W9-ATJ-190

Conversions (*continued*)

1 gram	$= 0.602 \ldots \times 10^{24}$ amu
	$= 2.20 \ldots \times 10^{-3}$ lb_m
1 gram/cm^3	$= 62.4 \ldots$ lb_m/ft^3
	$= 1000$ kg/m^3
	$= 1$ Mg/m^3
1 inch	$= 0.0254 \ldots$ m
1 joule	$= 0.947 \ldots \times 10^{-3}$ Btu
	$= 0.239 \ldots$ cal, gram
	$= 6.24 \ldots \times 10^{18}$ eV
	$= 0.737 \ldots$ ft·lb_f
	$= 1$ watt·sec
1 joule/meter^2	$= 8.80 \ldots \times 10^{-5}$ Btu/ft^2
1 [joule/(m^2·s)]/[°C/m]	$= 1.92 \ldots \times 10^{-3}$ [Btu/(ft^2·s)]/[°F/in.]
1 kilogram	$= 2.20 \ldots$ lb_m
1 megagram/meter^3	$= 1$ g/cm^3
	$= 10^6$ g/m^3
	$= 1000$ kg/m^3
1 meter	$= 10^{10}$ Å
	$= 10^9$ nm
	$= 3.28 \ldots$ ft
	$= 39.37$ in.
1 micrometer	$= 10^{-6}$ m
1 nanometer	$= 10^{-9}$ m
1 newton	$= 0.224 \ldots$ lb_f
1 ohm·inch	$= 0.0254 \ldots$ Ω·m
1 ohm·meter	$= 39.37$ Ω·in.
1 pascal	$= 0.145 \ldots \times 10^{-3}$ lb_f/in^2
1 poise	$= 0.1$ Pa·s
1 pound (force)	$= 4.44 \ldots$ newtons
1 pound (mass)	$= 0.453 \ldots$ kg
1 pound/foot^3	$= 16.0 \ldots$ kg/m^3
1 pound/inch^2	$= 6.89 \ldots \times 10^{-3}$ MPa
1 watt	$= 1$ J/s
1 (watt/m^2)/(°C/m)	$= 1.92 \ldots \times 10^{-3}$ [Btu/(ft^2·s)]/[°F/in.]

SI prefixes

giga	G	10^9
mega	M	10^6
kilo	k	10^3
milli	m	10^{-3}
micro	μ	10^{-6}
nano	n	10^{-9}

Fifth Edition

Elements of
Materials Science
and Engineering

Fifth Edition

Elements of Materials Science and Engineering

LAWRENCE H. VAN VLACK

University of Michigan
Ann Arbor, Michigan

Addison-Wesley Publishing Company
Reading, Massachusetts · Menlo Park, California
Wokingham, Berkshire · Amsterdam
Don Mills, Ontario · Sydney

This book is in the
Addison-Wesley Series in Metallurgy and Materials Engineering

Consulting Editor
Morris Cohen

Sponsoring Editor: Thomas Robbins
Managing Editor: Martha H. Stearns
Production Manager: Cheryl Wurzbacher
Production and Copy Editor: Marion E. Howe
Text and Cover Designer: Marie E. McAdam
Illustrators: Dick Morton
              Oxford Illustrators, Ltd.
Art Coordinator: Dick Morton
Manufacturing Supervisor: Ann DeLacey

Library of Congress Cataloging in Publication Data

Van Vlack, Lawrence H.
  Elements of materials science and engineering.

  Includes index.
  1. Materials. 2. Solids. I. Title.
TA403.V35 1984   620.1′1   83-25648
ISBN 0-201-08086-9

Reprinted with corrections, May 1985

Copyright © 1985, 1980, 1975, 1964, 1959 by Addison-Wesley Publishing Company, Inc.

All rights reserved. No part of this publication may be reproduced, stored in a retrieval system, or transmitted, in any form or by any means, electronic, mechanical, photocopying, recording, or otherwise, without the prior written permission of the publisher. Printed in the United States of America.

BCDEFGHIJ-DO-898765

To Today's Students, who will be Tomorrow's Engineers

Preface

As ably described by Professor Cohen in the Foreword, the central theme of *Materials Science and Engineering* is the concept that the properties and behavior of a material are closely related to the internal structure of that material. As a result, in order to modify properties, appropriate changes must be made in the internal structures. Also, if processing or service conditions alter the structure, the characteristics of the material are altered.

This new edition of *Elements of Materials Science and Engineering*, like the four previous editions, develops that structure ⇔ properties ⇔ performance relationship. It builds on basic college chemistry and physics courses, utilizing their principles and drawing upon those backgrounds to develop additional principles as required. Liberal use is made of engineering applications to illustrate the above relationships.

This edition has changes calculated to serve the student better. (1) The text is divided into three parts to meet academic needs. (2) There has been a modest increase in attention to polymeric, ceramic, electronic, and composite materials, as encouraged by many of those 180 schools who participated in the "Survey of the Introductory Materials Course" (*Journ. Materials Education*, Vol. 5, No. 3, 1983). (3) The review and study sections at the ends of the chapters have been enlarged as an aid to instruction.

This edition contains three parts. The first part addresses the phases of materials—their structure, properties, and performance. This part (through Chapter 9) is generally applicable to all engineering disciplines. Chapters include atomic order and disorder; metals, polymers, and ceramics; and semiconductors and nonconducting materials. Part I may constitute the introductory course if there are time restrictions. Part II considers multiphase materials with attention to phase relationships, heat treatments, composites, and service performance. It will have specific interest in those disciplines that emphasize design. Part III introduces the widely used materials of cast irons (cheapest metal), concrete (greatest tonnage), and wood (greatest volume) on an optional basis. None of these three materials is simple; but all can be approached on the structure–property basis.

Chapter 6 (polymeric materials) has been rewritten with an added section on processing. Chapter 9 (magnetic, dielectric, and optional materials) is new, as is Chapter 12 (composites). Fracture toughness has been added as a section in Chapter 13 (service performance) because of its importance to product reliability.

Study aids that were also present in the fourth edition include the chapter Previews, approximately 200 Example problems at the end of sections, a Summary at the end of each chapter, and more than 500 Study Problems. Added to the fifth edition are Quiz Samples (and answers), and a glossary of Terms and Concepts at the end of each chapter.

The study problems are separated into two categories. The first is a trial category (unshaded). These study problems parallel example problems closely, or relate directly to equations. Their purpose is to allow the students to "get their feet wet" before proceeding to the problems (shaded) that require more analysis and/or greater integration with previous topics.

As in previous editions, some sections and topics may be deleted by the instructor in deference to time or to sequel courses. These sections are marked with a bullet (•), and are not prerequisite to subsequent unmarked sections.

Unfortunately, individual acknowledgements are impossible for the hundreds of students at The University of Michigan, and to the dozens of instructors at other schools, each of whom have contributed in their way to this new edition. All comments have been seriously considered, and most have been used.

I would be remiss, however, if I did not acknowledge Professor Bigelow for his critical comments on the previous edition, and thank Professor M. Cohen (MIT) for his encouragement and numerous suggestions for this fifth edition. Likewise, the feedback of classroom responses by Professors Filisko, Hosford, Hucke, Leslie, and Tien was valued most highly. Behind the scenes of the revision, and always cooperative, were Mrs. Ardis Vukas in my office, and Marion Howe and Dick Morton of Addison-Wesley. It has indeed been a pleasure to work with them again. They made the work go smoothly. Fran's encouragement and patience have been generous and indispensable.

Ann Arbor, Michigan   L.H.VV.
September 1984

Contents

• See Preface for bullet (•) notation.

Foreword

Materials and Society

Morris Cohen

Materials are all about us; they are engrained in our culture and thinking as well as in our very existence. In fact, materials have been so intimately related to the emergence of civilizations that they have given names to the Stone, Bronze, and Iron Ages. Both naturally occurring and manufactured materials have become such an integral part of our lives that we often take them for granted, and yet materials rank with food, living space, energy, and information as basic resources of the human race. Materials are indeed the working substance of our society; they play a crucial role not only in our way of life but also in the well-being and security of nations.

But what are materials? How do we understand, manipulate, and use them? Materials are, of course, a part of the matter in the universe, but more specifically *they are substances whose properties make them useful in structures, machines, devices, or products.* For example, these categories include metals, ceramics, semiconductors, superconductors, polymers (plastics), glasses, dielectrics, fibers, wood, sand, stone, and many composites. The production and processing of these materials into finished goods account for about one fifth of the jobs and gross national product in the United States.

Since the human body might be regarded as a structure or machine or device, we could also embrace foods, drugs, biomatter, fertilizers, etc., among the classes of materials, but it is presently customary to leave these materials to the life and agricultural sciences. For similar reasons, even though fossil fuels, water, and air likewise fall within the broad definition of materials, they are usually dealt with in other fields.

The materials of the world's population can be visualized to flow in a vast *materials cycle*—a global cradle-to-grave system. Raw materials are taken from the earth by mining, drilling, excavating, or harvesting; then converted into bulk materials like metal ingots, crushed stone, petrochemicals, and lumber; and subsequently fabricated into engineering materials, like electric wire, structure steel, concrete, plastics, and plywood, for meeting end-product requirements in society. Eventually, after due performance in the service of humanity, these materials find their way back to earth as scrap, or preferably re-enter the cycle for reprocessing and further use before their ultimate disposal. In this huge international activity, almost one third of the world's raw materials are processed or consumed in the United States.

An important aspect of the materials–cycle concept is that it reveals many strong interactions among materials, energy, and the environment, and that all three must be taken into account in national planning and technological assessment. These considerations are becoming especially critical because of mounting shortages in energy and materials just at a time when the inhabitants of this planet are manifesting deeper concern for the quality of their living space. As a case in point, if scrap aluminum can be effectively recycled, it will require only about one twentieth the energy needed for an equivalent tonnage of primary aluminum from the ore, and the earth would be that much less scarred by the associated removal operations.

Consequently, the materials cycle is a system that intertwines natural resources and human needs. In an all-embracing way, materials form a global connective web that ties nations and economies not only to one another on this planet but also to the very substance of nature.

Clearly, then, in the development of human knowledge, one is not surprised to find a *science and engineering of materials* taking its place among all the other bodies of inquiry and endeavor that extend the reach of the human race. Simply stated, *materials science and engineering* (MSE) *is concerned with the generation and application of knowledge relating the composition, structure, and processing of materials to their properties and uses.* As suggested in Fig. I, there is a linkage that interrelates the structure, properties, processing, function, and performance of materials. MSE operates as a knowledge-conduction band that stretches from basic science and fundamental research (on the left) to societal needs and experience (on the right). The countercurrent flow of scientific understanding in one direction and empirical information in the other direction intermix very synergistically in MSE.

If we wish to highlight the *materials science* part of this spectrum, we focus on understanding the nature of materials, leading to theories or descriptions that explain how structure relates to composition, properties, and behavior. On

MATERIALS SCIENCE AND ENGINEERING

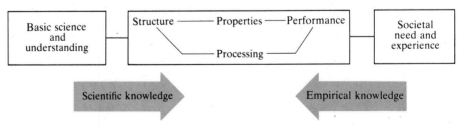

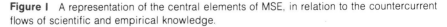

Figure I   A representation of the central elements of MSE, in relation to the countercurrent flows of scientific and empirical knowledge.

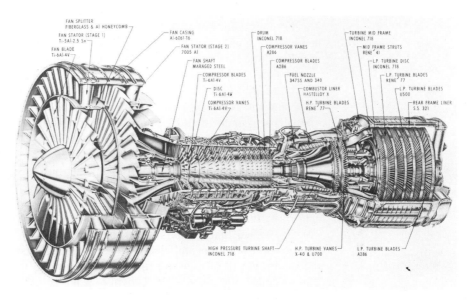

Figure II Materials in a jet engine. The jet engine is a complex system designed for converting fuel energy into motion. Here a major goal is to improve the ratio of thrust to weight. Each of the designated materials is selected and processed to serve an assigned function in harmony with all the other operating materials. (Taken from "Materials and Man's Needs," Committee on the Survey of Materials Science and Engineering, National Academy of Sciences, Washington, DC, 1974; courtesy of the General Electric Company.)

the other hand, the *materials engineering* part of the spectrum deals with the synthesis and use of both fundamental and empirical knowledge in order to develop, prepare, modify, and apply materials to meet specified needs. It is evident that the distinction between materials science and materials engineering is primarily one of viewpoint or center of emphasis; there is no line of demarcation between the two domains, and we find increasing logic in adopting the combined name of *materials science and engineering*. Actually, this textbook joins the two, both in its title and coverage, by using the term *materials science* generically to include many aspects of *materials engineering*.

Figures II and III illustrate the selection and function of materials in two very different applications. For societal use, all such applications of materials require numerous professional judgments concerning performance, reliability, maintainability, economic trade-offs, and environmental considerations. In the jet engine shown, each material is carefully chosen to provide an appropriate combination of properties that will serve the intended function. Examples include superalloys for turbine blades, heat-resistant alloys for combustion liners, fatigue-resistant titanium alloys for compressor vanes, high-strength steels for gears and shafts, wear-resistant alloys for bearings, and light-weight aluminum alloys for casings. And all of these materials must operate as an integrated whole in the overall system.

Another materials system is the electro-optical device shown schematically in Fig. III. This is a semiconductor laser, which is essentially composed of a suitably doped material such as gallium arsenide or indium phosphide. When electrons drop through the energy gap into the holes in the valence band, part of the energy thus released can stimulate other electrons to undergo this process, and the resulting photon emissions produce laser light. The laser-beam intensity is enhanced and sustained by multiple reflections from mirrors at the left and right faces of the device. Lasers of various types are now widely employed for intense localized heating purposes, e.g., welding, surface treatment, and surgery, as well as for transmitting pulsed signals along optical fibers in telecommunications. Electronic materials, whose useful properties depend on electrical, optical, and magnetic phenomena, constitute a rapidly growing field of activity in MSE.

So we see that MSE is a purposeful enterprise, reaching down into the micro-world of atoms and electrons and tying the condensed state of matter to the macro-world of material function and service to meet societal problems. The circular chart in Fig. IV portrays a large section of human knowledge, extending from basic sciences at the core, through applied sciences in the middle ring, to various engineering fields in the outer rim. In the center, we show physics and chemistry flanked by mathematics and mechanics; and on

Figure III Semiconductor laser (schematic). Photons of light are emitted when electrons combine with electron holes at the *p–n* junction of properly doped gallium arsenide or comparable semiconductors. The emitted light is reflected at the two ends, returning to stimulate the combination of additional electron-hole pairs. These build up an intense coherent beam that exits into an optical train for a variety of sophisticated applications.

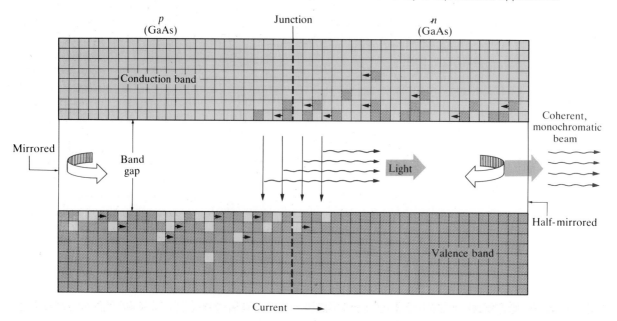

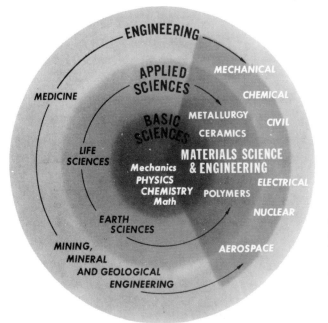

Figure IV Materials science and engineering. This text draws from introductory science principles of physics and chemistry to relate *composition and structure* of materials to the *properties and service behavior* that are important to engineers. More than one fourth of our national technical effort involves the development and technology of materials. This effort draws from the physical sciences and all branches of engineering.
(Courtesy National Academy of Science.)

moving out radially, we pass through various applications-oriented disciplines. The part of this map to be visualized as materials science and engineering is the shaded sector at the right, which may be compared to other sectors designating the life sciences and the earth sciences. In its broad sense, MSE is a multidiscipline that embraces (but does not replace!) some disciplines (e.g., metallurgy and ceramics) and some subdisciplines (e.g., solid-state physics and polymer chemistry), and also overlaps several engineering disciplines.

There are, of course, many scientists and engineers who are materials specialists—metallurgists, ceramists, and polymer chemists—and who are wholly involved in materials science and engineering. Equally important, government data, when analyzed, reveal that one of every six hours of professional work done by *all other engineers* directly involves materials and their utilization. The time fractions are even higher for chemists and physicists. As a result, the equivalent of one-half million of the nearly two million scientists and engineers in this country contribute to this major segment of our national product and our national well-being.

Thus, materials science and engineering constitutes a framework in which professionals in many disciplines work creatively to probe nature's processes and, at the same time, advance knowledge in response to the pull of human needs.

Cambridge, Massachusetts   Morris Cohen
September 1984

Chapter One

Introduction to
Materials Science
and Engineering

PREVIEW

Materials have always been an integral part of human culture and civilization; for example, the stone, bronze, and iron ages of the past. Likewise, today's advanced technologies involve sophisticated materials, since all of them utilize devices, products, and systems that must consist of materials. The engineer's expertise is in adapting *materials and energy for society's needs.*

The theme of this text is that *the properties of materials depend upon their internal structure.* In turn, the properties influence the *performance* of a material, both during manufacture and in ultimate service. To change the performance of a material, its internal structure must be modified. Conversely, if service conditions alter the structure, the engineer must anticipate what changes will occur in properties and performance.

The final section of this introductory chapter reviews selected properties that are associated with general physics. They will give us a basis for making our first structure ⇔ property ⇔ performance ties. For convenience, we commonly categorize materials as metals, polymers, and ceramics.

CONTENTS

This chapter, like all the chapters of this text, is concluded with a Review and Study section that includes (1) a *Summary,* (2) a glossary of key *Terms and Concepts,* (3) *Discussion* topics, (4) *Quiz Samples,* and (5) *Study Problems.*

STUDY OBJECTIVES

1. To become aware of the close association of materials with past cultures, and with current technological advances.

2. To become alert to the concept that materials possess an internal structure. (The various details of structure will be the focus of later chapters.)

3. To recognize most materials as being in one of three main categories—metals, polymers, and ceramics; and to know the basis of distinction between these.

4. To review selected properties from general physics, so that we will have an introductory basis for the structure ⇔ property ⇔ performance ties.

5. To become familiar with terms and concepts that relate materials and engineering.

1-1

MATERIALS AND CIVILIZATION

Among the various attributes of the human species is the ability to make things. Of course, all objects, tools, components, and engineering systems require the use of materials to meet their purposes. The role of materials has been sufficiently important in the progress of civilization so that the anthropologists and historians have identified early cultures by the more significant material. These include the *stone,* the *bronze,* and the *iron* ages of the past. Such identification could be carried into the present were it not for the fact that we are not limited to one predominant material but have a multitude of sophisticated materials—plastics, silicon, titanium, high-technology ceramics, optical fibers, etc. (Fig. 1–1.1).

With improvements in materials by the artisans and technologists across history, it was possible to make improved products. Clothing became more protective against the elements or more attractive; tools were made more available to relieve toil; homes were designed for more comfort; weapons became more destructive and lethal in an effort to obtain security against the opponent's advanced weapons; vehicles achieved longer ranges and quicker transport and gave more ready access to food, supplies, and enjoyment. Not only were functional products improved with the adaption of materials, but also the arts and crafts were influenced by the artisans' abilities to work (process) these materials into objects that were desired.

In early civilizations, as today, materials affected a broad spectrum of other human activities. During the Stone Age, the sites of the better flint deposits became the sites of villages and the crossroads of primitive barter. In ensuing centuries, the requirements for materials to meet societal needs brought forth exploration, political downfalls, as well as wars. The transition from the Bronze Age to the Iron Age, which started some 3500 years ago, introduced additional nontechnical changes. Iron was a "democratic material" in these early civilizations, because its widespread availability produced an

Figure 1–1.1   There are several hundred varieties of materials in this automobile engine. (The reader is asked to identify those that are apparent.) The design engineer selects each material on the basis of fabricability, properties, service behavior, cost, and availability. Improvements in materials will be required during the coming decade as weight limitations gain importance, as the engine is redesigned for more efficient energy conversion, and as the supply of raw materials tightens. (Courtesy of the Ford Motor Company.)

impact on the common person in every clan through the introduction of implements, tools, and utensils. No longer were the benefits of available metals limited to the affluent. Probably the most significant feature of this new democratic material was that the numbers of users exploded. More craftspeople were required in the materials trades. Advances now occurred in centuries and decades rather than in millennia. This acceleration continues to the present day.

Closer to modern times, we can cite the American and British railroad systems as an example of the symbiosis of materials technology and socioeconomic changes. As early as 1830, attempts were made to use steam power for land transportation. Rails were necessary for several obvious reasons; however, rails at that time were merely straps of soft wrought iron nailed to planks. An economical metal was not available with the required characteristics. Were it not for the developments of Kelley (U.S.) and of Bessemer (U.K.) in steel production, the railroad system could not have developed to open the West of the United States and to industrialize England. Conversely, were it not for the industrial and agricultural demand for transportation, the incentives, the capital, and the technology for steel development would have been missing.

A current example of the interplay of technological and nontechnical areas involves silicon. Economically, it has introduced a multibillion dollar industry. Communication has been facilitated at all distances from hearing aids to extra-terrestrial telemetry. Our day-to-day life is becoming altered via the availability of in-home entertainment, and by the introduction of computers at all levels of personal records. Changes are not solely technical.

1–2

MATERIALS AND ENGINEERING

Many prospective engineering students have asked, "What do engineers do?" The simplest, most general answer could be

Engineers adapt materials and/or energy for society's needs.

More specifically, engineers design products and systems, make them, and monitor their usage. Every product is made with materials; and energy is involved in production, in use, and even in communication. This association with design, production, and usage is the reason that engineering students take a course in materials as part of their undergraduate curricula.

The various engineering disciplines focus on their own subset of products and systems. As examples, the M.E. student gives attention to automobiles, power generating equipment, and other mechanical products, while the naval architect considers ships for marine transport, and the E.E. will construct circuits ranging from megawatt distribution systems to nanowatt computer signals. The petroleum refinery, designed by the Ch.E.; the bridge girder, by the C.E.; and the space shuttle, by the Aerospace Engineer (Fig. 1–2.1) are further examples. All of these products must be made out of materials. The materials that are specified must have the correct properties—first for production, and subsequently for service. They must not fail in use, producing a liability. Costs of the processed materials must be acceptable, and nowadays the choices of materials must be compatible with the environmental standards—from raw material sources, through manufacturing steps and product usage, to eventual discard. This means that those newer technical fields, such as "environmental engineering" and "public-policy engineering" must also give consideration to materials.

The age of technology is also an age of materials barriers. Nowhere is this more dramatically revealed than in energy conversion. The development of the solar cell (Fig. 1–2.2) requires that the engineer perfect the process for semi-conductor purification and crystal growth still further than the achievements to

(a)

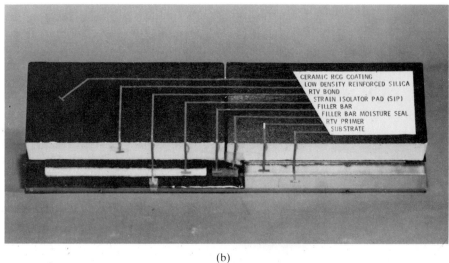

CERAMIC RCG COATING
LOW DENSITY REINFORCED SILICA
RTV BOND
STRAIN ISOLATOR PAD (SIP)
FILLER BAR
FILLER BAR MOISTURE SEAL
RTV PRIMER
SUBSTRATE

(b)

Figure 1–2.1 Materials system (space shuttle). (a) Touch down of first shuttle flight (1981). The shuttle was covered with 30,757 insulating surface tiles. (b) Reusable surface insulation (cross section). The success of the flight was related to the performance of the surface tiles, which were designed for reuse. (Courtesy NASA and J. D. Buckley, Langley Research Center.)

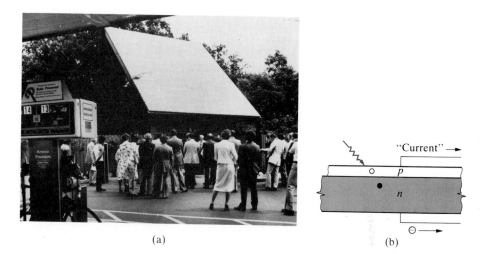

(a)   (b)

Figure 1–2.2 Solar energy (developmental). (a) Five kilowatt, photovoltaic array. (b) Solar cell (schematic). This demonstration unit consists of 72 panels, 65 cm × 130 cm each, and feeds 150 batteries, which in turn power 15 gasoline pumps and station lights. Progress toward commercial feasibilities depends on advances in materials technology. (Courtesy of Solarex Corp.)

date. Coal-to-watts conversion becomes more efficient at higher steam temperatures. Thus, we should not be surprised that as soon as materials were developed for 400°C (750°F) the design engineers then asked for 450°C (840°F) materials. Those are now available. The next step and the following step will always call for more and more stringent demands. We will never overcome the final materials barriers. However, design engineers and the materials engineers are always working in that direction.

Materials substitution is another arena of engineering activity. Aluminum substitution for steel is not automatic in order to develop lighter weight, more fuel-efficient cars. Aluminum and steel are not processed identically, so there must be changes in capital equipment. They have different rigidities, which must be considered in the technicalities of design. The substitution of stainless steel for regular steel could appear to be rational for longer life of bridges or ocean-going boats until the engineer compares costs, fabricability, and the sources of our raw materials with their international complications. The role of materials in modern society is manifold.

Some engineers will not encounter materials in their first assignment after graduating, and probably a few of those engineers will not advance to more responsible positions where they consider materials; however, sooner or later the majority of engineers will be involved in materials selection or performance, either directly or with colleagues in their technical organizations. Materials (and energy) are seldom avoided by today's engineer.

1-3

STRUCTURE ⇔ PROPERTIES ⇔ PERFORMANCE

Every known type of material must be considered for use by the engineer. Admittedly, some are not used widely because of availability, initial properties, cost, or service performance. Others, like iron and steel, paper, concrete, water, and vinyl plastics, find extensive uses. Of course, some of these categories can be subdivided. For example, it has been suggested that there are as many as 2000 types of steel, 5000 types of plastics, and 10,000 kinds of glass. Whether or not these numbers are realistic, the engineer has a large number of choices for the product being designed.

To the above numbers, we must add the capability of modifying properties of materials, as in the case of gears. They must be weak and soft to be machinable during manufacture, but strong and wear-resistant for service in the power train of automobiles, or of the heavy construction equipment. Thus, the task of knowing the properties and behavior of all types of available materials becomes enormous. In addition, several hundred new varieties of materials appear on the market each month. This means that individual engineers cannot hope to be familiar with the properties and performance of all types of materials in their numerous forms. However, the engineer can learn principles for guidance in the selection and application of materials.

Internal Structure and Properties

Since it is obviously impossible for the engineer or scientist to have detailed knowledge of the many thousands of materials already available, as well as to keep abreast of new developments, he or she must have a firm grasp of the underlying principles that govern the properties of *all* materials. The principle that is of most value to engineers and scientists is *properties of a material originate from the internal structures of that material.* This is analogous to saying that the operation of a TV set or other electronic product (Fig. 1–3.1) depends upon the components, devices, and circuits within that product. Anyone can twirl knobs, but electronic technicians must understand the internal circuits if they are going to repair a TV set efficiently; and the electrical engineer and the physicist must know the characteristics of each circuit element if they are to design or improve the performance of the final product.

The *internal structures* of materials involve atoms and the way atoms are associated with their neighbors in crystals, molecules, and microstructures. In the following chapters we shall devote much attention to these structures, because technical persons must understand them if they are going to produce and use materials, just as mechanical engineers must understand the operation of an internal-combustion engine if they are going to design or improve a car for the demands of the next decade.

Properties and Processing

Materials must be processed to meet the specifications that the engineer requires for the product being designed. The engineer calls this *production.* The most familiar processing steps of production simply change the shape of the material, e.g., machining, or forging. Of course, properties are important for

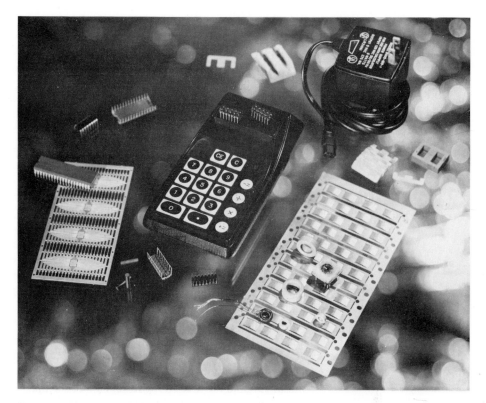

Figure 1–3.1 Electronic calculator. The performance of a calculator depends on the arrangement of the components of its internal circuit. Likewise, the behavior of a material depends upon the structure of its internal components. We shall see that these arrangements involve electron structures around atoms, the coordination of atoms, the structure of crystals, and the microstructure of adjacent crystals. The engineer and scientist can select and modify these internal structures to meet design needs, just as the circuit designer modifies electrical components. To do this intelligently, we must know the relationships between structure and properties of materials. (Courtesy of the Arnold Engineering Company.)

easy processing. Extremely hard materials immediately destroy the edge of a cutting tool, and soft materials like lead can "gum-up" saw blades, grinding wheels, and other tools. Likewise, strong materials are not amenable to plastic deformation, particularly if they are also nonductile, i.e., brittle. For example, it would be prohibitively expensive to produce sheet steels for most car fenders with anything other than the softest types.

Processing commonly involves more than simply changing the shape of the material by machining or by plastic deformation. Not uncommonly, the manufacturing process changes the properties of a material. For example, a wire is strengthened and hardened if it is drawn through a die to decrease its diameter. Typically, this hardening is not desired in a copper wire to be used as an electrical conductor; conversely, the engineer depends upon this strengthening that develops during processing the steel wire used in a steel-belted, radial tire. Whether desired or not, the properties should be expected to be modified whenever the manufacturing process changes the internal structure of the material. The internal structure of a material is altered when it is deformed; hence, there is a change in properties.

Thermal processing may also affect the internal structure of a material. Such processing includes annealing, quenching from elevated temperatures, and a number of other heat treatments. Our purpose will be to understand the nature of the structural changes, so that we, as engineers, may specify appropriate processing steps.

Properties and Service Behavior

A material in the completed product possesses a set of properties—strength, hardness, conductivity, density, color, etc.—chosen to meet the design requirements. It will retain these properties indefinitely, *provided* there is no change in the internal structure of the material. However, if the product encounters a service condition that alters the internal structure, we must expect the properties and behavior of the material to change accordingly. This explains why rubber gradually hardens when exposed to light and the air; why aluminum cannot be used in many locations of a supersonic plane; why a metal can fatigue under cyclic loading; why a drill of ordinary steel cannot cut as fast as a drill of high-speed steel; why a magnet loses its polarity in an *rf* field and why a semiconductor can be damaged by nuclear radiation. The list is endless. The consequence to the engineer is that consideration must be given not only to the initial demand but also to those service conditions that will alter the internal structure and hence the properties of a material.

The Engineering Approach

The method of the engineer is to understand the underlying principles of the particular assignment, then develop a solution. Since the solution to optimum materials selection for product and system *performance* involves the *properties* of the materials, the engineer must understand the internal *structures* that govern those properties.

Figure 1–3.2 associates these three factors for materials selection. This text will focus on the structure ⇔ property relationships. Detailed property ⇔ performance attentions will be received in sequel courses within the various engineering curricula. However, this text will cite numerous examples in which properties are important for performances, both during the raw material-to-product processing or during the subsequent service life.

During processing

STRUCTURE ⟺ PROPERTIES ⟺ PERFORMANCE

During service

Figure 1–3.2 The materials spectrum. The structure must be controlled to assure desired properties, which in turn influence the performance, both during the processing steps of production and in service. Conversely, a processing or service situation that changes the properties does so because the internal structure has been altered.

1–4

**TYPES OF
MATERIALS**

It is convenient to group materials into three main types: *metals, polymers* (or *plastics*), and *ceramics*. In reality, these are three idealized categories, since most materials have intermediate characteristics.

Metals

These materials are characterized by their high thermal and high electrical conductivity. They are opaque and usually may be polished to a high luster (Fig. 1–4.1). Commonly, but not always, they are relatively heavy and deformable.

What accounts for the above characteristics? The simplest answer is that metals owe their behavior to the fact that the valence electrons are not bound, but can leave their "parent" atoms. (Conversely, electrons are not free to roam to the same extent in polymers and ceramics.) Since some of the electrons are independent in metals, they can quickly transfer an electric charge and thermal energy. The opacity and reflectivity of a metal arises from the response of these unbound electrons to electromagnetic vibrations at light frequencies. It is another result of the partial independence of some of the electrons from their parent atoms.

**Polymers (commonly
called plastics)**

Plastics (Fig. 1–4.2) are noted for their low density and their use as insulators, both thermal and electrical. They are poor reflectors of light, tending to be transparent or translucent (at least in thin sections). Finally, some of them are flexible and subject to deformation. This latter characteristic is used in manufacturing.

Unlike metals, which have some migrant electrons, the *nonmetallic elements* of the upper right corner of the periodic table (inside back cover) have an *affinity* to *attract* or *share additional electrons*. Each electron becomes associated with a specific atom (or pair of atoms). Thus, in plastics, we find only limited electrical and thermal conductivity because all the thermal energy must

Figure 1–4.1 Metallic materials (stainless steel utensils for commercial kitchens). Metals possess ductility for the required processing. The luster and conductivity of metals arise from the behavior of the valence electrons. (Courtesy of the Vollrath Company.)

Figure 1–4.2   Plastic housing for a desktop computer (Z-100). Molding requirements for manufacture plus properties for service performance dictate the choice of the polymer product. (Reprinted by permission of Zenith Data Systems.)

be transferred from hot to cold regions by atomic vibrations, a much slower process than the electronic transport of energy that takes place in metals. Furthermore the less mobile electrons in plastics are more able to adjust their vibrations to those of light and therefore do not absorb light rays.

Materials that contain *only* nonmetallic elements share electrons to build up large molecules. These are often called *macromolecules*. We shall see in Chapter 6 that these large molecules contain many repeating units, or *mers,* from which we get the word *polymers.*

Ceramics

Simply stated, ceramics are *compounds that contain metallic and nonmetallic elements*. There are many examples of ceramic materials, ranging from the cement of concrete (and even the rocks themselves) to glass, to electrical insulators (Fig. 1–4.3), and to oxide nuclear fuel elements of UO_2, to name but a few.

Each of these materials is relatively hard and brittle. Indeed, hardness and brittleness are general attributes of ceramics, along with the fact that they tend to be more resistant than either metals or polymers to high temperatures and to severe environments. The basis for these characteristics is again the electronic behavior of the constituent atoms. Consistent with their natural tendencies, the metallic elements release their outermost electrons and give them to the nonmetallic atoms, which retain them. The result is that these electrons are immobilized, so that the typical ceramic material is a good insulator, both electrically and thermally.

Equally important, the positive metallic ions (atoms that have lost electrons) and the negative nonmetallic ions (atoms that have gained electrons) develop strong attractions for each other. Each *cation* (positive) surrounds itself with *anions* (negative). Considerable energy (and therefore considerable force) is usually required to separate them. It is not surprising that ceramic materials tend to be hard (mechanically resistant), refractory (thermally resistant), and inert (chemically resistant).

Figure 1–4.3 Ceramic products (high voltage insulators). All ceramic products contain compounds of metallic and nonmetallic elements; in this insulator they are oxides of magnesium, aluminum, and silicon. (R. Russell, and *Brick and Clay Record*.)

1–5

REVIEW OF SELECTED PROPERTIES OF MATERIALS

Materials are characterized by their properties. They may be strong, ductile, or have a high electrical resistivity. Conversely, they may have good conductivity, be brittle, or be soft. None of the adjectives just cited are quantitative as such. Engineers must be more specific if they are to know which of two materials is the stronger; furthermore, how strong? For example, is the steel in a bridge beam strong enough to carry the designed load? Or does the ceramic substrate underlying a 256-K computer chip have sufficient thermal conductivity to dissipate the nanowattage developed by each of the megabits per second in the microprocessors of a sophisticated computer?

It will not be our purpose in this section to detail all properties pertinent to engineering. It will be necessary, however, to define a few of the principal properties. This will give us a basis to see the relationship that was cited in Section 1–3 between the structure of materials and their properties. This section includes conductivity (and resistivity) plus a group of mechanical properties, many of which were encountered in basic physics courses. Those students who do not require a review may proceed immediately to Chapter 2. Other properties will be presented later in the text as attention is required for specific property–structural relationships.

Conductivity (and resistivity)

The resistance of a wire in a circuit is a function of the size of the wire as well as the material:

$$R = \rho \left(\frac{L}{A} \right). \tag{1–5.1}$$

The resistance, R, increases with increased length, L, and reduced cross-sectional area, A. The proportionality constant is the *electrical resistivity, ρ*, and is a property of the material. The dimensional unit for resistivity is an ohm·m.

Electrical conductivity, σ, is the reciprocal of the resistivity

$$\sigma = \frac{1}{\rho}. \tag{1–5.2}$$

Thus, it has ohm^{-1}·m^{-1} as its most commonly used dimensional units. However, some people use mhos/m. As yet, the SI units, siemans/m, have not been widely adopted.

In SI units, *thermal conductivity, k*, is expressed as power per unit area (watts/m^2) along a temperature gradient (K/m). These units reduce to W/m·K. We do not use the reciprocal of thermal conductivity (which would be a thermal resistivity) as a labeled property of a material.*

Stress and Strain

Stress, s, is a measure of force per unit area. Thus, it has the units of newtons/m^2, or pascals. The level of stresses encountered in engineering applications

---

* The dimensions of thermal conductivity in English units are generally an energy flux along a temperature gradient that uses (Btu/ft^2·s)/(°F/in.) as units.

are commonly in the megapascal range; hence MPa units are widely used.*

Strain, e, is the dimensional response to stress (*not* a synonym). It is expressed as a fraction (or as percent), and is therefore dimensionless. It is positive under *tensile* stresses and negative when *compressive* stresses are applied.

Elastic strain is reversible; that is, when the stress is removed, the strain disappears. Elastic strain is commonly a linear function of stress obeying Hooke's law of physics. The coefficient, or modulus relating unidirectional strain and stress, is the *modulus of elasticity, E* (or *Young's modulus*):

$$E = \frac{s}{e} = \frac{F/A}{\Delta L/L}. \qquad (1\text{–}5.3)$$

In Chapter 5, we shall encounter two additional elastic moduli.

Plastic strain occurs when the stress exceeds a critical value that initiates a permanent displacement of atoms from their neighbors. Thus, plastic strain is not reversible.

Strength is the critical stress required to initiate failure. However, as indicated in Table 1–5.1, the engineer has more than one way of defining failure. If failure is defined as the onset of permanent deformation (e.g., the bending, but not breaking, of an antenna for a car radio), the stress for the initial plastic deformation is involved. We speak of this critical stress as the *yield strength, S_y.* On the other hand, if the maximum tolerable tensile load before eventual *fracture* is the design parameter, the engineer uses the ultimate tensile strength, or simply the *ultimate strength, S_u.*

Hardness is the resistance of a material to penetration. A variety of procedures are used to measure hardness, as indicated in the footnote of Table 1–5.1. Hardness values are sometimes expressed in empirical units, and in other data as kg/mm^2. In most engineering usage, it is best to consider these values as relative indexes. However, hardness and strength values parallel each other in ductile materials. Although that relationship is generally empirical, it is useful for the engineer, since hardness values are much easier to obtain and generally nondestructive. For example, the hardness of a railroad rail can be measured *in situ,* with no damage to the rail; a tensile specimen would have to be cut out of the rail for machining and testing.

Ductility is the permanent strain that accompanies fracture. Here again, the engineer has more than one way of defining this strain. *Elongation* is the linear strain at fracture, $\Delta L_f/L$. It is commonly expressed as a percent, % El. It is imperative that a gage length be identified because the plastic strain almost invariably becomes localized. (See Fig. 1–5.1.) A second measure of ductility is

---

* This text will include certain English units parenthetically—particularly those pertaining to mechanical properties. Fortunately, the major companies of the U.S. are making shifts toward SI units. However, many senior engineers possess a wealth of experience within the nonmetric framework. Therefore, it is often desirable for the engineer to be "bilingual" in respect to units if we are to capitalize on their expertise.

TABLE 1–5.1

Mechanical Properties of Materials

Property, or characteristic	Symbol	Definition (or comments)	Common units	
			SI	English
Stress	s	Force/unit area (F/A)	pascal* ($N\dagger/m^2$)	psi* $lb_f/in.^2$
Strain	e	Fractional deformation ($\Delta L/L$)	—	—
Elastic modulus	E	Stress/elastic strain	pascal	psi
Strength		Stress at failure		
Yield	S_y	Resistance to initial plastic deformation	pascal	psi
Ultimate	S_u	Maximum strength (based on original dimensions)	pascal	psi
Ductility		Plastic strain at failure		
Elongation	e_f	($L_f - L_o$)/L_o	§	§
Reduction of area	R of A	($A_o - A_f$)/A_o	§	§
Toughness		Energy for failure by fracture	joules	ft-lb
Hardness‡		Resistance to plastic indentation	Empirical units	

* 1 pascal (Pa) = 1 newton/m² = 0.145×10^{-3} psi; 1000 psi = 6.894 MPa.

† A load of 1 kg mass produces a force F of 9.8 newtons (N) by gravity.

‡ Three different procedures are commonly used to determine hardness values:

    Brinell (BHN): A large indenter is used. The hardness is related to the diameter (1 to 4 mm) of the indentation.

    Rockwell (R): A small indenter is used. The hardness is related to the penetration depth. Several different scales are available, based on the indenter size and the applied load.

    Vickers (DPH): A diamond pyramid is used. A very light load may be used to measure the hardness in a microscopic area.

§ Dimensionless. Based on original (o) and final or fractured (f) measurements (usually expressed as percent).

the *reduction of area*, R of A, at the point of fracture:

$$R \text{ of } A = \frac{A_o - A_f}{A_o}.$$ (1–5.4)

Values for these two measures parallel each other; both are high in a ductile material and are nil in a nonductile material. However, there is no established mathematical relationship because the final plastic deformation is highly localized.

The Stress–Strain Curve

An established procedure for displaying the above property values is through the stress–strain curve (Fig. 1–5.2). Observe these features. The initial ds/de is linear because only elastic strain is developed. Beyond the yield strength, S_y, at the *proportional limit,* plastic strain overshadows the elastic strain.* Secondly,

* Elastic strain continues to increase as the stress is increased.

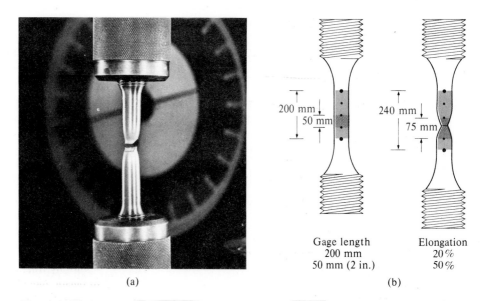

Gage length
200 mm
50 mm (2 in.)

Elongation
20%
50%

(a)   (b)

Figure 1–5.1 Tensile tests. (a) Completed test on a round specimen. (Courtesy of the U.S. Steel Corp.) (b) Elongation versus gage length. Since final deformation is localized, an elongation value is meaningless unless the gage length is indicated. For routine testing, a 50-mm (2-in.) gage length is common. (In inches, 2.0 in. → 3.0 in., and 7.9 in. → 9.5 in., for 50% and 20% elongation, respectively.)

the stress required for deformation increases as plastic strain occurs. Hardening occurs.

In the typical s–e curve, there is a maximum if we continue to calculate the stress on the basis of the original area, A_o. Beyond that maximum the curve drops, because necking (Fig. 1–5.1) reduces the cross-sectional area and thereby reduces the ability of the specimen to support a load. The curve abruptly ends with fracture. The calculated stress, at fracture, called the breaking strength, S_b, is thus less than the ultimate strength, S_u, since both are based on the original area. The breaking strength, S_b, has little if any practical value. The yield strength, S_y, and the strength, S_u, are very useful.

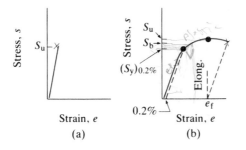

Strain, e
(a)

Strain, e
(b)

Figure 1–5.2 Stress–strain diagrams. (a) Nonductile material with no plastic deformation (example: cast iron). (b) Ductile material without marked yield point (example: aluminum). S_b = breaking strength; S_u = ultimate strength; S_y = yield strength; e_f = elongation (strain before fracture); X = fracture.

Toughness is the energy for failure by fracture (Table 1–5.1). Its units are joules (or ft·lbs). In brittle materials, which lack plastic deformation and therefore are nonductile, the toughness can be calculated from the area under the stress–strain curve. The result is joules per unit volume. Such a calculation has less meaning for ductile materials, again because the strain is localized in the necked region and the energy absorption is not uniformly distributed.

We shall pay more detailed attention to toughness in Chapter 13 since it is a critical service parameter.

Example 1–5.1

A copper wire has a diameter of 0.9 mm. (a) What is the resistance of a 30-cm wire? (b) How many watts are expended if 1.5 volts dc are applied across 30 m of this wire?

Procedure. Part (a) relates resistance to resistivity. Therefore, use data from Appendix C (ρ = 17 ohm·nm), and Eq. (1–5.1). Part (b) draws on elementary science, from which watts = volts × amps, ($P = EI = E^2/R$), and volts = amps × ohms, ($E = IR$).

Calculation

a)
$$R = \frac{(17 \times 10^{-9}\ \text{ohm·m})(0.3\ \text{m})}{(\pi/4)(9 \times 10^{-4}\ \text{m})^2} = 0.008\ \text{ohm.}$$

b)
$$P = \frac{E^2}{R} = \frac{E^2 A}{\rho L} = \frac{(1.5\ \text{V})^2(\pi/4)(9 \times 10^{-4}\ \text{m})^2}{(17 \times 10^{-9}\ \text{ohm·m})(30\ \text{m})} = 2.8\ \text{watts.} \blacktriangleleft$$

Example 1–5.2

Which part has the greater stress: (a) a rectangular aluminum bar of 24.6 mm × 30.7 mm (0.97 in. × 1.21 in.) cross section, under a load of 7640 kg, and therefore a force of 75,000 N (16,800 lb$_f$); or (b) a round steel bar whose cross-sectional diameter is 12.8 mm (0.505 in.), under a 5000-kg (11,000-lb) load?

Procedure. Stress is force per unit area. Also $f = ma$, with a being $g = 9.8$ m/s^2.

Units. $\dfrac{\text{Newtons}}{(\text{m})(\text{m})}$ = Pascals.   **Units.** $\dfrac{\text{Pounds}}{(\text{in.})(\text{in.})}$ = psi.

Calculation

a) $\dfrac{(7640)(9.8)}{(0.0246)(0.0307)} = 100$ MPa;   a) $\dfrac{16,800}{(0.97)(1.21)} = 14,300$ psi;

b) $\dfrac{(5000)(9.8)}{(\pi/4)(0.0128)^2} = 380$ MPa.   b) $\dfrac{11,000}{(\pi/4)(0.505)^2} = 55,000$ psi. $\blacktriangleleft$

Example 1–5.3

A 50-mm (1.97 in.) gage length is marked on a copper rod. The rod is strained so that the gage marks are 59 mm (2.32 in.) apart. Calculate the strain.

Units. $\dfrac{(\text{mm} - \text{mm})}{\text{mm}} = \dfrac{\text{mm}}{\text{mm}} = \dfrac{\text{percent}}{100}$.

Calculation. $\dfrac{59 - 50}{50} = 0.18$ mm/mm.   (or 18%).

Comment. Measurements in inches give the same answer. $\blacktriangleleft$

Example 1–5.4   If the average modulus of elasticity of the steel used is 205,000 MPa (30,000,000 psi), how much will a wire 2.5 mm (0.1 in.) in diameter and 3 meters (10 ft) long be extended when it supports a load of 500 kg (1100 lb$_f$ and 4900 N)?

Procedure. Modulus of elasticity = stress/strain; or, $e = s/E$.

Units. $\text{m/m} = \dfrac{\text{N/m}^2}{\text{pascals}}.$ $\quad\quad\quad\quad\quad$ $\text{in./in.} = \dfrac{\text{lb/in}^2}{\text{psi}}.$

Calculation

$$\text{Strain} = \frac{4900/(\pi/4)(0.0025)^2}{205,000 \times 10^6} \quad\quad \text{Strain} = \frac{1100/(\pi/4)(0.1)^2}{30,000,000}$$

$$= 0.005 \text{ m/m}. \quad\quad\quad\quad\quad\quad = 0.005 \text{ in./in.}$$

$$\text{Extension} = (0.005 \text{ m/m})(3 \text{ m}) \quad\quad \text{Extension} = (0.005 \text{ in./in.})(120 \text{ in.})$$
$$= 15 \text{ mm}. \quad\quad\quad\quad\quad\quad\quad = 0.6 \text{ in.} \blacktriangleleft$$

Review and Study

SUMMARY

1. Materials have been dominant forces in the cultures of advancing *civilizations*. In fact, we speak of the Stone Age, the Bronze Age, and the Iron Age. Furthermore, materials were critical for the industrial revolution, and they have a major role in the emergence of current "high technology."

2. The products and systems of all engineering descriptions must be made of materials. Every engineering advance encounters materials barriers. Sooner or later the majority of engineering graduates become involved with materials selection or performance.

3. There is a *structure–property–performance* relationship that provides a systematic approach to materials selections and behavior. Just as the internal circuiting dictates the capabilities of an electrical product, the *internal structure* of a material controls its properties and service performance.

4. It is convenient to categorize materials as metals, polymers, and ceramics. *Metallic* elements readily lose or release electrons. *Nonmetallic* elements accept or share electrons and thus produce macromolecular polymers. *Ceramic* materials are compounds of metallic and nonmetallic elements.

5. To facilitate the understanding of structure–property relationships, the reader may want to review the following items from physics: resistivity, conductivity, stress, strain, Hooke's law, Young's modulus, and elastic and plastic deformation. These are germane to the engineering properties of *strength, ductility,* and *toughness*.

TERMS AND CONCEPTS

Ceramics   Materials consisting of compounds of metallic and nonmetallic elements.

Conductivity   Transfer of thermal or electrical energy along a potential gradient.

Ductility   Total permanent strain prior to fracture—measured as elongation or as reduction of area.

Elongation (El.) Total plastic strain before fracture, measured as the percent of axial strain. (A gage length must be stated.)

Engineering The adaptation of materials and/or energy for society's needs. (Other definitions are also possible.)

Materials (engineering) Substances for use in technical products. (These include metals, ceramics, polymers, semiconductors, glasses, and natural substances like wood and stone; but generally exclude food, clothing, drugs, and related substances.)

Metals Materials characterized by their high electrical and thermal conductivities because their electrons are mobile. Metallic elements are located in the left and lower portions of periodic tables as conventionally presented.

Modulus of elasticity Stress per unit elastic strain. (*Young's modulus* is the most commonly encountered elastic modulus.)

Performance (materials) The behavior (and alteration) of materials during manufacturing and/or service.

Plastics Materials composed primarily of polymers.

Polymers Large molecules of nonmetallic elements composed of many repetitive units (mers). Commonly called plastics.

Property Characteristic quality of a material. (Quantitative values are preferable.)

Proportional limit The limit of the proportional range of the stress–strain curve. (Elastic strain increases beyond this "limit" but is masked by plastic strain.)

Reduction of area (R of A) Total plastic strain before fracture, measured as the percent decrease in cross-sectional area.

Resistivity (ρ) Reciprocal of electrical conductivity (usually expressed in ohm·m).

Strain (e) Unit deformation. Elastic strain is recoverable; plastic strain is permanent.

Strength (S) Stress to produce failure. Yield strength, S_y, is the stress to initiate the first plastic deformation; ultimate tensile strength, S_u, is the maximum stress on the basis of the original area.

Stress (s) Force per unit area. Nominal stress is based on the original design area; true stress is based on the actual area.

Stress–strain curve Plot of stress as a function of strain. (Stress is normally plotted on the basis of original area.)

Toughness A measure of the energy required for mechanical failure.

Young's modulus (E) Modulus of elasticity (axial). Stress per unit elastic strain.

FOR CLASS DISCUSSION

A_1 (a) The techniques of the artisan have often led to new materials. Give examples. (b) The needs for raw materials have led to exploration. Give examples. (c) New materials have advanced health care. Give examples.

B_1 Indicate three critical materials requirements that are encountered in your engineering field (i.e., M.E., C.E., E.E., etc.).

C_1 A car fender could be made out of armor plate so that it would not crumple in a collision. Give at least three reasons why it is not designed that way.

D_1 Examine a piece of wood. Describe some of its structural characteristics and indicate how they affect the properties.

E_1 List various metals that are used to make a kitchen stove; various ceramics; various polymers.

F_1 Categorize the following into the three types of materials of Section 1–4: bronze, portland cement, "rubber" cement, wood, window glass, Bakelite, rust, ethylene glycol (antifreeze).

G_1 The values for the thermal expansion and thermal conductivity coefficients are identical whether we use °C or the absolute temperature, K. In later chapters we will have problems where the choice of °C or K does make a difference. Why does it not matter here?

H_1 Compare and contrast: strength, ductility, and toughness.

I_1 "The yield strength is the stress where the strain switches from elastic to plastic." Comment on the validity of this statement.

J_1 Examine the prongs of an extension cord for a power saw. Why should there be specifications for (a) yield strength, (b) modulus of elasticity, (c) resistivity, and (d) corrosion resistance?

K_1 The percent reduction in area cannot be calculated from percent elongation (nor vice versa). Why is this true, when both are measures of ductility?

QUIZ SAMPLES

1A   Materials characterized three early stages of civilization. These were __**__ , __**__ , and __**__ .

1B   __**__ is considered by historians to be a "democratic material" because of its widespread availability for developing civilizations.

1C   As a generality, one may state that "Engineers adapt __**__ and/or __**__ for society's needs."

1D   Metallic elements readily __**__ valence electrons, while nonmetallic elements accept or __**__ electrons. Ceramic materials contain __**__ of metallic and non-metallic elements.

1E   The Young's modulus for a metal is 100,000 MPa; therefore, it takes a stress of __**__ to introduce 0.1% elastic strain.

1F   Stress is __**__ per unit __**__ , while strain is __**__ per unit __**__ , and a pascal is a __**__ per m^2.

1G   The elastic modulus of a rubber band is __** er__ than that of a copper wire of the same dimensions, because it takes smaller __**__ to produce 1% strain.

1H   Two measures of ductility are __**__ and __**__ . These are strains that precede __**__ .

1I   The __**__ strength is the stress required to initiate plastic deformation. Above this stress, the strain is both __**__ and plastic.

1J   The __**__ strength is the stress calculated as maximum load per unit of __**__ area.

1K   In a ductile material the breaking strength, S_b, is less than the ultimate strength, S_u, because __**__ · · · .

STUDY PROBLEMS*

1–2.1   Take the cord of a household appliance, such as a toaster or coffee-maker. List the materials used and the probable reason for their selection.

1–2.2   Examine an incandescent light bulb closely. How many different types of materials can you name? What thermal and electrical characteristics are required of each?

1–2.3   (a) Dismantle a cheap pen. List the materials that are used. (b) For each material, cite service conditions to which attention must be given in the material selection.

1–2.4   (a) Go to your car (or one owned by a friend). List all the materials you can identify. (Do not list names of parts.) (b) Compare your list with lists by your classmates. (c) Discuss the attributes that were important for their selection. (d) Suggest plausible substitutes for each. Why was the substitute *not* used?

1–4.1   Examine Appendix C. (a) Make generalizations about the comparative properties of metals, polymers, and ceramics. (b) From your own experience, cite exceptions to the generalizations you just made.

1–4.2   Take your list from S.P. 1–2.4 above and categorize the materials as (a) metals, (b) polymers (plastics), or (c) ceramics.

1–5.1   Copper has a resistivity of 17×10^{-9} ohm·m. (a) What is the end-to-end resistance of a copper strip 2 cm long by 5 mm wide $\times$ 1 mm thick? (b) The conductivity?

Answer: 68 microohms

1–5.2   (a) Using the data from Appendix C, determine the ratios of thermal to electrical conductivities, k/σ, for each metal. (b) Based on your calculations, make a generalization about the relationship between electrical and thermal conductivities.

1–5.3   A rod of copper should not be stressed to more than 70 MPa (10,200 psi) in tension. What diameter is required if it is to carry a load of 2000 kg (4400 lbs)?

Answer: 19 mm (or 0.74 in.)

1–5.4   What is the elastic strain in a copper rod that is stressed 70 MPa (10,200 psi)? (Moduli of elasticity and other properties of common metals are given in Appendix C.)

1–5.5   (a) A steel bar 12.7 mm in diameter supports a load of 7000 kg (and therefore a force of 68,600 N). What is the stress placed on the bar? (b) If the bar of part (a) has a modulus of elasticity of 205,000 MPa, how much will the bar be strained with this load?

Answer: (a) 540 MPa   (b) 0.0026

---

* Some of Study Problems, S.P., that follow each chapter are introductory in nature. They either mimic Examples closely, or come directly from equations. These can be used for "starters." Other Study Problems (shaded) require more analysis and synthesis by the student. Their procedures have not been spelled out but must be developed by the student through the use of accumulated knowledge and technical judgment. The engineering student should be able to solve these problems.

1-5.6 A wire of a magnesium alloy is 1.05 mm (0.04 in.) in diameter. According to Appendix C, its modulus of elasticity is 45,000 MPa (6.5×10^6 psi). Plastic deformation starts with a load of 10.5 kg (which is _____ N), or 23 lb_f. The total strain is 0.0081 after loading to 12.1 kg (26.6 lb_f). (a) How much permanent strain has occurred with a load of 12.1 kg (26.6 lb_f)? • (b) Rework this problem with English units.

1-5.7 Aluminum (6151 alloy) has a modulus of elasticity of 70,000 MPa (10^7 psi) and a yield strength of 275 MPa (40,000 psi). (a) How much load can be supported by a 2.75-mm (0.108-in.) wire of this alloy without yielding? (b) If a load of 44 kg (97 lb) is supported by a 30.5-m (100-ft) wire of this size, what is the total extension?

Answer: (a) 1630 N, or 167 kg (366 lb_f)   (b) 32 mm (1.27 in.)

1-5.8 A 6.4-mm (0.25-in.) diameter 1020 steel bar 1.83 m (6 ft) long supports a weight of 4450 N (1000 lb_f). What is the difference in length if the bar is changed to a 70–30 Monel?

1-5.9 A 1040 steel wire has a diameter of 0.89 mm (0.035 in.). Its yield strength is 980 MPa (142,000 psi) and its ultimate strength is 1130 MPa (164,000 psi). An aluminum alloy is also available with a yield strength of 255 MPa (37,000 psi), and a strength of 400 MPa (58,000 psi). (a) How much (%) heavier, or lighter, will an aluminum wire be than a steel wire to support a 40-kg (88-lb) load with the same elastic deformation as the steel wire? (b) How much (%) heavier, or lighter, must an aluminum wire be to support the same maximum load without permanent deformation? (c) Without breaking? (*Hint:* Establish a ratio of masses without calculating the actual areas.)

Answer: (a) +0.7%   (b) +32%   (c) −3%

QUIZ CHECKS

1A	stone, bronze, iron		**1G**	lower
1B	iron			stresses
1C	materials, energy		**1H**	elongation, reduction of area
1D	lose			fracture
	share		**1I**	yield
	compounds			elastic
1E	100 MPa		**1J**	ultimate
1F	force, area			original
	deformation, length		**1K**	the area decreases before breaking
	newton			(necking occurs)

---

* See footnote on p. 22.

Part One

Structure
and Properties
of Solid Phases

The next eight chapters examine the structures and properties of single-phase materials. Each of the materials that is cited will have only one basic structure. We will consider materials that possess a mixture of two or more structures in Parts II and III.

The coordinations of atoms with their neighbors will be the topic in Chapter 2. This includes bonding and the geometric arrangements of adjacent neighbors. This immediately leads to our first generalizations about properties (Section 2–6). The subject of Chapter 3 is crystals. They have a long-range repeating pattern that gives identical coordinations to all similar atoms of a material. However, there are always some imperfections in a crystal structure; and they have major consequences on certain properties. These are the topic of Chapter 4. Significant among the atomic disorders in solids is the ability of the atoms to diffuse and relocate. It is this diffusion that produces the structural changes discussed in Part II.

The delocalized electrons of metals, the covalent bonds of polymers (plastics), and the need for the atoms of ceramic compounds to always have unlike neighbors make metals, polymers, and ceramics noticeably different. Therefore, we will examine their structure–property relationships separately in Chapters 5, 6, and 7. However, regardless of these conventional distinctions, the same basic principles apply to each. Only the degree of response makes the properties differ.

Electron-controlled reactions will be the topics of Chapters 8 and 9. Conductivity, both metallic and semiconduction, will be considered first. This will require an introduction to the band nature of materials. Magnetic, dielectric, and optical properties will be related to structure in Chapter 9. This will utilize the background of the atomic structures that was encountered in earlier chapters.

Chapter Two

Atoms and Atomic Coordination

PREVIEW

In all solids, atoms are held together by bonds. They provide strength and related electrical and thermal properties to the solids. For example, strong bonds lead to high melting temperatures, high moduli of elasticity, shorter interatomic distances, and lower thermal expansion coefficients.

In this chapter, we shall draw upon the concepts introduced in general chemistry. We will need to examine the role of the valence electrons in the primary bonds—*ionic, covalent,* and *metallic*—in sufficient detail so we will be able to anticipate their effect on interatomic distances and atomic coordination.

Generalizations are presented that permit us to make our initial correlations between structure and properties.

CONTENTS

STUDY OBJECTIVES

1. To review the principal characteristics of atomic structure as presented in general chemistry.

2. To view molecules as groups of atoms having strong *intra*molecular bonds, but weak *inter*molecular bonds; this leads to variations in bond energies, bond lengths, and bond angles.

3. To know that the maximum coordination number for atoms is based (a) on the number of shared electron pairs in covalently bonded materials, and (b) on the size ratio in ionically bonded materials.

4. To develop a qualitative picture of how the forces and energies of bonds vary with interatomic distances (as presented in Fig. 2–4.2) to give us a basis for interatomic distance, modulus of elasticity, thermal expansion, etc., in later chapters.

5. To understand various factors that affect atom size. You should be able to rationalize the size differences in Table 2–4.1.

6. To develop the concept of delocalized electrons, particularly in metals since they involve the energy bands and energy gaps that must be used in Chapter 8.

7. To know certain broad generalizations that relate selected properties to bond energies.

8. To communicate with others on topics that utilize the terms and concepts of the chapter.

2-1

INDIVIDUAL ATOMS AND IONS

The atom is the basic unit of internal structure for our studies of materials. The initial concepts involving individual atoms are familiar to most of the readers. They include *atomic number, atomic mass,* and the relationships in the *periodic table.* We shall also give attention to the energy levels that are established by the electrons of the atoms.

Atomic Mass Unit

Since atoms are extremely small in comparison to our day-to-day concepts of mass, it is convenient to use the *atomic mass unit,* amu, as the basis for many calculations. The amu is defined as one twelfth of the atomic mass of carbon-12, the most common isotope of carbon. There are $0.6022 \ldots \times 10^{24}$ amu per g. We will use this conversion factor (called *Avogadro's number N*) in various ways. Since natural carbon contains approximately one percent C^{13} along with 98.9% C^{12}, the average atomic mass of a carbon atom is $12.011 \ldots$ amu. This is the value presented in the *periodic table* (Fig. 2–1.1) and in tables of selected

Periodic Table of the Elements

1A	IIA	IIIB	IVB	VB	VIB	VIIB	VIII	VIII	VIII	IB	IIB	IIIA	IVA	Va	VIA	VIIA	0
1 **H** 1.00797																1 **H** 1.0079	2 **He** 4.0026
3 **Li** 6.939	4 **Be** 9.012											5 **B** 10.811	6 **C** 12.011	7 **N** 14.007	8 **O** 15.9994	9 **F** 18.998	10 **Ne** 20.183
11 **Na** 22.990	12 **Mg** 24.312											13 **Al** 26.98	14 **Si** 28.086	15 **P** 30.97	16 **S** 32.064	17 **Cl** 35.453	18 **Ar** 39.95
19 **K** 39.102	20 **Ca** 40.08	21 **Sc** 44.96	22 **Ti** 47.90	23 **V** 50.94	24 **Cr** 52.00	25 **Mn** 54.94	26 **Fe** 55.85	27 **Co** 58.93	28 **Ni** 58.71	29 **Cu** 63.54	30 **Zn** 65.37	31 **Ga** 69.72	32 **Ge** 72.59	33 **As** 74.92	34 **Se** 78.96	35 **Br** 79.91	36 **Kr** 83.80
37 **Rb** 85.47	38 **Sr** 87.62	39 **Y** 88.91	40 **Zr** 91.22	41 **Nb** 92.91	42 **Mo** 95.94	43 **Tc** 99	44 **Ru** 101.07	45 **Rh** 102.91	46 **Pd** 106.4	47 **Ag** 107.87	48 **Cd** 112.40	49 **In** 114.82	50 **Sn** 118.69	51 **Sb** 121.75	52 **Te** 127.60	53 **I** 126.90	54 **Xe** 131.30
55 **Cs** 132.91	56 **Ba** 137.34	57–71 **La series***	72 **Hf** 178.49	73 **Ta** 180.95	74 **W** 183.85	75 **Re** 186.2	76 **Os** 190.2	77 **Ir** 192.2	78 **Pt** 195.1	79 **Au** 196.97	80 **Hg** 200.59	81 **Tl** 204.37	82 **Pb** 207.19	83 **Bi** 208.98	84 **Po** 210	85 **At** 210	86 **Rn** 222
87 **Fr** 223	88 **Ra** 226	89– **Ar series†**															

La series* (57–71):

58 **Ce** 140.12	59 **Pr** 140.91	60 **Nd** 144.24	61 **Pm** 145	62 **Sm** 150.35	63 **Eu** 151.96	64 **Gd** 157.25	65 **Tb** 158.92	66 **Dy** 162.50	67 **Ho** 164.93	68 **Er** 167.26	69 **Tm** 168.93	70 **Yb** 173.04	71 **Lu** 174.97

Ar series† (89–):

90 **Th** 232.04	91 **Pa** 231	92 **U** 238.03	93 **Np** 237	94 **Pu** 242	95 **Am** 243	96 **Cm** 247	97 **Bk** 249	98 **Cf** 251	99 **Es** 254	100 **Fm** 253	101 **Md** 256	102 **No** 253	103 **Lw** 257

Metals — Nonmetals

Figure 2–1.1 Periodic table of the elements, showing the atomic number and atomic mass (in amu). There are 0.602×10^{24} amu per gram; therefore the atomic masses are grams per 0.602×10^{24} atoms. Metals readily release their outermost electrons. Nonmetals readily accept or share additional electrons.

30

elements (Appendix B). Those atomic masses encountered most commonly by the reader include:

	Precision values	For most calculations
H	1.0079 . . . amu (or 1.0079 g/(0.602 × 10^{24} atoms))	1 amu (or 1 g/mole)
C	12.011 . . .	12
O	15.9994 . . .	16
Cl	35.453 . . .	35.5
Fe	55.847 . . .	55.8

These values may be rounded off as indicated for all but the most precise of our calculations.

The *atomic number* indicates the number of electrons associated with each neutral atom (and the number of protons in the nucleus). Each element is unique with respect to its atomic number. Appendix B lists selected elements, from hydrogen, with an atomic number of one, to uranium (92). It is the electrons, particularly the outermost ones, that affect most of the properties of engineering interest: (1) they determine the chemical properties; (2) they establish the nature of the interatomic bonding, and therefore the mechanical and strength characteristics; (3) they control the size of the atom and affect the critical conductivity of materials; and (4) they influence the optical characteristics. Consequently, we shall pay specific attention to the distribution and energy levels of the electrons around the nucleus of the atom.

Periodic Table

The *periodicity of elements* is emphasized in chemistry courses. We shall not repeat those characteristics here except to observe that the periodic table (Fig. 2–1.1) arranges the atoms of sequentially higher atomic numbers so that the vertical columns, called *groups,* possess atoms of similar chemical and electronic characteristics. In brief, those elements at the far left of the periodic table are readily ionized to give positive ions, *cations*. Those in the upper right corner of the periodic table more readily share or accept electrons. They are *electronegative*.

Electrons

Since electrons are components of all atoms, their negative electrical charge is commonly regarded as unity. In physical units, this charge is equal to 0.16 × 10^{-18} A·s per electron (or 0.16 × 10^{-18} coul/electron).

The electrons that accompany an atom are subject to very rigorous rules of behavior because they have the characteristics of standing waves during their movements in the neighborhood of the atomic nucleus. Again the reader is referred to introductory chemistry texts; however, let us summarize several

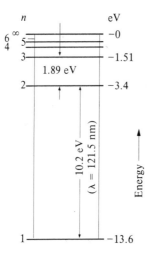

Figure 2–1.2 Energy levels for electrons (hydrogen). The electron of hydrogen normally resides in the lowest energy level. (At this level, it would take 13.6 eV, or 2.2 × 10⁻¹⁸ J, to separate the electron from the nucleus.) Electrons can be given additional energy, but only at specific levels. Gaps exist between these levels. These are forbidden energies.

features here. With individual atoms, electrons have specific energy states— *orbitals*. As shown in Fig. 2–1.2, the available electron energy states around a hydrogen atom can be very definitely identified.* To us, the important consequence is that there are large ranges of intermediate energies *not* available for the electrons. They are forbidden because the corresponding frequencies do not permit standing waves. Unless excited by external means, the one electron of a hydrogen atom will occupy the lowest orbital of Fig. 2–1.2.

Figure 2–1.3 shows schematically the energies of the lowest orbitals for sodium. Each orbital can contain no more than two electrons. These must be of opposite spins. Again, there are forbidden energy gaps between the orbitals that are unavailable for electron occupancy.

In our considerations, the topmost occupied orbital will have special significance since it contains the *valence electrons*. These electrons may be removed by a relatively small electric field, to give us the positive *cations* mentioned a few paragraphs ago. The energy requirements are called the *ionization energies*. In a later section we will see that these outermost or valence electrons are *delocalized* in metallic solids and free to move throughout the metal rather than remaining bound to individual atoms. This provides the basis for electrical and thermal conductivity.

When the valence orbitals are not filled, the atom may accept a limited number of extra electrons within these unfilled energy states, to become a negative ion, *anion*. These electronegative atoms with unfilled valence orbitals may also share electrons. This becomes important in covalent bonding and will be reviewed in the next section.

---

* This is done by spectrographic experiments.

Single atom

Figure 2–1.3   Energy levels for electrons (sodium). Since a sodium atom possesses eleven electrons, and only two electrons may occupy each level (orbital), several orbitals must be occupied. Gaps exist between them. It takes 5.1 eV (0.82×10^{-18} J) to remove the uppermost (valence) electron from sodium.

Example 2–1.1

Sterling silver contains approximately 7.5 w/o* copper and 92.5 w/o silver. What are the a/o* copper and a/o silver?

Procedure. Select a mass basis and calculate the number of atoms of each type of element present. Atomic masses are found in Fig. 2–1.1 and Appendix B.

Calculation. Basis: 10,000 amu alloy = 9250 amu Ag + 750 amu Cu.

Ag: 9250 amu Ag/(107.87 amu Ag/atom) = 85.75 atoms = 88 a/o

Cu: 750 amu Cu/(63.54 amu Cu/atom) = 11.80 atoms = 12 a/o.

Total atoms = 97.55   ◀

Example 2–1.2

The mass of small diamond is 3.1 mg. (a) How many C^{13} atoms are present if carbon contains 1.1 a/o of that isotope? (b) What is the weight percent of that isotope?

Procedure. (a) Determine the total number of carbon atoms present from the atomic masses; then 1.1 percent (on an atomic basis). (b) From the numbers of atoms in (a) and the atomic masses, calculate the mass (or weight) fraction.

Calculation

a)

$$\frac{0.0031 \text{ g}}{(12.011 \text{ g}/0.6022 \times 10^{24} \text{ atoms})} = 1.55 \times 10^{20} \text{ C atoms};$$

$$(1.55 \times 10^{20})(0.011) = 1.7 \times 10^{18} \text{ C}^{13} \text{ atoms}.$$

b) Basis: 3.1 mg.

$$\frac{\text{Mass}_{13}}{\text{Mass}_{\text{total}}} = \frac{(1.7 \times 10^{18})(13 \text{ amu})}{(1.55 \times 10^{20})(12.011 \text{ amu})} = 1.2 \text{ w/o}.$$

Comment. Since the mass of the C^{13} is greater than the average atom, the weight percent will be greater than the atom percent.   ◀

_____

* Weight percent, w/o; atom percent, a/o; linear percent, l/o; volume percent, v/o; mole percent, m/o; etc. In condensed phases (*solids and liquids*) weight percent is implied unless specifically stated otherwise. In *gases*, v/o or m/o are implied unless specifically stated otherwise.

Example 2–1.3

From Appendix B, indicate the orbital arrangements for a single iron atom, and for Fe^{2+} and Fe^{3+} ions.

Answers. The argon core is $1s^2 2s^2 2p^6 3s^2 3p^6$; therefore,

$$Fe: 1s^2 2s^2 2p^6 3s^2 3p^6 3d^6 4s^2. \tag{2–1.1}$$

During ionization, the $4s$ electrons are removed first; therefore:

$$Fe^{2+}: 1s^2 2s^2 2p^6 3s^2 3p^6 3d^6; \tag{2–1.2}$$

$$Fe^{3+}: 1s^2 2s^2 2p^6 3s^2 3p^6 3d^5. \blacktriangleleft \tag{2–1.3}$$

Example 2–1.4

Ten grams of nickel are to be electroplated on a steel surface with an area of $0.8953 \ m^2$. The electrolyte contains Ni^{2+} ions. (a) How thick will the nickel plate be? (b) What amperage is required if this is to be accomplished in 50 minutes?

Procedures. (a) Set up your own equation based on $\rho = m/V = m/At$. (b) Determine the number of nickel atoms. Two electrons are required per Ni^{2+}, each with 0.16×10^{-18} A·s (or 0.16×10^{-18} C). The (amp)(sec) product is obtained from these.

Calculation

a)
$$10 \ g/(8.9 \times 10^6 g/m^3)(0.8953 \ m^2) = 1.25 \times 10^{-6} \ m \qquad \text{(or } 1.25 \ \mu m\text{)}.$$

b)
$$\left[\frac{10 \ g \ Ni}{58.71 \ g \ Ni/0.6 \times 10^{24} \ atoms} \right] \left[\frac{(2 \ el/atom)(0.16 \times 10^{-18} \ A \cdot s/el)}{(3000 \ s)} \right] = 10.9 \ \text{amperes}.$$

Comment. By now, the student should be alert to the data available in the Appendixes. $\blacktriangleleft$

2–2

MOLECULES

The more common examples of molecules include compounds such as H_2O, CO_2, CCl_4, O_2, N_2, and HNO_3. Other small molecules are shown in Fig. 2–2.1. Within each of these molecules, the atoms are held together by strong attractive forces that have covalent bonds, although ionic bonds may be present. (*Covalently* bonded atoms *share* electron pairs; *ionic* bonds arise from coulombic attractions between *positive* and *negative* ions.) Unlike the forces that hold atoms together *within* the molecule, the bonds *between* molecules are weak and consequently each molecule is free to act more or less independently. These observations are borne out by the following facts: (1) Each of these molecular compounds has a low melting and a low boiling temperature compared with other materials. (2) The molecular solids are soft because the molecules can slide past each other with small stress applications. (3) The molecules remain intact in liquids and gases.

The molecules listed above are comparatively small; other molecules have large numbers of atoms. For example, pentatriacontane (shown in Fig. 2–2.2c) has over 100 atoms, and some molecules contain many thousand. Whether the molecule is small like CH_4 or much larger than that shown in Fig. 2–2.2(c), the

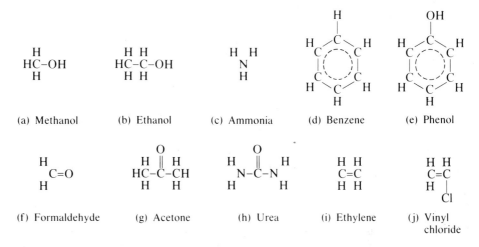

(a) Methanol   (b) Ethanol   (c) Ammonia   (d) Benzene   (e) Phenol

(f) Formaldehyde   (g) Acetone   (h) Urea   (i) Ethylene   (j) Vinyl chloride

Figure 2–2.1 Small molecules. Each carbon is surrounded by four bonds, each nitrogen by three, each oxygen by two, and each hydrogen and chlorine by one.

distinction between the strong *intra*molecular and the weaker *inter*molecular bonds still holds.

Other materials such as metals, MgO, SiO_2, and phenol-formaldehyde plastics have continuing three-dimensional structures of strong bonds. The difference between the structures of molecular materials and those with strong (primary) bonds continuing in all three dimensions produces major differences in properties. These differences will be considered in subsequent chapters.

Bond Lengths and Energies

The strength of bonds between atoms in a molecule, of course, depends on the kind of atoms and the other neighboring bonds. Table 2–2.1 is a compilation of bond lengths and energies for those atom couples most commonly encountered in molecular structures. The energy reported is the amount required to break one mole (Avogadro's number) of bonds. For example, 370,000 joules of energy are required to break 0.602×10^{24} C–C bonds, or $370,000/(0.602 \times 10^{24})$ joules per bond. This same amount of energy is released (-0.61×10^{-18} J/bond) if one of these C–C bonds is formed. Only the sign is changed.

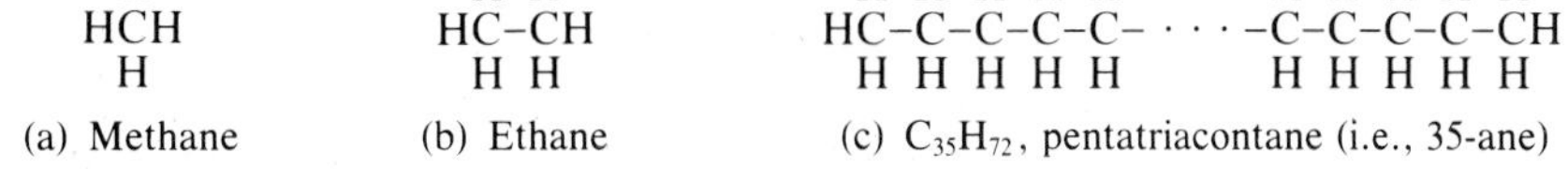

```
    H            H  H          H H H H H        H H H H H
   HCH          HC–CH          HC–C–C–C–C– · · · –C–C–C–C–CH
    H            H  H          H H H H H        H H H H H
(a) Methane   (b) Ethane    (c) C35H72, pentatriacontane (i.e., 35-ane)
```

Figure 2–2.2 Examples of molecules. Molecules are discrete groups of atoms. Primary bonds hold the atoms together within the molecule. Weaker, secondary forces attract molecules to each other.

TABLE 2–2.1		
Bond Energies and Lengths		

Bond	Bond energy*		Bond length, nm
	kJ/mole†	kcal/mole†	
C–C	370‡	88‡	0.154
C=C	680	162	0.13
C≡C	890	213	0.12
C–H	435	104	0.11
C–N	305	73	0.15
C–O	360	86	0.14
C=O	535	128	0.12
C–F	450	108	0.14
C–Cl	340	81	0.18
O–H	500	119	0.10
O–O	220	52	0.15
O–Si	375	90	0.16
N–H	430	103	0.10
N–O	250	60	0.12
F–F	160	38	0.14
H–H	435	104	0.074

* Approximate. The values vary with the type of neighboring bonds. For example, methane (CH_4) has the above value for its C–H bond; however the C–H bond energy is about 5% less in CH_3Cl, and 15% less in $CHCl_3$.

† Energies per 0.602×10^{24} bonds.

‡ All values are negative for forming bonds (energy is released), and positive for breaking bonds (energy is required).

Bond Angles

The chemist recognizes hybrid orbitals in certain covalent compounds, where the s and p orbitals are combined. The most important hybrid for us to review is the sp^3 orbital. Four equal orbitals are formed instead of having distinct $2s$ and $2p$ orbitals that occur in individual atoms (e.g., the individual sodium of Fig. 2–1.3). Methane (CH_4), carbon tetrachloride (CCl_4), and diamond provide examples of sp^3 orbitals that connect four identical atoms to the central carbon. Therefore, we find them equally spaced around the central carbon at 109.5° from each other.* Geometrically, this is equivalent to placing the carbon at a cube center and pointing the orbitals toward four of the eight corners (Fig. 2–2.3a). However, if the orbitals do not bond identical atoms to the central carbon, these time-averaged angles are distorted slightly, as shown for CH_3Cl (Fig. 2–2.3b).

* The angle 109.5° is a time-averaged value. Any particular H–C–H angle in CH_4 will vary rapidly as a result of thermal vibrations.

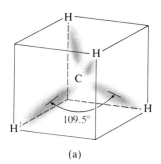

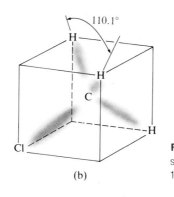

(a)

(b)

Figure 2–2.3 Bond angles. (a) Methane, CH₄, is symmetrical with each of the six angles equal to 109.5°. (b) Chloromethane, CH₃Cl, is distorted.

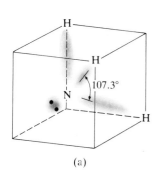

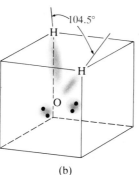

(a)

(b)

Figure 2–2.4 Bond angles. (a) Ammonia, NH₃, and (b) water have angles between the 109.5° of Fig. 2–2.3(a) and 90°. Ammonia has one lone-pair of electrons; water, two.

Greater distortions occur in hybrid orbitals when some of the electrons occur as *lone pairs* rather than in the covalent bond. This is particularly evident in NH₃ and H₂O (Fig. 2–2.4) where 107.3° and 104.5° are the time-averaged values for H–N–H and H–O–H, respectively.

One of the bond angles most frequently encountered in the study of materials is the C–C–C angle of the hydrocarbon chains (Fig. 2–2.5). While this will differ slightly, depending upon whether hydrogen or some other side radical is present, we may assume for our purposes that the C–C–C angle is close to 109.5°.

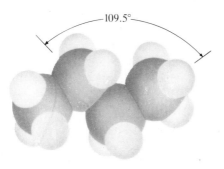

Figure 2–2.5 Bond angles (butane). Although we commonly draw straight chains (Figs. 2-2.2c and 2-2.7b), there is a C-C-C bond angle of about 109°.

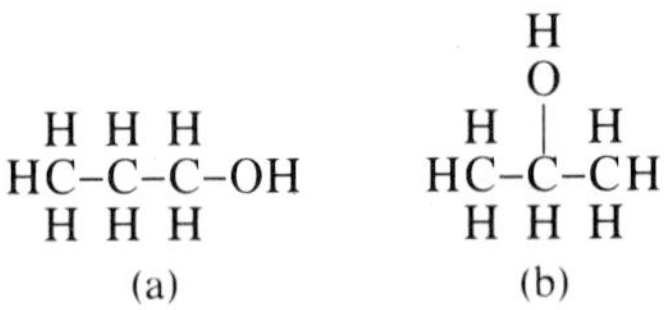

H
O
H H H   H | H
HC-C-C-OH   HC-C-CH
H H H   H H H
(a)   (b)

Figure 2–2.6 Isomers of propanol. (a) Normal propyl alcohol. (b) Isopropyl alcohol. The molecules have the same composition but different structures. Consequently, the properties are different. Compare with polymorphism of crystalline materials (Section 3–4).

Isomers

Many molecular compounds possess more than one atomic arrangement. This is illustrated in Fig. 2–2.6 for propyl and isopropyl alcohol. Variations in the structure of molecules with the same composition are called *isomers*. Differences in structure affect the properties of molecules. For example, the melting and boiling temperatures for normal propyl alcohol are $-127°C$ and $97.2°C$, respectively, whereas for isopropyl alcohol the corresponding temperatures are $-89°C$ and $82.3°C$.

Example 2–2.1

How much energy is given off when 70 g of ethylene (Fig. 2–2.7a) react to give polyethylene (Fig. 2–2.7b)?

Procedure. With each added C_2H_4 molecule, one $C=C$ is eliminated, and two $C-C$ bonds are formed. From Table 2–2.1, these changes involve $+680$ kJ per mole and $2(-370$ kJ/mole) of energy, respectively ($+$, for energy to break the bond; $-$, for energy released as a bond is formed). A mole includes 0.602×10^{24} bonds.

Calculation

$$\frac{+680,000 \text{ J}}{0.602 \times 10^{24} \text{ molecules}} - \frac{2(370,000 \text{ J})}{0.602 \times 10^{24} \text{ molecules}} = -9.96 \times 10^{-20} \text{ J/C}_2\text{H}_4,$$

$$70 \text{ g } (0.602 \times 10^{24} \text{ amu/g})/(28 \text{ amu/C}_2\text{H}_4) = 1.5 \times 10^{24} \text{ C}_2\text{H}_4,$$

$$(-9.96 \times 10^{-20} \text{ J/C}_2\text{H}_2)(1.5 \times 10^{24} \text{ C}_2\text{H}_4) = -150,000 \text{ J},$$

or

$$-150,000 \text{ J } (0.239 \text{ cal/J}) = -36 \text{ kcal}.$$

Comment. The reaction of Fig. 2–2.7 is the basic reaction for making large vinyl-type molecules that are used in plastics (Chapter 6). ◀

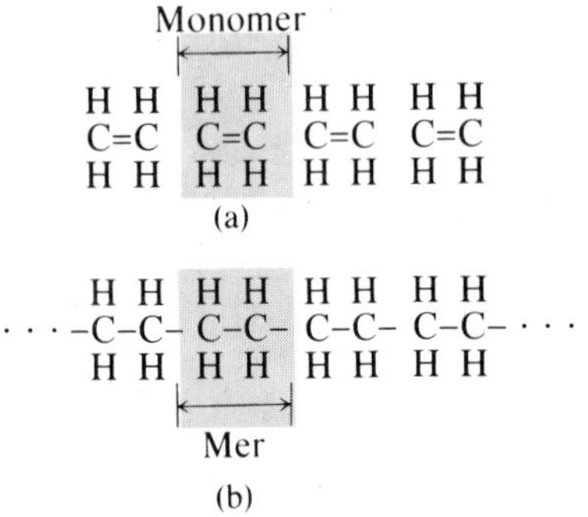

Monomer

H H   H H   H H   H H
C=C   C=C   C=C   C=C
H H   H H   H H   H H
(a)

H H   H H   H H   H H
····-C-C- C-C- C-C- C-C-····
H H   H H   H H   H H

Mer

(b)

Figure 2–2.7 Addition polymerization of ethylene. (a) Monomers of ethylene. (b) Polymer containing many C_2H_4 mers, or repeat units. The original double bond of the ethylene monomer is broken to form two single bonds and thus connect adjacent mers.

Example 2–2.2   Show sketches (carbon atoms only) of the various isomers of hexane, C_6H_{14}.

Sketches

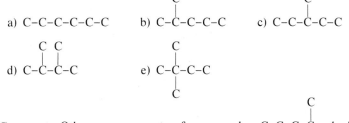

a) C–C–C–C–C–C

b) C–C–C–C–C (with C branch on second carbon)

c) C–C–C–C–C (with C branch on third carbon)

d) C–C–C–C (with two C branches)

e) C–C–C–C (with C branches on second carbon)

Comment. Other arrangements, for example, C–C–C–C (with C branch), duplicate one of the above alternatives (in this case, sketch *b*). ◀

2–3

ATOMIC COORDINATION

In the nearly unanimous majority of liquids and solids, atoms are closely associated with their neighbors. We saw this on an atomic scale in molecules. The oxygen atom of H_2O does not behave independently of its two hydrogen neighbors. For example, they must coordinate their thermal vibrations. The same is true for the carbon and hydrogens in methane (Fig. 2–2.3), where each carbon is *coordinated* with four hydrogens and each hydrogen is directly coordinated with one carbon.

Covalent Coordination

When atoms share a pair of electrons, they are said to have *covalent bonds*. Typically, these bonds are very strong. The hardness of diamond comes from the fact that each carbon atom is covalently bonded with four neighboring atoms, and each neighbor is bonded with an equal number of atoms to form a rigid 3-D structure (Fig. 2–3.1). Likewise, the *intra*molecular bonds in CH_4,

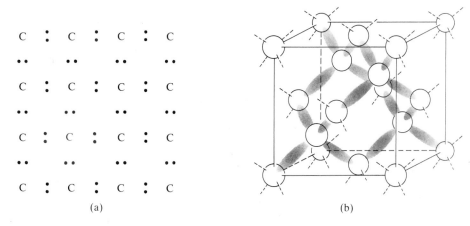

Figure 2–3.1 Diamond structure of carbon. The strength of the covalent bonds is what accounts for the great hardness of diamond. (a) Two-dimensional representation with a pair of electrons shared by neighboring atoms. (b) Three-dimensional representation with the bonds shown as the region of high electron probability (shaded).

TABLE 2–3.1			
Covalent Bonds per Atom*			
H	1	S	2
F	1	N	3
Cl	1	C	4
O	2	Si	4

* Also, the maximum coordination number.

polyethylene, and other polymeric (plastic) molecules arise from these covalent bonds of shared electrons.

Since carbon has four valence electrons and the sp^3 valence orbitals can accommodate a total of eight electrons (4 pairs), carbon is limited to four covalent bonds. Likewise, since nitrogen has five valence electrons, three neighbors produce the limiting eight electrons for the orbitals. Similar attention to oxygen, chlorine, etc., bring about the numbers that are familiar from introductory chemistry (Table 2–3.1). In methane, each of the carbon's four covalent bonds is with a separate hydrogen atom. In ethylene (C_2H_4 of Fig. 2–2.1), the double bond dictates only three neighbors. In acetylene, $H-C\equiv C-H$, the carbon atom has only two neighbors to account for the four covalent bonds. However, the *maximum coordination number* (i.e., number of immediate neighbors) for covalently bonded carbon is four. The other numbers of Table 2–3.1 also represent the maximum coordination numbers for those atoms. Lower coordination numbers are possible, if double bonds are present.

Ionic Coordination

The coordination numbers for atoms that are bonded by the coulombic attractions between positive and negative ions (rather than by sharing electrons covalently) are governed by different rules. Each positive ion (generally smaller) develops an *ionic bond* with as many surrounding larger negative ions as space will allow. This leads to limitations set by the ionic radii.

As shown in Fig. 2–3.2(a), each positive ion is coordinated (in two dimensions) with four neighbors. The attractions between ions with unlike charges

Figure 2–3.2 Ionic coordination (2-dimensional). (a) Coordination with $r/R > 0.41$. The smaller cation is coordinated with four anions (CN = 4 for 2-D). (b) With $r/R < 0.4$. The positive ion does not have maximum contact with all four neighboring negative ions. Likewise, there is repulsion between the contacting anions. (c) When $r/R < 0.4$, then CN = 3 is favored (2-D).

(a)   (b)   (c)

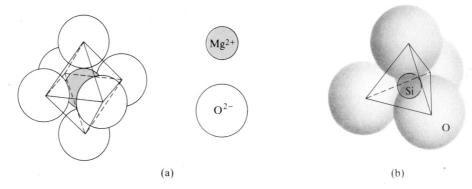

(a) (b)

Figure 2–3.3 Coordination numbers for ionic bonding (3-dimensional). (a) A maximum of six oxygen ions (O^{2-}) can surround each magnesium ion (Mg^{2+}). (b) The coordination number of Si^{4+} among O^{2-} is only four because the ion-size ratio is less than 0.4 (Table 2–3.2).

are greater than the repulsions between ions with like charges. With larger negative ions (or smaller positive ions) as shown in Fig. 2–3.2(b), the contact and repulsion of the anions prevents a strong attraction from developing between the smaller cation and all of the negative neighbors. Thus, a smaller coordination number is more stable (Fig. 2–3.2c). The maximum coordination number is dictated by the radii, r/R, where r is the radius of the smaller ion (almost always positive), and R is the radius of the larger anion.

Figure 2–3.2 showed 2-D sketches. The same concept is applicable to three dimensions, where a coordination number (CN) of six is shown in Fig. 2–3.3(a). The calculation of the minimum r/R ratio for CN = 6 is obtained in Example 2–3.1 from the sketch in Fig. 2–3.4(a), where the r/R ratio permits anion contact and the beginning of strong repulsion between the larger negative ions.

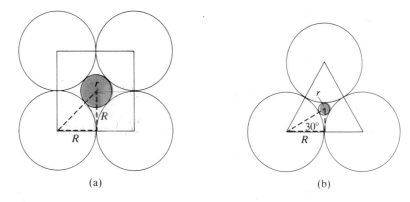

(a) (b)

Figure 2–3.4 Coordination calculations. (a) Minimum r/R for six-fold coordination. (b) Minimum r/R for three-fold coordination. (Compare with Example problems and Fig. 2–3.3.)

TABLE 2–3.2		
Coordination Numbers versus Minimum Radii Ratios		
Coordination number	**Radii ratios, r/R***	**Coordination geometry**
3-fold	≥ 0.15	
4	≥ 0.22	
6	≥ 0.41	
8	≥ 0.73	
12	1.0	——

* r—smaller radius; R—larger radius.

Other limiting r/R ratios are given in Table 2–3.2. As with covalent bonding, these limits are for *maximum coordination numbers;* and thus, minimum radii ratios. It is not uncommon to have fewer than the maximum number of neighbors.*

Extended Bonding

Every carbon atom in diamond demands the same coordination with its neighbors, namely, CN = 4. Thus, the structure of Fig. 2–3.1(b) continues far beyond those 15–20 atoms that were sketched. In a 1-carat diamond (0.2 g), there are 10^{22} carbon atoms. All of the 2×10^{22} bonds† between neighboring atoms are strong (primary). There are no weak (secondary) intermolecular bonds like we find between molecules (Fig. 2–2.2). More specifically, diamond is not molecular, because it has an extended, 3-D structure of primary (in this case covalent) bonds.

Ionic bonds (which also are primary bonds) involve the positive and negative ions of two or more elements and so they occur only in compounds. Since each cation is surrounded by negative ions, which in turn are coordinated with other positive ions, these compounds also develop extended, 3-D structures. This is indicated for NaCl in two dimensions (only) in Fig. 2–3.5. (We will see NaCl again, but in three dimensions in Fig. 3–1.1.) Since a grain of salt that

---

* In some cases, there must be fewer neighbors in order to balance charges.

† There are 2×10^{22} bonds because CN = 4 for carbon in diamond, but each bond is shared with a neighbor. Therefore, there are $\frac{1}{2}$ bonds per carbon.

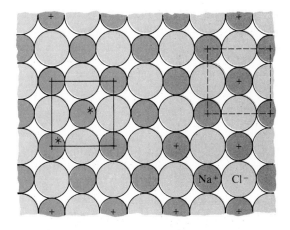

Figure 2–3.5 NaCl (2-dimensional). The ionic bonds between the positive, Na⁺, and negative, Cl⁻, ions extend in all directions. There are no isolated NaCl molecular pairs. (See Fig. 3–1.1 for a 3-D sketch of NaCl.)

weighs only 0.001 g has $\approx 10^{19}$ Na$^+$ ions and 10^{19} Cl$^-$ ions, and each ion is joined with its neighbors by primary ionic bonds, it is evident that a "molecule" of NaCl does not exist as a separate entity. There are no weak, or secondary, intermolecular bonds like we find between molecules (Fig. 2–2.2). Thus, although it is convenient to identify the molecular weight of NaCl (58.5 g) for calculation purposes, we cannot relate properties to an NaCl "molecule." It does not exist within our definition in Section 2–2.

Example 2–3.1

Show the origin of 0.41 as the minimum ratio for a coordination number of six.

Procedure. The minimum ratio of possible sizes to permit a coordination number of six is sketched in Fig. 2–3.4(a). From Fig. 2–3.3(a), note that the fifth and sixth ions sit above and below the center ion of Fig. 2–3.4(a).

Solution

$$(r + R)^2 = R^2 + R^2, \qquad r = \sqrt{2}\,R - R, \qquad \text{and } \frac{r}{R} = 0.41. \quad \blacktriangleleft$$

Example 2–3.2

Zircon can be grown as a single crystal and displayed as a gem. Its composition is $ZrSiO_4$. One such zircon weighs 0.3 g (1.5 carats) after cutting. How many zirconium atoms are present?

Procedure. Being a single crystal, each of the atoms is coordinated with primary bonds. There are no separable $ZrSiO_4$ "molecules"; however, we can use a mole of $ZrSiO_4$ for calculation purposes.

Calculation. From Fig. 2–1.1,

$$1 \text{ mole } ZrSiO_4 = 91.2 + 28.1 + 4(16) = 183.3 \text{ g},$$

and 1 mole $ZrSiO_4$ contains 0.6×10^{24} $ZrSiO_4$ and 0.6×10^{24} Zr.

$$0.3 \text{ g } (0.6 \times 10^{24} \text{ } ZrSiO_4/183.3 \text{ g}) = 10^{21} \text{ } ZrSiO_4$$
$$= 10^{21} \text{ Zr.}$$

Added information. Do *not* report 0.9819967×10^{21} as an answer. The source data had only one significant figure (0.3). Our final answer cannot have greater accuracy. The recommended practice is to round off the *final* answer (not the intermediate values) to the number of significant figures possessed by the least accurate *non-integer* datum (+ one). (This added figure has occasional value in revealing a possible bias within the preceding digit.) ◀

2–4

INTERATOMIC DISTANCES

The sketch of Fig. 2–3.1(b) for diamond shows a specific distance between adjacent carbon atoms. We also encountered atomic and ionic radii when we discussed the coordination number of atoms. There are attractive forces that pull atoms together into solids (and liquids). These attractive forces give strength to the solid. Of course, we make use of this strength and other accompanying properties. But what keeps the atoms from being drawn still closer together? It should be apparent from the preceding figures and the discussion that there is much vacant "space" in the volume surrounding the nucleus of an atom. The existence of this space is evidenced by the fact that neutrons can move through the fuel and the other materials of a nuclear reactor, traveling among many atoms before they are finally stopped (see Fig. 13–4.1).

The space between atoms is caused by interatomic repulsive forces, which exist in addition to the interatomic attractive forces described in Sections 2–2 and 2–3. Mutual repulsion results primarily because the close proximity of two atoms places too many electrons into interacting locations. The equilibrium distance is that distance at which the repulsive and the attractive forces are equal. An analogy may be made between the interatomic distances among atoms and the spacing between the two ring magnets of Fig. 2–4.1. (In this example, the magnets are aligned to give repulsion rather than attraction.) Of course the forces in this analogy are not identical to those between atoms; however, the principle of force balances is comparable. The top ring magnet is pulled by a force (gravity) toward the lower ring magnet (which in this case is fixed by the thumb). Since the force of gravity is essentially constant over the distance considered here, the top magnet falls to the point where it is repulsed by an equal magnetic force, of opposite direction. Because the repulsive force increases as an inverse function of the distance, equilibrium distance is achieved. Note that the magnets remain separated by space. (A nonmagnetic material may move through this space, just as a neutron (neutrons have no charge) may move among atoms in a solid.)

Coulombic Forces

The ionic bond will be used to illustrate the balance between attractive and repulsive forces in materials. The coulombic force F_C developed between two point charges is related to the quantity of the two charges $Z_1 q$ and $Z_2 q$, and their separation distance a_{1-2} as follows:

$$F_C = -k_0 (Z_1 q)(Z_2 q)/a_{1-2}^2, \qquad (2\text{–}4.1)$$

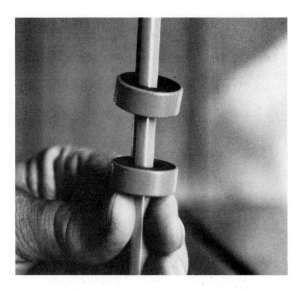

Figure 2–4.1 Balance of forces (ceramic ring magnets). Downward force on upper ring is caused by gravity. Upward force is caused by magnetic repulsion. Space remains between the two magnets at the equilibrium position. (Of course the forces in this analogy are not identical to those between atoms; however, the principle of opposing forces is comparable.)

where Z is the valence (+ or −) and q is 0.16×10^{-18} coulomb. The proportionality constant k_0 depends on the units used when we are considering adjacent ions.*

Electronic Repulsive Forces

The repelling force F_R between the electronic fields of two atoms or ions is also an inverse function with distance, but to a higher power:

$$F_R = -bn/a_{1-2}^{n+1}. \qquad (2\text{–}4.2)$$

Both b and n are empirical constants, with n equal to approximately 9 in ionic solids. Comparing, $F_C \propto a^{-2}$, and $F_R \propto a^{-10}$. Thus, the attractive forces predominate at greater distances of atomic separation, and the repulsive forces predominate at closer interatomic spacings (Fig. 2–4.2a). The equilibrium spacing, $o\text{–}a'$, is a natural result when

$$F_C + F_R = 0. \qquad (2\text{–}4.3)$$

A tension force is required to overcome the predominant forces of attraction if the spacing is to be increased. Conversely, a compressive force has to be applied to push the atoms closer together against the rapidly increasing electronic repulsion.

The equilibrium spacing is a very specific distance for a given pair of atoms, or ions. It can be measured to five significant figures by x-ray diffraction (Chapter 3), if temperature and other factors are controlled. It takes a large force to stretch or compress that distance as much as one percent. (Based on Young's modulus, a stress of 2000 MPa (300,000 psi) is required for iron.) It is for this

---

* With SI units, k_0 is 9×10^9 V·m/C, since $k_0 = 1/4\pi\epsilon_0$.

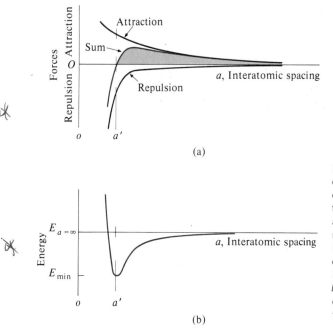

(a)

(b)

Figure 2–4.2 Interatomic distances. (a) The equilibrium spacing o–a' is the distance at which the attractive forces equal the repulsive forces. (b) The lowest potential energy occurs when o–a' is the interatomic distance. Since $E = \int F\, da$, the shaded area of (a) equals the depth of the energy trough in (b).

reason that the *hard ball* provides a usable model for atoms for many purposes where strength or atom arrangements are considered.*

Bonding Energy

The sum of the above two forces provides us with a basis for bonding energies (Fig. 2–4.2b). Since the product of force and distance is energy,

$$E = \int_{\infty}^{a} (F_C + F_R)\, da. \tag{2–4.4}$$

We will use infinite atomic separation as our energy reference, $E_{a=\infty} = 0$. As the atoms come together, energy is *released* in an amount equal to the shaded area of Fig. 2–4.2(a). The amount of energy released is shown in Fig. 2–4.2(b). Note, however, that at o–a', where $F = 0 = dE/da$, there is a minimum of energy because energy would have to be supplied to force the atoms still closer together. The depth of this *energy well*, $E_{a=\infty} - E_{min}$, represents the bonding

---

* The hard-ball model is not suitable for all explanations of atomic behavior. For example, a neutron (which doesn't have a charge) can travel through the space among the atoms without being affected by the electronic repulsive forces just described. Likewise, atomic nuclei can be vibrated vigorously by increased thermal energy, with only a small expansion of the *average* interatomic spacing. Finally, by a momentary distortion of their electrical fields, atoms can move past one another in a crowded solid. (See diffusion in Chapter 4.)

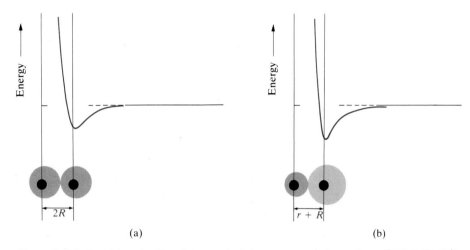

Figure 2–4.3 Bond lengths. The distance of minimum energy between two adjacent atoms is the bond length. It is equal to the sum of the two radii. (a) In a pure metal, all atoms have the same radius. (b) In an ionic solid, the radii are different because the two adjacent ions are never identical.

energy, because that much energy would be released $(-)$ as two atoms are brought together (at 0 K). The data of Table 2–2.1 list such values for covalent bonds.*

The schematic representations of Fig. 2–4.2 will be useful to us on various occasions in subsequent sections, when we pay attention to elastic moduli, thermal expansion, theoretical strengths, melting and vaporization temperatures, etc.

Atomic and Ionic Radii

The equilibrium distance between the centers of two neighboring atoms may be considered to be the sum of their radii (Fig. 2–4.3). In metallic iron, for example, the mean distance between the centers of the atoms is 0.2482 nm (or 2.482 Å) at room temperature. Since both atoms are the same, the radius of the iron atom is 0.1241 nm.

Several factors can change this distance between atom centers. The first is temperature. Any increase in energy above the minimum point shown in Fig. 2–4.2(b) will increase the mean distance because the energy trough is asymmetric. This increase in the mean spacing between atoms accounts for the thermal expansion of materials.

Ionic valence also influences interatomic spacing. The ferrous iron ion (Fe^{2+}) has a radius of 0.074 nm, which is smaller than that of the metallic iron

---

* The attraction forces of covalent bonds are more complicated than those of the coulombic forces that were presented in Eq. (2–4.1). However, a comparable energy well exists.

TABLE 2–4.1

Selected Atomic Radii

| Element | Metallic atoms | | Ions | | | Covalent bonds |
	CN*	Radius, nm	Valence	CN*	Radius, nm†	½ bond length,‡ nm
Carbon						Single 0.077
						Double 0.065
						Triple 0.06
Silicon			4+	6	0.042	Single 0.117
			4+	4	0.038	
Oxygen			2−	8	0.144	Single 0.075
			2−	6	0.140	Double 0.065
			2−	4	0.127	
			2−	2	~0.114	
Chlorine			1−	8	0.187	Single 0.099
			1−	6	0.181	
Sodium	8	0.1857	1+	6	0.097	
Magnesium	12	0.161	2+	6	0.066	
Aluminum	12	0.1431	3+	6	0.051	
			3+	4	0.046	
Iron	8	0.1241	2+	6	0.074	
	12	~0.127	3+	6	0.064	
Copper	12	0.1278	1+	6	0.096	

* CN = coordination number, i.e., the number of immediate neighbors. For ions, $1.1\,R_{CN=4} \approx R_{CN=6} \approx 0.97\,R_{CN=8}$.

† These values vary slightly with the system used. Patterned after Ahrens.

‡ From Table 2–2.1.

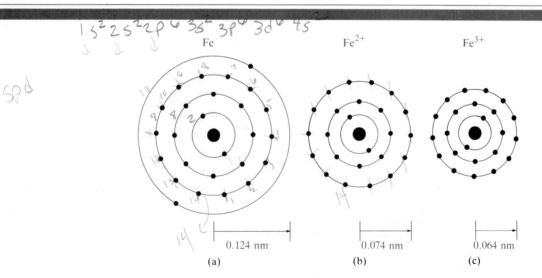

Figure 2–4.4   Atom and ion sizes (schematic). (a) Both iron atoms and iron ions have the same number of protons (26). (b) If two electrons are removed, the remaining 24 electrons and adjacent negative ions are pulled closer to the 26-proton nucleus. (c) A ferric ion holds its 23 electrons still closer to the nucleus.

atom (Table 2–4.1 and Appendix B*). Since the two outer valence electrons of the iron ion have been removed (Fig. 2–4.4), the remaining 24 electrons are pulled in closer to the nucleus, which still maintains a positive charge of 26. A further reduction in interatomic spacing is observed when another electron is removed to produce the ferric ion (Fe^{3+}). The radius of this ion is 0.064 nm or only about one half that of metallic iron.

A negative ion is larger than its corresponding atom. Since there are more electrons surrounding the nucleus than there are protons in the nucleus, the added electrons are not as closely attracted to the nucleus as were the original electrons.

A third factor affecting the size of the atom or ion is the number of adjacent atoms. An iron atom has a radius of 0.1241 nm when it is in contact with eight adjacent iron atoms, which is the normal arrangement at room temperature. If the atoms are rearranged to place this one iron atom in contact with twelve other iron atoms, the radius of each atom is increased slightly, to ≈ 0.127 nm. With a large number of adjacent atoms, there is more electronic repulsion from neighboring atoms, and consequently the interatomic distances are increased (Table 2–4.1).

We can apply a rule-of-thumb for the comparative radii of ions, since the radii are $\approx 10\%$ larger for CN = 6 than for CN = 4. Conversely, the radii are $\approx 3\%$ smaller for CN = 6 than for CN = 8. (See the footnotes to Table 2–4.1 and to Appendix B.)

We generally do not speak of atomic radii in covalently bonded materials because the electron distributions may be far from spherical (Fig. 2–2.3a). Furthermore, the limiting factor in atomic coordination is not the atom size, but rather the number of electron pairs available. Even so, we may make some comparisons of interatomic distances (bond lengths) when we look at Table 2–4.1. In ethane with a single C–C bond, this nucleus-to-nucleus distance is 0.154 nm as compared with 0.13 nm for a C=C bond and 0.12 nm for the C≡C bond. This change is to be expected, since the bonding energies are greater with the multiple bonds (Table 2–2.1).

● **Example 2–4.1**

MgO and NaCl are comparable, except that Mg^{2+} and O^{2-} ions are divalent and Na^+ and Cl^- ions are monovalent. Therefore, the Mg–O interatomic distance is 0.21 nm while the latter is 0.28 nm. Compare the coulombic attractive forces ($\rightarrow \leftarrow$) that are developed at these two distances for the two pairs of ions.

Procedure. This example is simply a plugging of Eq. (2–4.1); however, it lets one observe the differences in the coulombic attraction forces as the charges and inter-atomic distances are changed. The constant, k_0, has the value shown in the footnote following Eq. (2–4.1).

---

* The metallic radii used in this book are from the ASM *Metals Handbook*. The ionic radii are patterned after Ahrens.

● See Preface for the bullet, ●, notation.

Calculation

$$F_{Mg \to \leftarrow O} = -(9 \times 10^9 \text{ V·m/C}) \left(\frac{(+2)(-2)(0.16 \times 10^{-18} \text{ C})^2}{(0.21 \times 10^{-9} \text{ m})^2} \right)$$

$$= 20.9 \times 10^{-9} \text{ J/m}.$$

Similarly,

$$F_{Na \to \leftarrow Cl} = 2.9 \times 10^{-9} \text{ J/m}.$$

Comments. The opposing electronic repulsive forces (Eq. 2–4.2) will be -20.9 nJ/m and -2.9 nJ/m at these equilibrium distances. Thus, from Eq. (2–4.2), the empirical constant b is $\approx 0.4 \times 10^{-105}$ Jm^9 for MgO, and $\approx 10^{-105}$ Jm^9 for NaCl (assuming $n = 9$). ◀

● **Example 2–4.2**

Compare the energy of the $Mg^{2+} \to \leftarrow O^{2-}$ bond with the energy of the $Na^+ \to \leftarrow Cl^-$ bond.

Procedure. Combine Eqs. (2–4.1) and (2–4.4); then integrate from ∞ to a. We obtain

$$E = (k_0 Z_1 Z_2 q^2/a) + (b/a^n). \qquad (2\text{–}4.5)$$

Calculation. Use data from Example 2–4.1.

$$E_{Mg-O} = \frac{(9 \times 10^9 \text{ V·m/C})(-4)(0.16 \times 10^{-18} \text{ C})^2}{0.21 \times 10^{-9} \text{ m}} + \frac{0.4 \times 10^{-105} \text{ Jm}^9}{(0.21 \times 10^{-9} \text{ m})^9}$$

$$= -4.4 \times 10^{-18} \text{ J} + 0.5 \times 10^{-18} \text{ J}$$

$$= -3.9 \times 10^{-18} \text{ J};$$

$$E_{Na-Cl} = -0.8 \times 10^{-18} \text{ J} + 0.1 \times 10^{-18} \text{ J}$$

$$= -0.7 \times 10^{-18} \text{ J}.$$

Comments. We integrated from ∞ to a, rather than from a to ∞, since our reference energy is at infinite separation. The negative values indicate that energy is given off as the ions approach each other from a distance. In this energy range, it is common to use electron volts (1 joule = 6.24×10^{18} eV); therefore, the two calculated energies are -24 eV and -4.4 eV, respectively.

Added information. The greater bonding energy for MgO than for NaCl is reflected by the difference in their *melting temperatures*, 2800°C and 800°C, respectively. Likewise, the *elastic modulus*, which is the stress required for unit strain is nearly 12 times as great for MgO as for NaCl. Finally, the *thermal expansion coefficient*, α, is less for the more strongly bonded MgO than for the NaCl. ◀

2–5

ELECTRON MOVEMENTS AMONG COORDINATED ATOMS

Although we may represent electrons on paper as dots (e.g., Fig. 2–3.1a), we usually think of electrons as being in continuous motion. The popular representation of an atom is a nucleus with "planetary" electrons. Even this is too simple. The physicist chooses to analyze electron movements with wave equations, because the electrons have corresponding characteristics. We will not

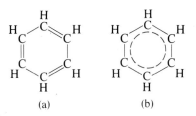

Figure 2–5.1 Delocalized electrons. (a) Benzene ring (simplified). (b) Benzene ring (preferred). The σ orbitals between the carbon atoms are stereospecific. Electrons in the other orbitals (π) can move from one side of the molecule to the other in response to internal and external electrical fields; they are delocalized.

develop wave equations. We will recall, however, from Section 2–1 that the electrons of individual atoms "have the characteristics of standing waves during their movements in the neighborhood of the atomic nucleus." This rationalized the energy levels of the orbitals in Figs. 2–1.2 and 2–1.3. More importantly, it revealed why electrons are restrained to discrete levels, with intervening forbidden energies.

Our attention in this section will be focused on the valence electrons, since they play a more prominent role in the conductive and dielectric properties of materials.

Delocalized Electrons

The covalent bond involves a pair of electrons that are *stereospecific;* that is, they reside in a closely defined region of space dictated by the two adjacent atoms.* The chemist calls the bonds arising from these localized electrons the σ bond.

Many molecules have only σ bonds between atoms for example, CH_4 of Fig. 2–2.3(a). In addition, the familiar molecular ring in benzene, C_6H_6, has electrons that become delocalized. Recall from Table 2–3.1, that each carbon atom requires four bonds. Thus, a simple representation of benzene is that of Fig. 2–5.1(a). A better representation is that shown in Fig. 2–5.1(b), because the six electrons that are not used for the σ bonds develop standing waves across the whole molecule. These electrons, which are called π electrons, respond to an electrical field by moving toward the side of the molecule nearer the positive electrode. However, they cannot leave the molecule (except under unusually catastrophic conditions). There are as many wave patterns for these delocalized electrons *as there are carbon atoms* in the ring. (In this case, six.)

Now, let us consider *graphite* (Fig. 2–5.2). The hexagonally arrayed layers, or sheets, of carbon atoms in graphite possess delocalized electrons (as well as electron pairs for σ bonds between specific atoms). Under neutral conditions, these delocalized electrons establish standing waves throughout the whole 2-dimensional layer of carbon atoms. There are as many possible wave patterns (*states*) for these delocalized electrons as there are carbon atoms in the sheet. (In this case, 10^x.) Furthermore, these delocalized electrons can respond to

---

* Even so, it is impossible to pin-point their location exactly, because of the *uncertainty principle*. However, the probability of the position is greatest along the line between the two atoms (the shaded areas of Figs. 2–2.3, 2–2.4, and 2–3.1(b)).

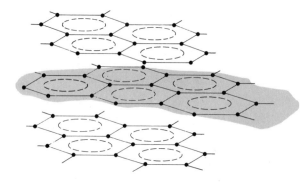

Figure 2–5.2 Delocalized electrons in graphite layers. Each layer contains "multiple benzene rings" (Fig. 2–5.1). The conductivity is more than 100 times greater in the parallel direction than in the perpendicular direction.

electric fields by moving across the sheets of graphite in a wavelike pattern. In fact, conductivity becomes possible if a positive electrode is present to remove electrons from one edge of the layer, and a negative electrode is available to supply electrons to the other edge.*

Typical *metals* have delocalized electrons that can move in three dimensions. It is thus common to speak of an electron "cloud" or "gas" because the outer, least strongly bonded electrons are able to move throughout the metal structure. The orbitals for metals are sketched schematically in Fig. 2–5.3 for sodium. The energy levels in multiatomic sodium differ from that of the single-atom orbitals that were first shown in Fig. 2–1.3. The prime change between Fig. 2–5.3(a) and 2–5.3(b) is that the upper valence orbital has split into as many levels, or states, as there are atoms in the system. Note that the average energy of the electrons in the valence band of Fig. 2–5.3(b) is below the energy of the 3s orbital for the individual atom. This accounts for the bonding in metals; in brief, energy would have to be supplied to overcome the metallic bond and separate the atoms from one another and to reestablish the individual atomic orbitals. Qualitatively speaking, we find very strong bonds holding tungsten atoms together. Its melting and boiling temperatures are very high (~3400 and 5900°C, respectively); also, it has an extremely high modulus of elasticity (50,000,000 psi, or 345,000 MPa; see Appendix C for comparisons). In contrast, the bonds in sodium are low as evidenced by its melting point (97.8°C) and soft behavior. Both have delocalized electrons for electrical and thermal conduction.

Mean Free Path

The delocalized electrons, just discussed, have wave characteristics as they move throughout a material. Waves proceed with a minimum of interruption when traveling through a periodic structure, i.e., a structure with a pattern that has uniform repetition. Any irregularity in the repetitive structure through

---

* Although the electrons can readily move within the graphite layers, there is not a comparable mechanism to move them from one layer to another. Therefore, electrical conductivity is very *anisotropic*, i.e., very directional.

which the wave travels will deflect the wave. Thus, if an electron had been traveling toward the positive electrode, a displaced atom or a foreign atom could cause it to be reflected toward the negative electrode. Of course, a second deflection or reflection could redirect the electron so that it picks up some of its original vector.

As the electrons move toward the positive electrode, they continuously acquire additional momentum and hence more velocity. Extra energy is necessary. As they move toward the negative electrode, they continuously lose momentum, and hence velocity; they must drop to lower energy states. Thus, the distance between deflections and reflections determines the *net*, or *drift velocity*. This representative distance in a material is called the mean free path. With longer mean free paths, greater drift velocities, $\bar{v}$, are attained in a given electric field, $\mathscr{E}$; that is, $d\bar{v}/d\mathscr{E}$ is higher. This value of $d\bar{v}/d\mathscr{E}$ is called the charge *mobility*, μ, which directly relates to the electrical *conductivity*, σ:

$$\sigma \propto \mu. \qquad (2\text{–}5.1)$$

The preceding sequence—mean free path → drift velocity → mobility → conductivity—will let us anticipate the effect of various structural factors upon the electrical behavior of a material (Chapters 5 and 8). For example, an increased temperature introduces greater thermal agitation among the atoms of a metal. As a consequence, the structure is less regular and the mean free path is shortened; in turn, the drift velocity, electron mobility, and electrical conductivity are reduced. In addition, the thermal conductivity is lowered, because most of the thermal energy is carried through metals by the movements of the mobile valence electrons. The above concepts are of great help to the modern engineer.

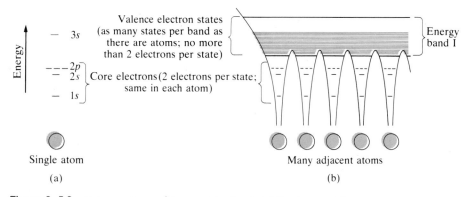

Figure 2–5.3   Valence electrons in metal (sodium). (a) Single atom (Fig. 2–1.3). (b) Multiatomic sodium metal. The valence electrons are delocalized into an energy band. These electrons are able to move throughout the metal and thus provide conductivity. The valence electrons fill only the bottom half of the band. Their average energy is lower than that of the 3s electrons with individual atoms.

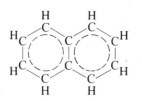

Figure 2–5.4   Naphthalene (Example 2–5.1). A carbon is at each hexagon corner. This leaves ten delocalized electrons.

Example 2–5.1

The structure of naphthalene is a double benzene ring with two carbons of each ring in common. (a) How many delocalized electrons are there? (b) How many wave patterns (electron states) are available to the delocalized electrons?

Procedure. Sketch the molecule. Determine the formula. Account for the valence electrons.

Answer. See Fig. 2–5.4.

a) Formula: $C_{10}H_8$.
   Valence electron balance
       Total valence electrons: $(4/C)(10\ C) + (1/H)(8H) = 48$;
   Electrons in bonds: C–H $= 16$
                           C–C (σ) $= 22$
                           Remainder (delocalized) $= 10$.

b) Electron states = number of carbon atoms = 10.

• **Added information.** Each state can contain two electrons (of opposite spin). Therefore, a total of 20 delocalized electrons could be accommodated (were it not for a resulting imbalance of charges in the molecule). ◀

2–6

GENERALIZATIONS BASED ON ATOMIC COORDINATIONS

Several engineering properties may now be associated *qualitatively* with the atomic relationships discussed in this chapter. The following generalizations will receive greater elaboration in subsequent chapters.

   Density is controlled by atomic weight, atomic radius, and coordination number. The latter is a significant factor because it controls the atomic packing (Chapter 3).

   Melting and *boiling temperatures* can be correlated with the depth of the energy trough shown in Fig. 2–4.2(b). Atoms have minimum energy (at the bottom of the trough) at a temperature of absolute zero. Increased temperatures raise the energy until the atoms are able to separate themselves one from another.

   Strength is influenced by the height of the total force or sum curve of Fig. 2–4.2(a). That force, when related to the cross-sectional area, gives the stress required to separate atoms. (As we shall see in Section 5–4, materials can deform through a process other than direct separation of the atoms. However, the amount of stress required to deform them is still governed by the interatomic forces.) Also, since larger interatomic forces of attraction imply deeper energy troughs, we observe that materials with high melting points are often the *harder* materials: e.g., diamond, Al_2O_3, TiC. In contrast, in materials with weaker bonds there is a correlation between softness and low melting point: e.g., lead, plastics, ice, and grease. Apparent exceptions to these generalizations can arise when more than one type of bond is present, as in graphite and clay.

The *modulus of elasticity* can be calculated from the slope of the sum curve of Fig. 2–4.2(a) because at the equilibrium distance, where the net force is zero, dF/da relates stress to strain. As long as the strain or change in interatomic distance is less than one percent, the modulus of elasticity remains essentially constant. Extreme compression or extreme tension respectively raise or lower the modulus of elasticity.

Thermal expansions of materials with comparable atomic packing vary inversely with their melting temperatures. This indirect relationship exists because the higher-melting-point materials have deeper and therefore more symmetrical energy troughs. Thus their mean interatomic distances increase less with a given change in thermal energy. Examples of several metals include Hg, melting temperature equals $-39°C$, coefficient of linear expansion equals 40×10^{-6} m/m·°C; Pb, 327°C, 29×10^{-6} m/m·°C; Al, 660°C, 22×10^{-6} m/m·°C; Cu, 1084°C, 17×10^{-6} m/m·°C; Fe, 1538°C, 12×10^{-6} m/m·°C; W, 3387°C, 4.2×10^{-6} m/m·°C.

Electrical conductivity is very dependent on the nature of the atomic bonds. Both ionically and covalently bonded materials are poor conductors, because electrons are not free to leave their host atoms. On the other hand, the delocalized electrons of metals easily move along a potential gradient. Semiconductors will be considered in Chapter 8; however, we can note here that their conductivity is controlled by the freedom of movement of their electrons.

Thermal conductivity is high in materials with metallic bonds, because delocalized electrons are efficient carriers of thermal as well as electrical energy (Section 5–1).

Review and Study

SUMMARY

1. We must use atomic weights in our study of materials. The *atomic mass unit,* amu, is useful. It is defined as one twelfth of the mass of the carbon-12 isotope. It is also equal to one gram divided by Avogadro's number—0.6022×10^{24}. Thus, *atomic weights* may be expressed either as amu's, or as g/mole. Since we will make

various references to the *periodic table* (Fig. 2–1.1), it is also printed inside the back cover of the book. (This makes an easy reference for atomic weights.) Electrons have the characteristics of standing waves in individual atoms. The resulting *orbitals* possess unique energies. Intervening energies are precluded, and large energy gaps exist.

2. *Molecules* are groupings of coordinated atoms that possess strong bonds among themselves (*intramolecular*), but weak *intermolecular* bonds. *Bond lengths* and *bond energies* may be measured. Energy must be supplied (+) to break bonds; it is released (−) when atoms are joined. In general, and with other factors equal, stronger bonds are shorter. Molecular compounds with more than one structure are called *isomers*. Because their structures are different, their properties are different.

3. All engineering materials for devices, components, and systems involve many, many coordinated atoms. The *coordination number,* CN, of each atom is the number of "contacting" neighbors of that atom. The maximum coordination number for a covalently bonded atom is dictated by the number of valence electrons that the atom possesses (Table 2–3.1). The maximum coordination number for ionic bonds is restricted by the ratio of their sizes (Table 2–3.2). Atoms (and ions) can be coordinated with fewer than the maximum number of neighbors; however, in general, these atoms possess more energy and, therefore, are less stably bonded. Many materials have only primary (strong) interatomic bonds. They do not group their atoms into molecules separated from each other by weaker bonds. Examples that were cited included the compounds of NaCl and MgO. While it is convenient to speak of their molecular weights and use them for calculation purposes, these compounds do not form molecules in the structural sense.

4. The distances between atoms may be measured very precisely in most materials. These *interatomic distances* are established as a balance between the attractive forces that pull the atoms together and the repulsive forces that arise when too many electrons get into close proximity. The equilibrium spacing for the attractive and repulsive forces is also the spacing of minimum energy so that an *energy well* can be identified and calculated. The *atomic radius* is one half of the interatomic distance between like atoms. *Ionic radii* are less readily determined since the interatomic distance involves the sum of two different radii. Radii are a function of temperature, valence, and the coordination number. (For ions, $1.1\,R_{CN=4} = R_{CN=6} = 0.97\,R_{CN=8}$.)

5. Electron movements within materials are restricted to the standing waves of orbitals, unless they are *delocalized*. The conductivity of metals exists because the valence electrons are free to move in three dimensions within an *energy band* that is only partially filled. Since the delocalized electrons of graphite are restricted to the plane of the carbon atoms, the conductivity of graphite is *anisotropic*. The *mean free path* for valence electron movements through metals is longest in perfectly periodic structures. Any irregularity, such as a missing atom or an impurity atom, deflects the wavelike progress of the electron, reducing the *net* or *drift velocity,* $\bar{v}$. In turn, the *electron mobility,* $\mu = d\bar{v}/d\mathcal{E}$ is lowered, and the *conductivity,* σ, is reduced because the latter is proportional to μ.

6. It is possible to make some initial generalizations based on atomic coordinations. *Density* is controlled by atomic weight, atomic radius, and coordination number.

Melting temperatures are correlated with the depth of the energy trough. Materials that have higher interatomic forces of attraction commonly possess greater *strengths, hardnesses,* and *moduli of elasticity.* Other factors equal, materials with strong bonds and high melting temperatures have lower *thermal expansion coefficients.* Finally, *electrical* and *thermal conductivities* are greater in those materials with delocalized valence electrons.

TERMS AND CONCEPTS

Anion   Negative ion.

Atomic mass unit (amu)   One twelfth of the mass of C^{12}; gram/$(0.602 \ldots \times 10^{24})$.

Atomic number   The number of electrons possessed by an uncharged atom. The number of protons per atom.

Atomic radius (elements)   Half of interatomic distance.

Atomic weight   Atomic mass expressed in atomic mass units, or in grams per mole.

Avogadro's number (N)   Number of amu's per gram; hence, the number of molecules per mole.

Bond angle   Angle between stereospecific bonds in molecules, or covalent solids.

Bond energy   Energy required to separate two chemically bonded atoms. Generally expressed as energy per mole of 0.6×10^{24} bonds.

Bond length   Interatomic distance of stereospecific bonds.

Cation   Positive ion.

Compound   A phase composed of two or more elements in a given ratio.

Conductivity   Transfer of thermal or electrical energy along a potential gradient.

Coordination number (CN)   Number of closest ionic or atomic neighbors.

Coulombic forces   Forces between charged particles, particularly ions.

Covalent bond   A bond between two atoms resulting from sharing an electron pair.

Delocalized electrons   Valence electrons independent of their atom of origin.

Drift velocity ($\bar{v}$)   Net velocity of electrons in an electric field.

Electric field ($\mathscr{E}$)   Voltage gradient, volts/m.

Electron charge (q)   The charge of 0.16×10^{-18} coul (or 0.16×10^{-18} amp·sec) carried by each electron.

Electronic repulsion   Repelling force of too many electrons in the same vicinity. Counteracts the attractive bonding forces.

Energy band   Band of permissible energy levels for valence electrons.

Energy well   Potential energy minimum between atoms.

Ion   An atom that possesses a charge because it has had electrons added or removed.

Ionic bond   Atomic bonding by coulombic attraction of unlike ions.

Ionic radius   Semiarbitrary radius assigned to ions. Varies with coordination number. (See Appendix B footnote.)

Isomers   Molecules with the same composition but different structures.

Lone pair   Electrons in a nonconnecting sp^3 orbital.

Mean free path   Mean distance traveled by electrons or by elastic waves between deflections or reflections.

Metallic bond   Interatomic bonds in metals characterized by delocalized electrons in energy bands.

Mobility (μ)   The drift velocity $\bar{v}$ of an electric charge per unit electric field, $\mathscr{E}$, (m/sec)/(volt/m). Alternatively, the diffusion coefficient of a charge per volt, (m²/sec)/volt.

Mole   Mass equal to the molecular weight of a material; 0.6×10^{24} molecules.

Molecular weight   Mass of one molecule (expressed in amu), or mass of 0.6×10^{24} molecules (expressed in grams). Mass of one formula weight.

Molecule   Finite groups of atoms bonded by strong attractive forces. Bonding between molecules is weak.

Orbital   Wave probabilities of atomic and molecular electrons.

Periodic table   See Fig. 2–1.1, or inside back cover.

Primary bond   Strong (>200 kJ/mole) interatomic bonds of the covalent, ionic, or metallic type.

Stereospecific   Covalent bonding between specific atom pairs (in contrast to omnidirectional coulombic attractions).

Thermal expansion coefficient (α)   (change in dimensions)/(change in temperature).

Valence electrons   Electrons from the outer shell of an electron.

FOR CLASS DISCUSSION

A_2 Distinguish between atomic number and atomic mass.

B_2 Explain why electrons can have only selected energy levels within an atom, and all other levels are forbidden.

C_2 The "hard-ball" model is widely used for metal atoms and ions but is less applicable to molecules. Why?

D_2 A small diamond has 10^{20} atoms. How many covalent bonds?

E_2 The energy of the C≡N bond is not listed in Table 2–2.1. However, its value is probably 915 kJ/mole; greater than that; less than that. Explain your choice.

F_2 Why can a neutron move through materials, even though the atoms "touch"?

G_2 Barium is not listed in Appendix B. On the basis of the periodic table (Fig. 2–1.1) and other data from Appendix B, predict the radius of the Ba^{2+} ion; Ba atom; Ag^{2+} ion.

H_2 Why is there a lower limit, but not an upper limit, on the radii ratios for CN = 6?

I_2 List the factors that affect the coordination number for ionic compounds; covalent compounds.

J_2 Plot the α_L vs. T_m data of Section 2–6. Add other metals to this list from data in a handbook. Explain your results in your own words.

K_2 Obtain the melting points for the pure metals (99+%) listed in Appendix C. Plot these values against their moduli of elasticity. Discuss the patterns of your results.

L_2 Gold and nickel are among the elements listed in Appendix B but are not among the materials in Appendix C. Provide an estimate of the thermal expansion coefficient and elastic modulus of each.

M_2 Graphite is very anisotropic, i.e., its properties vary markedly with direction. Explain why the conductivity of graphite is more than 100 times greater in the horizontal direction of Fig. 2–5.2 than in the vertical direction.

QUIZ SAMPLES

2A The mass of a chromium atom is $\underline{+-}$.

(a) 52g,   (b) 86×10^{-24} g,   (c) 52 amu,   (d) 52 amu/0.6×10^{24}
(e) None of these, rather _____

2B The charge per electron is $\underline{**}$ A·s; where the unit of $\underline{**}$ is equal to one ampere·second.

2C A molecule has strong $**$-molecular bonds and weaker $**$-molecular bonds.

2D Among the molecules that contain less than a dozen atoms, $\underline{+-}$ have (has) two carbon atoms each.

(a) methanol   (b) ethanol   (c) formaldehyde   (d) ethylene
(e) vinyl chloride   (f) None of these, rather _____

2E Two molecular compounds with identical compositions are called $\underline{**}$.

2F A polymer contains many identical repeating units called $\underline{**}$.

2G Of the following atoms, $\underline{+-}$ coordinate with other atoms in molecules by only a single covalent bond.

(a) hydrogen   (b) sulfur   (c) bromine
(d) None of these,   rather _____

2H An Mn^{2+} ion has a radius, r, of ~0.08 nm. It can develop a CN = 6 with anions that have $R = \underline{+-}$.

(a) 0.20 nm   (b) 0.15 nm   (c) 0.10 nm
(d) None of these, rather _____

STUDY PROBLEMS

2–1.1 (a) What is the mass of an aluminum atom? (b) The density of aluminum is 2.70 Mg/m^3 (2.70 g/cm^3); how many atoms per mm^3?

Answer: (a) 4.48×10^{-23} g/atom   (b) 6.02×10^{19} atoms/mm^3

2–1.2 A copper wire weighs 1.312 g, is 2.15 mm in diameter and 40.5 mm long. (a) How many atoms are present per mm^3? (b) Calculate its density.

2–1.3 A solder contains 60 w/o tin and 40 w/o lead. What are the atom percents of each element?

Answer: 72 a/o Sn; 28 a/o Pb

2–1.4 (a) How many iron atoms are there per gram? (b) What is the volume of a grain of metal containing 10^{20} iron atoms?

*2–1.5 (a) Use data from the appendixes to determine the mass of a single silver atom. (b) How many atoms are there per mm^3 of silver? (c) Based on its density, what is the volume of a grain of silver that contains 10^{21} atoms? (d) Assume the silver atoms are spheres ($R_{Ag} = 0.1444$ nm), and ignore the space among them. What is the volume occupied by the 10^{21} atoms? (e) What volume percent of the space is occupied?

Answer: (a) 1.79×10^{-22} g/Ag   (b) 5.86×10^{19} Ag/mm^3 (c) 17 mm^3
(d) 12.6 mm^3 (e) 74%

---

* See footnote on p. 22.

2–1.6   Change Study Problem 2–1.5 to nickel and answer the same questions.

2–1.7   (a) Al_2O_3 has a density of 3.8 Mg/m^3 (3.8 g/cm^3). How many atoms are present per mm^3? (b) Per gram?

Answer: (a) 1.12×10^{20} atoms/mm^3 (b) 2.95×10^{22} atoms/g

2–1.8   A cubic volume of MgO that is 0.42 nm along each edge contains 4 Mg^{2+} ions and 4 O^{2-} ions. What is the density of MgO?

2–1.9   A compound contains 33 a/o Cu and 67 a/o Al. What is the weight percent of each?

Answer: 54 w/o Cu; 46 w/o Al

2–1.10   Silver was plated onto a brass surface (1610 mm^2) until it was 7.5 μm thick. (a) How much silver (Ag^+) is required? (b) How many amperes were required to do this in 5 minutes?

2–1.11   An electroplated surface of sterling silver (92.5 w/o Ag; 7.5 w/o Cu) is to be plated on some silverware. How many amperes are required to plate 1 milligram per second? (Ag^+ and Cu^{2+}.)

Answer: 1.05 amp.

• 2–1.12   Indicate the orbital arrangements for a single atom (a) of chlorine; (b) of potassium.

2–2.1   Determine the molecular weight for each of the molecules of Fig. 2–2.1.

Answer: (a) 32   (b) 46   (c) 17   (d) 78   (e) 94   (f) 30   (g) 58   (h) 60   (i) 28 (j) 62.5

• 2–2.2   An organic compound contains 62.1 w/o carbon, 10.3 w/o hydrogen, and 27.6 w/o oxygen. Name a possible compound.

• 2–2.3   Refer to Fig. 2–2.3. Methyliodide (CH_3I) has a similar structure; however, the H—C—H angles equal 111.4°. What is the H—C—I angle in CH_3I? (This is a problem in trigonometry; but it will illustrate the distortion in a polar molecule.)

Answer: 107.5°

2–2.4   Sketch three of the four possible isomers of butanol (C_4H_9OH).

2–2.5   Sketch the structure of the various possible isomers for octane, C_8H_{18}.

2–3.1   Show the origin of 0.73 in Table 2–3.2.

Answer: $2(r + R) = \sqrt{3}(2R)$

2–3.2   (a) What is the radius of the smallest cation that can have a six-fold coordination with O^{2-} ions without distortion? (b) Eight-fold coordination?

2–3.3   Show the origin of 0.22 in Table 2–3.2. (*Hint:* Place the four large ions of Fig. 2–3.3(b) at four corners of a cube and the small ion at the cube center.) (See Fig. 2–2.3a.)

Answer: $2(r + R)/\sqrt{3} = 2R/\sqrt{2}$

2-3.4 Look ahead to Fig. 3-1.1. (a) What is the coordination number for each Na^+ ion? (b) For each Cl^- ion? (Assume that the structure continues with the same pattern beyond the present sketch.)

2-4.1 Refer to Example 2-4.1 and its comments. What are the attractive and repulsive forces between Mg^{2+} and O^{2-} ions (a) at $a = 0.2$ nm? (b) At 0.22 nm?

Answer: (a) 23 nJ/m, -35 nJ/m   (b) 19 nJ/m, -13 nJ/m

2-4.2 Plot the net force $(F_C + F_R)$ versus interatomic distance for $Mg^{2+} \rightarrow \leftarrow O^{2-}$ from 0.19 nm to 0.23 nm, that is, $a_{Mg-O} \pm 0.02$ nm. (Use data from Example 2-4.1 and Study Problem 2-4.1 as needed.)

2-4.3 The radius of an Na^+ ion is 0.097 nm when CN = 6. (a) What is it with CN = 4? (b) CN = 8?

Answer: (a) 0.088 nm   (b) 0.100 nm

2-4.4 All of the ions in CsCl have CN = 8. What is the anticipated center-to-center distance between the Cs^+ and the Cl^- ions?

2-4.5 Six O^{2-} ions surround an Mg^{2+} ion. Consider the ions to be hard balls with the radii shown in Table 2-4.1. What is the distance between the *surfaces* of the O^{2-} ions?

Answer: gap = 0.01 nm

2-4.6 Six Cl^- ions surround an Na^+ ion. What is the distance between the *surfaces* of the Cl^- spheres based on the data of Table 2-4.1?

2-4.7 (a) From Appendix B, cite three divalent cations that can have CN = 6 with S^{2-}, but not CN = 8. (b) Cite two divalent ions that can have CN = 8 with F^-.

Answer: (b) Ca, Hg, Pb

QUIZ CHECKS

2A	(b), (c)	2E	isomers
2B	0.16×10^{18} A·s coulomb	2F	mers
		2G	(a), (c)
2C	intramolecular intermolecular	2H	(b), (c)
2D	(b), (d), (e)		

Chapter Three

Atomic Order
in Solids

PREVIEW

After considering atom-to-atom bonding, our next step along the structural scale is to look at the *long-range patterns of atomic order*. This is made relatively simple in crystalline solids because unit cells are formed that repeat in each of the three dimensions. Each unit cell has all of the geometric characteristics of the total crystal.

It will be the purpose of this chapter to look specifically at the atomic arrangements in a few of the more simple structures (bcc, fcc, and hcp) and establish a credibility for their existence through density calculations.

The reader should become familiar with the notations for unit cell locations, crystal directions, and crystal planes, because they will be used subsequently to relate the crystal structure to the properties and behavior of materials.

STUDY OBJECTIVES

1. To develop the concept of the unit cell so that you may use the unit cell to visualize (a) the atomic arrangements that exist, (b) the long-range order that is present along various directions and planes, and (c) the packing densities in 1-, 2-, and 3-dimensions.

2. To identify the ordering patterns that are encountered in some of the simpler metals, specifically bcc and fcc, and also other patterns such as hcp and bct.

3. To be able (a) to make calculations relating the atomic radius in metals (i) to unit cell size, (ii) to atomic packing factors, and (iii) to densities, and (b) to modify the calculations appropriately for simple compounds such as NaCl.

4. To visualize crystal directions and planes from their indices, so that we may discuss anisotropic properties later without redescribing the geometric relationships each time.

5. To know that many solids can change crystal structure as a result of temperature and/or pressure changes; and that such changes require bond breaking, atom movements, and new bond formations, all of which take time.

• 6. To review the physics of wave motions so that you can understand the basis of x-ray diffraction, which is the very precise method that can be used to obtain interplanar spacings, atomic radii, etc.

3–1

CRYSTALS

Essentially all metals, a major fraction of ceramic materials, and certain polymers crystallize when they solidify. To many readers, a crystal implies a faceted, transparent, sometimes precious material. Crystals, however, have a more basic characteristic, one that we must consider if we are to understand the internal structures of metals and other materials.

Crystals possess a *periodicity* that produces *long-range order*. By this we mean that the local atomic arrangement is repeated at regular intervals millions of times in the three dimensions of space.

The order found in crystals may be illustrated on a local scale by the atomic coordinations sketched in Fig. 3–1.1: (1) each Na^+ ion has only Cl^- ions as immediate neighbors, and each Cl^- ion has only Na^+ ions as immediate neighbors; (2) the distance between immediate neighbors in NaCl is fixed, that is, $(r_{Na^+} + R_{Cl^-})$ always equals 0.097 nm plus 0.181 nm, or 0.278 nm; (3) neighbors

---

• See Preface for bullet, •, notation.

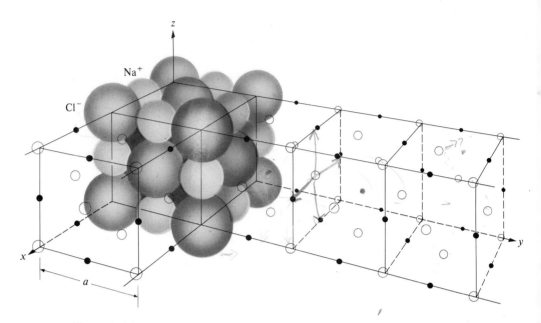

Figure 3–1.1 Crystal structure (NaCl). The atomic (or ionic) coordination is repeated to give a long-range periodicity. In this particular structure, the centers of all cube faces duplicate the pattern at the cube corners. There is also an Na⁺ ion at the center of each cell.

to any individual ion are always found in the identical directions as are neighbors for other corresponding ions.

While all of the above local relationships are important, greater significance rests in the fact that an extension of these atomic (or ionic) coordinations in three dimensions produces the characteristic long-range periodicity cited above. This is indicated by the colored overprint in Fig. 3–1.1, which suggests infinite extrapolation. The atoms (or ions) of a small volume called the *unit cell* (u.c.) are repeated at specific intervals. All unit cells in a crystal are identical. If we describe one, all are described. This will simplify for us later analyses and descriptions of internal structures.

Crystal Systems

Three-dimensional periodicity, which is characteristic of crystals, can be realized by several different geometries. The unit cell in Fig. 3–1.1 is cubic, in which the three dimensions are equal and at right angles. Such crystals are said to belong to the cubic system.

Before we look at other crystal systems, we should select a frame of reference. By convention, we orient the x-, y-, and z-axes with the origin in the lower, left, rear corner. The axial angles are labeled α, β, and γ as shown in Fig. 3–1.2(a). Also, by convention, we label the unit cell dimensions as a, b, and c, respectively, for the three axial directions (Fig. 3–1.2b). If necessary to facilitate calculations, the origin and orientation may be changed. However, this fact should be indicated to the reader.

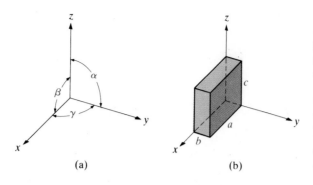

Figure 3–1.2 Crystal axes and unit-cell dimensions. The conventional orientation and labels are sketched. If necessary, the point of origin and the orientation may be changed to facilitate a particular problem; however, your reader should be advised of the change.

(a)　　　　　　　　(b)

Variations in the axial angles and in the relative size of a-, b-, and c-dimensions lead to seven (and only seven) crystal systems. These are presented in Table 3–1.1.

The *cubic system,* which is the most symmetrical system, will be encountered most widely in this text. Other systems that will be of importance to us are sketched in Fig. 3–1.3. That figure shows unit cells of the *tetragonal,* the *orthorhombic,* and the *hexagonal* systems.

Example 3–1.1

The unit cell of chromium is cubic and contains two atoms. Use the data of Appendix B to determine the dimension of the chromium unit cell.

Solution. The density of chromium is 7.20 Mg/m^3.

$$\text{Mass per unit cell} = 2 \text{ Cr}(52.0 \text{ g})/(0.602 \times 10^{24} \text{ Cr})$$
$$= 172.76 \times 10^{-24} \text{ g,}$$
$$\text{Volume} = a^3 = (172.76 \times 10^{-24} \text{ g})/(7.20 \times 10^6 \text{ g/m}^3)$$
$$= 23.994 \times 10^{-30} \text{ m}^3.$$
$$a = 0.2884 \times 10^{-9} \text{ m} \qquad \text{(or 0.2884 nm).}$$

Comment. This same value is obtained experimentally by x-ray diffraction (Section 3–8). ◀

TABLE 3–1.1

Crystal Systems (Cf. Fig. 3–1.2.)

System	Axes	Axial angles
Cubic	$a = b = c$	$\alpha = \beta = \gamma = 90°$
Tetragonal	$a = b \neq c$	$\alpha = \beta = \gamma = 90°$
Orthorhombic	$a \neq b \neq c$	$\alpha = \beta = \gamma = 90°$
Monoclinic	$a \neq b \neq c$	$\alpha = \gamma = 90° \neq \beta$
Triclinic	$a \neq b \neq c$	$\alpha \neq \beta \neq \gamma \neq 90°$
Hexagonal	$a = a \neq c$	$\alpha = \beta = 90°; \gamma = 120°$
Rhombohedral	$a = b = c$	$\alpha = \beta = \gamma \neq 90°$

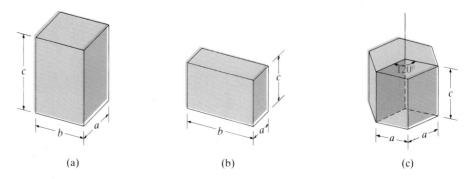

(a)   (b)   (c)

Figure 3–1.3 Noncubic unit cells. (a) Tetragonal: $a = b \neq c$; angles $= 90°$. (b) Orthorhombic: $a \neq b \neq c$; angles $= 90°$. (c) Hexagonal: $a = a \neq c$; angles $= 90°$ and $120°$.

3–2

CUBIC CRYSTALS

Body-centered Cubic Metals

Iron has a cubic structure. At room temperature, there is an iron atom at each corner of the unit cell, and another atom at the body center of the unit cell (Fig. 3–2.1). Iron is the most common metal with this body-centered cubic (bcc) structure, but not the only one. Chromium and tungsten, among others listed in Appendix B, also have body-centered cubic metal structures.

Every iron atom in this body-centered cubic metal structure is surrounded by eight adjacent iron atoms, whether the atom is located at a corner or at the center of the unit cell. Therefore, every atom has the same surroundings (Fig. 3–2.1a). There are two metal atoms per bcc unit cell. One atom is at the center of the cube and eight octants are located at the eight corners (Fig. 3–2.2).

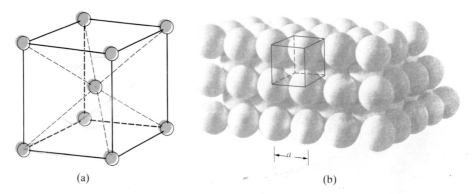

(a)   (b)

Figure 3–2.1 Body-centered cubic structure of a metal. Part (a) is a schematic view showing the location of atom centers. (b) Model made from hard balls. (G. R. Fitterer. Reproduced by permission from B. Rogers. *The Nature of Metals*, 2d ed., American Society for Metals, and the Iowa State University Press, Chapter 3.)

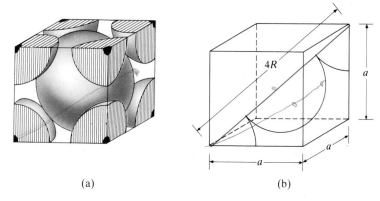

(a)   (b)

Figure 3–2.2 Body-centered cubic metal structure. This structure has two metal atoms per unit cell, and an atomic packing factor of 0.68. Since atoms are in "contact" along the body diagonals, Eq. (3–2.1) applies.

The materials with a body-centered cubic metal structure have atom contact along the body diagonal (b.d.) of the unit cell. Thus, from Fig. 3–2.2(b), we may write

$$(\text{b.d.})_{\text{bcc metal}} = 4R = a_{\text{bcc metal}}\sqrt{3}, \qquad (3\text{–}2.1a)$$

or

$$a_{\text{bcc metal}} = 4R/\sqrt{3}. \qquad (3\text{–}2.1b)$$

where a is the *lattice constant*.

We can develop the concept of the atomic *packing factor* (PF) of a bcc metal by assuming spherical atoms (hard-ball model) and calculating the volume fraction of the unit cell that is occupied by the atoms:

$$\text{Packing factor} = \frac{\text{Volume of atoms}}{\text{Volume of unit cell}}. \qquad (3\text{–}2.2)$$

Since there are two atoms per unit cell in a bcc metal, and assuming spherical atoms,

$$\text{PF} = \frac{2[4\pi R^3/3]}{a^3} = \frac{2[4\pi R^3/3]}{[4R/\sqrt{3}]^3} = 0.68.$$

Face-centered Cubic Metals

The atomic arrangements in copper (Fig. 3–2.3) are not the same as those in iron, although both are cubic. In addition to an atom at the corner of each unit cell of copper, there is one at the center of each face, but none at the center of the cube.

This face-centered cubic (fcc) structure is as common among metals as is the body-centered cubic structure. Aluminum, copper, lead, silver, and nickel possess this structure (as does iron at elevated temperatures).

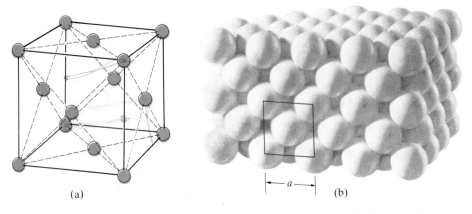

(a)   (b)

Figure 3–2.3 Face-centered cubic metal structure. Part (a) is a schematic view showing location of atom centers. (b) Model made from hard balls. (G. R. Fitterer. Reproduced by permission from B. Rogers, *The Nature of Metals*, 2d ed., American Society for Metals, and the Iowa State University Press, Chapter 3.)

The fcc metal structure has four atoms per unit cell. The eight corner octants total one atom, and each of the six face-centered atoms add one half of an atom for a net count of four atoms per unit cell. Since the atoms make contact along the face diagonal (f.d.), we may write

$$(\text{f.d.})_{\text{fcc metal}} = 4R = a_{\text{fcc metal}}\sqrt{2}, \qquad (3\text{–}2.3a)$$

or for the lattice constant,

$$a_{\text{fcc metal}} = 4R/\sqrt{2}. \qquad (3\text{–}2.3b)$$

As shown in Example 3–2.1, the packing factor for an fcc metal is 0.74, which is greater than the 0.68 for a bcc metal. This is to be expected since each atom in a bcc metal has only eight neighbors. Each atom in an fcc metal has 12 neighbors. This may be checked in Fig. 3–2.4 where the front face-centered

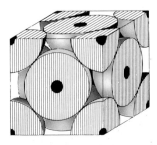

(a)   (b)

Figure 3–2.4 Face-centered cubic metal structure. This structure has four metal atoms per unit cell, and an atomic packing factor of 0.74. Atoms are in "contact" along the face diagonals; therefore, Eq. (3–2.3) applies. Note, however, that Eq. (3–2.3) does *not* apply to other fcc structures, e.g., those of Fig. 3–2.5.

atom has four adjacent neighbors, four neighbors in contact on the backside, and four additional neighbors that are not shown sitting in front.

Other fcc Structures

The *metal structure* of Figs. 3–2.3 and 3–2.4 is not the only fcc structure. The *NaCl structure* of Fig. 3–1.1 is also fcc. The centers of every face are identical in all respects with the corners. In the compound NaCl, where unlike atoms "touch," the dimensions of the unit cell are obtained from the sum of the two radii:

$$a_{fcc\ NaCl} = 2(r_{Na^+} + R_{Cl^-}). \tag{3–2.4}$$

This is apparent in Fig. 3–2.5(a). (Note that there is a gap between the Cl^- ions because of the size of the Na^+ ion. The Cl^- ions do *not* "touch.")

The *diamond-cubic structure* is also fcc. Figure 3–2.5(b) is a resketch of Fig. 2–3.1(b). Both show covalent bonds rather than spherical atoms or ions in contact; however, that does not matter for our purpose, which is to observe that the centers of all faces are identical in all respects to the corners. There are, however, eight atoms per unit cell for an average of two atoms per fcc position. Experiments show that the C–C interatomic distance is 0.154 nm (center-to-center) in this structure. A close examination of Fig. 3–2.5(b) reveals that the body diagonal (b.d.) is four times this value. Thus

$$(b.d.)_{diamond} = 4(C–C) = (a_{diamond})\sqrt{3}, \tag{3–2.5a}$$

or

$$a_{diamond} = 4(0.154\ nm)/\sqrt{3}, \tag{3–2.5b}$$

to give a value of 0.356 nm for a lattice constant, a.

Two additional fcc structures will be examined in Chapter 7.

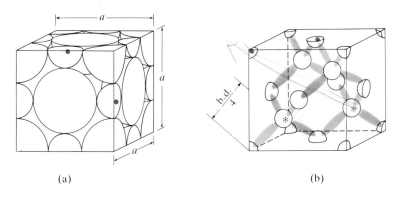

(a)   (b)

Figure 3–2.5 Additional fcc structures. Translations of $\pm\frac{a}{2}$ in two of the three coordinate directions lead to identical positions in fcc structures. (a) NaCl structure (cf. Fig. 3–1.1). $a = 2(r + R)$. (b) Diamond cubic (cf. Fig. 2–3.1b). Each unit cell contains 8 ($\frac{1}{8}$) carbon atoms at the corners, 6 ($\frac{1}{2}$) carbon atoms at the face centers, and four wholly within the cell for a total of eight carbon atoms, $a\sqrt{3} = 4(C–C) = $ b.d.

Example 3–2.1   Calculate (a) the atomic packing factor of an fcc metal (Fig. 3–2.4); (b) the ionic packing factor of fcc NaCl (Fig. 3–2.5a).

Procedure. Observe in Fig. 3–2.4(a) that there are four atoms per unit cell in an fcc metal structure. The corresponding sketch modification of Fig. 3–1.1 in Fig. 3–2.5(a) shows 4 Na^+ and 4 Cl^- ions per unit cell. Note, however, that the lattice constants relate to the radii differently in the fcc metal structure (Eq. 3–2.3) and in the fcc NaCl structure (Eq. 3–2.4).

Calculation

a) From Eqs. (3–2.2) and (3–2.3),

$$PF = \frac{4(4\pi R^3/3)}{a^3} = \frac{16\pi R^3(2\sqrt{2})}{(3)(64R^3)} = 0.74.$$

b) From Eqs. (3–2.2) and (3–2.4), Fig. 3–2.5(a), and radii from Appendix B,

$$PF = \frac{4(4\pi r^3/3) + 4(4\pi R^3/3)}{(2r + 2R)^3} = \frac{16\pi(0.097^3 + 0.181^3)}{3(8)(0.097 + 0.181)^3} = 0.67.$$

Comments. It is apparent, from this example, that the packing factor is independent of atom size, if only one size is present. In contrast, the relative sizes do affect the packing factor when more than one type of atom is present. The face-centered cubic metal structure has the highest packing factor (0.74) that is possible for a pure metal, and thus this structure could also be called a *cubic close-packed* (ccp) structure. As we might expect, many metals have this structure. We shall see in the next section that a hexagonal close-packed metal structure also has a packing factor of 0.74. ◀

Example 3–2.2   Copper has the fcc metal structure with an atom radius of 0.1278 nm. Calculate its density and check this value with the density listed in Appendix B.

Procedure. Since $\rho = m/V$, we must obtain the mass per unit cell and the volume of each unit cell. The former is calculated from the mass of four atoms (Fig. 3–2.4); the latter requires the calculation of the lattice constant, a, using Eq. (3–2.3) for an fcc metal structure.

Calculation

$$a = \frac{4}{\sqrt{2}} (0.1278 \text{ nm}) = 0.3615 \text{ nm}.$$

$$\text{Density} = \frac{\text{Mass/Unit cell}}{\text{Volume/Unit cell}} \qquad (3\text{–}2.6a)$$

$$= \frac{(\text{Atoms/Unit cell})(\text{g/atom})}{(\text{Lattice constant})^3}. \qquad (3\text{–}2.6b)$$

$$\text{Density} = \frac{4[63.5/(0.602 \times 10^{24})]}{(0.3615 \times 10^{-9} \text{ m})^3} = 8.93 \text{ Mg/m}^3 \qquad (\text{or } 8.93 \text{ g/cm}^3).$$

The experimental value listed in Appendix B is 8.92 Mg/m^3. ◀

Example 3–2.3

Calculate the volume of the unit cell of LiF, which has the same fcc structure as NaCl (Fig. 3–1.1).

Procedure. Although LiF has an fcc structure, we cannot use the geometry of Fig. 3–2.4(b), since the fluoride ions do not touch, as did the metal atoms. Rather, according to Eq. (3–2.4), a is twice the sum of r_{Li^+} and R_{F^-}. (Check Fig. 3–2.5(a) again.)

Calculation. From Appendix B,

$$a = 2(0.068 + 0.133) \text{ nm,}$$

$$a^3 = 0.065 \text{ nm}^3 \qquad \text{(or } 65 \times 10^{-30} \text{ m}^3\text{).}$$

Comment. We use ionic radii since LiF is an ionic compound. Furthermore, each ion has six neighbors; therefore, the radii of Appendix B do not have to be modified. ◀

3–3

HEXAGONAL CRYSTALS

The sketches of Figs. 3–3.1(a) and 3–3.1(b) are two representations of *hexagonal* unit cells. The angles within the base are 120° (and 60°). The rhombic representation of (b) is the more fundamental, because it is the smallest repeating volume; however, the hexagonal representation, which actually contains three of the rhombic cells, is used more often because it shows the hexagonal (6-fold) symmetry characteristics that give the name to this unit cell.

Hexagonal Close-packed Metals

The hexagonal structure formed by several metals, including magnesium, titanium and zinc, is shown in Fig. 3–3.2. This hexagonal structure is called the *hexagonal close-packed* metal structure (hcp). It is characterized by the fact that each atom in one layer is located directly above or below interstices among three atoms in the adjacent layers. Consequently, each atom touches three atoms in the layer below its plane, six atoms in its own plane, and three atoms in the layer above for CN = 12. There is an average of six atoms per unit cell in the hcp metal structure of Fig. 3–3.2 (or two per unit cell if we use the related rhombic representation).

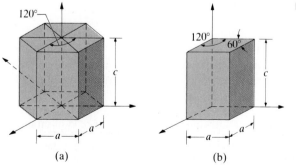

(a) (b)

Figure 3–3.1 Hexagonal cells. (a) Hexagonal representation (4 axes). (b) Rhombic representation (3 axes). The two are equivalent geometrically, with $a \neq c$, $\gamma = 120°$, and $\alpha = \beta = 90°$; however, $V_a = 3 V_b$.

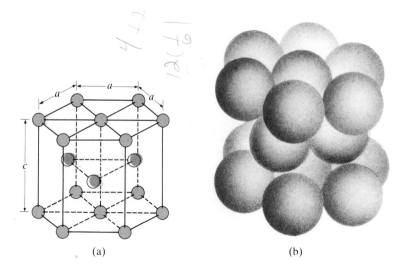

(a)

(b)

Figure 3–3.2 Hexagonal close-packed metal structure. (a) Schematic view showing the location of atom centers. (b) Model made from hard balls.

Since both hcp metals and fcc metals have a coordination number of 12, they have the same atomic packing factors, that is, 0.74.

Example 3–3.1

The atomic packing factor of magnesium, like all hcp metals, is 0.74. What is the volume of its unit cell, which is shown in Fig. 3–3.2(a)?

Procedures. From Appendix B, magnesium has $\rho = 1.74$ Mg/m³ (or 1.74 g/cm³); its atomic mass = 24.31 amu, and $\bar{R}_{Mg} = 0.161$ nm. From Fig. 3–3.2(a), 12/6 + 2/2 + 3 = 6 atoms/unit cell.

Two approaches are possible. (a) Since the packing factor is given as 0.74, calculate the volume of six atoms per unit cell. (b) Since $\rho = m/V$, calculate the mass of six atoms per unit cell, then the volume per unit cell.

Calculation

a)
$$V_{u.c.} = 6 \text{ atoms } (4\pi/3)(0.161 \text{ nm})^3/0.74$$
$$= 0.14 \text{ nm}^3 \qquad\qquad (\text{or } 1.4 \times 10^{-28} \text{ m}^3).$$

b)
$$m_{u.c.} = 6 \text{ atoms } (24.31 \text{ g})/(0.602 \times 10^{24}) = 2.42 \times 10^{-22} \text{g};$$

$$V_{u.c.} = \frac{m}{\rho} = (2.42 \times 10^{-22} \text{ g})/(1.74 \times 10^6 \text{ g/m}^3)$$
$$= 1.4 \times 10^{-28} \text{ m}^3 \qquad\qquad (\text{or } 0.14 \text{ nm}^3).$$

Comments. The average radius, $\bar{R}$, of the magnesium atom is 0.161 nm. However, x-ray diffraction data show that the Mg atoms are compressed almost 1 percent to become oblate spheroids. (See Comments with Example 3–3.2 and the footnote to Appendix B.) ◀

• Example 3–3.2

Assume spherical atoms. What is the c/a ratio of an hcp metal?

Derivation. Refer to Fig. 3–3.2(a) and consider the three central atoms plus the one at the center of the top. This is an equilateral tetrahedron with edges $a = 2R$. From geometry,

$$h = a\sqrt{2/3}.$$

$$c = 2h = 2a\sqrt{2/3} = 1.63a. \tag{3–3.1}$$

Comments. The c/a ratios of hcp metals depart somewhat from this figure: Mg, 1.62; Ti, 1.59; Zn, 1.85. This means we must envision magnesium and titanium atoms as slightly compressed spheres, and zinc atoms as elongated spheroids. ◀

● **Example 3–3.3**

The unit-cell volume of hcp titanium at 20°C is 0.106 nm³, as sketched in Fig. 3–3.2(a). The c/a ratio is 1.59. (a) What are the values of c and a? (b) What is the radius of the Ti in a direction that lies in the base of the unit cell?

Procedure. Work from a, the dimension of the bottom edge; then calculate the unit cell volume from the height ($= 1.59 a$) and the basal area, A.

Calculation

a)

$$A = 6(1/2)(a)(a \sin 60°) = 2.60 \, a^2.$$

$$\text{Volume} = (1.59a)(2.60a^2) = 4.13a^3 = 0.106 \text{ nm}^3.$$

$$a^3 = 0.02566 \text{ nm}^3; \quad \text{and} \quad a = 0.2950 \text{ nm}.$$

$$c = 1.59(0.295 \text{ nm}) = 0.469 \text{ nm}.$$

b)

$$a = 2R_{\text{base}}; \quad \therefore R_{\text{base}} = 0.1475 \text{ nm}.$$

Comments. The average radius is 0.146 nm (Appendix B). The atoms of titanium (and the unit cell) are slightly compressed in the c-direction. (See Comments with Example 3–3.2.) ◀

3–4
POLYMORPHISM

We recall from Section 2–2 that two molecules may possess different structures even though their compositions are identical. We called those molecules *isomers*. An analogous situation that will be extremely important to us occurs in crystalline solids. *Polymorphs* are two or more distinct types of crystals that have the same composition.* The most familiar example is the dual existence of graphite and diamond as two polymorphs of carbon.

The prime example of polymorphism in metals will be iron, since our whole ability to heat-treat steel and modify its properties stems from the fact that, as iron is heated, it changes from the bcc to the fcc metal structure. Furthermore, the change is reversible as iron cools. At room temperature bcc iron has a coordination number of 8, an atomic packing factor of 0.68, and an atomic radius of 0.1241 nm. Pure iron changes to fcc at 912°C, at which point its coordination number is 12, its atomic packing factor is 0.74, and its atomic radius is 0.129 nm. [At 912°C (1673°F) the atomic radius of bcc iron, due to thermal expansion, is 0.126 nm.]

* When this is found in elemental solids such as metals, the term *allotropes* is sometimes used.

Many other materials have two or more polymorphic forms. In fact some, such as SiC, have as many as 20 crystalline modifications; however, this is unusual. Invariably, polymorphs have differences in density and other properties. In succeeding chapters we shall be interested in the property variations and in the time required to change from one crystal modification (phase) to another.

Example 3–4.1

Iron changes from the bcc to the fcc metal structure at 912°C (1673°F). At this temperature the atomic radii of the iron atoms in the two structures are 0.126 nm and 0.129 nm, respectively.

a) What is the percent of volume change, v/o, as the structure changes?

b) Of linear change, l/o?

(*Note:* As indicated in Section 2–4 and Table 2–4.1, the higher the coordination number the larger the radius.)

Procedure. Use a common basis for the two structures. The simplest is four iron atoms; therefore, *two* unit cells of bcc iron and *one* unit cell of fcc iron. Also, $a_{bcc}\sqrt{3} = 4R_{bcc}$, and $a_{fcc}\sqrt{2} = 4R_{fcc}$.

Calculation

a)
$$2V_{bcc} = 2a_{bcc}^3 = 2[4(0.126 \text{ nm})/\sqrt{3}]^3 = 0.0493 \text{ nm}^3.$$

$$V_{fcc} = a_{fcc}^3 = [4(0.129 \text{ nm})/\sqrt{2}]^3 = 0.0486 \text{ nm}^3.$$

$$\frac{\Delta V}{V} = \frac{0.0486 - 0.0493}{0.0493} = -0.014 \qquad \text{(or } -1.4 \text{ v/o change)}$$

b)
$$(1 + \Delta L/L)^3 = 1 + \Delta V/V,$$

$$\Delta L/L = \sqrt[3]{1 - 0.014} - 1 = -0.0047 \qquad \text{(or } -0.47 \text{ l/o change).}$$

Comment. Iron expands thermally until it reaches 912°C, where there is an abrupt shrinkage; further heating continues the expansion (Fig. 4–5.2). ◀

Example 3–4.2

The densities of ice and water at 0°C are 0.915 and 0.9995 Mg/m³ (or g/cm³), respectively. What is the percent volume expansion during the freezing of water?

Procedure. Determine the volumes of 1 Mg of each.

$$\text{Volume ice} = 1.093 \text{ m}^3;$$

$$\text{Volume liquid} = 1.0005 \text{ m}^3.$$

Calculation

$$\Delta V/V = (1.093 \text{ m}^3 - 1.0005 \text{ m}^3)/1.0005 \text{ m}^3$$
$$= +0.093 \qquad \text{(or } 9.3 \text{ v/o).}$$

Comments. We are familiar with the major changes that occur during freezing and thawing. In principle, polymorphic changes are similar. That is, there is a change in structure within the solid. This brings about changes in volume, density, and almost every other physical property. ◀

3–5

UNIT CELL GEOMETRY

The general convention that we will follow is to orient a crystal so its x-axis points toward us, the y-axis points to our right as we face it, and the z-axis upward. Thus, by convention the origin is at the left, lower, rear corner of the unit cell.* The opposing directions are negative.

Points Within Unit Cells

Every point within a unit cell can be identified in terms of the coefficients along the three coordinate axes. Thus the origin is 0,0,0. Since the center of the unit cell is at $\frac{a}{2},\frac{b}{2},\frac{c}{2}$, the indices of that location are $\frac{1}{2},\frac{1}{2},\frac{1}{2}$. These *coefficients of locations* are always expressed in unit cell dimensions. Thus, the far corner of the unit cell is 1,1,1 regardless of the crystal system—cubic, tetragonal, orthorhombic, etc.

A translation from any selected site within a unit cell by an integer multiple of lattice constants (a, b, and/or c) leads to an identical position in another unit cell. Thus, in the two-dimensional lattice of Fig. 3–5.1, the two points labeled * are separated by translations of $3b$ (parallel to y) and $2c$ (parallel to z). This example is obviously noncubic (or nonsquare). However, integer multiples lead to replication in all crystal systems.

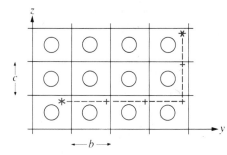

Figure 3–5.1 Unit translations. A translation of an integer number of unit dimensions, i.e., lattice constants, leads to a site that is identical in every respect to the original site.

Replication is also realized by $\pm\frac{a}{2}$, $\pm\frac{b}{2}$, $\pm\frac{c}{2}$ translations in body-centered unit cells (or into adjacent unit cells). Thus, a point with indices of 0.3, 0.4, 0.2 duplicates the site at 0.8, 0.9, 0.7; or one at 0.8, −0.1, 0.7 in the next unit cell to the left. (See Fig. 3–5.2.)

In face-centered unit cells, replication of points accompanies translations of $\pm\frac{a}{2}$, $\pm\frac{b}{2}$, 0; of $\pm\frac{a}{2}$, 0, $\pm\frac{c}{2}$; and of 0, $\pm\frac{b}{2}$, $\pm\frac{c}{2}$ (in addition to the $\pm a$, $\pm b$, $\pm c$ translations). As a result, the $\frac{3}{4},\frac{1}{4},\frac{1}{4}$ site is identical to the $\frac{1}{4},\frac{3}{4},\frac{1}{4}$ site. This may be checked out in the diamond cubic structure of Fig. 3–2.5(b) with the two atoms containing stars.

---

* We are not locked into this convention if there is reason to change. However, we will assume this convention applies in our discussions *unless* informed otherwise. Conversely, your answers should comply with the convention *unless* you advise your reader of the reorientation that you have used.

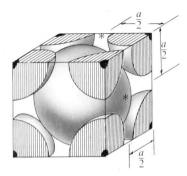

Figure 3–5.2 Translation in bcc. A translation of $\pm\frac{a}{2}, \pm\frac{a}{2}, \pm\frac{a}{2}$ from any point leads to an identical site, for example, * to * (from $0,\frac{1}{2},1$ to $\frac{1}{2},1,\frac{1}{2}$).

Before proceeding, a word of caution is in order. Figure 3–5.3 shows a cubic structure that is *not* body-centered. It is simple cubic. It does not meet the requirement of a bcc structure, namely, translations of $\pm\frac{1}{2}$, $\pm\frac{1}{2}$, $\pm\frac{1}{2}$ reproduce the original site. In Fig. 3–5.3, Zn is at 0,0,0; Cu is at $\frac{1}{2},\frac{1}{2},\frac{1}{2}$. The sites are occupied by different atoms! Only full unit cell translations lead to replication; hence, this cubic structure is "simple."

Example 3–5.1

a) Sketch a noncubic unit cell and show the locations of points that have the following coefficients:

$$0,0,0; \quad 0,0,\tfrac{1}{2}; \quad \tfrac{1}{2},\tfrac{1}{2},\tfrac{1}{2}; \quad \tfrac{1}{2},\tfrac{1}{2},0; \quad 1,1,0; \quad 1,1,1; \quad 1,1,2.$$

b) Assume the cell is orthorhombic (all axial angles = 90°), and a = 0.270 nm, b = 0.403 nm, and c = 0.363 nm. What is the translation distance t between points at $0,-1,0$ and at $1,1,2$?

Solution

a) See Fig. 3–6.1(a).

b) Since the axial angles are 90°,

$$t = \sqrt{[(1 - 0)(0.270 \text{ nm})]^2 + [(1 - (-1))(0.403 \text{ nm})]^2 + [(2 - 0)(0.363 \text{ nm})]^2}$$
$$= 1.118 \text{ nm}.$$

Comment. This procedure for locating points within unit cells does not limit us to the reference unit cell. ◀

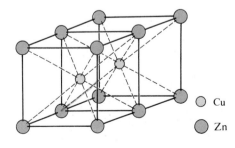

Figure 3–5.3 Simple cubic structure (β' brass). Each copper atom is coordinated with eight zinc atoms; and each zinc atom is coordinated with eight copper atoms. The structure is simple cubic, because the center site does not duplicate the corner sites. The prototype for this structure is CsCl (Fig. 7–2.1).

○ Cu

◉ Zn

Example 3–5.2   Consider the following four points within a unit cell:

$$\tfrac{1}{4},\tfrac{1}{4},\tfrac{3}{4}; \qquad \tfrac{7}{8},\tfrac{3}{4},\tfrac{1}{5};$$

$$\tfrac{1}{8},\tfrac{1}{8},\tfrac{1}{8}; \qquad 0.53,\ 0.25,\ 0.17.$$

a) Identify a second identical site *within* the reference unit cell (bcc) for each of the four points.

b) Identify three identical sites *within* an fcc unit cell for *each* of the above four points.

Answers

a) The translations in body-centered cells are $\pm 0.5a,\ \pm 0.5a,\ \pm 0.5a$. Therefore,

For $\tfrac{1}{4},\tfrac{1}{4},\tfrac{3}{4}$: $(0.25 + 0.5),\ (0.25 + 0.5),\ (0.75 - 0.5) = \tfrac{3}{4},\tfrac{3}{4},\tfrac{1}{4}.$

For $\tfrac{1}{8},\tfrac{1}{8},\tfrac{1}{8}$: $(0.125 + 0.5),\ (0.125 + 0.5),\ (0.125 + 0.5) = \tfrac{5}{8},\tfrac{5}{8},\tfrac{5}{8}.$

For $\tfrac{7}{8},\tfrac{3}{4},\tfrac{1}{5}$: $(0.875 - 0.5),\ (0.75 - 0.5),\ (0.20 + 0.5) = \tfrac{3}{8},\tfrac{1}{4},0.7.$

For $0.53,\ 0.25,\ 0.17$: $(0.53 - 0.5),\ (0.25 + 0.5),\ (0.17 + 0.5) = 0.03,\ 0.75,\ 0.67.$

b) The translations in face-centered cells are permutations of $\pm 0.5a,\ \pm 0.5a,$ and 0.

For $\tfrac{1}{4},\tfrac{1}{4},\tfrac{3}{4}$:	$\tfrac{3}{4},\tfrac{3}{4},\tfrac{3}{4}$	$\tfrac{3}{4},\tfrac{1}{4},\tfrac{1}{4}$	$\tfrac{1}{4},\tfrac{3}{4},\tfrac{1}{4}.$
For $\tfrac{1}{8},\tfrac{1}{8},\tfrac{1}{8}$:	$\tfrac{5}{8},\tfrac{5}{8},\tfrac{1}{8}$	$\tfrac{5}{8},\tfrac{1}{8},\tfrac{5}{8}$	$\tfrac{1}{8},\tfrac{5}{8},\tfrac{5}{8}.$
For $\tfrac{7}{8},\tfrac{3}{4},\tfrac{1}{5}$:	$\tfrac{3}{8},\tfrac{1}{4},\tfrac{1}{5}$	$\tfrac{3}{8},\tfrac{3}{4},0.7$	$\tfrac{7}{8},\tfrac{1}{4},0.7.$
For $0.53,\ 0.25,\ 0.17$:	$0.03,\ 0.75,\ 0.17$	$0.03,\ 0.25,\ 0.67$	$0.53,\ 0.75,\ 0.67.$

Comment. These translations within body-centered and face-centered cells are independent of whether the crystal is cubic, orthorhombic (or tetragonal). ◄

Example 3–5.3   Calculate the distances in NaCl between the center of a sodium ion and the center of
a) its nearest neighbor;
b) its nearest positive ion;
c) its second nearest Cl^- ion;
d) its third nearest Cl^- ion;
e) the nearest site that is identical.

Procedure. Refer to Fig. 3–1.1. From Appendix B, the radii of Na^+ and Cl^- are 0.097 nm and 0.181 nm, respectively (since each has CN = 6).

Calculations

$$a = 2(0.097 + 0.181\ nm) = 0.556\ nm.$$

a) Distance $= a/2 = 0.278$ nm.
b) Distance $= \sqrt{(a/2)^2 + (a/2)^2} = 0.393$ nm.
c) Distance $= \sqrt{(a/2)^2 + (a/2)^2 + (a/2)^2} = 0.482$ nm.
d) Distance $= \sqrt{(a/2)^2 + a^2} = 0.622$ nm.
e) Same as (b), since the nearest Na^+ ions are at identical positions. The translation is $\tfrac{a}{2},\tfrac{a}{2},0.$ ◄

Example 3–5.4   Cesium iodide (CsI) has the structure shown in Fig. 3–5.3. What is its packing factor if the radii are 0.172 nm and 0.227 nm, respectively?

Procedure. The body diagonal of the unit cell is equal to $(2r + 2R)$. Each unit cell has 1 Cs^+ and 1 Cl^- ion.

Calculation

$$a = [2(0.172 + 0.227)]/\sqrt{3} = 0.461 \text{ nm.}$$

$$\text{Packing factor} = (4\pi/3)(0.172^3 + 0.227^3)/(0.461)^3 = 0.72.$$

Comments. The packing factor is greater than that obtained in the calculation following Eq. (3–2.2) because a bcc metal does not contain a second (different) atom in the center.

These radii differ from the radii of Appendix B by a factor of 0.97, since the data in the appendix are for CN = 6. Here, CN = 8. ◄

3–6

CRYSTAL DIRECTIONS

When we correlate various properties with crystal structures in subsequent chapters, it will be necessary to identify specific crystal directions, because many properties are directional. For example, the elastic modulus of bcc iron is greater parallel to the body diagonal than to the cube edge. Conversely, the magnetic permeability of iron is greatest in a direction parallel to the edge of the unit cell.

Indices of Directions

All parallel directions use the same label or index. Therefore, to label a direction, pick the parallel ray that passes through the origin, which is at 0,0,0. The direction is labeled by the coefficient of a point on that ray. However, since there are an infinite number of points on any ray, we specifically choose the point with the lowest set of integers (Fig. 3–6.1). Thus the [111] direction passes from 0,0,0 through 1,1,1. Note, however, that this direction also passes through $\frac{1}{2},\frac{1}{2},\frac{1}{2}$ (and 2,2,2). Likewise the [112] passes through $\frac{1}{2},\frac{1}{2},1$; but for simplicity's sake, we use the integer notation. Observe that we enclose the direction indices in square brackets $[uvw]$, and use the letters u, v, and w for the indices arising from the three principal directions, x, y, and z, respectively. Parallel directions always have the same indices. Finally, note that we may have negative coefficients, which we designate with an overbar; a $[11\bar{1}]$ direction will have a component in the minus z direction. (We do *not* use commas between the indices.)

Angles Between Directions

In certain calculations (e.g., resolved shear stresses), it will be necessary to calculate the angle between two different crystal directions. For most of the calculations we will encounter, this may be performed by simple inspection. Thus, in Fig. 3–6.1, the angle between [110] and [112] directions (i.e., [110] ⋪ [112]) is arctan $2c/\sqrt{a^2 + b^2}$. If that unit cell had been cubic rather than orthorhombic, so that $a = b = c$, the angle would have been arctan $2a/a\sqrt{2}$, or

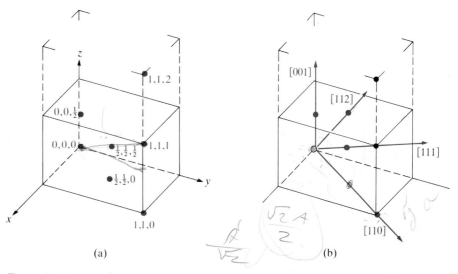

Figure 3–6.1 Orthorhombic unit cell. (a) Indices of points. (b) Indices of directions. As stated in Section 3–5, the origin is commonly but not necessarily located at the lower, left, rear corner. Likewise, as discussed in this section, we use square brackets as closures to indicate crystal directions, [uvw]. We will be using parentheses (hkl) to enclose indices for crystal planes (Section 3–7). Commas are omitted within all closures (and closures are not used for the indices of points).

arccos $a\sqrt{2}/a\sqrt{6}$. In fact, with cubic crystals (*only*), we can determine cos [uvw] ∢ [u'v'w'] by the *dot product*. This latter procedure will be useful to us since most of our calculations in this introductory text will involve these symmetric cubic crystals.

• **Linear Density**

The repeating distance between identical positions within a crystal differs from direction to direction and from structure to structure. For example, in the [111] direction of a bcc metal, there is repetition every 2R (which is $a\sqrt{3}/2$). The repeating distance is $a\sqrt{2}$ in the [110] direction of bcc, but $a/\sqrt{2}$ in fcc. These can be checked in Figs. 3–2.1 and 3–2.3.

Conversely, the reciprocals of these repetition distances are the *linear densities*. Thus, in the [110] direction of aluminum, which is fcc with a equal to 0.405 nm, the linear density is one lattice point per [$a/\sqrt{2}$], which inverts to $\sqrt{2}/(0.405 \times 10^{-6}$ mm), equaling 3.5×10^6/mm.

$$\text{Linear density} = \frac{\text{Number}}{\text{Unit length}}. \qquad (3–6.1)$$

We will observe in Chapter 5 that deformation occurs most readily in those directions with the greatest linear density, and therefore the shortest repeating distance.

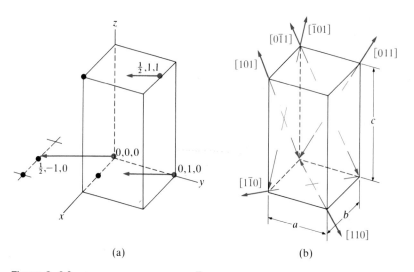

Figure 3–6.2 Crystal directions. (a) [1$\bar{2}$0] (see Example 3–6.1). (b) ⟨101⟩ (see Example 3–6.2b).

In most metals, there is one atom per lattice point; therefore, the linear density of atoms equals the linear density of identical sites. However, Example 3–6.4 shows that the two may differ in some crystals.

Families of Directions

In the cubic crystal, the following directions are identical except for our arbitrary choice of the x-, y-, and z-labels on the axes:

$$[111] \quad [11\bar{1}] \quad [1\bar{1}1] \quad [\bar{1}11]$$

$$[\bar{1}\bar{1}\bar{1}] \quad [\bar{1}\bar{1}1] \quad [\bar{1}1\bar{1}] \quad [1\bar{1}\bar{1}]$$

Any directional property* will be identical for these four opposing pairs.† Therefore, it is convenient to identify this *family of directions* as ⟨111⟩ rather than writing the eight separate indices. Note that the closure symbols are angle brackets ⟨ ⟩; and again no commas are used.

Example 3–6.1

All parallel directions possess the same directional indices. Sketch rays in the [1$\bar{2}$0] direction that pass through locations (a) 0,0,0; (b) 0,1,0; and (c) $\frac{1}{2}$,1,1.

Sketch. See Fig. 3–6.2(a). ◀

Example 3–6.2

The ⟨101⟩ family of directions include what individual directions (a) in a cubic crystal? ● (b) in a tetragonal crystal?

---

* For example, Young's modulus, magnetic permeability, index of refraction.
† See the comment with Example 3–6.2.

Answers

a) In cubic, $a = b = c$. Therefore,

$$[110], [101], [011], [1\bar{1}0], [10\bar{1}], [01\bar{1}].$$

b) In tetragonal cells $a = b \neq c$ (Fig. 3–6.2b). Therefore, only the u and v indices of $\langle uvw \rangle$ are interchangeable; the w index is not:

$$[101], [011], [\bar{1}01], [0\bar{1}1].$$

Comments. The $[\bar{1}01]$ and the $[10\bar{1}]$ directions are commonly considered to be the two senses of one direction, and not two separate directions. However, if desired, we could list the negative indices of each of the above directions, and double the number. ◄

Example 3–6.3

a) What is the angle between the [111] and [001] directions in a *cubic* crystal?

b) $[111] \not{\angle} [\bar{1}\bar{1}1]$?

Solution. From the *cubic* analog of Fig. 3–6.1(b),

a)

$$\cos [111] \not{\angle} [001] = a/a\sqrt{3},$$

$$[111] \not{\angle} [001] = 54.75°$$

b) Observe that [001] bisects $[111] \not{\angle} [\bar{1}\bar{1}1]$; therefore,

$$[111] \not{\angle} [\bar{1}\bar{1}1] = 2(54.75°) = 109.5°;$$

or, by the dot product (*since the crystal is cubic*):

$$\cos [111] \not{\angle} [\bar{1}\bar{1}1] = -\tfrac{1}{3},$$

$$[111] \not{\angle} [\bar{1}\bar{1}1] = 109.5°.$$

Comment. Compare the results here with Fig. 2–2.3(a) where all four of the bond angles are 109.5°. ◄

• Example 3–6.4

The lattice constant a is 0.357 nm for diamond, which is cubic. (a) What is the linear density of lattice points in the $[1\bar{1}1]$ direction? (b) Of atoms?

Observations. Refer to Fig. 3–2.5(b). Locate the origin at the rear, *right,* lower corner of the unit cell. From there, the $[1\bar{1}1]$ direction passes through the front, left, upper corner. En route, it passes through the center of a carbon atom (the atom on the right with an identifying star). The repeating distance is from 0,0,0 to 1,1,1, or $a\sqrt{3}$. A trip along this route produces *two* atoms for every $a\sqrt{3}$ step.

Calculations

$$a\sqrt{3} = 0.357\sqrt{3} = 0.618 \text{ nm.}$$

a)

$$\text{Linear density} = \frac{1}{(0.618 \times 10^{-6} \text{ mm})} = 1.6 \times 10^6/\text{mm.}$$

b)

$$\frac{\text{Two atoms}}{(0.618 \times 10^{-6} \text{ mm})} = \frac{3.2 \times 10^6}{\text{mm}}.$$

Comment. Based on our relocated origin, the atom (identified with a star) is at $\tfrac{1}{4}, -\tfrac{1}{4}, \tfrac{1}{4}$. This location can be checked by relating it to its four neighbors at

$$0,0,0; \quad \tfrac{1}{2},0,\tfrac{1}{2}; \quad \tfrac{1}{2},-\tfrac{1}{2},0; \quad 0,-\tfrac{1}{2},\tfrac{1}{2}. \quad ◄$$

3–7

CRYSTAL PLANES

A crystal contains planes of atoms; these influence the properties and behavior of a material. Thus it will be advantageous to identify various planes within crystals.

The lattice planes most readily visualized are those that outline the unit cell, but there are many other planes. The more important planes for our purposes are those sketched in Figs. 3–7.1, 3–7.2, and 3–7.3. These are labeled (010), (110), and (111) respectively, where the numbers within the parentheses (*hkl*) are called the *Miller indices*.

Miller Indices

We may use the plane with the darker tint in Fig. 3–7.4 to explain how (*hkl*) numbers are obtained. The plane intercepts the *x*-, *y*-, and *z*-axes at **1***a*, **1***b*, and **0.5***c*. The Miller indices are simply the reciprocals of these intercepts: (112). The plane of lighter tint in Fig. 3–7.4 is the (111) plane, since it intercepts the axes at **1**a, **1***b*, and **1***c*. Returning to the earlier figures, we have

Figure	Plane	Intercepts	Miller indices
3–7.1(a)	Middle	∞a, **1***b*, ∞c	(010)
3–7.2(a)	Left	**1***a*, **1***b*, ∞c	($\bar{1}$10)
3–7.3(a)	Middle	$-$**1***a*, **1***b*, **1***c*	($\bar{1}$11)

Note that a *minus intercept* is handled readily with an overbar. Furthermore, observe that we use parentheses (*hkl*) to denote planes (and no commas), in order to avoid confusion with individual directions that were denoted in Section 3–6 with square brackets: [*uvw*]. All parallel planes are identified with the same indices.

A more rigorous definition of *Miller indices* is now possible. They are *the **reciprocals** of the three intercepts that the plane makes with the axes, cleared of fractions and of common multipliers.*

Planes lying on an axis or through the origin would appear to present a problem because their intercepts are not uniquely definable. However, since all parallel planes possess the same indices, we can readily handle the problem by

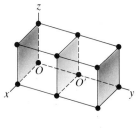

(a)

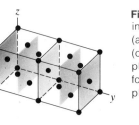

(c)

Figure 3–7.1 (010) planes in cubic structures. (a) Simple cubic. (b) Bcc. (c) Fcc. (Note that the (020) planes included for bcc and fcc are comparable to (010) planes.)

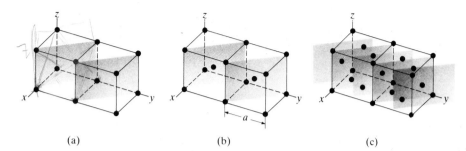

Figure 3–7.2 (110) planes in cubic structures. (a) Simple cubic. (b) Bcc. (c) Fcc. (The (220) planes included for fcc are comparable to (110) planes.)

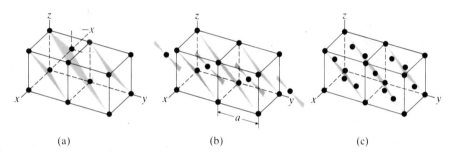

Figure 3–7.3 ($\bar{1}$11) planes in cubic structures. (a) Simple cubic. (b) Bcc. (c) Fcc. Negative intercepts are indicated by bars above the index. The ($\bar{2}$22) planes included for bcc are comparable ($\bar{1}$11) planes.)

shifting the origin. Consider Fig. 3–7.1(a) where all three shaded planes possess the same geometric features and should be labeled identically. With the origin at O, the middle plane has (010) as its indices. The right plane is calculated to be $(0\frac{1}{2}0)$, but cleared of fractions, as required, it becomes (010). That is, $(0\frac{1}{2}0) \times 2 = (010)$. The left plane cannot be calculated with O as an origin. However, shift

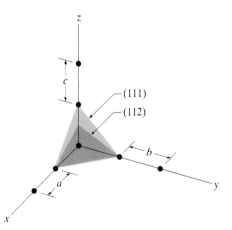

Figure 3–7.4 Miller indices. The (112) plane cuts the three axes at 1, 1, and $\frac{1}{2}$ unit distances.

the origin to O'. The left plane is now $(0\bar{1}0)$. However, $(0\bar{1}0) \times (-1) = (010)$. This is permissible, because a multiplication simply produces a parallel plane. Multiplication by a negative number transposes the plane to the other side of the origin.

Families of Planes

Depending on the crystal system, two or more planes may belong to the same family of planes. In the cubic system, an example of multiple planes includes the following, which constitute a *family of planes,* or *form:*

$$\frac{(100)\quad(010)\quad(001)}{(\bar{1}00)\quad(0\bar{1}0)\quad(00\bar{1})} = \{100\}.*$$

The collective notation for a family of planes is $\{hkl\}$. Figure 3–1.1 indicates for us the form of the $\{100\}$ family that has the six planes listed above. Each face is identical except for the consequences of our arbitrary choice of axis labels and directions. The reader is asked to verify that the $\{111\}$ family includes eight planes, and the $\{110\}$ family includes 12 planes, when all of the permutations and combinations of individual planes are included.

• Indices for Planes in Hexagonal Crystals (*hkil*)

The three Miller indices (hkl) can describe all possible planes through any crystal. In hexagonal systems, however, it is commonly useful to establish four axes, three of them coplanar (Fig. 3–3.1a). This leads to four intercepts and $(hkil)$ indices.† The fourth index i is an additional index that mathematically is related to the sum of the first two:

$$h + k = -i. \tag{3–7.1}$$

These optional $(hkil)$ indices are generally favored because they reveal hexagonal symmetry more clearly. Although partially redundant, they are used almost exclusively, in preference to the equivalent (hkl) indices, in scientific papers.

Planar Densities of Atoms

When we consider plastic deformation later, we will need to know the density of atoms on a crystal plane. Example 3–7.3 shows how we may calculate these by means of the relationship:

$$\text{Planar density} = \frac{\text{Atoms}}{\text{Unit area}}. \tag{3–7.2}$$

As we did in dealing with linear density, we may also calculate the planar density of atoms or of lattice points.

Study Aids (crystal directions and planes)

If your available time for study was too short to understand the origin and significance of crystal indices, you are strongly urged to use the paperback

---

* These six planes of the $\{100\}$ family enclose space; hence the term, *form.* However, it is permissible to state that the $\{100\}$ family has only three members—(100), (010), and (001)—since pairs like (100) and $(\bar{1}00)$ are parallel, but simply on the opposite side of the origin. Likewise, the $\{111\}$ family may be identified with either four or eight planes. (See Example 3–7.7.)

† Called *Miller–Bravais* indices.

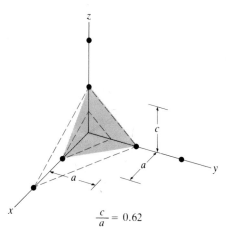

Figure 3–7.5 Noncubic intercepts (tetragonal structure). The shaded (111) plane cuts the three axes of any crystal at equal unit distances. However, since c may not equal a, the actual intercepting distances are different. (The dashed lines refer to Example 3–7.2.)

$$\frac{c}{a} = 0.62$$

Study Aids for Introductory Materials Courses.* Directions and Miller indices are presented in greater detail by a series of annotated sketches. A separate section presents the four indices used for hexagonal crystals. Optional sketches address the topics of families of direction and planes and the use of dot and cross products.

Example 3–7.1

Sketch a (111) plane through a unit cell of a simple tetragonal crystal having a c/a ratio of 0.62.

Solution. Figure 3–7.5 shows this plane (shaded). The (111) plane cuts the three axes at unit distances. However, the unit distance along the z-axis is shorter than the unit distances on the x- and y-axes. ◀

Example 3–7.2

Sketch two planes with (122) indices on the axes of Fig. 3–7.5.

Solution. The intercepts will be the reciprocals of (122)—$1a$, **0.5**b, and **0.5**c. This is shown by the inner set of dashed lines. A plane with intercepts—$2a$, $1b$, and $1c$—is parallel and therefore has the same indices (outer set of dashed lines). ◀

Example 3–7.3

(a) How many atoms per mm² are there on the (100) planes of lead (fcc)? (b) On the (111) planes?

Solution. Pb radius = 0.1750 nm (from Appendix B),

$$a_{Pb} = \frac{4R}{\sqrt{2}} = \frac{4(0.1750 \text{ nm})}{1.414} = 0.495 \text{ nm.}$$

a) Figure 3–7.6 shows that the (100) plane contains two atoms per unit-cell face.

$$(100): \qquad \frac{\text{atoms}}{\text{mm}^2} = \frac{2 \text{ atoms}}{(0.495 \times 10^{-6} \text{ mm})^2} = 8.2 \times 10^{12} \text{ atoms/mm}^2.$$

* *Study Aids for Introductory Materials Courses*, Reading, Mass.: Addison-Wesley, 1977.

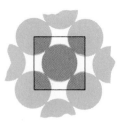

Figure 3–7.6 Sketch for Example 3–7.3(a). A (100) plane in an fcc metal structure has two atoms per a^2.

b) Figure 3–7.7(b) shows that the (111) plane contains three one-sixth atoms in the triangular shaded area. That area is $bh/2 = R^2\sqrt{3}$.

$$(111): \quad \frac{\text{atoms}}{\text{nm}^2} = \frac{\frac{3}{6}}{\frac{1}{2}bh} = \frac{\frac{3}{6}}{\frac{1}{2}(2)(0.1750 \text{ nm})(\sqrt{3})(0.1750 \text{ nm})}$$
$$= 9.4 \text{ atoms/nm}^2 = 9.4 \times 10^{12} \text{ atoms/mm}^2. \quad \blacktriangleleft$$

Example 3–7.4

A plane includes points at 0,0,0 and $\frac{1}{2},\frac{1}{4},0$ and $\frac{1}{2},0,\frac{1}{2}$. What are its Miller indices?

Solution. Make a sketch (Fig. 3–7.8).

Since the plane passes through the origin, shift the origin, for example, 1 unit in the x-direction. The intercepts are now $-1a$, $0.5a$, and $+1a$; therefore, $(1\bar{2}1)$.

Comment. Had we shifted the origin one unit upward, the intercepts would have been $+1, -\frac{1}{2}, -1$ to give us $(1\bar{2}\bar{1})$. This is parallel to $(\bar{1}21)$, and therefore equivalent. $\quad \blacktriangleleft$

Example 3–7.5

What direction is the line of intersection of the (111) and (112) planes?

Answer. Refer to Fig. 3–7.4: $[1\bar{1}0]$, (or $[\bar{1}10]$).

Comment. For our purposes, the line of intersection is most readily obtained by inspection. However, it may also be obtained as a cross product of the indices of the two planes. $\quad \blacktriangleleft$

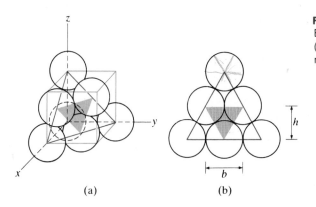

Figure 3–7.7 Sketch for Example 3–7.3(b). (a) Exposed (111) plane. (b) Rotation for measurement.

(a)   (b)

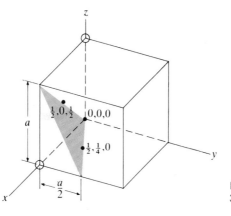

Figure 3–7.8 The ($\bar{1}$21) plane (see Example 3–7.4).

• **Example 3–7.6**

Refer to Fig. 3–3.1(a). Provide an index for the plane facing the reader.

Answer

Axis:	a_1	a_2	a_3	c
Intercepts:	**1**	**∞**	**−1**	**∞**

Therefore, the ($hkil$) indices are ($10\bar{1}0$), since we use reciprocals.

Comment. If we had used the alternate presentation of Fig. 3–3.1(b), the (hkl) indices would be (100). The additional index, i, is related to h and k by Eq. (3–7.1). ◀

Example 3–7.7

List the individual planes that belong to the {111} family in cubic crystals.

Answer

$$(111) \quad (\bar{1}11) \quad (1\bar{1}1) \quad (11\bar{1});$$
$$(\overline{111}) \quad (1\bar{1}\bar{1}) \quad (\bar{1}1\bar{1}) \quad (\bar{1}\bar{1}1).$$

Comment. In reality, there are only four planes (rather than 8) because those of the second set are parallel to those of the first set. However, unlike the negative directions (Example 3–6.2), we commonly list the redundant planes because the eight planes (in this case) enclose a volume and produce a crystal *form*. The form for {111} is a bipyramid. ◀

• **3–8**

X-RAY DIFFRACTION

Excellent experimental verification for the crystal structures that we have been discussing is available through x-ray diffraction. When these high-frequency electromagnetic waves are selected to have a wavelength slightly greater than the *interplanar* spacings of crystals, they are diffracted according to very exacting physical laws. The angles of diffraction let us decipher crystal structures

---

• See Preface for bullet, •, notation.

with a high degree of accuracy. In turn, one can readily determine the interplanar spacings (and therefore atomic radii) in metals to four significant figures, and even more precisely, if necessary. Let us first examine the spacings between planes. Then we will turn to diffraction.

Interplanar Spacings

Recall, from Section 3–7, that all parallel planes bear the same (hkl) notation. Thus the several (110) planes of Fig. 3–7.2 have still another (110) plane that passes directly through the origin. As a result, if we measure a *perpendicular* distance from the origin to the next adjacent (110) plane, we have measured the interplanar distance, d. We will observe that in the simple cubic structures of Figs. 3–7.1(a), 3–7.2(a), and 3–7.3(a), the interplanar distances are $a, (a\sqrt{2})/2$, and $(a\sqrt{3})/3$ for d_{010}, d_{110}, and $d_{\bar{1}11}$, respectively. That is, there is one spacing per cell edge a for d_{010}; two spacings per face diagonal, $a\sqrt{2}$, for d_{110}; and three spacings per body diagonal, $a\sqrt{3}$, for $d_{\bar{1}11}$. We may formulate a general rule for d-spacings in *cubic* crystals:

$$d_{hkl} = \frac{a}{\sqrt{h^2 + k^2 + l^2}}, \qquad (3\text{–}8.1)$$

where a is the lattice constant and h, k, and l are the indices of the planes.* The interplanar spacings for noncubic crystals may be expressed with equations that are related to Eq. (3–8.1), but take into account the variables of Table 3–1.1. We will not consider diffraction of noncubic crystals.

Bragg's Law

When x-rays encounter a crystalline material, they are diffracted by the planes of the atoms (or ions) within the crystal. The *diffraction angle* θ depends upon the wavelength λ of the x-rays and the distance d between the planes:

$$n\lambda = 2d \sin \theta. \qquad (3\text{–}8.2)$$

Consider the parallel planes of atoms in Fig. 3–8.1, from which the wave is diffracted. The waves may be "reflected" from the atom at H or H' and remain in phase at K. However, x-rays are reflected not only from the surface plane but also from the adjacent subsurface planes. If these reflections are to remain in phase and be *coherent,* the distance $MH''P$ must equal one or more integer wavelengths of the rays. The value n of Eq. (3–8.2) is the integer number of waves that occur in the distance $MH''P$.

Diffraction Analyses

The most common procedure for making x-ray diffraction analyses utilizes very fine powder of the material in question. It is mixed with a plastic cement and formed into a very thin filament that is placed at the center of a circular camera

---

* The reciprocal nature of Miller indices permits this type of simplified calculation. Likewise, the cross product of the Miller indices for two intersecting planes gives the direction indices for the line of intersection. Finally, in order for a direction to lie in a plane, the dot product of the indices for that direction and for the plane must equal zero. In brief, there is purpose in using the "down-side up" Miller indices.

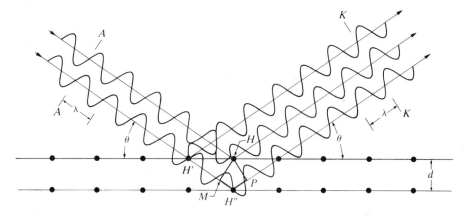

Figure 3–8.1 X-ray diffraction. *MH″P,* which is 2*d* sin *θ*, must equal λ to keep the emerging rays in phase.

(Fig. 3–8.2). A collimated beam of x-rays is directed at the powder. Since there is a very large number of powder particles with essentially all possible orientations, the diffracted beam emerges as a cone of radiation at an angle 2θ from the initial beam. (Observe in Fig. 3–8.1 that the diffracted beam is 2θ away from the initial beam.)

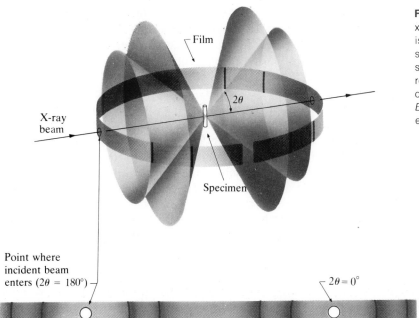

Figure 3–8.2 The exposure of x-ray diffraction patterns. Angle 2θ is precisely fixed by the lattice spacing *d* and the wavelength λ as shown in Eq. (3–8.2). Every cone of reflection is recorded in two places on the strip of film. (B. D. Cullity, *Elements of X-ray Diffraction,* 1st ed., Addison-Wesley.)

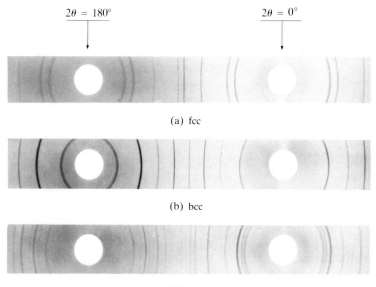

$2\theta = 180°$ $2\theta = 0°$

(a) fcc

(b) bcc

(c) hcp

Figure 3–8.3 X-ray diffraction patterns for (a) copper, fcc, (b) tungsten, bcc, and (c) zinc, hcp. The crystal structure and the lattice constants may be calculated from patterns such as these. (B. D. Cullity, *Elements of X-ray Diffraction*, 2d ed., Addison-Wesley.)

The diffraction cone exposes the film strip in the camera at two places; each is 2θ from the straight-through exit port. There is a separate cone (or pair of *diffraction lines*) for each d_{hkl} value of interplanar spacings. Thus, the diffraction lines may be measured and the d-spacings calculated from Eq. (3–8.2). All fcc metals will have a similar set of diffraction lines, but with differing 2θ values, since they have different lattice constants; for example, $a_{Cu} = 0.3615$ nm; $a_{Al} = 0.4049$ nm; $a_{Pb} = 0.495$ nm, etc. Thus, we can differentiate between various fcc metals.

In Fig. 3–8.3, we see x-ray diffraction films for copper (fcc), tungsten (bcc) and zinc (hcp). It is immediately apparent that the sequences of diffraction lines are different for the three types of crystals. Since we lack space here to explain these differences, we will have to simply observe that, with different "fingerprints," it is possible not only to determine the size of the lattice constants, with utmost precision, but also to identify the crystal lattice. X-ray diffraction is an extremely powerful tool in the study of the internal structure of materials.

Example 3–8.1

X-rays of an unknown wavelength are diffracted 43.4° by copper (fcc) whose lattice constant a is 0.3615 nm. Separate determinations indicate that this diffraction line for copper is the first-order ($n = 1$) line for d_{111}.

a) What is the wavelength of the x-rays?

b) The same x-rays are used to analyze tungsten (bcc). What is the angle, 2θ, for the second-order ($n = 2$) diffraction lines of the d_{010} spacings?

Solution

a) Since $2\theta = 43.4°$, and by using Eqs. (3–8.1) and (3–8.2),

$$(1)\lambda = 2\left[\frac{0.3615 \text{ nm}}{\sqrt{1^2 + 1^2 + 1^2}}\right]\sin 21.7°$$

$$= 0.1543 \text{ nm}.$$

b) From Appendix B,

$$R_w = 0.1367 \text{ nm}.$$

From Eq. (3–2.1),

$$a_w = 4(0.1367 \text{ nm})/\sqrt{3} = 0.3157 \text{ nm}.$$

From Eqs. (3–8.1) and (3–8.2),

$$\sin\theta = 2(0.154 \text{ nm})(\sqrt{0 + 1 + 0})/(2)(0.3157 \text{ nm}),$$
$$2\theta = 58.4°.$$

Comments. The second-order diffraction of d_{010} is equivalent to the first-order diffraction of a d_{020} spacing; i.e., the perpendicular distance from the origin to a plane that cuts the x-, y-, and z-axes at ∞, $\frac{1}{2}$, and ∞, respectively. This may be checked with Eqs. (3–8.1) and (3–8.2).

The d_{hkl}-spacings for all planes of the same form are equal. For example, $d_{100} = d_{010} = d_{001}$ in the cubic system. ◀

Review and Study

SUMMARY

In Chapter 2 we considered the coordination of atoms with their neighbors. This chapter carries our understanding of internal structure one step further by looking at longer-range geometric order. Since all atoms of one kind have similar coordination requirements, it is not surprising that we find the same patterns repeated time and time again throughout the material.

1. Crystals are solids with a long-range periodicity. Crystals can be categorized into seven *crystal systems* based on the angles between the reference axes, and the periodicity of the pattern. Structures that are body-centered and face-centered cubic will receive much, but not all, of our attention.

2. In *body-centered cubic* structures, the pattern repeats itself by translations of $\pm\frac{a}{2}$, $\pm\frac{b}{2}$, $\pm\frac{c}{2}$. In *face-centered cubic* structures, any two of the above three translations produce replication. Packing factors and densities may be calculated from the atom radius in metal structures where a single type of atom is present. Several face-centered cubic structures were encountered. So far, these include the *fcc metal structure*, the *NaCl structure*, and the *diamond cubic structure*.

3. Nearly 30 percent of the metallic elements form a *hexagonal close-packed* metal structure. Like fcc metals, the packing factor for hcp metals is 0.74, and CN = 12.

4. An element or compound may have more than one crystal structure. These are called *polymorphs*. Although identical in composition, polymorphs do not have identical properties since their structures differ. The two polymorphs of iron—bcc and fcc—have major technical significance.

5. *Points* within unit cells are indexed by their axial coefficients. For example, $\frac{1}{2},\frac{1}{2},1$ is the center of the top face of a unit cell, regardless of the crystal system.

6. Many properties vary with *crystal direction*. Directions are indexed as a ray passing through the origin and point u, v, w. Such a direction is identified as [uvw]. The values of u, v, and w should be integers with lowest common multiples. Thus, a ray from 0,0,0, to the center of the top face is labeled [112] since it extends to the point: 1,1,2. All parallel directions possess the same indices. Families of directions $\langle uvw \rangle$ include all directions that are identical except for our arbitrary choice of coordinate references.

7. *Crystal planes* are labeled by *Miller indices*. They are the *reciprocals* of the three intercepts that the plane makes with the three axes, cleared of fractions and of common multipliers. For example, a (120) plane cuts the x and y axes at a and $b/2$, respectively, but is parallel with the z axis (cuts it at ∞). Parallel planes carry the same indices.

   We will give the greatest, but not exclusive, attention to the (100), (110), and (111) planes; therefore, the reader should be familiar with the following facts:

   i) Any plane of the {100} family parallels two of the coordinate axes and cuts the third;

   ii) Any plane of the {110} family parallels one axis and cuts the other two with equal intercept coefficients; and

   iii) The planes of the {111} family cut all three axes with equal intercept coefficients.

• 8. Crystal structures are determined by *x-ray diffraction*. The *interplanar spacings* can be calculated to four and more significant figures by measuring the diffraction angles. These are the bases for determining interatomic distances and calculating atomic radii.

TERMS AND CONCEPTS

Anisotropic Having different properties in different directions.

Atomic radius, R Half of the interatomic distance.

• **Bragg's law** Diffraction law for periodic structures (Eq. 3–8.2).

Body-centered cubic (bcc) The center of the cube is identical to the cube corners.

Crystal A solid with long-range repetitive pattern in the three coordinate directions.

Crystal direction [uvw] A ray from an arbitrary origin through a selected unit-cell location. The indices are the lattice coefficients of that location.

Crystal plane (hkl) A two-dimensional array of ordered atoms. (See also Miller indices.)

Crystal system   Categorization of unit cells by axial and dimensional symmetry (Table 3–1.1).

Diamond cubic structure   The fcc structure of an element with CN = 4. (See Fig. 3–2.5(b).)

• **Diffraction lines**   The diffracted beam from a crystal surface. Detected photographically, or by a Geiger-type counter.

• **Diffraction (x-ray)**   Deviation of an x-ray beam by regularly spaced atoms.

Face-centered cubic (fcc)   The centers of cube faces are identical to each other and to the cube corners.

Family of directions ⟨*uvw*⟩   Crystal directions that are identical except for our arbitrary choice of axes.

Family of planes {*hkl*}   Crystal planes that are identical except for our arbitrary choice of axes.

Hexagonal close-packed metal (hcp)   A hexagonal metal with CN = 12, and PF = 0.74.

• **Interplanar spacing,** d_{hkl} Perpendicular distance between two adjacent crystal planes with the same index.

Lattice   The space arrangement of the periodicity in a crystal.

Lattice constants   Edge dimensions of the unit cell.

Linear density   Items, e.g., atoms, per unit length.

Long-range order   A repetitive pattern over many atomic distances.

Miller indices (*hkl* **)**   Index relating a plane to the reference axes of a crystal. *Reciprocals* of axial intercepts, cleared of fractions and of common multipliers.

NaCl structure   The fcc structure of an AX compound with CN = 6. (See Figs. 3–2.5(a) and 7–2.3.)

Orthorhombic   A crystal with three unequal but perpendicular axes.

• **Planar density**   Items, e.g., atoms, per unit area.

Polymorphism   A composition with more than one crystal structure.

Repetition distance   Translation vector between identical lattice points.

Simple cubic (sc)   A cubic unit cell with lattice points at the corners only.

Tetragonal   Two of three axes equal; all three at right angles.

Translation   Vector displacement between lattice points.

Unit cell   A small (commonly the smallest) repetitive volume that comprises the complete lattice pattern of a crystal.

FOR CLASS DISCUSSION

A₃   What is meant by long-range order?

B₃   List the several crystal systems and indicate their distinguishing characteristics.

C₃   The corner site of a cubic unit cell is associated with how many unit cells? The corner site of a rectangular area of a plane belongs to how many "unit areas"?

D₃   From Fig. 3–2.4, show that CN = 12 for fcc.

E₃   Examine wallpaper that has an intricate design and pick a point at random. Find three nearby identical positions. What are the translation distances between the identical positions?

F₃   Among 3-D crystals, we find bcc and bct (i.e., body-centered tetragonal). Also there is an fcc, but no fct. Explain why.

G₃   What is the ratio of volumes for the two presentations of the hexagonal unit cell in Fig. 3–3.1?

H₃   Compare the size and "shape" of the holes that are centered at $\frac{1}{2},\frac{1}{2},0$ and at $\frac{1}{2},\frac{1}{4},0$ of a bcc metal.

I₃   Compare the size and "shape" of the holes that are centered at $\frac{1}{2},0,0$ and $\frac{1}{4},\frac{1}{4},\frac{1}{4}$ of an fcc metal.

J₃   Diamond requires a higher pressure, to be stable, than does graphite. What does this indicate about the densities of diamond and graphite?

K₃   Cobalt changes during heating from hcp to fcc at 1120°C. How will its volume be altered (a) compared with iron which contracts on heating from the low to the high temperature forms (Example 3–4.1), and (b) with titanium that contracts on cooling (Study Problem 3–4.3), when it changes structure from the high to the low temperature form?

L₃   From Fig. 3–2.4, show that fcc iron could be categorized as a body-centered tetragonal (bct) with a c/a ratio of 1.414. Why do we use fcc rather than bct?

M₃   Gray tin has the structure of diamond (Fig. 3–2.5b). White tin has a body-centered tetragonal structure with atoms at

$$0,0,0 \quad \tfrac{1}{2},\tfrac{1}{2},\tfrac{1}{2} \quad \tfrac{1}{2},0,\tfrac{1}{4} \quad 0,\tfrac{1}{2},\tfrac{3}{4}.$$

Sketch the unit cell. Show that the second and third sites listed above are not equivalent, but the third and fourth sites are equivalent.

N₃   How many directions are in the ⟨012⟩ family of a cubic crystal? (We usually consider [uvw] and [$\overline{uvw}$] to be the same, since a reversal of all indices simply follows the same ray but in the opposite sense; that is, [$\overline{uvw}$] = $-$[uvw].)

• **O₃**   Repeat N₃ but for a tetragonal crystal. (Recall $c \neq a$) in tetragonal; therefore we cannot permutate the 2 with the 0 and 1, as we can in the cubic system.)

P₃   The hcp unit cell of Fig. 3–3.2(b) can be redrawn as a rhombic lattice. (Cf. Fig. 3–3.1b.) At how many locations is a point repeated per unit cell? (Do the points that you picked have identical neighbors in identical directions and at identical distances? If not, they are not equivalent.)

Q₃   Which of the following directions lie in the (110) plane?

$$[112] \quad [1\bar{1}0] \quad [001] \quad [1\bar{1}2] \quad [8\bar{8}9]$$

• **R₃**   Show that the dot product of [uvw] and (hkl) equals zero when the direction lies within the plane.

• **S₃**   The [111] direction is normal to the (111) plane in a *cubic* crystal, but not in a tetragonal crystal. Why?

• **T₃**   With Fig. 3–7.2(c), prove to yourself that second-order ($n = 2$) diffraction for d_{110} equals first-order ($n = 1$) diffraction for d_{220}.

U₃   The atomic packing factor is independent of atom size in metals. Under what conditions would the packing factor for a sand within a container by independent of sand size?

V_3 As stated in Section 3–1, the majority of metals are cubic; however, few molecular solids are cubic. Suggest a reason why.

• W_3 We consider (020) to be comparable to (010); but we differentiate between d_{020} and d_{010}. Why?

QUIZ SAMPLES

3A   The unit cells of crystals in the __+−__ system(s) have equal dimensions in *two* of the three axial directions.

  (a) cubic   (b) tetragonal   (c) monoclinic   (d) hexagonal
(e) rhombohedral

3B   The crystal system(s) not listed in 3A is (are) __**__ .

3C   The fcc structure(s) that possess(es) a packing factor of 0.74 is (are) __+−__ .

  (a) metal   (b) NaCl   (c) diamond   (d) CaF_2 of Fig. 7–3.1.

3D   MgO has the same structure as NaCl. Therefore, a crystal has __**__ times as many Mg^{2+} ions as it has unit cells.

3E   A polymorphic change from hcp to __**__ will introduce no volume change, while a change from bcc to __**__ or __**__ produces a volume contraction.

3F   In the unit cell of a bcc crystal, a point at 0.6, 0.2, 0.45 is identical to a point at __**__ .

3G   Based on data in Appendix B, the distance from $0,\frac{1}{2},0$ to $0,0,\frac{1}{2}$ in the unit cell of aluminum is __+−__ nm.

  (a) 0.143   (b) $0.143\sqrt{2}$ (c) 0.143(2)   (d) $4(0.143)/\sqrt{2}$ (e) $0.405/\sqrt{2}$
(f) None of the above, rather _____   (g) Impossible to calculate.

3H   The [102] direction that passes through $\frac{1}{2},\frac{1}{2},\frac{1}{2}$ also passes through __+−__ .

  (a) 1,0,2   (b) $\frac{1}{2},0,1$ (c) $-1,0,-2$ (d) 0,0,0   (e) None of the above, rather _____   (f) Impossible to answer, since it does not pass through the origin.

3I   The (112) plane that contains location 0,0,1 also contains locations __+−__ .

  (a) 1,0,0   (b) $0,0,\frac{1}{2}$ (c) $1,0,\frac{1}{2}$ (d) None of the above, rather _____
(e) Impossible to determine since the z intercept is not $\frac{c}{2}$.

3J   The plane that passes through the origin and locations $\frac{1}{2},\frac{1}{2},\frac{1}{2}$ and 1,0,1 is the __**__ plane.

3K   Among the following directions, __+−__ is (are) parallel to the $(11\bar{1})$ plane.

  (a) $[1\bar{1}0]$ (b) [011]   (c) $[8\bar{7}1]$ (d) [112]   (e) $[\bar{2}1\bar{1}]$
(f) None of the above, rather _____

3L   Gray tin has the structure of diamond (Fig. 3–2.5b), but with $a = 0.649$ nm. There are __**__ tin atoms per mm in the [111] direction; but the distance between identical points in the [111] direction is __**__ nm.

• **3M**   The perpendicular distance between (111) planes in copper is __**__ nm; this distance is __**__ nm between (110) planes and __**__ nm between adjacent (001) planes.

• **3N**   The x-ray wavelength in Fig. 3–8.3(a) is 0.156 nm; therefore, the largest inter-planar spacing is __+−__ nm. (Assume $n = 1$.)

(a) 0.21 nm   (b) 0.156/(2 sin 21.9°)   (c) 0.156/(2 sin 43.8°)   (d) 0.11 nm
(e) None of the above, rather _____   (f) Impossible to calculate unless we have data from App. B.

STUDY PROBLEMS

3–1.1   The unit cell of aluminum is cubic with $a = 0.4049$ nm. From its density, calculate how many atoms there are per unit cell.

Answer: 4

3–1.2   There are two atoms per unit cell in zinc. What is the volume of the unit cell?

3–1.3   Tin is tetragonal with $c/a = 0.546$. There are 4 atoms/unit cell. (a) What is the unit cell volume? (b) What are the lattice dimensions?

Answer: $c = 0.318$ nm, $a = 0.582$ nm

3–1.4   Gold foil is 0.08 mm thick and 670 mm^2 in area. (a) It is cubic with $a = 0.4076$ nm. How many unit cells are there in the foil? (b) What is the mass of each unit cell if the density is 19.32 Mg/m^3 (= 19.32 g/cm^3)?

3–2.1   Calculate the atomic packing factor of tungsten (bcc with $a = 0.3157$ nm).

Answer: 0.68

3–2.2   The volume of the unit cell of chromium in Example 3–1.1 is 24×10^{-30} m^3, or 0.024 nm^3. Based on the data used in that example problem, plus PF$_{bcc\ metal} = 0.68$, calculate the radius of the chromium atom to verify that of Appendix B.

3–2.3   Silver is fcc and its atomic radius is 0.1444 nm. How large is the side of its unit cell?

Answer: 0.4084 nm

3–2.4   Nickel is fcc with a density of 8.9 Mg/m^3 (=8.9 g/cm^3). (a) What is the volume per unit cell based on the density? (b) From your answer in part (a), calculate the radius of the nickel atom. Check this with the radius listed in Appendix B.

3–2.5   Titanium is bcc at high temperatures and its atomic radius is 0.145 nm. (a) How large is the edge of the unit cell? (b) Calculate the density.

Answer: (a) 0.335 nm   (b) 4.24 Mg/m^3 (=4.24 g/cm^3)

3–2.6   Lead is fcc and its atomic radius is 1.750×10^{-10} m. What is the volume of its unit cell?

3–2.7   Magnesium oxide (MgO) has the same structure as NaCl (Fig. 3–1.1). (a) What is its lattice constant a? (b) Its density? (*Note:* MgO is an ionic compound; therefore, we must use ionic radii rather than atomic radii.)

Answer: 3.8 Mg/m^3 (=3.8 g/cm^3)

3–2.8   Barium is bcc with a density of 3.6 Mg/m^3 (=3.6 g/cm^3). (a) Calculate the center-to-center distance between closest atoms. (b) What is the edge dimension of the unit cell? (Its atomic number is 56 and its atomic weight is 137.3 amu.)

3–2.9 Refer to Fig. 3–1.1. (a) What is the distance between the centers of the closest Cl^- ions? (b) What is the distance from the center of the Na^+ ion to the center of the *second-closest* Cl^- ion? (c) What is the distance between the surfaces of the closest Cl^- ions? The closest Na^+ ions?

Answer: (a) 0.39 nm   (b) 0.48 nm   (c) 0.03 nm, 0.2 nm

3–2.10 Calculate the density of NaCl.

3–2.11 (a) How many atoms are there per mm^3 in solid strontium? (b) What is the atomic packing factor? (c) It is cubic. What is its structure? (Atomic number = 38; atomic mass = 87.62 amu; atomic radius = 0.215 nm; ionic radius = 0.127 nm; density = 2.6 Mg/m^3.)

Answer: (a) $1.78 \times 10^{19}/mm^3$ (b) 0.74   (c) fcc metal

3–2.12 (a) How many atoms are there per mm^3 in solid tantalum? (b) What is the atomic packing factor? (c) It is cubic. What is its structure? (Atomic number = 73; atomic mass = 180.95 amu; atomic radius = 0.1429 nm; ionic radius = 0.068 nm; density = 16.6 Mg/m^3.)

3–3.1 From the data for cobalt (hcp) in Appendix B, calculate the volume of the unit cell (Fig. 3–3.2).

Answer: 0.066 nm^3

* **3–3.2** Zinc has an hcp structure. The height of the unit cell is 0.494 nm. The centers of the atoms in the base of the unit cell are 0.2665 nm apart. (a) How many atoms are there per hexagonal unit cell? (Show reasoning.) (b) What is the volume of the hexagonal unit cell? (c) Is the calculated density greater or less than the actual density of 7.135 Mg/m^3?

3–3.3 The atomic mass of zirconium is 91.22 amu, and the average atomic radius of this hcp metal is 0.16 nm. (a) What is the volume associated with each atom (sphere + interstitial space)? (b) From the answer in part (a), calculate the density.

Answer: (a) 23.2×10^{-30} m^3 (or 0.0232 nm^3)   (b) 6.5 Mg/m^3 (=6.5 g/cm^3)

3–3.4 Magnesium is hcp with nearly spherical atoms having a radius of ~0.161 nm. Its c/a ratio is 1.62. (a) Calculate the volume of the unit cell from these data. (b) Check your answer by using it to calculate the density of magnesium.

3–4.1 The lattice constant a for diamond (Fig. 3–2.5b) is 0.357 nm. What percent volume change occurs when it transforms to graphite (ρ = 2.25 Mg/m^3, or 2.25 g/cm^3)?

Answer: 56 v/o expansion

3–4.2 The volume of a unit cell of bcc iron is 0.02464 nm^3 at 912°C. The volume of a unit cell of fcc iron is 0.0486 nm^3 at the same temperature. What is the percent change in density as the iron transforms from bcc to fcc?

3–4.3 Titanium is bcc in its high-temperature form. The radius increases 2% when the bcc changes to hcp during cooling. What is the percentage volume change? (*Recall* that there will be a change in atomic packing factor.)

Answer: −2.5 v/o (= −0.8 l/o)

---

* See footnote on p. 22.

3-4.4   Metallic tin has a tetragonal structure with $a = 0.5820$ nm and $c = 0.3175$ nm and with four atoms per unit cell. Another form of tin (gray) has the cubic structure of diamond (Fig. 2–3.1b) with $a = 0.649$ nm. What is the volume change as tin transforms from gray to metallic?

3-5.1   Refer to Fig. 3–2.1. We may identify the structure of a bcc metal with only two locations. What are they?

Answer: $0,0,0$ and $\frac{1}{2},\frac{1}{2},\frac{1}{2}$. (The remaining atoms are at redundant translations.)

3-5.2   Refer to Fig. 3–2.3. We may identify the structure of an fcc metal with four locations. What are they?

3-5.3   Copper is fcc and has a lattice constant of 0.3615 nm. What is the distance between the $1,0,0$ and $0,\frac{1}{2},\frac{1}{2}$ locations?

Answer: 0.4427 nm

3-5.4   MgO has the same structure as NaCl (Fig. 3–1.1). Its lattice constant is $\sim$0.42 nm. What is the center-to-center distance from the Mg^{2+} ion at $0,0,\frac{1}{2}$ and (a) its nearest Mg^{2+} neighbor? (b) its second-nearest Mg^{2+} neighbor?

3-5.5   When a copper atom is located at the origin of an fcc unit cell, a small interstitial hole is centered at $\frac{1}{4},\frac{1}{4},\frac{1}{4}$. Where are there other equivalent holes?

Answer: $\frac{3}{4},\frac{3}{4},\frac{1}{4}$ $\frac{3}{4},\frac{1}{4},\frac{3}{4}$ $\frac{1}{4},\frac{3}{4},\frac{3}{4}$ for a total of four per unit cell if we count $\frac{1}{4},\frac{1}{4},\frac{1}{4}$.

3-5.6   Repeat Study Problem 3–5.5, but start with $\frac{3}{4},\frac{3}{4},\frac{3}{4}$ rather than $\frac{1}{4},\frac{1}{4},\frac{1}{4}$.

3-5.7   Each atom in bcc iron has eight nearest neighbors. (a) How many *second* nearest neighbors are there? (b) If a_{bcc} for iron is 0.286 nm, what are the center-to-center distances to their second nearest neighbors?

Answer: (a) 6   (b) 0.286 nm

3-5.8   Refer to Study Problem 3–5.5, and Fig. 3–2.3. Observe that locations $\frac{1}{4},\frac{1}{4},\frac{1}{4}$ and $\frac{3}{4},\frac{3}{4},\frac{3}{4}$ both lie among four immediate neighbors. Are they equivalent locations? In other words, why was $\frac{3}{4},\frac{3}{4},\frac{3}{4}$ not among the answers for Study Problem 3–5.5?

3-5.9   The structure of β' brass is simple cubic as shown in Fig. 3–5.3. Estimate its density if the radii of the Cu and Zn atoms are 0.13 nm and 0.14 nm, respectively.

Answer: 7.1 Mg/m^3 (=7.1 g/cm^3)

3-6.1   (a) A ray in the [122] direction passes through the origin. Where does it leave the reference unit cell? (b) Another ray in a parallel [122] direction leaves the reference unit cell at $1,1,1$. Where did it enter the reference unit cell?

Answer: (a) $\frac{1}{2},1,1$ (b) $\frac{1}{2},0,0$

3-6.2   A ray in the [111] direction passes through location $\frac{1}{2},\frac{1}{2},0$. What are two other locations along its path?

3-6.3   Draw a line from $\frac{1}{2},\frac{1}{2},0$ to the center of the *next* unit cell, which could be indexed $1,\frac{3}{2},\frac{1}{2}$. What is the direction?

Answer: [021]

3–6.4 A line is drawn through a tetragonal unit cell from location $\frac{1}{2},0,0$ to $\frac{1}{8},\frac{3}{4},\frac{3}{4}$. What is the index of that direction?

3–6.5 (a) In a cubic crystal, what is the tangent of the angle between the [110] direction and the [111] direction? (b) The [001] direction and the [112] direction?

Answer: (a) $a/(a\sqrt{2}) = 0.71$ (b) $(a\sqrt{2})/(2a) = 0.71$

3–6.6 (a) In a cubic crystal, what is cos [113] ≮ [110]? (b) What is sin [001] ≮ [112]?

3–6.7 What is the angle between [111] and [1$\bar{1}$1] in a cubic crystal? (*Hint:* [101] bisects this angle.)

Answer: 70.5°

3–6.8 (a) What is the angle between [101] and [$\bar{1}$01] in a cubic crystal? (b) In a tetragonal crystal where $c/a = 0.55$?

• **3–6.9** What is the linear density of atoms in (a) the ⟨100⟩ directions of copper (fcc, $a = 0.361$ nm)? (b) The ⟨100⟩ directions of iron (bcc, $a = 0.286$ nm)?

Answer: (a) 2.77×10^6 Cu/mm (b) 3.50×10^6 Fe/mm

• **3–6.10** What is the linear density of atoms in (a) the [221] direction of copper? (b) The [221] direction of iron? (See Study Problem 3–6.9 for lattice constants.)

3–6.11 Draw a line from a point at $\frac{1}{4},\frac{1}{4},\frac{3}{4}$ to the center of the next unit cell in the rear ($-x$ direction). (a) What is the [uvw] direction between these two points? (b) The unit cell is tetragonal with $a = 0.31$ nm and $c = 0.33$ nm. What is the distance between the two points in part (a)?

Answer: (a) [$\bar{3}1\bar{1}$] (b) 0.259 nm

3–6.12 Draw a line from the $\frac{1}{2},0,\frac{1}{2}$ location of aluminum through the center of the base of next unit cell (x-direction). (a) What is the direction? •(b) What is the linear density of atoms in that direction? (c) On a sketch, show a parallel direction that passes through the 0,0,1 location. Indicate two other unit cell locations through which this parallel line passes.

3–6.13 (a) How many directions are there in the ⟨210⟩ family of a cubic crystal? (b) Of a tetragonal crystal?

Answer: (a) 12 (b) 4

3–6.14 How many directions are there in (a) the ⟨100⟩ family of a cubic crystal? (b) The ⟨110⟩ family? (c) The ⟨111⟩ family?

3–7.1 A plane intercepts the crystal axes at $a = $ **0.5** and $b = $ **1**. It is parallel to the z-axis. What are the Miller indices?

Answer: (210)

3–7.2 A plane intercepts axes at $a = $ **2**, $b = $ **1**, and $c = $ **1**. What are the Miller indices?

3–7.3 Give the indices for a plane with intercepts at $a = -2$, $b = \frac{2}{3}$, and $c = \frac{3}{2}$.

Answer: $(\bar{3}94)$

3–7.4 What are the indices for a plane with intercepts at $a = \frac{1}{2}$, $b = -\frac{3}{2}$, and $c = \frac{1}{3}$?

3–7.5 What are the axial intercepts for a (211) plane that passes through the point, 0,2,0?

Answer: $a = 1$, $b = 2$, $c = 2$

3–7.6 A (120) plane contains a point at 1,1,1. Where does it intercept the three axes?

3–7.7 What are the axial intercepts for a (111) plane that passes through the center of the unit cell, that is, $\frac{1}{2},\frac{1}{2},\frac{1}{2}$?

Answer: $a = 1.5$, $b = 1.5$, $c = 1.5$

3–7.8 A $(33\bar{1})$ plane contains points at 0,0,0 and $-1,1,0$. What is another crystal location (of many) that lies in this plane?

3–7.9 A plane includes points at 0,0,0 and 0,1,0 and $\frac{1}{2},1,\frac{1}{2}$. What are its Miller indices?

Answer: $(\bar{1}01)$, or $(10\bar{1})$

3–7.10 (a) How many atoms are there per mm^2 in the (100) plane of nickel? (b) The (110) plane? (c) The (111) plane?

3–7.11 (a) How many atoms are there per mm^2 in the (111) plane of iron? (b) The (210) plane of silver?

Answer: (a) $7 \times 10^{12}/mm^2$ (b) $5.4 \times 10^{12}/mm^2$

3–7.12 (a) Sketch a unit cell of silver (fcc) and shade the (012) plane. (b) What is the planar density of atoms?

3–7.13 (a) Sketch a unit cell of chromium (bcc) and shade the (102) plane. (b) What is the planar density of atoms?

Answer: (b) $5.4 \times 10^{12}/mm^2$

3–7.14 (a) How many atoms are there per mm^2 in the (110) plane of diamond? (b) The (111) plane? (See Fig. 3–2.5(b); $a = 0.357$ nm.)

3–7.15 (a) What is the line of intersection between the (110) plane and $(1\bar{1}1)$ plane of a cubic crystal? (b) Of a tetragonal crystal?

Answer: (a) $[\bar{1}12]$ (b) $[\bar{1}12]$

3–7.16 (a) What is the line of intersection between the (112) plane and the (010) plane? (b) The (112) plane and the (110) plane?

3–7.17 What are the $(hkil)$ indices of a plane in a hexagonal crystal that intercepts axes at $a_1 = 1$, $a_2 = 1$, and $c = 0.5$?

Answer: $(11\bar{2}2)$

3–7.18 Which of the atoms sketched in Fig. 3–3.2(a) lie in the $(10\bar{1}1)$ plane that intercepts the vertical axis at $c = 1$? (Be very careful. It is easy to mislead yourself in answering this question.)

• **3-7.19** (a) List the planes that belong to the {100} family in tetragonal crystals. (b) The {001} family.

Answer: (a) $(100)(010)(\bar{1}00)(0\bar{1}0)$ (b) $(001)(00\bar{1})$

3-7.20 List the twelve planes of the {110} family in cubic crystals.

3-7.21 (a) What ⟨110⟩ directions lie in the (111) plane of copper? (b) In the $(1\bar{1}1)$ plane?

Answer: (a) $[1\bar{1}0]$, $[\bar{1}01]$, $[01\bar{1}]$, and their inverses   (b) $[110]$, $[\bar{1}01]$, $[011]$, and their inverses

3-7.22 (a) What ⟨111⟩ directions lie in the (110) plane of iron? (b) In the $(1\bar{1}0)$ plane?

• **3-8.1** The lattice constant for a unit cell of aluminum is 0.4049 nm. (a) What is d_{220}? (b) d_{111}? (c) d_{200}?

Answer: (a) 0.1431 nm   (b) 0.2337 nm   (c) 0.2024 nm

• **3-8.2** Nickel is face-centered cubic with an atom radius of 0.1246 nm. (a) What is the d_{200} spacing? (b) The d_{220} spacing? (c) The d_{111} spacing?

• **3-8.3** X-rays with a wavelength of 0.058 nm are used for calculating d_{200} in nickel. The diffraction angle 2θ is 19°. What is the size of the unit cell? ($n = 1$.)

Answer: 0.35 nm

• **3-8.4** A sodium chloride crystal is used to measure the wavelength of some x-rays. The diffraction angle 2θ is 10.2° for the d_{111} spacing of the chloride ions. What is the wavelength? (The lattice constant is 0.563 nm.)

• **3-8.5** The first line (lowest 2θ) of Fig. 3-8.3(b) is for the d_{110} spacing in tungsten. (a) Determine the diffraction angle, 2θ, graphically. (b) The radius of the tungsten atom is 0.1367 nm. What wavelength was used? (c) The second diffraction line is for what d_{hkl} spacing? ($n = 1$.)

Answer: (a) 41°   (b) 0.156 nm   (c) $h^2 + k^2 + l^2 = 4$, $\therefore d_{200}$

• **3-8.6** The first line (lowest 2θ) of Fig. 3-8.3(a) is for the d_{111} spacing of copper. (a) What is its 2θ value? (b) The radius of the copper atom is 0.1278 nm. Calculate the wavelength of the x-rays. (c) The second diffraction line is for what d_{hkl} spacing?

QUIZ CHECKS

3A	(b), (d)		**3I**	(c)
3B	orthorhombic, triclinic		**3J**	$(\bar{1}01)$ or $(10\bar{1})$
3C	(a)		**3K**	(a), (b), (c), (d), (e)
3D	four		**3L**	1.8×10^6/mm
3E	fcc			1.12 nm
	fcc or hcp		• **3M**	0.209 nm
3F	0.1, 0.7, 0.95			0.128 nm
3G	(c), (e)			0.181 nm
3H	(e): $\frac{1}{2}+n, \frac{1}{2}, \frac{1}{2}+2n$		• **3N**	(a), (b)

Chapter Four

Atomic Disorder
in Solids

PREVIEW

Nature is not perfect: inherently there is some disorder present. Thus the structures just described in Chapter 3 are subject to exceptions. Often these exceptions are minor: maybe one atom out of 10^{10} is out of place. Even so, they can become important. Imperfections account for the behavior of semiconductors, for the ductility of metals, and for the strengthening of alloys. Imperfections account for the color of sapphire. They also permit the movement of atoms during heat treating, so that new structures and enhanced properties may be realized.

This chapter considers *impurities* first, then *crystalline imperfections*. Some solids are so disordered that we do not detect crystallinity. Finally we will look at *atomic vibrations and movements* within solids.

STUDY OBJECTIVES

1. To modify the knowledge you obtained in Chemistry on liquid solutions to new concepts on solid solutions. (These include substitutional and interstitial solid solutions and nonstoichiometry.)

2. To identify imperfections as 0-dimensional (vacancies and interstitials), as 1-dimensional (dislocations), or as 2-dimensional (boundaries and surfaces), and to explain the accompanying energy requirements in terms of interatomic distances.

3. To be able to index quantitatively the grain boundary area within a solid (and to be familiar with the empirical ASTM "grain-size index").

4. To differentiate between the changes that accompany (a) the solidification of a crystalline material, and (b) the solidification of an amorphous (noncrystalline) material.

5. To know the common factors that affect the diffusivity of atoms. These are significant in manufacturing processes and heat treatments.

6. To be familiar with the terms and concepts introduced in this chapter.

4–1

IMPURITIES IN SOLIDS

We admire "the real thing"; thus we commonly prefer pure wool, refined sugar, and like to think of 24-carat gold. Although these ideals may be noble, there are instances where, because of cost, availability, or properties, it is desirable to have impurities present. An example is *sterling silver,* which contains 7.5% copper and only 92.5% silver (Example 2–1.1). This material, which we rate highly, could be refined to well over 99% purity.* It would cost more; however, it would be of inferior quality. Without altering its appearance, the 7.5% Cu makes the silver stronger, harder, and therefore more durable—at a lower cost!

Of course, we must consider the properties pertinent to our design. Zinc added to copper produces *brass,* again at a lower cost than the pure copper. Brass is harder, stronger, and more ductile than copper. On the other hand, brass has lower electrical conductivity than copper, and so we use the more expensive pure copper for electrical wiring and similar applications where conductivity is important.

Alloys are combinations of two or more metals into one material. These combinations may be *mixtures* of two kinds of crystalline structures (e.g., bcc iron and Fe_3C in a steel bridge beam). Alternatively, alloys may involve *solid*

---

* As stated in the footnote with Example 2–1.1, weight percent (w/o) is implied in liquids and solids *unless specifically stated otherwise.*

solutions, which will be exemplified in the next section by brass. Although the term of alloy is generally not used specifically, various combinations of two or more oxide components may be incorporated advantageously into ceramic products (e.g., in the sparkplug insulator of Fig. 7–7.1(b)). Likewise, the computer housing of Fig. 1–4.2 contains a combination of several types of molecules.

4–2

SOLID SOLUTIONS IN METALS

Solid solutions form most readily when the *solvent* and *solute* atoms have similar sizes and comparable electron structures. For example, the individual metals of brass—copper and zinc—have atomic radii of 0.1278 nm and 0.139 nm, respectively. They both have 28 subvalent electrons and they each form crystal structures of their own with a coordination number of 12. Thus, when zinc is added to copper, it substitutes readily for the copper within the fcc metal structure, until a maximum of nearly 40 percent of the copper atoms has been replaced. In this solid solution of copper and zinc, the distribution of zinc is entirely random (see Fig. 4–2.1).

Substitutional Solid Solutions

The solid solution described above is called a *substitutional* solid solution because the zinc atoms substitute for copper atoms in the crystal structure. This type of solid solution is quite common among various metal systems. The solution of copper and nickel to form cupronickel is another example. Any fraction of the atoms in the original copper structure may be replaced by nickel. Copper–nickel solid solutions may range from no nickel and 100 percent copper, to 100 percent nickel and no copper. There is no *solubility limit*. All copper–nickel alloys are face-centered cubic.

On the other hand, there is a very definite limit to the amount of tin that may replace copper to form *bronze,* and retain the face-centered cubic structure of the copper. Tin in excess of the maximum amount of *solid solubility*

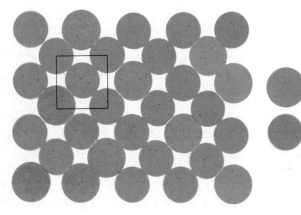

Figure 4–2.1 Random substitutional solid solution (zinc in copper, i.e., brass). The crystal pattern is not altered.

Zinc

Copper

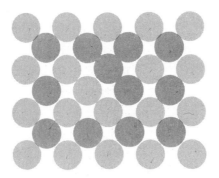

Figure 4–2.2 Ordered substitutional solid solution. The majority (but not all) of the atoms are coordinated among atoms unlike themselves. If ordering is complete, a compound is formed (Section 4–3).

must form another phase. This solubility limit will be considered in more detail in Chapter 10.

If there is to be extensive replacement in a substitutional type of solid solution, the atoms must be nearly the same size. Nickel and copper have a complete range of solutions because both of their individual structures are fcc and their radii are 0.1246 nm and 0.1278 nm, respectively. As the difference in size increases, less substitution can occur. Only 20 percent of copper atoms can be replaced by aluminum, because the latter has a radius of 0.1431 nm as compared with only 0.1278 nm for copper. Extensive solid solubility rarely occurs if there is more than about 15 percent difference in radius between the two kinds of atoms. There is further restriction in solubility when the two components have different crystal structures or valences.

The limiting factor is the *number* of substituted atoms rather than the *weight* of the atoms that are substituted. However, engineers ordinarily express composition as weight percent. It is therefore necessary to know how to convert weight percent to atomic percent, and vice versa (see Example 2–1.1).

• **Ordered Solid Solutions**

Figure 4–2.1 shows a *random substitution* of one atom for another in a crystal structure. In such a solution, the chance of one element occupying any particular atomic site in the crystal is equal to the atomic percent of that element in the alloy. In that case, there is no *order* in the substitution of the two elements.

However, it is not unusual to find an *ordering* of the two types of atoms into a specific arrangement. Figure 4–2.2 shows an ordered structure in which most dark "atoms" are surrounded by light "atoms." Such ordering is less common at higher temperatures, since greater thermal agitation tends to destroy the orderly arrangement.

Interstitial Solid Solutions

In another type of solid solution, illustrated in Fig. 4–2.3, a small atom may be located in the interstices between larger atoms. Carbon in iron is an example. At temperatures below 912°C (1673°F), pure iron occurs as a body-centered cubic structure. Above 912°C (1673°F), there is a temperature range in which iron has a face-centered cubic structure. In this face-centered structure, a

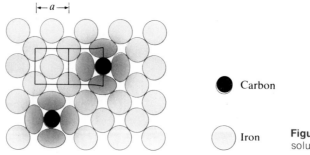

Carbon

Iron

Figure 4–2.3 Interstitial solid solution (carbon in fcc iron).

relatively large *interstice*, or "hole," exists in the center of the unit cell. Carbon, being an extremely small atom, can move into this hole to produce a solid solution of iron and carbon. At those temperatures where the iron has a body-centered cubic structure, the interstices between the iron atoms are much smaller. Consequently the solubility in body-centered cubic iron is very limited (Section 10–6).

Example 4–2.1

Bronze is a solid-solution alloy of copper and tin in which 3%, more or less, of the copper atoms are replaced by tin atoms. The fcc unit cell of copper is retained, but expanded a bit because the tin atoms have a radius of approximately 0.151 nm. (a) What is the weight percent in a 3 a/o tin bronze? (b) Assuming the lattice constant increases linearly with the atomic fraction of tin, what is the density of this bronze?

Procedure. Select a basis for calculation. One hundred atoms (25 fcc unit cells) is a natural; thus, 97 Cu + 3 Sn. (a) Determine the masses. (b) Since $\rho = m/V$, determine $\bar{a}$ for $V = a^3$, and use the calculated mass.

Calculation

a) Mass of copper: $97(63.54 \text{ amu}) = 6163 \text{ amu} = 0.945$ (or 94.5 w/o);

   Mass of tin: $3(118.69 \text{ amu}) = \underline{356} \text{ amu} = 0.054$ (or 5.4 w/o).

   Total mass = 6519 amu

b) Average radius = $0.97(0.1278 \text{ nm}) + 0.03(0.151 \text{ nm}) = 0.1285 \text{ nm}$;

   $a = (4)(0.1285)/\sqrt{2} = 0.3634 \text{ nm}$.

   $$\rho = \frac{6519 \text{ amu}/(0.602 \times 10^{24} \text{ amu/g})}{25(0.3634 \times 10^{-9} \text{ m})^3} = 9.0 \text{ Mg/m}^3 \quad (=9.0 \text{ g/cm}^3). \quad \blacktriangleleft$$

Example 4–2.2

The maximum solubility limit of tin in a Cu–Sn bronze is 15.8 w/o at 586°C. What is the atom percent tin?

Procedure. Select a mass basis for calculation, e.g., 100,000 amu. Therefore, 15,800 amu Sn + 84,200 amu Cu.

Calculation

Number of Cu atoms:	84,200/63.54 = 1325, which is 90.9 a/o Cu;
Number of Sn atoms:	15,800/118.69 = $\underline{\ \ 133}$, which is 9.1 a/o Sn.
	Total atoms = 1458 ◀

Example 4–2.3

At 1000°C, there can be 1.7 w/o carbon in solid solution with fcc iron (Fig. 4–2.3). How many carbon atoms will there be for every 100 unit cells?

Procedure. Since there are 100 unit cells, we have 400 Fe atoms, which account for 98.3 percent of the mass.

$$\text{Total mass} = (400\ \text{Fe})(55.85\ \text{amu/Fe})/0.983 = 22,726\ \text{amu}.$$
$$\text{Carbon atoms} = (22,726\ \text{amu})(0.017)/(12.01\ \text{amu/C atom}) = 32.$$

Comment. Note that the numbers were not rounded off to two significant figures until the final answer. Had we done so at 23,000 amu, we would have biased our final answer to the high side (33 carbon atoms).

Added information. The carbon atom sits at $\frac{1}{2},\frac{1}{2},\frac{1}{2}$ locations in about one third of the unit cells. Since the carbon atom is slightly larger than the hole, it is not possible to have carbon atoms at all equivalent locations. ◀

• 4–3

SOLID SOLUTIONS IN COMPOUNDS

Substitutional Solutions

These solid solutions can occur in ionic phases as well as in metals. In ionic phases, just as in the case of solid metals, atom or ion size is important. A simple example of an ionic solid solution is shown in Fig. 4–3.1. The structure is that of MgO (Fig. 3–1.1) in which the Mg^{2+} ions have been replaced by Fe^{2+} ions. Inasmuch as the radii of the two ions are 0.066 nm and 0.074 nm, complete substitution is possible. On the other hand, Ca^{2+} ions cannot be similarly substituted for Mg^{2+} because their radius of 0.099 nm is comparatively large.*

An additional requirement, which is more stringent for solid solutions of ceramic compounds than for solid solutions of metals, is that the valence charges on the replaced ion and the new ion must be identical. For example, it is difficult to replace the Mg^{2+} in MgO with an Li^+, although the two have identical radii, because there would be a net deficiency of charges. Such a substitution could be accomplished only if there were other compensating changes in charge. (See Section 7–4.)

Nonstoichiometric Compounds

Many compounds have exact ratios of elements (e.g., H_2O, CH_4, MgO, Al_2O_3, Fe_3C, to name but a few). They have a fixed ratio of atoms; thus they are *stoichiometric*. Bonds form between unlike atoms. As a result, the structure becomes even more perfectly ordered than we saw in Fig. 4–2.2.

---

* See Appendix B for ionic radii.

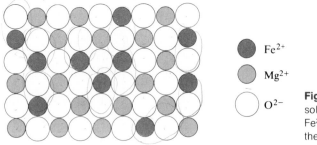

Figure 4–3.1 Substitutional solid solution in a compound. Fe^{2+} is substituted for Mg^{2+} in the MgO structure.

Other compounds deviate from specific integer ratios for the two (or more) elements that are present. Thus we find that "Cu_2Al" includes 31 a/o to 37 a/o Al (16 to 20 w/o Al) rather than being exactly $33\frac{1}{3}$ a/o Al. Likewise, at 1000°C "FeO" ranges from 51 to 53 a/o oxygen, rather than being exactly 50 a/o. We call these compounds *nonstoichiometric* because they do not have a fixed ratio of atoms.

Nonstoichiometric compounds always involve some solid solution. In the Cu_2Al example cited above, the atoms are nearly enough the same size and sufficiently comparable electronically so that with excess aluminum, some of the copper atoms are replaced by aluminum atoms (up to the maximum of 37 a/o Al). Conversely, in the presence of excess copper, the Cu/Al atom ratio reaches 69/31, by substituting a few copper atoms into the aluminum sites of Cu_2Al.

The nonstoichiometry of $Fe_{1-x}O$ arises from a different origin. The iron ions and the oxygen ions are too different to permit any measurable substitution. However, iron compounds always include some ferric (Fe^{3+}) ions along with the ferrous (Fe^{2+}) ions. Thus in order to balance charges there always must be more than 50 a/o oxygen. In fact, each two Fe^{3+} ions require an extra O^{2-} ion; or conversely, each pair of Fe^{3+} ions must be accompanied by a positive ion *vacancy* (Fig. 4–3.2). This is called a *defect structure* since there

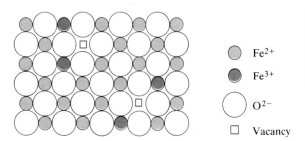

Figure 4–3.2 Defect structure ($Fe_{1-x}O$). This structure is the same as NaCl (Fig. 3–1.1) except for some iron ion vacancies. Since a fraction of the iron ions are Fe^{3+} rather than Fe^{2+}, the vacancies are necessary to balance the charge. The value of x ranges from 0.04 to 0.16, depending on temperature and the amount of available oxygen.

are irregularities in the atom packing. At 1000°C and with iron saturation, the composition is $Fe_{0.96}O$; with oxygen saturation, the composition is $Fe_{0.88}O$.

These defect structures will be important to us in later chapters when we discuss electrical conduction. Also, we will observe in this chapter that the atoms or ions can diffuse more readily through the crystalline solid when vacancies are present.

Example 4–3.1

An iron oxide (Fig. 4–3.2) contains 52 a/o oxygen and has a lattice constant of 0.429 nm. (a) What is the Fe^{2+}/Fe^{3+} ion ratio? (b) What is the density? (This structure is like that of NaCl, except for the vacancies, □. See Fig. 3–1.1.)

Procedure. (a) Choose a basis—100 atoms (= 52 O^{2-} + 48 iron ions). Make the sum of the charges equal zero, using y Fe^{3+} and $(48 - y)$ Fe^{2+}. (b) Based on Fig. 3–2.5(a), the 52 O^{2-} ions require 13 unit cells, but there are only 48 Fe ions present (and 4 □'s).

Calculation

a) Charge balance: $52(2-) + (y)(3+) + (48 - y)(2+) = 0$,

$$y = 8 \ Fe^{3+}; \qquad 48 - y = 40 \ Fe^{2+};$$

$$Fe^{2+}/Fe^{3+} = 5.$$

b) $\rho = \dfrac{[48(55.8) + 52(16)] \ \text{amu}/13 \ \text{u.c.}}{(0.602 \times 10^{24} \ \text{amu/g})(0.429 \times 10^{-9} \ \text{m})^3/\text{u.c.}} = 5.7 \ \text{Mg/m}^3 \ (= 5.7 \ \text{g/cm}^3).$ ◀

• Example 4–3.2

A β'-brass is nominally an intermetallic compound, CuZn, with the simple cubic structure shown in Fig. 3–5.3. It may also be called a *partially ordered solid solution,* particularly since it is nonstoichiometric with a range of 46 to 50 a/o zinc at 450°C. Assume 90% of the $\frac{1}{2},\frac{1}{2},\frac{1}{2}$ sites of Fig. 3–5.3 are occupied by copper atoms in a 46 a/o Zn–54 a/0 Cu alloy. What percent of the 0,0,0 sites are occupied by copper atoms?

Solution. Basis: 50 unit cells = 50 $\frac{1}{2},\frac{1}{2},\frac{1}{2}$ sites (and 50 0,0,0 sites)

$$= 100 \ \text{atoms} = 54 \ \text{Cu} + 46 \ \text{Zn}.$$

$\frac{1}{2},\frac{1}{2},\frac{1}{2}$ sites: $\qquad\qquad$ 45 Cu + 5 Zn;

0,0,0 sites: $\qquad\qquad$ 9 Cu + 41 Zn.

Therefore, 9 of the 50, or 18% of the 0,0,0 sites, are occupied by copper.

Comments. At low temperatures almost all of the neighbors of zinc atoms are copper atoms, and *vice versa*. However, as the temperature is increased, the atoms start to disorder. In this problem, at 450°C, the 0,0,0 sites are 82 Zn–18 Cu, while the $\frac{1}{2},\frac{1}{2},\frac{1}{2}$ sites are 10 Zn–90 Cu (all atom percents). Thus the two sites are not equivalent, and the structure is simple cubic.

Above 470°C, this alloy becomes fully random, with no preference for copper atoms to be surrounded by zinc atoms. Under those conditions, the substitutional solid solution (called β-brass rather than β'-brass) is bcc because the unit-cell centers and corners have equal probability for the same average composition. ◀

4–4

IMPERFECTIONS IN CRYSTALS

We have just seen a type of imperfection in crystals where a vacancy is required to accommodate a charge imbalance (Fig. 4–3.2). When imperfections such as vacancies involve one or a few atoms, we call them *point defects*. Other imperfections may be lineal through the crystal; hence, the term *line defects*. They become particularly significant when crystals are plastically deformed by shear stresses. In fact, a small number of these may cause metal crystals to deform 1000 times more readily than would be possible in their absence. When present in large numbers, these lineal imperfections increase the strength of the metal. Finally, other imperfections are two-dimensional and involve both external *surfaces* and internal *boundaries*.

Point Defects

The simplest point defect is a *vacancy,* which involves a missing atom (Fig. 4–4.1a) within a crystal. Such defects can be a result of imperfect packing during the original crystallization, or they may arise from thermal vibrations of the atoms at elevated temperatures (Section 4–6), because as thermal energy is increased there is an increased probability that individual atoms will jump out of their position of lowest energy. Vacancies may be single, as shown in Fig. 4–4.1(a), or two or more of them may condense into a di-vacancy (Fig. 4–4.1b) or a tri-vacancy.

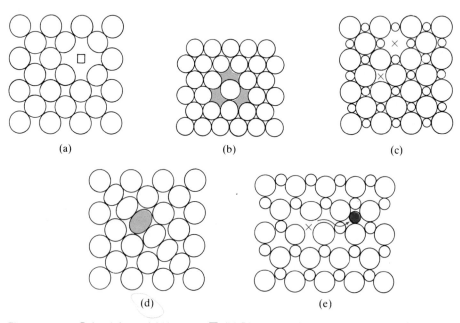

(a)   (b)   (c)

(d)   (e)

Figure 4–4.1 Point defects. (a) Vacancy, ☐. (b) Di-vacancy (two missing atoms). (c) Ion-pair vacancy (Schottky defect). (d) Interstitialcy. (e) Displaced ion (Frenkel defect).

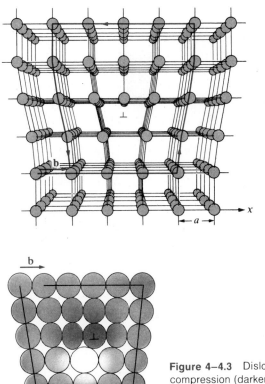

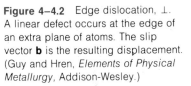

Figure 4–4.2 Edge dislocation, ⊥. A linear defect occurs at the edge of an extra plane of atoms. The slip vector **b** is the resulting displacement. (Guy and Hren, *Elements of Physical Metallurgy*, Addison-Wesley.)

Figure 4–4.3 Dislocation energy. Atoms are under compression (darker) and tension (light) adjacent to the dislocation.

Ion-pair vacancies (called Schottky imperfections) are found in compounds that must maintain a charge balance (Fig. 4–4.1c). They involve vacancies of pairs of ions of opposite charges. Ion-pair vacancies, like single vacancies, facilitate atomic diffusion (Section 4–7).

An extra atom may be lodged within a crystal structure, particularly if the atomic packing factor is low. Such an imperfection, called an *interstitialcy*, produces atomic distortion (Fig. 4–4.1d).

A *displaced ion* from the lattice into an interstitial site (Fig. 4–4.1e) is called a Frenkel defect. Close-packed structures have fewer interstitialcies and displaced ions than vacancies, because additional energy is required to force the atoms into the interstitial positions.

Line Defects (dislocations)

The most common type of line defect within a crystal is a dislocation. An *edge dislocation* (⊥) is shown in Fig. 4–4.2. It may be described as an edge of an extra plane of atoms within a crystal structure. Zones of compression and of tension accompany an edge dislocation (Fig. 4–4.3) so that there is a net increase in energy along the dislocation. The displacement distance for atoms

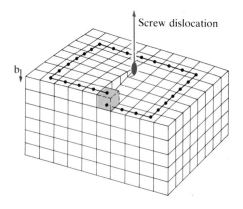

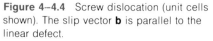

Figure 4–4.4 Screw dislocation (unit cells shown). The slip vector **b** is parallel to the linear defect.

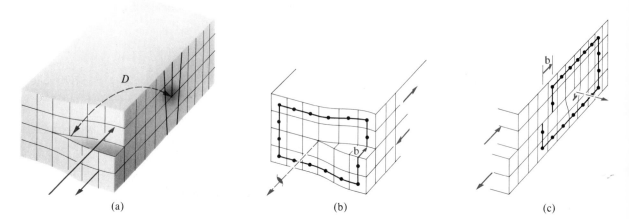

(a)   (b)   (c)

Figure 4–4.5 Dislocation formation by shear. (a) The *dislocation line*, *D*, expands through the crystal until displacement is complete. (b) This defect forms a screw dislocation where the line is parallel to the shear direction. (c) The linear defect is an edge dislocation where the line is perpendicular to the shear direction.

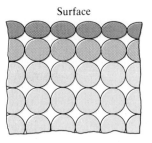

Surface

Figure 4–4.6 Surface atoms (schematic). Since these atoms are not entirely surrounded by others, they possess more energy than internal atoms.

around the dislocation is called the *slip vector,* **b**.* This vector is at right angles to the edge dislocation line.

A *screw dislocation* (ʂ) is like a spiral ramp with an imperfection line down its axis (Fig. 4–4.4). Its slip vector is parallel to the defect line. Shear stresses are associated with the atoms adjacent to the screw dislocation; therefore, extra energy is involved here as with the previously cited edge dislocations.

Dislocations of both types may originate during crystallization. Edge dislocations, for example, arise when there is a slight mismatch in the orientation of adjacent parts of the growing crystal so that an extra row of atoms is introduced or eliminated. As shown in Fig. 4–4.4, a screw dislocation provides for easy crystal growth because additional atoms and unit cells can be added to the "step" of the screw. Thus the term *screw* is very apt, because the step swings around the axis as growth proceeds.

Dislocations more commonly originate during deformation. We see this in Fig. 4–4.5, where shear is seen to introduce both edge dislocations and screw dislocations. Both lead to the same final displacement and are in fact related through the dislocation line that connects them.

Surfaces

Crystalline imperfections may extend in two dimensions as a boundary. The most obvious boundary is the external *surface*. Although we may visualize a surface as simply a terminus of the crystal structure, we should quickly appreciate the fact that atomic coordination at the surface is not fully comparable to the atoms within a crystal. The surface atoms have neighbors on only one side (Fig. 4–4.6); therefore they have higher energy and are less firmly bonded than the internal atoms. This energy may be rationalized with Fig. 2–4.2 by noting that if additional atoms were to be deposited onto the surface atoms, energy would be released just as it was for the combination of two individual atoms. We find our best visible evidence of this surface energy in the case of liquid drops that have spherical shape to minimize the surface area (and therefore the surface energy) per unit volume. Surface adsorption provides additional evidence of the energy differential at the surface.

Grain Boundaries

Although a material such as copper in an electric wire may contain only one phase, i.e., only one structure (fcc), it contains many crystals of various orientations. These individual crystals are called *grains*. The shape of a grain in a solid is usually controlled by the presence of surrounding grains. Within any particular grain, all of the unit cells are arranged with one orientation and one pattern. However, at the *grain boundary* between two adjacent grains there is a transition zone that is not aligned with either grain (Fig. 4–4.7).

Although we cannot see the individual atoms illustrated in Fig. 4–4.7, we can quite readily locate grain boundaries in a metal under a microscope, if the metal has been treated by *etching*. First the metal is smoothly polished so that a plane, mirrorlike surface is obtained, and then it is chemically attacked for a

---

* Also called a *Burgers vector.*

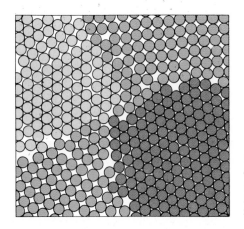

Figure 4–4.7 Grain boundaries. Note the disorder at the boundary. (Reproduced by permission from Clyde W. Mason, *Introduction to Physical Metallurgy*, American Society for Metals, Chapter 3.)

short period of time. The atoms at the surface of mismatch between one grain and the next will dissolve more readily than other atoms and will leave a line that can be seen with the microscope (Fig. 4–4.8); the etched grain boundary does not act as a perfect mirror as does the remainder of the grain (Fig. 4–4.9).

We may consider that the grain boundary is two-dimensional, although it may be curved, and actually has a finite thickness of 1 or 2 atomic distances. The mismatch of the orientation of adjacent grains produces a less efficient packing of the atoms along the boundary. Thus the atoms along the boundary have a higher energy than those within the grains. This accounts for the more rapid etching along the boundaries described above. The higher energy of the

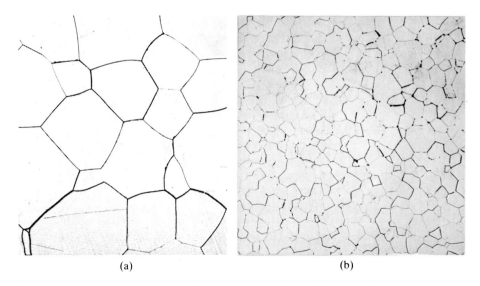

(a)   (b)

Figure 4–4.8 Grain boundaries. (a) Molybdenum (×250) (O. K. Riegger). (b) High-density periclase, MgO (×250) (R. E. Gardner and G. W. Robinson, Jr., *J. Amer. Ceram. Soc.*).

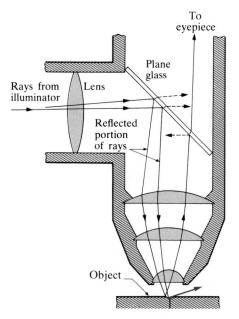

To
eyepiece

Plane
glass

Rays from
illuminator

Lens

Reflected
portion
of rays

Object

Figure 4–4.9 Grain boundary observation.
The metal has been polished and etched.
The corroded boundary does not reflect light
through the microscope; thus it appears as a
dark line in photomicrographs. (Reproduced
by permission from B. Rogers, *The Nature of
Metals*, 2d ed., American Society for Metals,
and Iowa State University Press, Chapter 2.)

boundary atoms is also important for the nucleation of polymorphic phase
changes (Section 3–4). The lower atomic packing along the boundary favors
atomic diffusion (Section 4–7), and the mismatch between adjacent grains modifies the progression of dislocation movements (Fig. 4–4.5). Thus the grain
boundary modifies plastic strain of a material.*

**Grain Boundary
Area and Grain Size**

It is obvious that the two *microstructures* of Fig. 4–4.8 are different. The grains
in Fig. 4–4.8(a) are larger than are the grains in Fig. 4–4.8(b). Conversely, this
MgO has more grain-boundary area than does the molybdenum. (Both are at
×250, i.e., their lineal dimensions have been magnified 250 times.)

Since grain boundaries affect a material in a number of ways, it is useful to
know the amount of grain-boundary surface per unit volume, S_V. This area can
be readily estimated if we place a line randomly across the microstructure. Of
course, this line will intercept more grain boundaries in a *fine-grained* material
than in a *coarse-grained* material. The relationship is

$$S_V = 2 P_L, \qquad (4-4.1)$$

where P_L is the number of points of intersections per unit length between the
line and the boundaries.

We shall not prove the above relationship, but we can demonstrate it in
Fig. 4–4.10 where a 50-mm circle has been placed randomly on the previous

---

* At normal temperatures the grain boundaries interfere with slip. Therefore a *fine-grained* material is stronger than a *coarse-grained* material. At elevated temperatures, the boundaries can
accommodate the dislocations. As a result the situation is reversed, and creep results (Section
5–7).

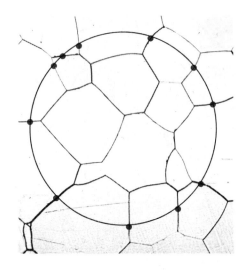

Figure 4–4.10 Grain-boundary area calculation. Since the magnification is ×250, the length of the circle is π(50 mm)/250, or 0.63 mm (0.025 in.). It intersects 11 boundaries in that distance. Therefore, there are 2(11/0.63 mm), or 35 mm², of boundary area per mm³ (see Eq. (4–4.1). (O. K. Riegger.)

photomicrograph of molybdenum. There are 11 intersections. Since the magnification is ×250, the circumference on the metal is actually $50\pi/250 = 0.63$ mm, and the value of P_L is 11/0.63 mm = 17.5/mm. Based on Eq. (4–4.1), the surface area per unit volume S_V is 35 mm²/mm³. This may be compared with the boundary area of the MgO in Fig. 4–4.8(b) in Example 4–4.1.

Although the boundaries are the microstructural features that relate to property behavior, it is common to refer to *grain size*. A method to determine a grain-size number has been standardized by the American Society for Testing and Materials (ASTM). Although empirical, it is a quantitative and reproducible index. This index uses **2** as a base:

$$N_{(0.0645 \text{ mm}^2)} = 2^{n-1}. \tag{4–4.2}$$

The term N is the number of grains observed in an area of 0.0645 mm² (1 in.² at ×100).* The value n is the *grain-size number* (G.S.#). Example 4–4.2 calculates the G.S.# for the molybdenum shown in Fig. 4–4.8(a). Figure 4–4.11 shows two of a series of grain-size nets that can be used for quick visual assignments of a grain-size number to ×100 photomicrographs. These grain-size numbers are pertinent to the heat treating of steels (Section 11–8) and for the transition temperatures of steels (Section 13–2).

Example 4–4.1

Estimate the grain-boundary area per unit volume in the MgO of Fig. 4–4.8(b).

Solution. Lay a 50-mm straight-edge at random across the figure and count the grain boundaries intersected. If, by repeating this for a total of 5 times, you get counts of 13, 17, 12, 14, and 12, you would have 68 counts per 250 mm. However, the magnification is ×250. Therefore, $P_L = 68$/mm.

---

* The procedure was originally standardized to use a microscope with ×100 lenses, and to count the grains within a 1 in. × 1 in. area (=0.0001 in², or 0.0645 mm²).

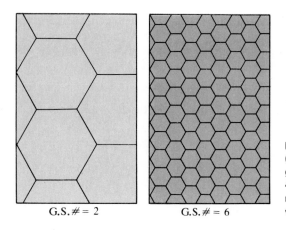

G.S.# = 2   G.S.# = 6

Figure 4–4.11   Grain-size numbers (ASTM comparison nets). A large grain-size number, G.S.# (Eq. 4–4.2), indicates more grains and more grain-boundary area per unit volume (×100).

Calculation

$$S_V = 2(68/\text{mm}) = {\sim}140/\text{mm} \qquad (\text{or } 140 \text{ mm}^2/\text{mm}^3).$$

Comments. There is about four times as much grain-boundary area per unit volume in the MgO of Fig. 4–4.8 as in the molybdenum. Both of these are *estimates* from samplings. However, with reasonable care, one can be accurate within ±10%, which is sufficient for most purposes. ◄

Example 4–4.2   Assign an ASTM G.S.# to the molybdenum of Fig. 4–4.8(a).

Procedure. Since the magnification is *not* ×100, we cannot obtain N directly by counting a 1 in.^{2} (645 mm^2) sample of the area. However, we can count the grains in the total area and correct for magnification.

Estimation. There are ~17 grains in an area of $(59 \text{ mm}/250)^2$. (See comments.)

$$\frac{{\sim}17}{0.056 \text{ mm}^2} = \frac{N}{0.0645 \text{ mm}^2};$$

$$N = {\sim}20 = 2^{n-1};$$

$$n = 5^+.$$

Comments. The photomicrograph in Fig. 4–4.8 is, itself, a sample and therefore subject to statistical variations. As a result, we should not expect utmost precision. Furthermore, it is seldom necessary to be more specific than ±0.5 in our estimates of the grain-size number.

   The number of grains in the area that was sampled is most readily obtained by (1) counting the grains that lie entirely within the area, (2) adding to the count half of the grains at the edges (since these are shared by adjacent areas), and (3) then adding a fourth of each of the four-corner grains. If you took another sample of the metal in Fig. 4–4.8(a), and your counts were as low as 15 or as high as 20, your grain-size estimate would still be $n = 5^+$. The count must double to shift from one G.S.# to the next. ◄

Example 4–4.3

Accurate measurements can be made to four significant figures of the density of aluminum. When cooled rapidly from 650°C, $\rho_{Al} = 2.698$ Mg/m^3. Compare that value with the theoretical density obtained from diffraction analyses where a was determined to be 0.4049 nm.

Procedure. Use the method for density determination that was used in Example 3–2.2.

Solution. Since the atomic weight is 26.98 amu and aluminum is fcc (Appendix B),

$$\rho_{\text{theor.}} = \frac{4(26.98 \text{ amu})/(0.6022 \times 10^{24} \text{ amu/g})}{(0.4049 \times 10^{-9} \text{ m})^3} = 2.700 \text{ Mg/m}^3;$$

$$\frac{\rho}{\rho_{\text{theor.}}} = \frac{2.698}{2.700} = 0.999; \quad \text{or} \sim 1 \text{ vacancy per 1000 atoms.}$$

Added Information. The match is close. Aluminum, like most metals just below their melting temperatures, has about 1 vacancy in 10^3 atoms. The equilibrium fraction drops to approximately 1 vacancy in 10^7 atoms at half of the absolute melting temperature. ◀

4–5

NONCRYSTALLINE MATERIALS

Long-range order is absent in some materials of major engineering and scientific importance. Included are all liquids, glass, the majority of plastics, and a few metals if the latter are cooled extremely rapidly from their liquid. In principle, we can view this lack of repetitive structure as a volume, or 3-dimensional, disorder, and as a continuation of our sequence: point defects, lineal defects, and 2-dimensional boundaries. We call these materials *amorphous* (literally "without form") in contrast to crystalline materials.

Liquids

For the most part, liquids are *fluids* (i.e., they *flow* under their own mass). However, just as "molasses in January" gets semi-rigid, various liquids of technical importance can become very viscous and even solid, without crystallizing.

First let us look at the disorder that occurs in a single-component metal as it approaches melting, then transforms into a liquid. We can use aluminum for our example. As was implied in the last section, the greater thermal energy at higher temperatures introduces not only greater thermal vibrations, but also some vacancies. Just short of the melting point, crystalline aluminum may contain up to 0.1% vacancies in its lattice. When the vacancies approach one percent in a close-packed structure, "turmoil reigns." The regular 12-fold coordination is destroyed and the long-range order of crystal structure disappears (Fig. 4–5.1).

This disorder of melting increases the volume of most materials* (Fig. 4–5.2). With this disorder, the number of nearest neighbors drops from 12 to

---

* It increases the volume of all those materials that are close-packed and that do not have directional bonds. A *few* materials with low packing factors and stereospecific bonds (Chapter 2) collapse into denser structures when they are thermally excited. Water is the prime example of this exception. (And are we lucky!!)

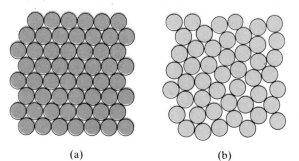

Figure 4–5.1   Melting (metal). (a) Crystalline metal with CN = 12. (The 6 in the plane plus 3 above and 3 below.) (b) Liquid metal. Long-range order is lost; CN < 12, and the average interatomic distance increases slightly.

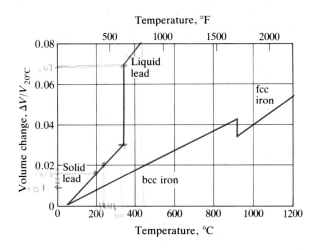

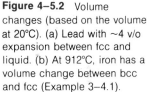

Figure 4–5.2   Volume changes (based on the volume at 20°C). (a) Lead with ~4 v/o expansion between fcc and liquid. (b) At 912°C, iron has a volume change between bcc and fcc (Example 3–4.1).

only 11 or 10; however, these are not in a regular pattern, so the space per atom and the average interatomic distance is increased a few percent. Each material requires a characteristic amount of energy to be melted. We call this energy the *heat of fusion*. Since the heat of fusion, ΔH_f, is the energy required to disorganize a mole of atoms, and the melting temperature, T_m, is a measure of the atomic bond strength, we find a general correlation between the two (Table 4–5.1).*

Glasses

As indicated earlier, glasses are sometimes considered to be very viscous liquids, inasmuch as they are noncrystalline. However, only a few liquids can actually form glasses. Therefore, in order to make a distinction, we must look at the structure of glass more critically.

At high temperatures glasses form true liquids. The atoms have freedom to move around and respond to shear stresses. When a commercial glass is supercooled from its liquid temperature, there is thermal contraction caused by

---

* For the correlation to be good, we must have comparable materials, e.g., among metals.

TABLE 4–5.1

Heats of Fusion of Metals

Metal	Melting temperature, °C (°F)	Heat of fusion, joule/mole*
Tungsten, W	3410 (6170)	32,000
Molybdenum, Mo	2610 (4730)	28,000
Chromium, Cr	1875 (3407)	21,000
Titanium, Ti	1668 (3034)	21,000
Iron, Fe	1538 (2800)	15,300
Nickel, Ni	1453 (2647)	17,900
Copper, Cu	1085 (1985)	13,500
Aluminum, Al	660.4(1221)	10,500
Magnesium, Mg	650 (1202)	9,000
Zinc, Zn	420 (788)	6,600
Lead, Pb	327.4 (621)	5,400
Mercury, Hg	−38.9(−38)	2,340

* joule/0.6 × 10^{24} atoms; 4.18 J = 1 cal.

atomic rearrangements that produce more efficient packing of the atoms. This volume change (Fig. 4–5.3) is typical of all liquid phases; however, with more extensive cooling, there is an abrupt change in the thermal expansion coefficient, dV/dT, of glasses. Below a certain temperature, called the glass transition temperature, or more simply the *glass temperature*, T_g, there are no further rearrangements of the atoms and the only volume change is a result of reduced thermal vibrations. This lower coefficient is comparable to the thermal

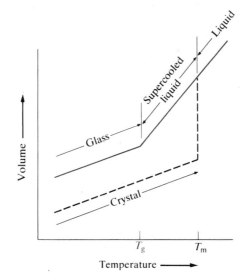

Figure 4–5.3 Volume changes in supercooled liquids and glasses. When a liquid is cooled, it contracts rapidly and continuously because with decreased thermal agitation, the atoms develop more efficient packing arrangements. In the absence of crystallization the contraction continues below T_m to the glass transition temperature, T_g, where the material becomes a rigid glass. Below T_g, no further rearrangements occur, and the only further contraction is caused by reduced thermal vibrations of the atoms in their established locations.

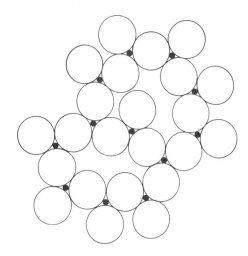

Figure 4–5.4 Structure of B_2O_3 glass. Although there is no long-range crystalline order, there is a short-range coordinational order. Each boron atom is among three oxygen atoms. Each oxygen atom is coordinated with two boron atoms.

expansion coefficient in crystals where thermal vibrations are the only factor causing volume changes, and no rearrangement occurs.

The term *glass* applies to those materials that have the expansion characteristics of Fig. 4–5.3. Glasses may be either inorganic or organic and are characterized by a *short-range order* (and an absence of long-range order). Figure 4–5.4 presents one of the simplest glasses (B_2O_3), in which each small boron atom fits among three larger oxygen atoms. Since boron has a valence of three and oxygen a valence of two, electrical balance is maintained if each oxygen atom is located between two boron atoms. As a result, a continuous structure of strongly bonded atoms is developed. Below the glass-transition temperature, where the atoms are not readily rearranged, the fluid characteristics are lost and a noncrystalline solid exists. Such a solid has a significant resistance to shear stresses and therefore cannot be considered a true liquid.

The temperature–volume characteristics of Fig. 4–5.3 were originally observed in silicate glasses (Section 7–5). However, it soon became apparent that these characteristics have major significance in polymeric materials. Thus in Chapter 6, we will look at the glass-transition temperature, T_g, closely. Below the T_g, polymers are hard and brittle and have low dielectric constants. Above the T_g, a plastic becomes flexible and even rubbery, with concurrent changes in its dielectric constant.

Phases

Reference has already been made to phases and will be made again numerous times throughout this text. We are now in a position to define a phase as that part of a material that is *distinct from others in structure and/or composition.* Consider "ice–water." While of the same composition, the ice is a crystalline solid with a hexagonal lattice; the water is a liquid. The *phase boundary* between the two locates a discontinuity in structure: they are *separate phases.* Or consider silver-plated copper. Both silver and copper are fcc; however, the silver atoms are sufficiently larger than the copper atoms* so that there is nearly complete composition discontinuity at room temperature. Thus, they form two separate phases.

Commonly, two phases of a material have distinct differences in both composition and structure, e.g., a plastic with a fiberglass reinforcement. In contrast, some phases lose their distinctiveness and dissolve, e.g., after dissolving in a cup of coffee, sugar is no longer a separate phase. The same is true of zinc, which, by itself hcp, dissolves in copper (fcc) to produce brass, a single-phase solid solution.

In terms of the above discussion, a *solution* (liquid or solid) is a phase with more than one *component*. A *mixture* is a material with more than one *phase*.

There are many crystalline phases because there are innumerable permutations and combinations of atoms, or groups of atoms. There are relatively few amorphous phases because, lacking long-range order, their atomic arrangements are less definite and permit a greater range of solution than do crystals.† There is only one gaseous phase. The atoms or molecules are far apart and randomly distributed; as a result additional vapor components may be introduced into one "structure." No discontinuities are observed in a gas other than at the atomic or molecular level.

Example 4–5.1

From Fig. 4–5.2, calculate the packing factor (a) of solid lead at 326°C; (b) of liquid lead at 328°C. (Assume the hard-ball radius of 0.1750 nm is retained.)

Procedure. Lead is fcc with a calculated PF of 0.74 (see Example 3–2.1a); thus an increase in volume of 2.9% at 326°C and 7% at 328°C decreases the PF accordingly.

Calculation

a)
$$PF_{326°C} = 0.74/1.029 = 0.719;$$

b)
$$PF_{328°C} = 0.74/1.07 = 0.692.$$

Comment. One could suggest that the radius increases by 1.0 l/o from 0.1750 nm at 20°C to 0.1767 nm at 326°C to give the 2.9 v/o expansion. If so, the packing factor at 326°C would remain at 74%, and the packing factor of the liquid becomes 0.712. One cannot argue against that suggestion, since one definition of radius is "one half of the closest interatomic distance." In either event, there is an abrupt discontinuity in packing efficiency at the melting temperature. ◀

4–6

ATOMIC VIBRATIONS

The atoms of a material become static only at 0 K (−273°C or −460°F). Under that condition the atoms settle down to their lowest energy positions among their neighbors (Fig. 2–4.2b). As the temperature is increased, the increased energy permits the atoms to vibrate into greater and shorter interatomic dis-

---

* $R_{Cu} = 0.1278$ nm; $R_{Ag} = 0.1444$ nm.

† The fiberglass and plastic just cited are both amorphous. While their structures are not sufficiently similar to produce a single phase, each can be a solvent for large quantities of solutes.

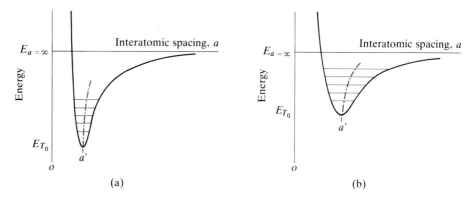

Figure 4–6.1 Energy and expansion. (a) Strongly bonded solid. (b) Weakly bonded solid. With equal additions of thermal energy, above absolute zero, T_0, the mean interatomic spacing changes less in a material with a deeper energy trough (cf. Fig. 2–4.2b). The expansion becomes more pronounced at higher temperatures (higher energy).

tances. However, as can be noted by the energy well of that earlier figure, the displacements in the two directions are not symmetric; i.e., for a given energy level (temperature), the atoms can move farther apart more readily than they can be pushed together. This produces a *thermal expansion* because the mean interatomic distance is increased.

Thermal Expansion In the vicinity of our ambient temperature, the heat capacity of many solids is essentially constant. Therefore, we can consider a succession of temperature intervals to the energy trough (Fig. 4–6.1a). With successive energy increases, the median interatomic spacing increases; hence, the thermal expansion we spoke of in the previous paragraph. Note two features, however. There is less dimensional change per ΔT in a strongly bonded (high-melting) material characterized by a deep energy trough than in a weakly bonded material (Fig. 4–6.1b). This is illustrated in Fig. 4–6.2 as a graphical presentation of the data

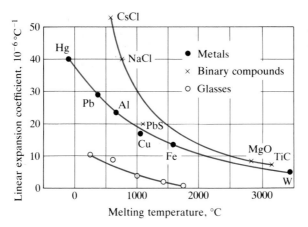

Figure 4–6.2 Melting temperatures and expansion coefficients (20°C). The more strongly bonded, higher melting materials have lower expansion coefficients (cf. Fig. 4–6.1). Comparisons must be made among materials with comparable structures.

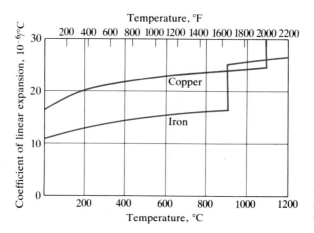

Figure 4–6.3 Thermal expansion versus temperature. The discontinuity for copper at 1084.5°C (1984°F) is a result of melting. Iron has a discontinuity because there is a bcc ↔ fcc rearrangement of the atoms at 912°C (1673°F).

given in the generalizations of Section 2–6. Figure 4–6.2 emphasizes that our generalization is limited to comparable materials.

The second feature of Fig. 4–6.1 is that as the temperature increases, the change in dimension becomes relatively more pronounced with the median curves bending to the right. This means that thermal expansion coefficients increase with temperature (Fig. 4–6.3), and expansion data in tables must be for identified temperatures. In Appendix C, the temperature is 20°C.

Thermal expansion is *isotropic* only in cubic and amorphous materials. In other crystals, the amount of expansion varies with orientation. Consider graphite (Fig. 2–5.2); the structure is markedly *anisotropic*. As a result, the thermal agitation expands the lattice more in the vertical direction than in the two horizontal directions.* This is not surprising because the bonds between layers are weaker than those within the layers.

Thermal Energy Distribution

The total kinetic energy, K.E., of a mole of gas increases in proportion to the temperature T, so that the equation

$$\text{K.E.} = \tfrac{3}{2}RT \qquad (4\text{–}6.1)$$

is appropriate. The R of this equation is the same gas constant encountered in introductory chemistry courses, where its units are commonly reported as 1.987 cal/mole·K. If we switch to joules of the SI units and pay attention to the individuals instead of moles (or 0.602×10^{24}), the value becomes 13.8×10^{-24} J/K†. This is called *Boltzmann's constant* and is identified by k. Thus, the

---

* At 700°C, for example, the thermal-expansion coefficient of graphite is only 0.8×10^{-6}/°C within the plane of the layers, but 29×10^{-6}/°C perpendicular to the layers. Both of these values decrease at lower temperatures, and in fact, the first coefficient passes through zero at 400°C and becomes slightly negative at room temperature. The second value ($\perp$ to layers) is 26×10^{-6}/°C at 20°C, so that the overall volume coefficient $\alpha_V \simeq \alpha_x + \alpha_y + \alpha_z$ is positive. (See Example 4–6.1.)

† $(1.987 \text{ cal/mole·K})(4.18 \text{ J/cal})/(0.602 \times 10^{24}/\text{mole}) = 13.8 \times 10^{-24}$ J/K.

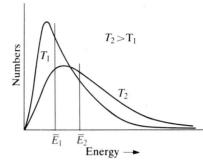

Figure 4–6.4 Energy distributions. Both the average energy $\bar{E}$ and the fraction with energies in excess of a specified level are increased as the temperature T is increased.

average K.E. of an individual molecule of gas is

$$\text{K.E.} = \tfrac{3}{2}kT. \tag{4–6.2}$$

However, this does not imply that all molecules of air in this room have this same energy. Rather, there will be a statistical distribution of energies as indicated in Fig. 4–6.4. At any particular instant of time, a very few molecules will have nearly zero energy; many molecules will have energies near to the average energy, and some molecules will have extremely high energies. As the temperature increases, there is (1) an increase in the average energy of the molecules, and (2) an increase in the number of molecules with energies in excess of any specified value.

The above applies to the kinetic-energy distribution of molecules in a gas. However, the same principle applies to the distribution of vibrational energy of atoms in a liquid or solid. Specifically, at any particular instant of time, a negligible number of atoms will have zero energy; many atoms will have energies near the average energy; and some atoms will have extremely high energies.

Our interest will be directed toward those atoms that have high energies. Very often we should like to know the probability of atoms possessing more than a specified amount of energy, e.g., what fraction of the atoms has energy greater than E of Fig. 4–6.5. These are the atoms that can introduce changes

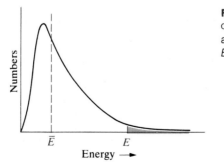

Figure 4–6.5 Energies. The ratio of the number of high-energy atoms (shaded) to total number of atoms is an exponential function $(-E/kT)$ when $E \gg \bar{E}$ (Eq. 4–6.3b).

within solids by breaking old bonds and joining new neighbors. The statistical solution to this problem has been worked out by Boltzmann as follows:

$$\frac{n}{N_{tot}} \propto e^{-(E-\bar{E})/kT},$$ (4–6.3a)

where k is the previously described Boltzmann's constant. The number n of atoms with an energy greater than E, out of the total number N_{tot} present, is a function of the temperature T. When E is considerably in excess of the average energy $\bar{E}$, the equation reduces to

$$\frac{n}{N_{tot}} = Me^{-E/kT},$$ (4–6.3b)

where M is the proportionality constant. The value of E is normally expressed as joules/atom; thus, k is 13.8×10^{-24} J/atom·K. However, conversions may be made from other units by means of Appendix A.

Example 4–6.1

At 700°C, the linear expansion coefficients, α, for graphite are as follows:

$$\alpha_\perp = 29 \times 10^{-6}/°C,$$

$$\alpha_\parallel = \alpha_\parallel = 0.8 \times 10^{-6}/°C$$

where $\alpha_\perp$ is the expansion coefficient perpendicular to the graphite layers of Fig. 2–5.2, and $\alpha_\parallel$ is the coefficient in the two directions parallel to the layers. What is the volume increase in graphite between 600°C and 800°C?

Solution. Since $V = L^3$, and $V + \Delta V = (L + \Delta L_\perp)(L + \Delta L_\parallel)^2$,

$$\Delta V/V = (\Delta L_\perp/L) + (2\Delta L_\parallel/L) + \cdots,$$

$$\alpha_V \Delta T \approx (\alpha_\perp + 2\alpha_\parallel)\Delta T = (29 + 1.6)(10^{-6}/°C)(200°C)$$

$$= 0.006 \qquad \text{(or 0.6 v/o)}.$$

Comments. In general,

$$\alpha_V \approx \alpha_x + \alpha_y + \alpha_z,$$ (4–6.4)

where the three subscripts at the right refer to the linear expansion coefficient of the three coordinate directions. ◀

Example 4–6.2

At 500°C (773 K), a diffusion experiment indicates that one out of 10^{10} atoms has enough *activation energy* to jump out of its lattice site into an interstitial position. At 600°C (873 K), this fraction is increased to 10^{-9}. (a) What is the activation energy required for this jump? (b) What fraction of the atoms has enough energy at 700°C (973 K)?

Procedure. The fraction of atoms with energies in excess of a value, E, is an exponential function of temperature. Therefore, we use Eq. (4–6.3b). (a) We have two data pairs to write two equations, each with two unknowns, E and $\ln M$. Solve simultaneously for the activation energy, E. (b) Use the values of E and $\ln M$ to calculate n/N at 700°C.

Calculation

$$\ln(n/N_{tot}) = \ln M - E/kT \qquad (4\text{–}6.5)$$

a) $\ln 10^{-10} = -23 = \ln M - (E/(13.8 \times 10^{-24} \text{ J/atom·K})(773 \text{ K})),$

and

$\ln 10^{-9} = -20.7 = \ln M - (E/(13.8 \times 10^{-24} \text{ J/atom·K})(873 \text{ K})).$

Solving simultaneously,

$$\ln M = -2.92, \quad \text{and} \quad E = 0.214 \times 10^{-18} \text{ J/atom};$$

or, in terms of a mole,

$$E = 129,000 \text{ J/mole} \qquad (=30,900 \text{ cal/mole}).$$

b) $\ln(n/N_{tot}) = -2.92 - ((0.214 \times 10^{-18} \text{ J/atom})/(13.8 \times 10^{-24} \text{ J/atom·K})(973 \text{ K})),$

$n/N_{tot} \approx 6 \times 10^{-9}.$

Comments. The relationship shown in Eq. (4–6.5) has a logarithmic value that is linear with reciprocal temperature, $1/T$,

$$y = C - Bx,$$

or

$$\ln \frac{n}{N} = C - \frac{E}{k} \left(\frac{1}{T}\right), \qquad (4\text{–}6.6)$$

where $y = \ln (n/N)$, and $x = (1/T)$. The slope B is E/k. Equation (4–6.6) is called an *Arrhenius* equation. It is widely encountered in any situation where the reaction of process is thermally activated. Examples include the intrinsic conductivity of semiconductors, the catalytic reaction of emission control, the diffusion processes in metals, the creep in plastics, and the viscosity of fluids. We will encounter this relationship again.

◀

4–7

ATOMIC DIFFUSION   As the temperature is increased and the atoms vibrate more energetically, a small fraction of the atoms will relocate themselves in the lattice. Example 4–6.2 related this fraction to temperature. Of course, the fraction depends not only on temperature, but also on how tightly the atoms are bonded in position. The energy requirement for an atom to change position is called the *activation energy*. This energy may be expressed as calories/mole, Q; as J/atom, E (Table 4–7.2); or as eV/atom.

Let us use Fig. 4–7.1 to illustrate activation energy schematically. A carbon atom is small ($r \approx 0.07$ nm) and can sit interstitially among a number of fcc iron atoms. If it has enough energy,* it can squeeze between the iron atoms to

---

* Approximately 34,000 cal/mole, 0.24×10^{-18} J/atom, or 1.5 eV/atom.

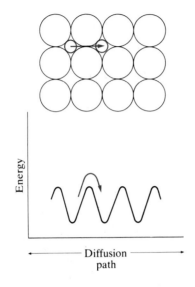

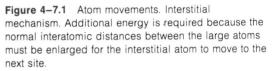

Energy

Diffusion
path

Figure 4–7.1 Atom movements. Interstitial mechanism. Additional energy is required because the normal interatomic distances between the large atoms must be enlarged for the interstitial atom to move to the next site.

the next interstice when it vibrates in that direction. At 20°C there is only a small probability that it will have that much energy. At higher temperatures, the probability increases (cf. Example 4–6.2).

Other diffusion mechanisms are sketched in Fig. 4–7.2. When all the atoms are the same size, or nearly so, the vacancy mechanism becomes predominant.

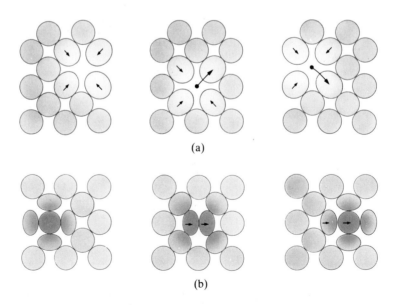

(a)

(b)

Figure 4–7.2 Diffusion mechanisms. (a) By vacancies. (b) By interstitialcies. The vacancies move in the opposite direction from the diffusing atoms. (In Chapter 8 we shall observe an analogy for the movements of electron holes.) The movements follow "random-walk" statistics.

The vacancies may be present, either as part of a defect structure (Fig. 4–3.2) or because of extensive thermal agitation (e.g., Example 4–4.3, where aluminum was just 10° below its melting point).

Self-diffusion

Let us consider diffusion in metallic nickel (fcc). In uniform surroundings, each of the twelve surrounding atoms (in the 3-dimensional crystal) have equal probability of moving into a vacancy that is present in nickel. Stated conversely, a vacancy in nickel has equal probability of moving into any of the neighboring lattice points.

Normally, no net diffusion is observed in pure nickel because atom movements are random, and the atoms are all identical. However, through the use of radioactive isotopes, it is possible to determine the diffusion of atoms in their own structure, i.e., *self-diffusion*. For example, radioactive nickel (Ni^{59}) can be plated onto the surface of normal nickel. With time, and as a function of temperature, there is progressive self-diffusion of the tracer isotopes into the adjacent nickel (Fig. 4–7.3). (And there is counter movement of the untagged atoms into the surface layer. The mechanisms of diffusion include those shown in Fig. 4–7.2, as well as along grain boundaries where the structure is more open (Fig. 4–4.7).)

The homogenization process shown in Fig. 4–7.3 must be interpreted as follows. Although there is equal probability that an individual atom (or vacancy) will move in any of the coordinate directions, a concentration gradient

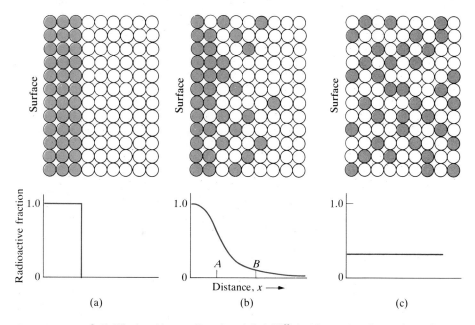

Figure 4–7.3 Self-diffusion. Here radioactive nickel (Ni^{63}) has been plated onto the surface of nonradioactive nickel. (a) Time, $t = t_0$. (b) Diffusion gradient, $t_0 < t < t_\infty$. (c) Homogenized, $t = t_\infty$.

favors a net movement of tracer atoms to the right in the nickel. At point A in Fig. 4–7.3(b), there are more tagged atoms than there are at point B. Thus, even with the same probability per atom for the individual tagged atoms at A to move right as there are for a tagged atom at B to move left, the difference in numbers produces the net movement as indicated. Greater uniformity develops until eventually (it may be a long time), the tagged atoms become equally dispersed throughout the crystal (as are the untagged atoms).

Diffusivity

The description of self-diffusion and counterdiffusion shown in Fig. 4–7.3 can also be applied for solute atoms in a solid solution. As an example, assume a concentration gradient where there is one carbon atom per 20 unit cells of fcc iron at point (1), and only one carbon atom per 30 unit cells at point (2), which is one millimeter away. Now, since there are random movements of carbon atoms at *each* point, we will find a net flux of carbon atoms from point (1) to point (2), simply because there are half again as many atoms jumping in the vicinity of point (1). (See Example 4–8.1.)

The diffusion *flux J* of atoms (expressed in atoms/m²·sec) is proportional to the concentration gradient, $(C_2 - C_1)/(x_2 - x_1)$. In mathematical terms,

$$J = -D \frac{dC}{dx}. \tag{4–7.1a}*$$

The proportionality constant D is called *diffusivity,* or diffusion coefficient. The negative sign indicates that the flux is in the down-hill gradient direction. The units are

$$\frac{\text{atoms}}{(\text{m}^2)(\text{sec})} = \left[\frac{\text{m}^2}{\text{sec}}\right]\left[\frac{\text{atoms/m}^3}{\text{m}}\right]. \tag{4–7.1b}$$

Diffusivity varies with the nature of the solute atoms, with the nature of the solid structure, and with changes in temperature. Several examples are given in Table 4–7.1. Some reasons for the various values of Table 4–7.1 are as follows:

1. Higher temperatures provide higher diffusivities, because the atoms have higher thermal energies and therefore greater probabilities of being activated over the energy barrier between atoms (Fig. 4–7.1).

2. Carbon atoms have a higher diffusivity in iron than do nickel atoms in iron because the carbon atom is a small one (Appendix B).

3. Copper atoms diffuse more readily in aluminum than in copper because the Cu–Cu bonds are stronger than the Al–Al bonds (as evidenced by their melting temperatures).

---

* This is called *Fick's first law*. There is also a Fick's second law,

$$\frac{\partial C}{\partial t} = D \left(\frac{\partial^2 C}{\partial x^2}\right), \tag{4–7.2}$$

that shows the rate at which the concentration will change with time. In fact, the values of $\partial C/\partial t$ and $\partial^2 C/\partial x^2$ were determined experimentally in the laboratory to calculate the values of D found in Table 4–7.1.

TABLE 4–7.1

Atomic Diffusivity*

Solute	Solvent (host structure)	Diffusivity, m²/sec	
		500°C (930°F)	1000°C (1830°F)
1. Carbon	fcc iron	(5×10^{-15})†	3×10^{-11}
2. Carbon	bcc iron	10^{-12}	(2×10^{-9})
3. Iron	fcc iron	(2×10^{-23})	2×10^{-16}
4. Iron	bcc iron	10^{-20}	(3×10^{-14})
5. Nickel	fcc iron	10^{-23}	2×10^{-16}
6. Manganese	fcc iron	(3×10^{-24})	10^{-16}
7. Zinc	Copper	4×10^{-18}	5×10^{-13}
8. Copper	Aluminum	4×10^{-14}	10^{-10} M‡
9. Copper	Copper	10^{-18}	2×10^{-13}
10. Silver	Silver (crystal)	10^{-17}	10^{-12} M
11. Silver	Silver (grain boundary)	10^{-11}	—
12. Carbon	hcp titanium	3×10^{-16}	(2×10^{-11})

* Calculated from data in Table 4–7.2.

† Parentheses indicate that the phase is metastable.

‡ M—Calculated, although temperature is above melting point.

4. Atoms have higher diffusivity in bcc iron than in fcc iron because the former has a lower atomic packing factor (0.68 versus 0.74). (We shall observe later that the fcc structure has larger interstitial holes; however, the passageways between the holes are smaller than in the bcc structure.)

5. The diffusion proceeds more rapidly along the grain boundaries because this is a zone of crystal imperfections and of lower packing (Fig. 4–4.7).

Diffusivities versus Temperature

The discussion of Section 4–6 related the distributions of thermal energy to temperature. Boltzmann was able to quantify this with Eq. (4–6.3), which showed that the number of atoms that have more than a specified amount of energy increases in proportion to an exponential function that includes that energy and the reciprocal of temperature. With diffusion, the *activation energy* for atom movements corresponds to the energy E of Boltzmann's equation. Thus,

$$D = D_0 e^{-E/kT}, \qquad (4\text{–}7.3)$$

where D_0 is the proportionality constant independent of temperature that includes M of Eq. (4–6.3b). The logarithm of the diffusivity is related to the reciprocal of the temperature, $1/T$:

$$\ln D = \ln D_0 - \frac{E}{kT}. \qquad (4\text{–}7.4a)$$

TABLE 4–7.2

Constants for Diffusivity Calculations* ($\ln D = \ln D_0 - Q/RT = \ln D_0 - E/kT$)†

Solute	Solvent (host structure)	D_0, m^2/sec	Q, cal/mole	E, J/atom
1. Carbon	fcc iron	0.2×10^{-4}	34,000	0.236×10^{-18}
2. Carbon	bcc iron	2.2×10^{-4}	29,300‡	0.204×10^{-18}
3. Iron	fcc iron	0.22×10^{-4}	64,000	0.445×10^{-18}
4. Iron	bcc iron	2.0×10^{-4}	57,500	0.400×10^{-18}
5. Nickel	fcc iron	0.77×10^{-4}	67,000	0.465×10^{-18}
6. Manganese	fcc iron	0.35×10^{-4}	67,500	0.469×10^{-18}
7. Zinc	Copper	0.34×10^{-4}	45,600	0.317×10^{-18}
8. Copper	Aluminum	0.15×10^{-4}	30,200	0.210×10^{-18}
9. Copper	Copper	0.2×10^{-4}	47,100	0.327×10^{-18}
10. Silver	Silver (crystal)	0.4×10^{-4}	44,100	0.306×10^{-18}
11. Silver	Silver (grain boundary)	0.14×10^{-4}	21,500	0.149×10^{-18}
12. Carbon	hcp titanium	5.1×10^{-4}	43,500	0.302×10^{-18}

* See J. Askill, *Tracer Diffusion Data for Metals, Alloys, and Simple Oxides,* New York: Plenum (1970), for a more complete listing of diffusion data.

† $R = 1.987$ cal/mole·K; $k = 13.8 \times 10^{-24}$ J/atom·K.

‡ Lower below 400°C.

The other term k is the same Boltzmann constant as in Eq. (4–6.3); that is, 13.8×10^{-24} J/atom·K.

Table 4–7.2 lists the values of D_0 and the activation energy for a number of diffusion reactions. Since the chemist prefers molar and calorie units, we can rewrite the above equation:

$$\ln D = \ln D_0 - \frac{Q}{RT}. \qquad (4\text{–}7.4b)$$

Energy is expressed as Q (cal/mole). The gas constant R is 1.987 cal/mole·K.

The twelve sets of data in Tables 4–7.1 and 4–7.2 are plotted in Fig. 4–7.4. These are Arrhenius-type plots.*

Study Aids (atom movements)

A series of 16 sketches are included in the paperback *Study Aids for Introductory Materials Courses.* The factors affecting the diffusivity are emphasized.

Example 4–7.1

There are 0.19 a/o copper at the surface of some aluminum and 0.18 a/o copper, 1.2 mm underneath the surface. What will the flux of copper atoms be from the surface inward at 500°C? (The aluminum is fcc, and $a = 0.4049$ nm.)

Procedure. We must obtain the concentration gradient of the copper atoms, since the flux is proportional to it. However, concentrations must be expressed in number/unit volume. Therefore, it is necessary to first determine the total number of atoms per m^3.

* See the comments after Example 4–6.2.

Diffusivity is a function of temperature according to Eq. (4–7.3), and the data in Table 4–7.2.

Calculation of the concentration gradient

$$\text{Atoms/m}^3 = 4/(0.4049 \times 10^{-9} \text{ m})^3$$
$$= 6 \times 10^{28}/\text{m}^3.$$

$$(dC/dx)_{Cu} = \frac{(0.0018 - 0.0019)(6 \times 10^{28}/\text{m}^3)}{0.0012 \text{ m}}$$
$$= -5 \times 10^{27} \text{ Cu/m}^4.$$

Calculation of the flux

$$\text{Diffusivity} = (0.15 \times 10^{-4} \text{ m}^2/\text{s}) \exp\left[\frac{-0.210 \times 10^{-18}\text{J}}{(13.8 \times 10^{-24} \text{ J/K})(773 \text{ K})}\right]$$
$$= 4 \times 10^{-14} \text{ m}^2/\text{s}.$$

$$\text{Flux} = -(4 \times 10^{-14} \text{ m}^2/\text{s})(-5 \times 10^{27} \text{ Cu/m}^4)$$
$$= 2 \times 10^{14} \text{ Cu/m}^2\cdot\text{s} \qquad\qquad (\text{or } 2 \times 10^8 \text{ Cu/mm}^2\cdot\text{s}).$$

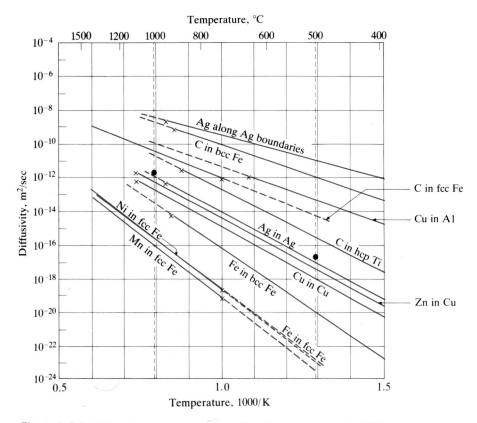

Figure 4–7.4 Diffusivity versus temperature. (See Tables 4–7.1 and 4–7.2.)

Comments. The value of copper diffusivity may also be obtained graphically from Fig. 4–7.4 (and for 500°C, from Table 4–7.1). Since the activation energy enters the exponent of the equation, its three significant figures do not carry over into the value of diffusivity. Usually diffusivity values are significant to only the first or second figure. ◄

Example 4–7.2

The diffusivity of aluminum in copper is 2.6×10^{-17} m²/s at 500°C and 1.6×10^{-12} m²/s at 1000°C. (a) Determine the values of D_0, Q, and E for this diffusion couple. (b) What is the diffusivity at 750°C?

Procedure. With two unknowns, D_0 and E (or Q), we need at least two data points for a solution. From these we can obtain the answers, either by calculation or graphically. With equal care, the two procedures are equally accurate, because they use the same initial data.

Calculation. With Eq. (4–7.4a),

a)
$$\ln (2.6 \times 10^{-17}) = \ln D_0 - \frac{E}{13.8 \times 10^{-24}\ (773)};$$

$$\ln 1.6 \times 10^{-12} = \ln D_0 - \frac{E}{13.8 \times 10^{-24}\ (1273)}.$$

Solving simultaneously, we get
$$D_0 = 4 \times 10^{-5} \text{ m}^2/\text{s}, \quad \text{and} \quad E = 0.3 \times 10^{-18} \text{ J/atom.}$$

Alternatively, with Eq. (4–7.4b),
$$Q = 43,000 \text{ cal/mole.}$$

b) From Eq. (4–7.4a),
$$\ln D = \ln 4 \times 10^{-5} - \left(\frac{0.3 \times 10^{-18} \text{ J}}{(13.8 \times 10^{-24} \text{ J/K})(1023 \text{ K})}\right);$$

and therefore,
$$D = 2.5 \times 10^{-14} \text{ m}^2/\text{s} \qquad (\text{or } 10^{-13.6} \text{ m}^2/\text{s}).$$

Graphically. The two data points are located by dots on Fig. 4–7.4. By interpolation at 1/1023 K,
$$\log_{10} D = -13.6 \qquad (\text{or } D = 2.5 \times 10^{-14} \text{ m}^2/\text{s}).$$

The slope is E/k; the intercept at $1/T = 0$ is log D_0.

Comments. Observe that the diffusivity of copper through aluminum is higher than for aluminum through copper. This is to be expected from our knowledge of the bond strength of the host metals:
$$(T_m \text{ of Cu}) > (T_m \text{ of Al}); \qquad \text{therefore,} \quad D_{Al/Cu} < D_{Cu/Al}. \ ◄$$

Example 4–7.3

A steel contains 8.5 w/o Ni at the center x of a grain of fcc iron, and 8.8 w/o Ni at the edge e of the grain. The two points are separated by 40 μm. What is the flux of atoms between x and e at 1200°C? ($a = 0.365$ nm.)

Procedure. Change to atom percent nickel; then calculate the number of Ni atoms/μm^3 at the two points. Finally, we need the diffusivity at 1200°C to calculate the flux of Ni atoms.

Calculation

$$(100 \text{ amu})(0.085)/(58.71 \text{ amu/Ni}) = 0.1448 = 8.1 \text{ a/o Ni};$$

$$(100 \text{ amu})(0.915)/(55.85 \text{ amu/Fe}) = \underline{1.638} = 91.9 \text{ a/o Fe}.$$

$$\text{Total atoms} = 1.783$$

A comparable calculation at e gives ~8.4 a/o Ni.

Per unit cell,

$$C_x = (4 \text{ atoms})(0.081)/(0.365 \times 10^{-3} \mu m)^3 = 6.66 \times 10^9/\mu m^3;$$

$$C_e = (4 \text{ atoms})(0.084)/(0.365 \times 10^{-3} \mu m)^3 = 6.91 \times 10^9/\mu m^3.$$

$$\ln D = \ln (0.77 \times 10^{-4} \text{ m}^2/\text{s}) - \left(\frac{0.465 \times 10^{-18} \text{ J}}{(13.8 \times 10^{-24} \text{ J/K})(1473 \text{ K})}\right);$$

$$D = 9 \times 10^{-15} \text{ m}^2/\text{s} = 9 \times 10^{-3} \mu m^2/\text{s}.$$

$$\text{Flux} = -(9 \times 10^{-3} \mu m^2/\text{s})\left(\frac{(6.66 - 6.91)(10^9/\mu m^3)}{40 \mu m}\right)$$

$$= 5.6 \times 10^4 \text{ atoms}/\mu m^2 \cdot \text{s} \qquad (\text{or } 5.6 \times 10^{16} \text{ atoms/m}^2 \cdot \text{s}). \blacktriangleleft$$

• 4–8

DIFFUSION PROCESSES

There are numerous applications of diffusion to technical processes. *Carburization* of steel may be familiar to the reader. In this process, a low-carbon steel (which is relatively tough, but soft) is heated in a carbon-containing atmosphere so that carbon diffuses into the steel, producing a hard, carbon-enriched *case* (Fig. 4–8.1). Another example among commercial diffusion processes is encountered in making semiconductors. Boron can be diffused into silicon to provide a *p*-type region of a junction device (Chapter 8).

We will not detail these processes here but will note that simultaneous consideration may be given to diffusivity and to processing time. If we multiply each side of Eq. (4–7.1) by time, we get an integrated flux:

$$\text{atoms/cm}^2 = Dt(-dC/dx). \qquad (4\text{–}8.1)$$

Thus, the total number of atoms entering or leaving a material through a unit area and down a fixed concentration gradient is proportional to Dt, the product of diffusivity and time. This is a convenient relationship because it lets us anticipate processing time as a function of temperature. For example, assume a 1.0-mm carburized case is formed on a steel (fcc iron) in 100 minutes at 800°C, where the diffusivity is ~2.4×10^{-12} m^2/s. With the same concentration gradient, it would take only 8 minutes at 1000°C where the diffusivity is 12.5 times as great (~3×10^{-11} m^2/s from Table 4–7.1).

Distance below surface, mm

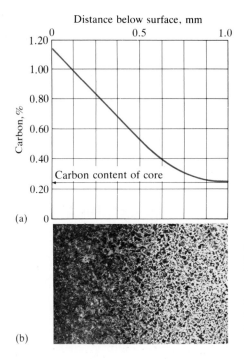

(a)

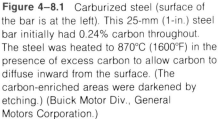

(b)

Figure 4–8.1 Carburized steel (surface of the bar is at the left). This 25-mm (1-in.) steel bar initially had 0.24% carbon throughout. The steel was heated to 870°C (1600°F) in the presence of excess carbon to allow carbon to diffuse inward from the surface. (The carbon-enriched areas were darkened by etching.) (Buick Motor Div., General Motors Corporation.)

The thickness x of a case increases with $\sqrt{Dt}$:*

$$x \propto \sqrt{Dt}, \qquad (4\text{–}8.2)$$

or

$$\frac{\sqrt{Dt}}{x} = \text{constant}.$$

From this and from the examples of the previous paragraph, a case that is ≈ 1.9 mm thick would develop in a 30-min period at 1000°C:

$$\frac{[(2.4 \times 10^{-12}\ \text{m}^2/\text{s})(6000\ \text{s})]^{1/2}}{1\ \text{mm}} = \frac{\sqrt{Dt}}{x} = \frac{[(3 \times 10^{-11}\ \text{m}^2/\text{s})(1800\ \text{s})]^{1/2}}{x_{1000°C,\ 30\ \text{min}}},$$

$$x_{1000°C,\ 30\ \text{min}} \approx 1.9\ \text{mm}.$$

Example 4–8.1

At the surface of a steel bar there is one carbon atom per 20 unit cells of iron. One millimeter behind the surface, there is one carbon atom per 30 unit cells. The diffusivity at 1000°C is 3×10^{-11} m²/s. The structure is fcc at 1000°C ($a = 0.365$ nm). How many carbon atoms diffuse through each unit cell per minute?

---

* This is derivable from Fick's second law (Eq. 4–7.2) in a footnote in the previous section.

Procedure. Determine the number of carbon atoms per unit volume at both locations to obtain the concentration gradient; then calculate the flux in terms of $(0.365 \text{ nm})^2$.

Calculation. Carbon concentrations are

$$C_2 = 1/[30(0.365 \times 10^{-9} \text{ m})^3]$$
$$= 0.68 \times 10^{27}/\text{m}^3;$$

$$C_1 = 1/[20(0.365 \times 10^{-9} \text{ m})^3]$$
$$= 1.03 \times 10^{27}/\text{m}^3.$$

From Eq. (4–7.1a),

$$J = -(3 \times 10^{-11} \text{ m}^2/\text{s})\left(\frac{(0.68 - 1.03)(10^{27}/\text{m}^3)}{10^{-3} \text{ m}}\right)$$

$$= 1.05 \times 10^{19}/\text{m}^2 \cdot \text{s} \qquad\qquad (\text{or} \sim 10/\text{nm}^2 \cdot \text{s}).$$

Each unit cell has an area of $(0.365 \times 10^{-9} \text{ m})^2$. Therefore,

$$J_{\text{u.c.}} = (10.5/\text{nm}^2 \cdot \text{s})(0.365 \text{ nm})^2 (60 \text{ s/min})$$
$$= 84 \text{ atoms/min.}$$

Comments. It is apparent that a piece of steel is not a dormant material; rather, numerous changes occur within it.

We use the above process to *carburize* steel. In Chapter 11, we will see how this can be used to advantage to modify the surface hardness of a steel. ◀

Example 4–8.2

Radioactive nickel has been diffused into the surface of iron for 20 minutes at 1200°C. The limit of detection of the radioactivity is at a position that is 1.5 mm behind the surface. A companion sample was heated for 1 hour at 1100°C. Will the limit of detection be greater or less than the 1.5 mm observed in the first sample?

Procedure. Since Dt/x^2 is a constant, we can calculate it where $x = 1.5$ mm; then calculate x for the values of D and t at 1200°C and 3600 s. The diffusivity, D, may be obtained from Eq. (4–7.3), or lifted from the graph of Fig. 4–7.4:

$$D_{1200°C} = 9 \times 10^{-15} \text{ m}^2/\text{s};$$

$$D_{1100°C} = 1.7 \times 10^{-15} \text{ m}^2/\text{s}.$$

Calculation

1200°C, 20 min.:
$$Dt/x^2 = (9 \times 10^{-15} \text{ m}^2/\text{s})(1200 \text{ s})/(0.0015 \text{ m})^2$$
$$= 4.8 \times 10^{-6};$$

1100°C, 60 min.:
$$4.8 \times 10^{-6} = (1.7 \times 10^{-15} \text{ m}^2/\text{s})(3600 \text{ s})/x^2$$
$$x = 0.0011 \text{ m} \qquad\qquad (\text{or } 1.1 \text{ mm}).$$

The limit of detection would be at 1.1 mm. ◀

Review and Study

SUMMARY

In Chapter 3 attention was given to the preciseness and regularity of the atomic structure of crystals. Perfect crystals serve as a basis for considering many properties and characteristics of materials, such as density, anisotropy, slip planes, phase stability, piezoelectricity, and semiconducting compounds. At the same time, crystals are not always perfect, and many important properties and behaviors of materials arise from the irregularities. We cited some of these in the preview of the chapter. We must consider the *disorder* in materials as well as their order.

1. *Impurities* are found in all materials unless special means are used to reduce them to a low level. An alloy is a metal with impurities that are intentionally present. An *alloy* is a combination of two or more metallic elements in one material. The combinations may be *mixtures* or *solid solutions*.

2. *Solid solutions* contain atoms (or molecules) of a second component as a solute. These may be present either interstitially among the solvent atoms or as a substitute replacement within the crystal lattice. In order for solid solution to be extensive in crystals, the solute and solvent must be comparable in size and electronic behavior.

- 3. Solid solutions also exist within compounds. Some *nonstoichiometric compounds* arise from incomplete ordering. Others containing transition elements exist as a result of mixed valences. The latter commonly produce a *defect* structure.

4. *Imperfections* are found in all crystals unless special means are used to reduce them to a low level. It is convenient to categorize them by geometry:

    i) *point defects*, that include vacancies and/or interstitials;

    ii) *linear defects*, commonly called dislocations; and

    iii) *boundaries*, which may be external surfaces or internal discontinuities between grains or phases. The latter may be treated as two-dimensional defects.

5. Liquids generally lack the long-range order that characterizes crystals. For some materials, it is possible to avoid crystallization and retain the amorphous character of a liquid into a solid (i) if the anticipated crystal structure is complex, or (ii) by rapid cooling. We label these amorphous solids by the term *glass*. Although lacking a freezing temperature, amorphous materials possess a glass temperture T_g, which will be very important to us when we study the behavior of plastics.

    A *phase* was defined as that part of a material that is distinct from other parts in structure and/or composition.

6. Thermal expansion results from increased atomic vibrations at higher temperatures. Furthermore, the expansion coefficient increases with increased temperature.

    At any instant of time, most atoms or molecules possess near-average energy. However, some will possess very little energy, and some will possess abnormally high energies. We are interested in the high end of this statistical distribution, because those atoms may be activated to break bonds, and to diffuse to new locations within the material. It is only through this mechanism that the internal structure (and hence the properties) of the material may be modified.

7. Diffusion occurs more readily (i) at high temperatures, (ii) when the diffusing atom is small (e.g., carbon in iron), (iii) when the packing factor of the host structure is low (e.g., bcc versus fcc), (iv) when the bonds of the host structure are weak (e.g., low-melting materials), and (v) when there are defects in the material (e.g., vacancies or grain boundaries).

● 8. Diffusion is used industrially to carburize steels and to form junctions of semiconductor devices. The thicknesses of diffusion surface layers, x, increases with $\sqrt{Dt}$.

TERMS AND CONCEPTS

● **Activation energy (E or Q)** Energy barrier that must be met prior to reaction.

Alloy A metal containing two or more elements.

Amorphous Noncrystalline, and without long-range order.

Anisotropic Having different properties in different directions.

Arrhenius equation Thermal activation relationship. (See comments with Example 4–6.2.)

ASTM grain-size number (G.S.#) Standardized grain counts. (See Eq. (4–4.2).)

Boltzmann's constant (k) Thermal energy coefficient ($13.8 \ldots \times 10^{-24}$ J/K).

Brass Alloys of copper (>50%) plus zinc.

Bronze An alloy of copper and tin (unless otherwise specified; e.g., an aluminum bronze is an alloy of copper and aluminum).

Component The basic chemical substances required to create a chemical mixture or solution.

Compound A phase composed of two or more elements in a given ratio.

Concentration gradient (dC/dx) Change in concentration with distance. Concentration is expressed in number per unit volume.

Defect structure Nonstoichiometric compounds that contain either vacancies or interstitials within the structure.

Diffusion The movement of atoms or molecules in a material.

Diffusion flux (J) Transport per unit area and time.

Diffusivity (D) Diffusion flux per unit concentration gradient.

Dislocation, edge ($\perp$) Linear defect at the edge of an extra crystal plane. The slip vector is perpendicular to the defect line.

Dislocation, screw ($\mathfrak{s}$) Linear defect with slip vector parallel to the defect line.

Energy distribution Spectrum of energy levels arising from thermal activation.

Fick's first law Proportionality between diffusion flux and concentration gradient.

Glass An amorphous solid below its transition temperature. A glass lacks long-range crystalline order, but normally has short-range order.

Glass transition temperature (T_g) Transition temperature between a supercooled liquid and its glassy solid.

Grain Individual crystal in a polycrystalline microstructure.

Grain boundary The zone of crystalline mismatch between adjacent grains.

Grain boundary area (S_v) Intergranular area/unit volume; for example, in²/in³ or mm²/mm³.

Grain size Statistical grain diameter in a random cross section. (Austenite grain size is reported as the number of former austenite grains within a standardized area. See Section 4–4.)

Heat of fusion (H_f) Energy per mole (or other stated unit) to melt a material.

Imperfection (crystal) Defect in a crystal. May be a point, a line, or a boundary.

Interstice   Unoccupied space between atoms or ions.

Ion vacancy   Unoccupied ion site within a crystal structure. The charge of the missing ion must be appropriately compensated.

Microstructure   Structure of grains and phases. Generally requires magnification for observation.

Mixture   Combination of two phases.

Nonstoichiometric compounds   Compounds with noninteger atom (or ion) ratios.

Phase   A physically homogeneous part of a materials system. (See Section 4–5.)

Point defect   Crystal imperfection involving one (or a very few) atoms.

Short-range order   Specific first-neighbor arrangement of atoms, but random long-range arrangements.

Slip vector (b)   Displacement distance of a dislocation. It is parallel to a screw dislocation and perpendicular to an edge dislocation.

Solid solution   A homogeneous crystalline phase with more than one chemical component.

Solid solution, interstitial   Crystals that contain a second component in their interstices. The basic structure is unaltered.

Solid solution, ordered   A substitutional solid solution with a preference by each of the components for specific lattice sites.

Solid solution, substitutional   Crystals with a second component substituted for solvent atoms in the basic structure.

Solubility limit   Maximum solute addition without supersaturation.

Solute   The minor component of a solution.

Solvent   The major component of a solution.

Sterling silver   An alloy of 92.5 Ag and 7.5 Cu. (This corresponds to nearly the maximum solubility of copper in silver.)

Stoichiometric compounds   Compounds with integer atom (or ion) ratios.

Surface   Boundary between a condensed phase and gas.

Thermal expansion coefficient (α)   (change in dimensions)/(change in temperature).

Vacancy ($\square$)   A normally occupied lattice site that is vacant.

FOR CLASS DISCUSSION

A_4 What is the difference between brass and bronze?

B_4 Distinguish between an alloy that is a mixture of phases and one that is a solid solution.

C_4 Distinguish between solvent and solute, as applied to solid solutions.

D_4 Distinguish between interstitial and substitutional solid solutions.

E_4 From Appendix B, select elements that have a favorable size for substitutional solid solution with metallic iron. Select those ions that might substitute for Fe^{2+} ions.

F_4 There is a good argument for considering that an ordered solid solution is a compound. Discuss.

G_4 Nonstoichiometric AlMg does not possess a significant number of vacancies; nonstoichiometric FeO does. Explain the difference.

H_4 Why do Schottky defects occur in pairs?

I₄ Explain the origin of surface energy; of grain-boundary energy.

J₄ Cite various ways grain boundaries affect the behavior of materials.

K₄ Magnification is always expressed on the basis of the lineal multiplication factor. With ×100, what is the area multiplication factor?

L₄ Compare and contrast the change that occurs at 380°C in $AuCu_3$ (Study Problem 4–2.6) with melting.

M₄ Why is energy given off during the freezing of water (80 cal/g, or 6 kJ/mole)?

N₄ Explain the glass temperature, T_g, to a friend who is not in this course.

• **O₄** Which will be higher, the *mean energy* or the *median energy* of the gas molecules in this room?

P₄ What phases are present after mixing 10 g of oil with 100 g of water? 10 g of salt with 100 g of water? Give an example of a materials "system" with three phases present.

Q₄ What is meant by activation energy? By phase?

• **R₄** The term D_0 in Eq. (4–7.3) includes M of Eq. (4–6.3b) but does not equal M. Suggest why.

S₄ Why must the T of Arrhenius-type equations be in absolute temperature, K? We can use °C for thermal expansion and thermal conductivity calculations.

T₄ Refer to Table 4–7.1. Why are the values higher for couple #2 than for couple #1? For couple #2 than for couple #4? For couple #11 than for couple #10? For couple #8 than for couple #9?

U₄ Self-diffusion is the movement of atoms within their own structure. Explain how this can occur. Suggest a way to measure it in the absence of a concentration gradient, through the use of radioactive isotopes.

QUIZ SAMPLES

4A For extensive substitutional solid solution, the solute and solvent atoms must have similar __+−__ .

(a) atomic diameters   (b) atomic weights   (c) electronic properties
(d) atomic numbers

4B A tin bronze contains 10 w/o Sn; the atomic percent, a/o, will be __**-er__ . An aluminum bronze contains 10 a/o Al; the weight percent, w/o, will be __**-er__ .

4C The maximum solubility of carbon in bcc iron is 0.02 w/o. This means 1 C/ __**__ unit cells.

4D A __**-metric__ compound does not have a fixed ratio of atoms.

4E Point defects within crystals include __+−__ .

(a) cation pairs   (b) hole-interstitial pairs   (c) anion vacancies   (d) split atoms   (e) Dislocations   (f) None of the above, rather _____

4F The slip vector of a screw dislocation is __**__ to the dislocation line. The adjacent atoms possess extra __**__ .

4G Two-dimensional imperfections include external __**__ , and boundaries between __**__ or __**__ .

4H A bigger ASTM grain-size number means __+−__ .

(a) Larger grain diameters   (b) more grains   (c) more grain-boundary area per unit volume   (d) None of the above, because this grain-size index has to be empirical, since the grains are not spherical.

4I G.S. #5 has __**__ times as many grains per square inch in a two-dimensional section as G.S. #3. The linear dimensions of the former are __**__ times as large.

4J A crystal changes to a liquid at the __**__ temperature; a glass changes to a(n) __**__ liquid at the glass-transition temperature.

4K The thermal expansion coefficient of a solid generally __**-creases__ as the temperature is raised. Furthermore, the volume expansion coefficient for a solid is generally __**-er__ than for a liquid.

4L In order to have a linear plot of thermal activation as a function of temperature, we must plot __+−__ .

(a) number, n, vs. °K   (b) log n vs. °K   (c) n vs. 1/°C   (d) n vs. °K^{-1} (e) None of the above, rather _____

4M Diffusivity is the coefficient relating flux to __**__ .

4N Other factors equal, diffusivity is greater when the __sol-**__ atom is small; the packing factor is __**-er__ ; the melting temperature is __**-er__ .

4O The dimensions of diffusivity are __**__ .

4P In order to have a linear plot of diffusivity as a function of temperature, we must plot __+−__ .

(a) D vs. °K   (b) ln D vs. °K   (c) log D vs. °K   (d) D vs. °K^{-1} (e) None of the above, rather _____

• **4Q** If carburization time is doubled, the carburized case will become __**__% thicker (other factors equal). The temperature is changed to increase the diffusivity by a factor of 10; the time should be only __**__% as long to give the same case thickness.

STUDY PROBLEMS   **4-2.1** An alloy contains 85 w/o copper and 15 w/o tin. Calculate the atomic percent (a/o) of each element.

Answer: 8.6 a/o Sn; 91.4 a/o Cu

4-2.2 There is 5 a/o magnesium in an Al–Mg alloy. Calculate the w/o magnesium.

4-2.3 Consider Fig. 4–2.3 to be an interstitial solution of carbon in fcc iron. What is the w/o carbon present?

Answer: 1.3 w/o carbon

4-2.4 Consider Fig. 4–2.1 to be a substitutional solid solution of copper and gold. What is the w/o Cu present if (a) Cu is the more prevalent atom? (b) Au is the more prevalent atom?

4-2.5 An alloy contains 80 w/o Ni and 20 w/o Cu in substitutional fcc solid solution. Calculate the density of this alloy.

Answer: 8.92 g/cm^3 (=8.92 Mg/m^3)

4-2.6 An alloy with 25 a/o gold and 75 a/o copper forms an fcc solid solution which is random above 380°C. Below that temperature, it becomes ordered with gold atoms at the corners of the unit cell and copper atoms at the center of each face. What is the weight percent gold?

4-2.7 Refer to Study Problem 4–2.6. (a) Estimate the lattice constant of that alloy. (b) Calculate its density.

Answer: (a) 0.373 nm   (b) 12.4 g/cm^3 (=12.4 Mg/m^3)

4-2.8 Silicon that is used as a semiconductor contains 10^{21} aluminum atoms per m^3. (a) What w/o Al is present? (b) How many unit cells of silicon per aluminum atom? (Si is cubic with 8 atoms/unit cell; a = 0.543 nm.)

4-3.1 (a) What is the w/o FeO in the solid solution of Fig. 4–3.1? (b) The w/o Fe^{2+}? (c) Of O^{2-}?

Answer: (a) 51 w/o   (b) 39.8 w/o   (c) 30.8 w/o

4-3.2 If all the iron ions of Fig. 4–3.1 were changed to Ni ions, what would be the w/o MgO?

4-3.3 What is the density of Fe$_{<1}$O, if the Fe^{3+}/Fe^{2+} ratio is 0.14? (Fe$_{<1}$O has the structure of NaCl; and ($r_{Fe} + R_O$) averages 0.215 nm.)

Answer: 5.73 Mg/m^3 (=5.73 g/cm^3)

4-3.4 An intermetallic compound of aluminum and magnesium varies from 52Mg–48Al to 56 Mg–44Al (weight bases). What are the atom ratios of these compositions?

4-4.1 Calculate the radius of the largest atom that can exist interstitially in fcc iron without crowding. (*Hint:* Sketch the (100) face of several adjacent unit cells.)

Answer: 0.053 nm

4-4.2 Calculate the radius of the largest atom that can fit interstitially into fcc silver without crowding.

4-4.3 Determine the radius of the largest atom that can be located in the interstices of bcc iron without crowding. (*Hint:* The center of the largest hole is located at $\frac{1}{2},\frac{1}{4},0$.)

Answer: 0.036 nm

4-4.4 (a) What is the coordination number of the interstitial site in Study Problem 4–4.1? (b) What structure would result if *every* such site were occupied by a smaller atom or ion?

4-4.5 (a) What is the coordination number for the interstitial site in Study Problem 4–4.3? (b) How many of these sites are there per unit cell?

Answer: (a) 4   (b) 12

4-4.6 In copper at 1000°C, one out of every 473 lattice sites is vacant. If these vacancies remain in the copper when it is cooled to 20°C, what will be the density of the copper?

4–4.7 The numbers of vacancies increase at higher temperatures. Between 20°C and 1020°C, the lattice constant of a bcc metal increased 0.5 l/o from thermal expansion. In the same temperature range, the density decreased 2.0%. Assuming there was one vacancy per 1000 unit cells in this metal at 20°C, estimate how many vacancies there are per 1000 unit cells at 1020°C.

Answer: 11/1000 unit cells

4–4.8 Estimate the grain-boundary area per unit volume for the iron in Fig. 10–8.2(a). The magnification is ×500.

4–4.9 (a) Assume that the G.S. #6 of Fig. 4–4.11 represents a two-dimensional cut through a polycrystalline solid. Estimate the corresponding grain-boundary area. (b) Repeat for G.S. #2.

Answer: (a) 60 mm²/mm³ (or 1500 in²/in³)   (b) 15 mm²/mm³ (or 380 in²/in³)

4–4.10 Joe Moe concluded that the ASTM grain size of a metal was #2. However, he had neglected to observe that the magnification was ×300 and not ×100. (a) What was the correct grain-size number? (b) Refer to Fig. 10–8.2(a). What is the ASTM G.S.#?

4–5.1 The density of liquid aluminum at its melting point is 2.37 Mg/m³. Assume a constant atom size; calculate its atomic packing factor.

Answer: 0.65

4–5.2 (a) Estimate the heat of fusion of sodium from Table 4–5.1. (b) Of silver.

4–5.3 Based on the slopes of the curves in Fig. 4–5.2, estimate the volume expansion coefficient, α_V, of (a) solid lead and (b) liquid lead at the melting point of lead.

Answer: (a) 100×10^{-6}/°C   (b) 120×10^{-6}/°C

4–6.1 An aluminum wire is stretched between two rigid supports at 35°C. It cools to 15°C. What stress is developed?

Answer: 31 MPa (or 4500 psi).

4–6.2 Estimate the linear expansion coefficient of bcc iron at 900°C from the data in Fig. 4–5.2. Compare your answer with the data in Fig. 4–6.3.

4–6.3 Quartz (SiO_2) is hexagonal. Mean values of its two thermal expansion coefficients are $\alpha_a = 15 \times 10^{-6}$/°C and $\alpha_c = 10 \times 10^{-6}$/°C. What is the percent volume expansion between 20°C and 570°C of a quartz crystal?

Answer: ≈2.2 v/o

4–6.4 Refer to Example 4–6.2. What fraction of the atoms have sufficient energy to jump out of their sites at 1000°C?

4–6.5 At 800°C, 1 out of 10^{10} atoms, and at 900°C, 1 out of 10^9 atoms have sufficient energy for movements within a solid. (a) What is the activation energy in J/atom? (b) In cal/mole?

Answer: 0.4×10^{-18} J/atom (or 57,600 cal/mol)

4-6.6 An activation energy of 2.0 eV (or 0.32×10^{-18} J) is required to form a vacancy in a metal. At 800°C there is one vacancy for every 10^4 atoms. At what temperature will there be one vacancy for every 1000 atoms?

4-7.1 A solid solution of copper in aluminum has 10^{26} atoms of copper per m³ at point X, and 10^{24} copper atoms per m³ at point Y. Points X and Y are 10 micrometers apart. What will be the diffusion flux of copper atoms from X to Y at 500°C?

Answer: 4×10^{17} atoms/m²·s

4-7.2 (a) What is the ratio of diffusivities for carbon in bcc iron to carbon in fcc iron at 500°C? (b) Of carbon in fcc iron to nickel in fcc iron at 1000°C? (c) Of carbon in fcc iron at 1000°C to carbon in fcc iron at 500°C? (d) Why are the ratios greater than 1?

4-7.3 The inward flux of carbon atoms in fcc iron is 10^{19}/m²·s at 1000°C. What is the concentration gradient?

Answer: -3.3×10^{29}/m⁴

4-7.4 There are 4 a/o carbon at the surface of the iron in Study Problem 4-7.3. What is the a/o carbon at 1 mm behind the surface? ($a = \sim 0.365$ nm at 1000°C.)

4-7.5 (a) Using the data of Table 4-7.2, calculate the diffusivity of copper in aluminum at 400°C. (b) Check your answer with Fig. 4-7.4.

Answer: 2×10^{-15} m²/s

4-7.6 Zinc is moving into copper. At point X there are 2.5×10^{17} Zn/mm³. What concentration is required at point Y (2 mm from X) to diffuse 60 Zn atoms/mm²·min at 300°C (Y to X)?

4-7.7 What is the weight percent of zinc in the copper at point X of Study Problem 4-7.6?

Answer: 0.3 w/o

4-7.8 A zinc gradient in a copper alloy is 10 times greater than the aluminum gradient in a copper alloy. Compare the flux of solute atoms/m²·s in the two alloys at 500°C. (The data for $D_{Al \, in \, Cu}$ are in Example 4-7.2.)

4-7.9 Refer to Study Problem 4-7.1. (a) What is the diffusion coefficient of copper in aluminum at 100°C? (b) What will be the diffusion flux of copper atoms from X to Y at 100°C?

Answer: (a) 3×10^{-23} m²/s   (b) 300 atoms/mm²·s

4-7.10 Aluminum is to be diffused into a silicon single crystal. At what temperature will the diffusion coefficient be 10^{-14} m²/s? ($Q = 73,000$ cal/mole and $D_0 = 1.55 \times 10^{-4}$ m²/s.)

4-7.11 The Ni-in-fcc Fe and the Fe-in-fcc Fe curves of Fig. 4-7.4 cross. (a) By calculation, determine the temperature where they cross. (b) Suggest why the two curves are nearly coincident.

Answer: (a) ~1200 K (~900°C); a more precise answer is not warranted.

4–7.12 The diffusion of carbon in tungsten has an activation energy of 0.78×10^{-18} J/atom (112,000 cal/mole) and D_0 of 0.275 m^2/s. Where does the diffusivity curve lie on Fig. 4–7.4? Suggest why it lies where it does in comparison to the other curves for carbon diffusion.

4–7.13 How much should the concentration gradient be for nickel in iron if a flux of 100 nickel atoms/mm^2·s is to be realized (a) at 1000°C? (b) at 1400°C?

Answer: (a) $-(5 \times 10^{23}/m^3)/m$ (b) $-(7 \times 10^{20}/m^3)/m$

4–7.14 Compare the diffusivities of iron (fcc), copper, and silver at 60% of their melting temperatures, $0.6\,T_m$.

4–8.1 A suitable carburized case on steel (fcc iron) was obtained in 1 hour at 820°C (1510°F). How long would it take to double the thickness of the case at the same temperature?

Answer: 4 hours

4–8.2 A suitable carburized case on steel (fcc iron) was obtained in 1 hour at 820°C (1510°F). How long would it take to obtain a case of the same thickness at 850°C (1560°F)?

4–8.3 Refer to Study Problem 4–8.1. (a) What temperature is required to double the thickness in one hour? (b) In two hours?

Answer: (a) 1200 K (927°C)   (b) 1144 K (871°C)

4–8.4 Refer to Example 4–8.1. How many atoms diffuse per minute through each cell at 950°C?

QUIZ CHECKS			
4A	(a), (c)	4J	melting
4B	lower		supercooled
	lower	4K	increases
4C	540		smaller
4D	nonstoichiometric	4L	(e): log n vs. °K^{-1}
4E	(b), (c)	4M	concentration gradient
4F	parallel	4N	solute
	energy		lower
4G	surfaces		lower
	grains or phases	4O	m^2/s
4H	(b), (c)	4P	(e): log D vs. °K^{-1}
4I	four	4Q	41%,
	two		10%

Chapter Five

Single-phase Metals

PREVIEW

Metals are somewhat simpler than the other two principal categories of materials—polymers and ceramics. This is true because a large number of metals contain only one kind of atom (or are a solid solution such as brass in which zinc proxies for copper without a change in structure). Therefore, we will examine how properties relate to the structure of metals first. We will then turn to polymers and ceramics in Chapters 6 and 7, respectively.

We must consider both crystalline structure and imperfections as we look at *elastic* and *plastic deformation*. A new feature will be presented; that is, the opportunity to anneal the metal and remove previously introduced structural imperfections. This can affect properties significantly.

STUDY OBJECTIVES

1. To familiarize yourself with the magnitudes of commonly encountered properties of annealed metals, such as copper and nickel, and of annealed alloys, such as brasses and Cu–Ni.

2. To describe various single-phase microstructures and to understand the basis for grain growth.

3. To extend your knowledge of elastic behavior beyond that of Section 1–5 to include (a) the three principal types of moduli (Young's, shear, and bulk), and (b) the variations with temperature, bond strength, and crystal orientation.

4. To know the origin of solution hardening, of strain hardening, and of annealing in terms of dislocations, and of recrystallization.

5. To apply your knowledge of solution hardening, strain hardening and recrystallization to the selection of materials and processes to meet specifications.

6. To understand the terms and concepts that pertain to metals and alloys.

5-1

SINGLE-PHASE ALLOYS

Many widely used metals have only a single phase. These may be commercially pure metals with only one component. Examples of such metals include copper for electric wiring, zinc for the coating on galvanized steel, and the aluminum used for housewares. However, in numerous other cases a second component is intentionally added to the metal in order to improve the properties. Any such combination of metals is called an *alloy*.

Alloys are single-phase metals if the solid solubility limit is not exceeded. Brass (a single-phase alloy of copper and zinc), bronze (a similar alloy of copper and tin), and copper–nickel alloys are typical of the single-phase alloys we shall study in this chapter. Multiphase, or polyphase, alloys contain additional phases because the solid solubility limit is exceeded. The majority of our steels, as well as many other metals, are multiphase alloys. They will be discussed in Part II of the text.

Properties of Single-phase Alloys

The properties of alloys are different from those of pure metals, as shown by Figs. 5–1.1 and 5–1.2 (pp. 154–155) for brass and Cu–Ni solid solutions. The increases in strength and hardness (Section 1–5) are due to the presence of solute atoms that interfere with the movements of dislocations in the crystals

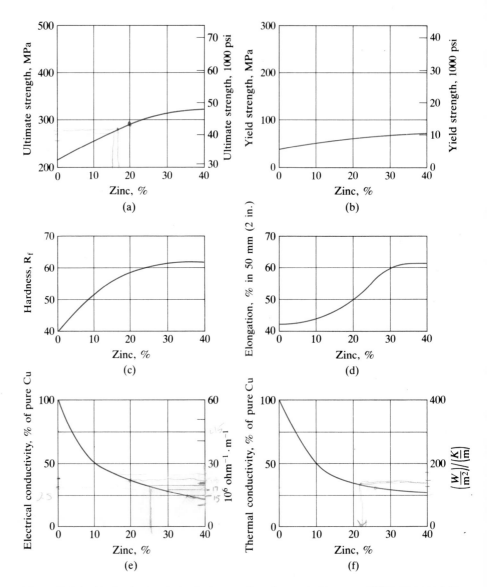

Figure 5–1.1 Mechanical and physical properties of annealed brasses (20°C). The solubility limit of zinc in fcc copper is near 40%. (Adapted from ASM data.)

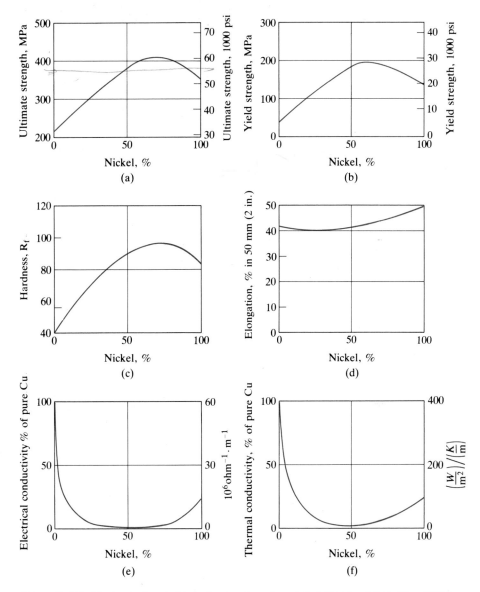

Figure 5–1.2 Mechanical and physical properties of annealed copper–nickel alloys (20°C). Copper and nickel form a complete series of solid solutions. (Adapted from ASM data.)

during plastic deformation. We shall observe later (Section 5–4) that this interference arises because the dislocations cannot readily move past the alloying atoms.

Very small amounts of impurities reduce the electrical conductivity of a metal, because foreign atoms introduce nonuniformities in the local potential within the crystal lattice. Therefore, the electrons experience more deflections and reflections, with a consequent reduction in the length of the mean free path (Section 2–5).

In a metal, electrons carry the majority of the energy for thermal conduction. Thus there is a correspondence between the thermal and electrical conductivity. (Compare (e) and (f) of Figs. 5–1.1 and 5–1.2.) In fact, it was pointed out some time ago that k/σ in most pure metals is about 7×10^{-6} watt·ohm/K at normal temperatures ($\sim 20°C$) when thermal conductivity k and electrical conductivity σ are expressed in (watts/m²)/(K/m) and ohm^{-1}·m^{-1}, respectively (Fig. 5–1.3). This ratio is proportional to temperature (in kelvin). As a result, we can write

$$k/\sigma = LT = k\rho. \tag{5–1.1}$$

The proportionality constant L (called the *Lorenz number*) is 2.3×10^{-8} watts·ohm/K². The electrical resistivity, ρ, is the reciprocal of electrical conductivity, σ. Since data are readily available on the electrical resistivity values for metals at various temperatures, the Lorenz number provides a convenient rule of thumb for less readily available thermal conductivity values.

Microstructures of Single-phase Alloys

Grains were defined in Section 4–4 as individual crystals. Materials with many grains are described as polygranular, or more commonly, *polycrystalline*. Adjacent crystals have dissimilar orientations so that a grain boundary is present

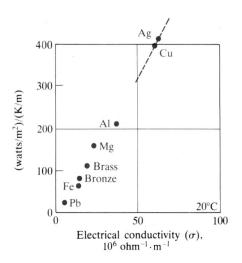

Figure 5–1.3   Metal conductivity (thermal versus electrical at 20°C). Since electrons also transport thermal energy, a good electrical conductor is a good thermal conductor. (The k/σ ratio is proportional to absolute temperature. This relationship applies best to pure metals. It does not include nonmetals.)

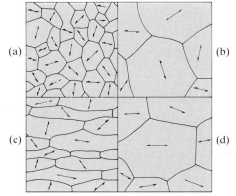

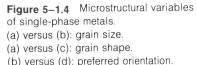

Figure 5–1.4 Microstructural variables of single-phase metals.
(a) versus (b): grain size.
(a) versus (c): grain shape.
(b) versus (d): preferred orientation.

(Fig. 4–4.7). The microstructures of single-phase metals can be varied by changes in *size, shape,* and *orientation* of the grains (Fig. 5–1.4). These aspects are not wholly independent, because the shape and size of grains are both consequences of grain growth. Likewise, grain shape is commonly dependent on the crystalline orientation of grains during growth.

Although it is common to speak of grain size in terms of diameter, few if any grains of a single-phase metal are spherical. Rather, the grains must completely fill space and also maintain a minimum of total boundary area. This was shown in Figs. 4–4.8 and 5–1.4(a), where the grains are described as being *equiaxed* because they have approximately equal dimensions in the three coordinate directions. In addition to equiaxed grains, other commonly encountered *grain shapes* may be platelike (Fig. 5–1.4c), columnar, or dendritic (i.e., treelike). No attempt will be made to systematize them in this book.

The *orientation* of grains within a metal is typically quite random (Fig. 5–1.4a). However, there are exceptions, that can be important from the standpoint of engineering properties. For example, the [100] directions of iron have a higher magnetic permeability than other directions. Therefore, if the grains within a polycrystalline transformer sheet are not random, but are processed to have a *preferred orientation* so that the [100] direction is preferentially aligned with the magnetic field, a significantly more efficient performance may be obtained from the transformer. The metallurgist has learned how to do this, with the result of billions of dollars worth of savings in electrical power distribution.

Example 5–1.1

A copper-nickel alloy must have an ultimate strength greater than 290 MPa (42,000 psi), and a thermal conductivity greater than 0.040 (W/mm²)/(°C/mm). Select an alloy from Fig. 5–1.2, bearing cost in mind.

Procedure. There are specification "windows," framed on one side by strength requirements, and bounded on the other side by conductivity specifications. We must define these bounds and pick the most appropriate composition within the specification window.

Solution

Ultimate strength, S_u	19% < x < 100% Ni.
Thermal conductivity, k	0% < x < 25% Ni, and 85% < x < 100% Ni.
Specification ranges	19%–25% Ni, and 85%–100% Ni.

Specify 80 Cu–20 Ni, since nickel is more expensive than copper.

Comments. If the coins taken from one's pocket are examined, the reader will realize that nickel is more expensive than copper.

Specifications seldom are set at the extreme edge of the range, because a slight variance in composition may exceed the specification range. ◀

Example 5–1.2

Copper has a resistivity of 47×10^{-9} ohm·m at 500°C. Estimate the $\sigma_{500°C}/\sigma_{20°C}$ and the $k_{500°C}/k_{20°C}$ ratios for copper.

Procedure. The values of $\sigma_{20°C}$ and $k_{20°C}$ are available from Fig. 5–1.1. Two methods are available for the thermal conductivity ratio. (i) Calculate k at 500°C from the resistivity and the Lorenz number. (ii) Estimate the k/σ ratio at 20°C from Fig. 5–1.1 and increase the ratio from 20°C to 500°C.

Calculation

$$\sigma_{20°C} = 60 \times 10^6 \text{ ohm}^{-1}\cdot\text{m}^{-1}; \quad k_{20°C} = 400 \text{ W/m·K}.$$

a) $\dfrac{\sigma_{500°C}}{\sigma_{20°C}} = \dfrac{1/(47 \times 10^{-9} \text{ ohm·m})}{(60 \times 10^6 \text{ ohm}^{-1}\cdot\text{m}^{-1})} = 0.35.$

b) $\quad k_{500°C} = LT/\rho$

$$= (2.3 \times 10^{-8} \text{ W·}\Omega\text{/K}^2)(773 \text{ K})/(47 \times 10^{-9} \text{ ohm·m})$$

$$= 378 \text{ W/m·K}$$

or $(k\rho)/773 \text{ K} = (k/\sigma)/293 \text{ K}.$

$$(k\rho)_{500°C} = \left(\frac{400 \text{ W/m·K}}{60 \times 10^6 \text{ ohm}^{-1}\text{m}^1}\right)\left(\frac{773 \text{ K}}{293 \text{ K}}\right) = 17.5 \times 10^{-6} \text{ W·}\Omega\text{/K}$$

$$k_{500°C} = \frac{17.5 \times 10^{-6} \text{ W·}\Omega\text{/K}}{47 \times 10^{-9} \Omega\text{·m}} = 375 \text{ W/m·K}.$$

$$k_{500°C}/k_{20°C} = (375/400) = 0.94.$$

Comment. Electrical conductivity decreases significantly with temperature, while thermal conductivity is more constant. In fact, in solid solution alloys, the thermal conductivity may increase with temperature. (See Study Problem 5–1.3.) ◀

Example 5–1.3

When a rod is loaded with 250 kg (550 lb), it must have an elastic strain of less than 0.001 m/m; it must not deform plastically; and it must have a resistance of less than 0.1 ohm/ m. What is the minimum diameter for a rod made from monel metal (70 Ni–30 Cu)? Made from brass (70 Cu–30 Zn)?

Procedure. Data from Appendix C and Figs. 5–1.1 and 5–1.2 include:

	70 Ni–30 Cu	70 Cu–30 Zn
S_y, MPa	195	70
ρ, ohm·m	480×10^{-9}	62×10^{-9}
E, MPa	180,000	110,000

Using $A = \pi d^2/4$,

Rearrange $S_y = F/A$ to give $d = \sqrt{4F/\pi S_y}$;

Rearrange Eq. (1–5.1) to give $d = \sqrt{(4\rho L)/(\pi R)}$;

Rearrange Eq. (1–5.3) to give $d = \sqrt{(4F)/(e\pi E)}$.

(handwritten:) $R = \rho\left(\dfrac{L}{A}\right)$

$E = \dfrac{S}{e} = \dfrac{F/A}{\kappa L/L}$

Calculation

		70 Ni–30 Cu	70 Cu–30 Zn
$<S_y$:	$d = \sqrt{4(250 \times 9.8\ \text{N})/\pi\ S_y}$	$d \geq 4.0$ mm	$d \geq 6.7$ mm
$<0.1\ \Omega/\text{m}$:	$d = \sqrt{4\rho(1\ \text{m})/\pi(0.1\ \Omega)}$	$d \geq 2.5$ mm	$d \geq 0.9$ mm
$e < 0.001$:	$d = \sqrt{4(250 \times 9.8\ \text{N})/(0.001\ \pi E)}$	$d \geq 4.2$ mm	$d \geq 5.3$ mm

The minimum diameters are shown in bold-face type. Anything smaller would not meet the three requirements. ◄

• 5–2

PROCESSING OF SINGLE-PHASE ALLOYS

Most metals are processed initially by melting. While molten, they will be chemically refined to remove unwanted impurities. If alloys such as brass or bronze are being produced, zinc or tin, respectively, are added to molten copper. Being molten, the zinc (or tin) readily dissolves and becomes uniformly distributed within the liquid metal.

Next, the typical metal is cast (poured) into a mold to solidify. The mold may be the shape of the final product, as in the case of a large bronze bell; or the solidified *casting* may require some machining for final shaping (Fig. 5–2.1). In other cases, the mold is an *ingot,* which is simply a solidified block of metal (or other material) that is subsequently deformed by *mechanical working* or *forming* to produce a rod, wire, tube, plate, forging, etc. During the mechanical processing, the shape is changed. As a result, stresses must be applied that are above the yield strength described in Chapter 1. Consequently, the mechanical processing on the initial large ingot is commonly performed at high temperatures at which the material is softer and more ductile. At those temperatures

Figure 5–2.1 Casting (cut-away section of a brass faucet). By solidifying the molten metal in a sand mold, the shape was obtained that required only a small amount of machining.

less energy is required for deformation and there is less chance for fracturing during processing. *Rolling, forging,* and *extrusion* (Fig. 5–2.2) are among the common hot deformation processes.

Following the primary deformation steps, it is often desirable to continue the processing at ambient temperatures. This is possible because the dimensions are now smaller and the required forces and energy reduced accordingly. Also furnaces and fuel costs are avoided. Further, most metals oxidize rapidly at high temperatures. This is generally avoided at low temperatures. Finally, we shall see in this chapter that strengths can commonly be increased if the deformation is performed while the metal is at normal temperatures. Wire *drawing, spinning* (Fig. 5–2.3), and *stamping* are among the common secondary deformation processes performed at ambient temperatures.

In order to understand the behavior of metals during either the processing steps or subsequent service, it will be necessary to be more complete in our knowledge of elastic deformation than we were in Chapter 1, and to visualize the mechanisms of plastic deformation. Finally, it will be necessary to show the consequences of elevated temperatures on materials because a metal may change its structure (and therefore properties) during high-temperature processes and/or service. These topics will follow in turn.

Figure 5–2.2 Mechanical working (primary). (a) Rolling. (b) Forging. (c) Extrusion. These processes can produce severe plastic deformations at high temperatures.

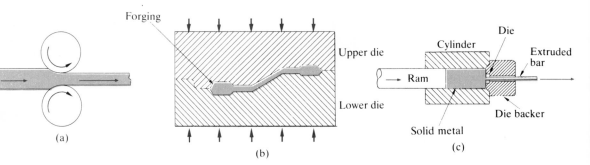

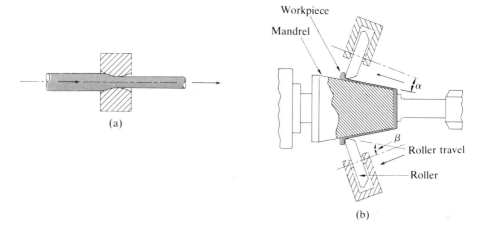

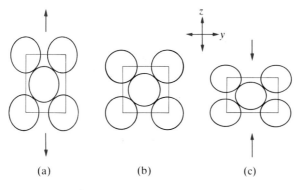

Figure 5–2.3 Mechanical working (secondary). (a) Wire drawing. (b) Spinning. Most metals are strengthened as they are deformed at ambient temperatures. Therefore, these processes are generally limited to products with smaller cross sections than the primary processes of Fig. 5–2.2.

5–3

**ELASTIC
DEFORMATION**

Elastic deformation occurs when a solid is stressed (Section 1–5). If the load is applied in tension, the piece becomes slightly longer; removal of the load permits the specimen to return to its original dimension. Conversely, when a load is applied in compression, the piece becomes slightly shorter. Elastic strain is a result of a slight elongation of the unit cell in the direction of the tensile load, or a slight contraction of the unit cell in the direction of the compressive load (Fig. 5–3.1).

Elastic Moduli

When only elastic deformation occurs, the strain is essentially proportional to the stress. This ratio between stress and strain is the *modulus of elasticity* (Young's modulus) and is a characteristic of the type of metal. The greater the

Figure 5–3.1 Elastic normal strain (greatly exaggerated). Atoms are not permanently displaced from their original neighbors. (a) Tension (+). (b) No strain. (c) Compression (−).

forces of attraction between atoms in a metal, the higher the modulus of elasticity.

Any lengthening or compression of the crystal structure in one direction, due to a uniaxial force, produces an adjustment in the dimensions at right angles to the force. In Fig. 5–3.1(a), for example, a small contraction is indicated at right angles to the tensile force. The negative ratio between the lateral strain e_y and the direct tensile strain e_z is called *Poisson's ratio v*:

$$v = -\frac{e_y}{e_z}.$$ (5–3.1)

Engineering materials may be loaded in shear as well as in tension (and compression). In shear loading, the two forces are parallel but are not aligned (Fig. 5–3.2b). As a result, the *shear stress, τ*, is that force, F_s divided by the sheared area, A_s:

$$\tau = F_s/A_s.$$ (5–3.2)

A shear stress produces an angular displacement, α. We define *shear strain, γ*, as the tangent of that angle, i.e., as x/y in Fig. 5–3.2(b). The recoverable or elastic shear strain is proportional to the shear stress:

$$G = \tau/\gamma,$$ (5–3.3)

where G is the *shear modulus*. Also called the modulus of rigidity, the shear modulus is different from the modulus of elasticity E; however, the two are related at small strains by

$$E = 2G(1 + v).$$ (5–3.4)

Since Poisson's ratio v is normally between 0.25 and 0.5, the value of G is approximately 35% of E.

A third elastic modulus is the *bulk modulus, K*. It is the reciprocal of the compressibility β of the material and is equal to the hydrostatic pressure P_h per

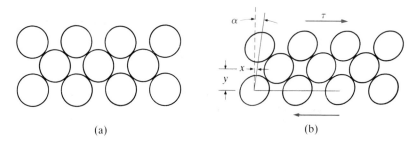

(a)   (b)

Figure 5–3.2 Elastic shear strain. Shear couples produce a relative displacement of one plane of atoms past the next. This strain is elastic so long as atoms keep their original neighbors. (a) No strain. (b) Shear strain.

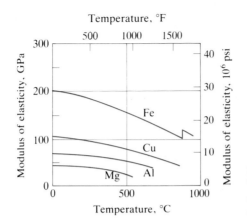

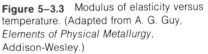

Figure 5–3.3   Modulus of elasticity versus temperature. (Adapted from A. G. Guy, *Elements of Physical Metallurgy*, Addison-Wesley.)

unit of volume compression, $\Delta V/V$:

$$K = \frac{P_h V}{\Delta V} = \frac{1}{\beta}.$$ (5–3.5)

The bulk modulus is related to the modulus of elasticity as follows:

$$K = \frac{E}{3(1 - 2\nu)}.$$ (5–3.6)

The reader may derive this equation in Study Problem 5–3.6.

Elastic Moduli versus Temperature

Elastic moduli decrease as temperature increases, as shown in Fig. 5–3.3 for four common metals. In terms of Fig. 2–4.2(a), a thermal expansion reduces the value of dF/da, and therefore the modulus of elasticity. The discontinuity in the curve for iron in Fig. 5–3.3 is due to the change from bcc to fcc at 912°C (1673°F). Not surprisingly, the more densely packed fcc polymorph requires greater stresses for a given strain, i.e., the elastic modulus increases for fcc. Also note from Fig. 5–3.3 that higher-melting metals have greater elastic moduli.

Elastic Moduli versus Crystal Direction

Elastic moduli are *anisotropic* within materials; that is, they vary with crystallographic direction. As an example, iron has an average modulus of elasticity of about 205 GPa (30,000,000 psi); however, the actual modulus of a crystal of iron varies from 280 GPa (41,000,000 psi) in the [111] direction to only 125 GPa (18,000,000 psi) in the [100] direction (Table 5–3.1). The consequence of any such anisotropy becomes significant in polycrystalline materials. Assume, for example, that Fig. 5–3.4(a) represents the cross section of a steel wire in which

TABLE 5–3.1					
Moduli of Elasticity (Young's Modulus)*					
	Maximum		**Minimum**		**Random**	
Metal	**GPa**	**10^6 psi**	**GPa**	**10^6 psi**	**GPa**	**10^6 psi**
Aluminum	75	11	60	9	70	10
Gold	110	16	40	6	80	12
Copper	195	28	70	10	110	16
Iron (bcc)	280	41	125	18	205	30
Tungsten	345	50	345	50	345	50

* Adapted from E. Schmid and W. Boas, *Plasticity in Crystals.* English translation, London: Hughes and Co.

the average stress is 205 MPa (30,000 psi). If the grains are randomly oriented, the elastic strain is 0.001, because the average modulus of elasticity is 205 GPa (30,000,000 psi). However, in reality, the stress varies from 125 MPa (18,000 psi) to 280 MPa (41,000 psi) as shown in Fig. 5–3.4(b), because grains have

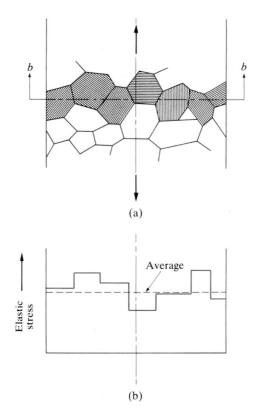

(a)

(b)

Figure 5–3.4 Stress heterogeneities (schematic). Elastic stresses vary with grain orientation, because the moduli of elasticity are not isotropic.

different orientations, but each is strained equally (0.001). Of course, this means that some grains will exceed their yield strength before other grains reach their yield strength.

Example 5-3.1

A plate of steel has a 100.0-cm × 100.0-cm square scribed on its surface. It is loaded in one direction (perpendicular to opposite edges of the square) with a 200-MPa (29,000-psi) stress. (a) What are the dimensions of the scribed area? (Poisson's ratio of steel = 0.29.)

Without removing the initial stress, a second tension stress of 410 MPa (60,000 psi) is applied at right angles to the first, i.e., perpendicular to the other edges of the square. (b) What are the new dimensions of the scribed area?

Procedure. Since no preferred orientation is indicated, we will assume the random grain value of Young's modulus (Table 5-3.1). The strains are additive when the two stresses are applied.

Calculation

a) With Eq. (1-5.3): e_z = 200 MPa/205,000 MPa = 0.000975;
   with Eq. (5-3.1): e_y = -0.29(0.000975) = -0.00028;
   1000 mm (1 + 0.000975) × 1000 mm (1 - 0.00028) = 1001.0 mm × 999.7 mm.

b) e_y = -0.00028 + 410 MPa/205,000 MPa = 0.00172;
   e_z = 0.000975 - 0.29(410/205,000) = 0.00040;
   1000 mm (1 + 0.0004) × 1000 mm (1 + 0.00172) = 1000.4 mm × 1001.7 mm.

Added information. We can write a general equation for elastic deformation in three dimensions from Eqs. (1-5.3) and (5-3.1):

$$e_x = \frac{s_x}{E} - \frac{\nu s_y}{E} - \frac{\nu s_z}{E}. \qquad (5-3.7) \blacktriangleleft$$

Example 5-3.2

What is the percentage volume change in iron if it is hydrostatically compressed with 1400 MPa (200,000 psi)? (Poisson's ratio = 0.29.)

Procedure. We need the bulk modulus ($= P_h/(\Delta V/V)$), which is obtainable from E and ν (Eq. 5-3.6). An alternative solution makes use of Eq. (5-3.7). Since $e_x = e_y = e_z$ ($= \Delta L/L$), and with $\Delta V/V \approx 3(\Delta L/L)$, we get the same answer.

Solution

$$K = (205,000 \text{ MPa})/3(1 - 0.58))$$
$$= 162,700 \text{ MPa}.$$
$$\Delta V/V = -1400 \text{ MPa}/162,700 \text{ MPa} = -0.86 \text{ v/o}.$$

Alternative solution

$$e_x = (1 - 2\nu)(-1400 \text{ MPa})/(205,000 \text{ MPa}) = -0.00287$$
$$\Delta V/V \approx 3e = -0.86 \text{ v/o}.$$

Comment. The approximation $\Delta V/V = 3(\Delta L/L)$ comes from $1 + \Delta V/V = (1 + \Delta L/L)^3$, and it is valid when the Δ's are small. $\blacktriangleleft$

5–4

PLASTIC DEFORMATION WITHIN SINGLE CRYSTALS

Cubic metals and their nonordered alloys deform predominantly by *plastic shear,* or *slip,* where one plane of atoms slides over the next adjacent plane. This is also one of the methods of deformation in hexagonal metals. Shear deformation occurs even when compression or tension stresses are applied, because these stresses may be resolved into shear stresses.

Critical Shear Stress, and Slip Systems

Slip occurs more readily along certain crystal directions and planes than along others. This is illustrated in Fig. 5–4.1, where a single crystal of an hcp metal was deformed plastically. The shear stress required to produce slip on a crystal plane is called the *critical shear stress* τ_c.

We can summarize the predominant sets of *slip systems* in several familiar metals in Table 5–4.1. A slip system includes the *slip plane* (*hkl*),* and a *slip direction* [*uvw*]. There are a number of slip systems, because of the multiple planes in a family of planes and multiple directions in a family of directions (Sections 3–6 and 3–7). Two facts stand out in Table 5–4.1.

1. The slip direction in each metal crystal is the direction with the highest linear density of lattice points, or the shortest distance between lattice points.

2. The slip planes are planes that have wide interplanar spacings (Section 3–8) and therefore, high planar densities (Section 3–7).

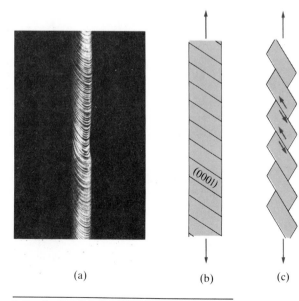

(a)   (b)   (c)

Figure 5–4.1 Slip in a single crystal (hcp). Slip paralleled the (0001) plane,* which contains the shortest slip vector. (Constance Elam, *Distortion of Metal Crystals,* Oxford: Clarendon Press.)

---

* See Section 3–7 for (*hkil*) indices of hexagonal crystals.

	TABLE 5–4.1			
	Predominant Slip Systems in Metals			

Structure	Examples	Slip planes	Slip directions	Number of independent slip systems
bcc	α-Fe, Mo, Na, W	$\{101\}$	$\langle \bar{1}11 \rangle$	12
bcc	α-Fe, Mo, Na, W	$\{211\}$	$\langle \bar{1}11 \rangle$	12
fcc	Ag, Al, Cu, γ-Fe, Ni, Pb	$\{111\}$	$\langle \bar{1}10 \rangle$	12
hcp	Cd, Mg, α-Ti, Zn	$\{000\bar{1}\}$*	$\langle 11\bar{2}0 \rangle$	3
hcp	α-Ti	$\{10\bar{1}0\}$*	$\langle 11\bar{2}0 \rangle$	3

* See Section 3–7 for ($hkil$) indices of hexagonal crystals.

Resolved Shear Stresses

The axial stress (tension or compression) that is required to produce plastic deformation depends not only on the critical shear stress for the slip system of each material, but also on the orientation of the applied stress with respect to the slip system (Fig. 5–4.2). For one thing, the shear area, A_s, will not equal the cross-sectional area, A, perpendicular to the tension or compression forces. Then, the forces must be resolved from the force F in the axial direction to the force F_s in the shear direction. The area increases as the plane is rotated away from the axis of applied stress; thus, $A_s = A/\cos \phi$, where ϕ is the angle between the axial direction and the normal to the slip plane (Fig. 5–4.2). The resolved force is $F_s = F \cos \lambda$, where λ is the angle between the two force directions. Since the shear stress, τ, is the shear force, F_s, per unit shear area, A_s,

$$\tau = F_s/A_s = (F \cos \lambda)/(A/\cos \phi) = s \cos \lambda \cos \phi. \qquad (5\text{–}4.1)$$

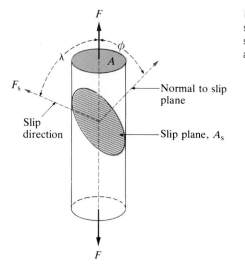

Figure 5–4.2 Resolved stresses. The axial stress s equals F/A. The resolved shear stress τ equals F_s/A_s, where $F_s = F \cos \lambda$, and $A_s = A/\cos \phi$. This leads to Eq. (5–4.1).

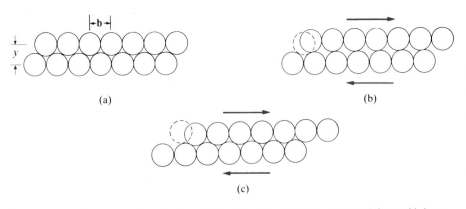

(a)

(b)

(c)

Figure 5–4.3 An assumed mechanism of slip (simplified). Metals actually deform with less shear stress than this mechanism would require.

This is *Schmid's law,* which relates the resolved shear stress, τ, to the axial stress, s, $(=F/A)$.

Slip occurs with the minimum axial force when both λ and ϕ are 45°. Under these conditions τ is equal to one half the axial stress F/A. The resolved shear stress is less in relation to the axial stress for any other crystal orientation, dropping to zero as either λ or ϕ approaches 90°.

Mechanism of Slip   Figure 5–4.3 shows a simplified mechanism of slip. If we attempted to calculate the strength of metals on this basis, the result would indicate that the strength of metals should be approximately $G/6$, where G is the shear modulus. Since metals are not that strong, a different slip mechanism must be operative. All experimental evidence supports a mechanism involving dislocation movements. If we use Fig. 5–4.4 as a model of a dislocation and place a shear stress along the horizontal direction, the dislocation can be moved (Fig. 5–4.5) with a shearing displacement within the crystal. (See also Fig. 4–4.5.) The shear stress required for this type of deformation is a small fraction of the previously cited value of $G/6$, and matches observed shear strengths.

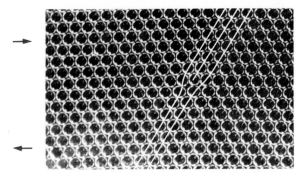

Figure 5–4.4 Edge dislocation. "Bubble-raft" model of an imperfection in a crystal structure. Note the extra row of atoms. (Bragg and Nye, *Proc. Roy. Soc. (London).)*

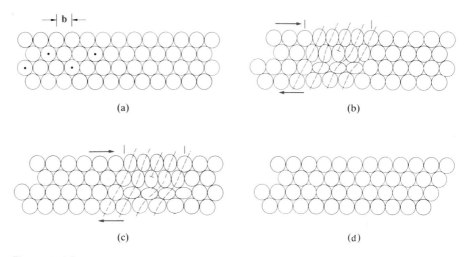

Figure 5–4.5 Slip by dislocations. In this model only a few atoms at a time are moved from their low-energy positions. Less stress is therefore required to produce slip than if all of the atoms moved at once as proposed in Fig. 5–4.3.

• The mechanism of slip requires the growth and movement of a dislocation line; therefore, energy is required. The energy, E, of a dislocation line is proportional to the length of the dislocation line, l, the product of the shear modulus, G, and the square of the unit slip vector $\mathbf{b}$ (Section 4–4):

$$E \propto lG\mathbf{b}^2. \qquad (5\text{–}4.2)$$

This automatically means that the easiest dislocations to generate and expand for plastic deformation are those with the shortest unit slip vector, $\mathbf{b}$ (Fig. 4–4.3), particularly since the $\mathbf{b}$-term is squared. The directions in a metal with the shortest slip vector will be the directions with the greatest linear density of atoms. The lowest value for the shear modulus, G, accompanies the planes that are farthest apart and hence have the greatest planar density of atoms. Thus, we may develop the rule of thumb that predicts the lowest critical shear ·stresses *on the most densely packed planes and in the most densely packed directions.**

Dislocation Movements in Solid Solutions

The energy associated with an edge dislocation (Fig. 4–4.3) is the same, whether the dislocation is located at (b) or (c) of Fig. 5–4.5. Therefore, no net energy is required for the movement between the two.† Such is not the case

---

* • Apparent exceptions to the rule generally arise because the planes and/or directions of easy slip are oriented to receive a low resolved shear stress. As an example, slip cannot occur on $(0\bar{1}1)[\bar{1}11]$ of a bcc metal from a force applied in the [011] direction because $\cos\phi = 0$. However, this leads to the possibility of slip in the $(211)[\bar{1}11]$ system, where $\cos\phi$ and $\cos\lambda$ are $1/\sqrt{3}$ and $\sqrt{\tfrac{2}{3}}$, respectively; and therefore $\tau = 0.47(F/A)$.

† This statement does not apply if (1) the movement includes an increase in the length of the dislocation, or (2) there is a pile-up of dislocations (Fig. 5–7.6).

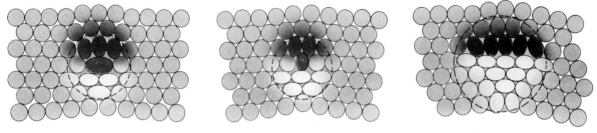

(a) Larger impurity atom (b) Smaller impurity atom (c) Same size atom

Figure 5–4.6 Solid solution and dislocations. An odd-sized atom decreases the stress around a dislocation. As a result, energy must be supplied and additional stress applied to detach the dislocation from the solute atom. This accounts for solution hardening.

when solute atoms are present. As shown in Fig. 5–4.6, when an impurity atom is present, the energy associated with a dislocation is less than it is in a pure metal. Thus, when a dislocation encounters foreign atoms, its movement is restrained because energy must be supplied to release it for further slip. As a result, solid-solution alloys always have higher strengths than do pure metals (Figs. 5–1.1 and 5–1.2). We call this *solution hardening.*

• **Intermetallic Compounds**

Phases such as Al_5Mg_3, Cu_2Al, $CuAl_2$, and β' brass (CuZn) receive mention elsewhere in this text. As is true with ceramics (Chapter 7) or any 3-D compound that contains neighbors of unlike atoms, these phases resist shear, because slip would destroy the preferred order among the unlike atoms (Fig. 4–2.2). They are, therefore, hard and also nonductile. As a result, they are brittle.

These hard phases cannot be used readily by themselves; however, they strengthen metals when they are present as very fine particles in alloys that are otherwise soft (Chapter 11).

Example 5–4.1

What force does it take in the $[1\bar{1}0]$ direction to have a resolved force of 130 N (29 lbs$_f$) in the [100] direction of a cubic crystal?

Solution

$$F_{[100]}/F_{[1\bar{1}0]} = \cos [1\bar{1}0] \not\ll [100].$$

$$F_{[1\bar{1}0]} = 130 \text{ N}/\cos 45° = 184 \text{ N}$$

$$= 184 \text{ N} \qquad \text{(or 41 lbs}_f\text{)}.$$

Comment. Draw a sketch of a cubic unit cell if you do not immediately see that the angle to be resolved is 45°. ◄

Example 5–4.2

The critical shear stress τ_c for the $\langle \bar{1}10 \rangle \{111\}$ slip systems of *pure* copper was found to be 1 MPa (145 psi). (a) What stress s must be applied in the [001] direction to produce slip in the [101] direction on the $(\bar{1}11)$ plane? (b) In the [110] direction on the $(\bar{1}11)$ plane?

Procedure. Draw a sketch to locate $[001] \not\sphericalangle [\bar{1}11]$ and $[001] \not\sphericalangle [101]$.

Calculations

a)

$$\cos \phi = \frac{\text{Edge of unit cell}}{\text{Body diagonal of unit cell}} = \frac{a}{a\sqrt{3}} = 0.577;$$

$$\cos \lambda = \frac{\text{Edge of unit cell}}{\text{Face diagonal of unit cell}} = \frac{a}{a\sqrt{2}} = 0.707;$$

$$s = \frac{F}{A} = \frac{\tau}{\cos \phi \cos \lambda} = \frac{1 \text{ MPa}}{(0.577)(0.707)} = 2.45 \text{ MPa} \quad (= 355 \text{ psi}).$$

b) By inspection, $[001] \not\sphericalangle [110] = 90°$, therefore $\cos \lambda = 0$, and $s = \infty$. (Slip cannot occur in this direction when the stress is applied in the [001] direction.)

Comment. The dot product is useful here but can only be used for *cubic* crystals.   ◀

● **Example 5–4.3**   With a sketch show the 12 slip systems that are included in the $\langle \bar{1}10 \rangle \{111\}$ slip systems of fcc metals.

Answers. See Fig. 5–4.7. Each of the four planes of the {111} family has three slip directions. Thus we have

$$[\bar{1}10](111) \quad [1\bar{1}0](11\bar{1}) \quad [110](1\bar{1}1) \quad [110](\bar{1}11)$$

$$[10\bar{1}](111) \quad [101](11\bar{1}) \quad [10\bar{1}](1\bar{1}1) \quad [101](\bar{1}11)$$

$$[0\bar{1}1](111) \quad [011](11\bar{1}) \quad [011](1\bar{1}1) \quad [01\bar{1}](\bar{1}11).$$

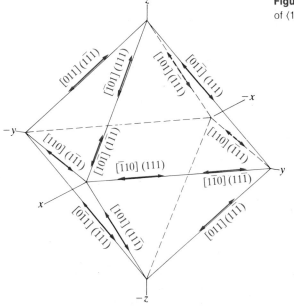

Figure 5–4.7 The 12 slip systems of $\langle 1\bar{1}0 \rangle \{111\}$. (See Example 5–4.3.)

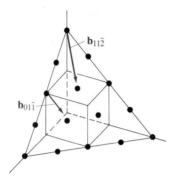

Figure 5–4.8   See Example 5–4.4.

Comment. The rear planes are parallel to the front planes and therefore involve the same slip systems, for example, $[110](\bar{1}11)$ is the same slip system as $[110](1\bar{1}1)$. Likewise, $[\bar{1}10](111)$ and $[1\bar{1}0](111)$ are the same slip system, but of opposite sense. ◀

• **Example 5–4.4**

Both $[01\bar{1}]$ and $[11\bar{2}]$ lie in the (111) plane of fcc aluminum. Therefore, both $[01\bar{1}](111)$ and $[11\bar{2}](111)$ slip are conceivable. (a) Make a sketch of the (111) plane and show the $[01\bar{1}]$ and $[11\bar{2}]$ unit slip vectors. (b) Compare the energies of the dislocation lines that have these two displacement vectors.

Solution

a) See sketch (Fig. 5–4.8).

b) Refer to Eq. (5–4.2):

$$\frac{E_{01\bar{1}}}{E_{11\bar{2}}} = \frac{lG\,\mathbf{b}_{01\bar{1}}^2}{lG\,\mathbf{b}_{11\bar{2}}^2}.$$

Both involve the same slip plane (111); therefore, the same shear modulus applies. Based on unit length l,

$$\frac{E_{01\bar{1}}}{E_{11\bar{2}}} = \left(\frac{\mathbf{b}_{01\bar{1}}}{\mathbf{b}_{11\bar{2}}}\right)^2 = \left(\frac{a/\sqrt{2}}{a\sqrt{6}/2}\right)^2 = \frac{1}{3};$$

$$E_{01\bar{1}} = \tfrac{1}{3}E_{11\bar{2}}.$$

Comment. With this $\tfrac{1}{3}$ ratio, slip occurs appreciably more readily with the first of the two potential slip systems for fcc metals. ◀

5–5

PROPERTIES OF PLASTICALLY DEFORMED METALS

Plastic deformation changes the internal structure of a metal; therefore, it is to be expected that deformation also changes the *properties* of a metal. Evidence of such property changes may be obtained through resistivity measurements. The distorted structure reduces the mean free path of electron movements (Section 2–5) and therefore increases the resistivity (Fig. 5–5.1).

A second property change that has much greater engineering significance is strength. Plastically deformed metals become stronger, as we saw in Chapter 1.

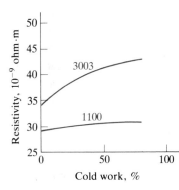

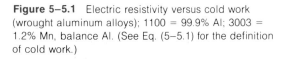

Figure 5–5.1 Electric resistivity versus cold work (wrought aluminum alloys); 1100 = 99.9% Al; 3003 = 1.2% Mn, balance Al. (See Eq. (5–5.1) for the definition of cold work.)

There, the *s/e* curve of Fig. 1–5.2(b) continued to rise after the yield strength was exceeded.

In Fig. 5–5.1, as well as in other cases, it is convenient to refer to the percent of *cold work* as an index of plastic deformation. Cold work is the amount of plastic strain introduced during processing, expressed by the percent decrease in cross-sectional area from deformation; that is,

$$\% \, CW = \left[\frac{A_o - A_f}{A_o} \right] 100, \qquad (5\text{–}5.1)$$

where A_o and A_f are the original and final areas respectively.

Strain Hardening

The copper of Fig. 5–5.2 has been plastically deformed. The deformation shows up as traces of slip planes on a previously polished surface. The effect of plastic deformation may also be revealed in an electron microscope. Figure

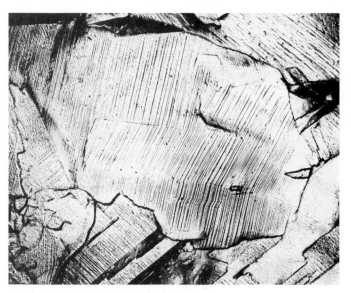

Figure 5–5.2 Plastically deformed, polycrystalline copper (×25). The traces of the slip planes are revealed at the polished surface of the metal. (National Bureau of Standards. Reproduced by permission from B. Rogers, *The Nature of Metals*, 2nd ed., American Society for Metals, and Iowa State University Press, Chapter 13.)

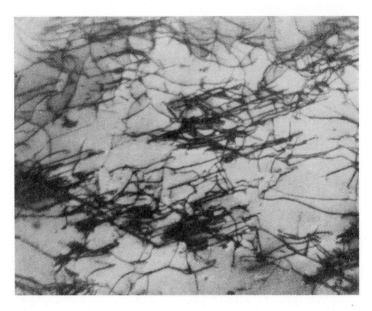

Figure 5–5.3 Dislocations in plastically deformed metal (stainless steel, ×30,000). Additional deformation introduces more dislocations. Also, additional dislocations interfere with further deformation. *Strain hardening* is the result. (M. J. Whelan.)

5–5.3 shows dislocations in a stainless steel that had been severely cold-worked. The entanglement of lines are dislocations, the numbers and lengths of which increase greatly with additional cold work. As seen in Fig. 5–5.4, the total length of dislocation lines markedly affects the shear stress. We may conclude that *although dislocations account for plastic deformation* (Fig. 5–4.5), *they interfere with the movements of other dislocations*. The dislocation

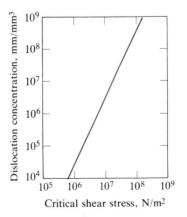

Figure 5–5.4 Dislocation density versus critical shear stress. An increased number of dislocations provides interference to dislocation movements. (Adapted from Wiedersich, *Jour. of Metals*.)

entanglements, or "traffic jams," increase the critical shear stress, τ_c, and therefore the strength of the material.

Data of the type presented in Fig. 5–5.4 are shown in a more practical engineering format in Figs. 5–5.5, 5–5.6, and 5–5.7.

The increase in hardness that results from plastic deformation is called *strain hardening*. Laboratory tests show that an increase in both tensile strength and yield strength accompanies this increase in hardness. On the other hand, strain hardening reduces ductility, because part of the deformation occurred during cold-working, before the gage marks (Fig. 1–5.1) are placed on the test bar. Thus, less ductility is observed during testing. The process of strain hardening increases yield strength more than ultimate strength (Fig. 5–5.7), and the two approach the true breaking strength as the amount of cold work is increased.

Example 5–5.1

An annealed iron rod was originally 12.3 mm (0.48 in.) in diameter. It is cold-worked by drawing it through a die with a hole that is 10.4 mm (0.41 in.) in diameter. (a) What was the ductility of the rod before drawing? (b) After drawing?

Procedure. An annealed metal has 0% CW. Calculate the cold work during drawing from the change in diameter. In both cases, obtain the ductility from Fig. 5–5.5(c).

Calculations

a) 55% Elongation (50 mm).

b) $$\text{Cold work} = \frac{(\pi/4)(12.3)^2 - (\pi/4)(10.4)^2}{(\pi/4)(12.3)^2} = 0.285 \qquad \text{(or 28.5\%)}.$$

From Fig. 5–5.5(c),

30% Elongation (50 mm).

Figure 5–5.5 Cold work versus mechanical properties (iron and copper). Cold work is the amount of plastic strain, expressed as the decrease in cross-sectional area (Eq. 5–5.1).

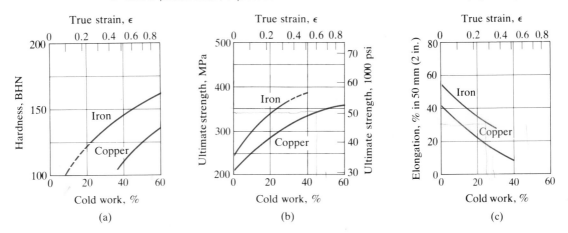

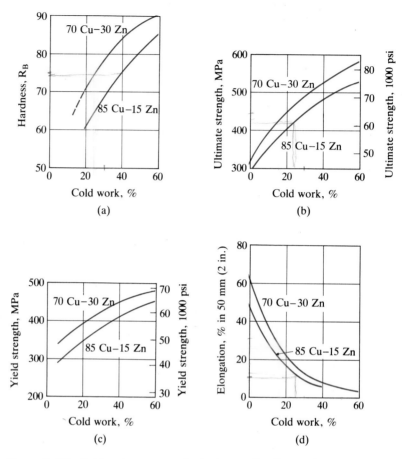

Figure 5–5.6 Cold work versus mechanical properties (brasses).

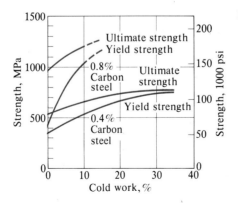

Figure 5–5.7 Cold work versus strength of plain-carbon steels.

Comment. A limited amount of extrapolation is generally permissible. Also one may interpolate between curves to obtain estimates of data for other alloys, for example, 80 Cu–20 Zn in Fig. 5–5.6, or 90 Cu–10 Zn, if Figs. 5–5.6 and 5–5.5 are used together. ◄

Example 5–5.2   A cold-worked copper or a brass may be used in an application with the specifications of $S_u \geq 345$ MPa ($\geq 50{,}000$ psi), and a ductility of greater than 20% elongation in 50 mm (2 in.). Choose a metal.

Answers. The cold-work data provide the following specification ranges:

	Copper	Brass (85–15)	Brass (70–30)
Figure	5–5.5	5–5.6	5–5.6
Ultimate strength	≥40% CW	≥9% CW	≥4% CW
Ductility (% Elong.)	≤26% CW	≤16% CW	≤23% CW
Range	—	9%–16% CW	**4%–23% CW**

Comments. The design engineer has a choice of brasses. Other factors equal, he or she would choose the higher-zinc brass (70 Cu–30 Zn) because zinc is cheaper than copper. Furthermore, the specification range of 4%–23% CW gives more flexibility in processing the metal. ◄

5–6

RECRYSTALLI-ZATION

This is a process of growing new crystals from previously deformed crystals.

Crystals that have been plastically deformed, like those shown in Figs. 5–5.2 and 5–5.3, have more energy than unstrained crystals because they are loaded with dislocations and point imperfections. Given a chance, the atoms will move to form a more perfect, unstrained array. Such an opportunity arises when the metal is heated, through the process called *annealing*. The greater thermal vibrations of the lattice at high temperatures permit a reordering of the atoms into less distorted grains. Figure 5–6.1 shows the progress of this *recrystallization*.

Brass was cold-worked 33% and therefore strain-hardened (Fig. 5–6.1a). Several samples were heated to 580°C (1076°F) for a few seconds. New grains, i.e., new fcc crystals of brass, are detected in the sample that was heated for only 3 seconds (Fig. 5–6.1b). The initial recrystallization started along traces of the earlier slip planes. In 4 seconds, the brass has been nearly half recrystallized (Fig. 5–6.1c), and it is completely recrystallized in 8 seconds (Fig. 5–6.1d). Were we to examine the new crystals of that last figure by an electron microscope, we would observe a greatly reduced number of dislocations. Also the hardness has dropped significantly from 165 BHN initially to less than 100 BHN.

(a) Cold-worked 33%   (b) 3 sec at 580 °C (1076 °F)

(c) 4 sec at 580 °C   (d) 8 sec at 580 °C

Figure 5–6.1 Recrystallization of strain-hardened brass (×40). (Courtesy of J. E. Burke, General Electric Co.)

Recrystallization Temperatures

The recrystallization process requires atom movements and rearrangements. These rearrangements for recrystallization occur more readily at high temperatures. In fact, we will observe in Fig. 5–6.2 that a marked decrease in strength has occurred in a sample held for one hour at 300°C (570°F). Initially cold-worked 75%, this sample was almost completely recrystallized in that period of time. In contrast, samples held for an hour at temperatures below 200°C (~400°F) retained almost all of their higher strength, which was obtained during the 75% cold work. We thus speak of a *recrystallization temperature*, T_R, in this case approximately 270°C where the microstructure and strength change drastically.*

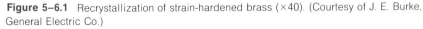

* There is a slight loss of strength and a major recovery of electrical conductivity at temperatures slightly below the recrystallization temperature. Called *recovery*, these lower-temperature changes come about because the point imperfections (vacancies, interstitials, etc., that were introduced by cold work) move to dislocation edges within the strained crystals. The point imperfections do not affect deformation significantly; therefore, only slight softening occurs during this recovery stage. The mean free paths of the electrons are significantly lengthened by the disappearance of the point defects; therefore, the resistivity drops with their removal.

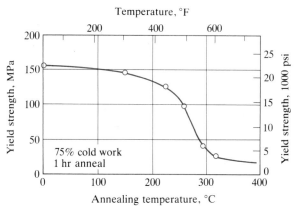

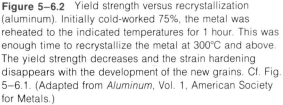

Figure 5–6.2 Yield strength versus recrystallization (aluminum). Initially cold-worked 75%, the metal was reheated to the indicated temperatures for 1 hour. This was enough time to recrystallize the metal at 300°C and above. The yield strength decreases and the strain hardening disappears with the development of the new grains. Cf. Fig. 5–6.1. (Adapted from *Aluminum*, Vol. 1, American Society for Metals.)

The temperature for recrystallization is dictated by several factors. A "ball-park" figure is that T_R lies between three tenths and six tenths of the absolute, K, melting temperature, that is, $0.3\,T_m$ to $0.6\,T_m$. The rationale for this generality is that the diffusivity D for self-diffusion is directly related to the melting temperature for the metal.* However, *time* is a second factor that affects recrystallization temperatures. For example, the recrystallization of a commercially pure aluminum alloy that had been cold-worked 75% may be completed in 1 minute at 350°C (623 K and 662°F), but requires 60 minutes at 300°C and 40 days at 230°C. We expect this, because the flux of atoms is proportional to the diffusivity (Eq. 4–7.1a), which in turn is a function of temperature (Eq. 4–7.3).

A third factor that affects the recrystallization temperature is the amount of *strain hardening*. As shown by the hardness data in Fig. 5–6.3, the recrystallization temperature drops from above 320°C (>600°F) for a 65–35 brass with 20% cold work to approximately 280°C (~535°F) for the same brass with 60% cold work (each with one hour of annealing time). An explanation takes cognizance of the fact that a highly strain-hardened metal has more stored energy in the form of vacancies and dislocations (Fig. 4–4.3) than does one with little cold work. With this energy available, it takes less thermal energy for atoms to rearrange themselves into an annealed grain. That is, recrystallization can occur at lower temperatures.

Finally, *pure* materials recrystallize at lower temperatures than solid solutions. Thus, highly pure electrical copper wire anneals much more readily than a comparably deformed brass wire, even though the melting temperature would predict a small opposite relationship. The rationale, which involves dislocation movements, will not be detailed here.

---

* A solution of Study Problem 4–7.14 gives diffusivity values at $0.6\,T_m$ for iron, copper, and silver of $10^{-17.5}$, $10^{-17.3}$, and $10^{-17.4}$ m²/s, respectively. (Each of these calculations were for fcc structures and, therefore, can be compared directly.)

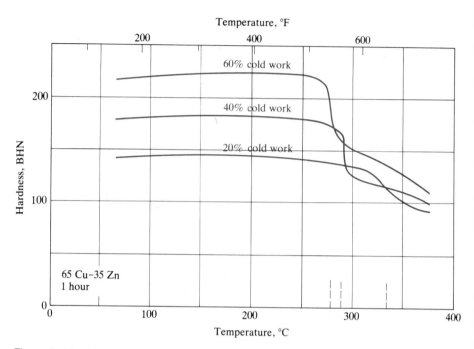

Figure 5–6.3 Softening during annealing (65Cu–35Zn brass). The hardnesses were determined at 20°C after heating to the indicated temperatures for 1 hour. The more highly strain-hardened brass softens at a lower temperature and with less thermal energy. (ASM data.)

Recrystallization Rates

The progress of recrystallization typically follows an S-shaped, or *sigmoidal,* curve. For example, the 3.25%-Si electrical steel of Fig. 5–6.4 shows undetectable recrystallization in the first 100 s at 600°C, but is 50% recrystallized in 20 min (1200 s). More than 2 hr ($\sim 10^4$ s) are required for completion at this temperature.

Recrystallization proceeds similarly at the other temperatures of Fig. 5–6.4, but with different time frames. Observe that the 50% recrystallization time is relatively easily identified because the reaction is most rapid at that point. When we relate the time of that reaction point to the recrystallization temperature, we observe an Arrhenius behavior (Fig. 5–6.4b). This is expected because atom movements govern the reaction, and they, in turn, are dependent upon thermal activation.

The mathematical relationship in Fig. 5–6.4(b) is

$$\ln t = C + B/T \tag{5–6.1}$$

where C and B are constants. We may relate this equation to Eq. (4–7.4) if we recognize that a fast reaction rate R (a short time, since $R = 1/t$) corresponds to rapid diffusion. Thus

$$\ln R = \ln R_0 - E/kT = -\ln t, \tag{5–6.2}$$

where C and B of Eq. (5–6.1) are $-\ln R_0$ and E/k, respectively.

Hot-working versus Cold-working of Metals

In the production operations described in Figs. 5–2.2 and 5–2.3, the distinction between *hot-working* and *cold-working* does not rest on temperature alone, but on the relationship of the processing temperature to the recrystallization temperature. Hot-working refers to shaping processes that are performed above the recrystallization temperature; cold-working refers to shaping processes that are performed below it. Thus the temperature for cold-working copper may be higher than that for hot-working lead.

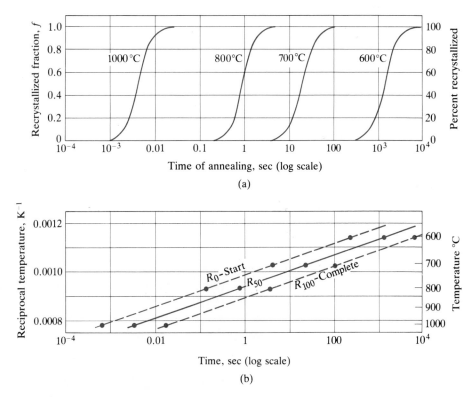

Figure 5–6.4 Recrystallization time (3.25% Si, electrical steel). (a) The rate increases (time shortens) at higher temperatures. (Adapted from data by G. R. Speich and R. M. Fisher, U.S. Steel Corp.) (b) Arrhenius plot. R_0: start of recrystallization; R_{50}: 50 percent recrystallized; R_{100}: recrystallization completed. (R_{50} is usually the more precise value, since R_0 and R_{100} are asymptotes.)

The choice of the recrystallization temperature as the point for distinguishing between hot- and cold-working is quite logical from the production point of view. Below the recrystallization temperature, the metal becomes harder and less ductile with additional deformation during processing. More power is required for deformation and there is a greater chance for cracking during the process. Above the recrystallization temperature, the metal will anneal itself during, or immediately after, the mechanical working process. Thus, it remains soft and relatively ductile during processing.

Engineering Significance of Cold-working and Annealing

Strain hardening by cold work is of prime importance of the design engineer. It permits the use of smaller parts with greater strength. Of course, the product must not be used at temperatures that will anneal the metals.

Cold work reduces the amount of plastic deformation that a metal can subsequently undergo during a shaping operation. The hardened, less ductile, cold-worked metal requires more power for further working and is subject to cracking. Therefore, *cold-work–anneal cycles* are used to assist production. Example 5–6.1 describes such a process.

The loss of ductility during cold-working has a useful side-effect in machining. With less ductility, the chips break more readily (Fig. 5–6.5), thus facilitating the cutting operation. As a result, the engineer may commonly specify cold-working for "screw-stock" metal, i.e., metals that require cutting in automatic screw machines.

Example 5–6.1

A 70–30 brass rod is required to have a diameter of 5.0 mm (0.197 in.), an ultimate strength of more than 420 MPa (61,000 psi), and a 50-mm elongation of more than 18%. The rod is to be drawn from a 9.0-mm (0.355-in.) diameter rod that had been annealed. Specify the final processing steps for making the 5.0-mm rod.

(a)   (b)

Figure 5–6.5 The cutting of metal turnings by a machine tool. The strain-hardened metal in (a) formed the more desirable chips, while the annealed metal in (b) formed continuous turnings. (Hans Ernst, Cincinnati Milacron, Inc.)

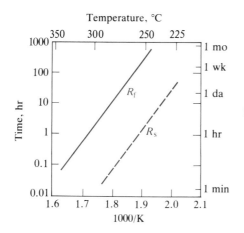

Figure 5–6.6 Recrystallization time versus temperature (aluminum, 75% cold-worked). Dashed line: start of recrystallization, R_s. Solid line: recrystallization finished (Example 5–6.2). An Arrhenius relationship (ln t versus $1/T$) is followed because recrystallization requires atom movements. (Adapted from *Aluminum*, Vol. 1, American Society for Metals.)

Procedure. To "meet the spec," there must be at least 16% CW for S_u, and no more than 24% CW for ductility (Figs. 5–5.6b,d). We can use 20%. The *last* step should use an annealed bar, diameter = d that has its area reduced 20% to diameter = 5.0 mm.

Calculation

$$CW = 0.20 = \frac{d^2\pi/4 - (5.0)^2\pi/4}{d^2\pi/4}$$

$$d = 5.6 \text{ mm} \qquad\qquad (= 0.22 \text{ in.})$$

Comments. Hot work from 9.0 mm to 5.6 mm (or cold work and anneal in one or more cycles). The rod should be in the annealed condition at 5.6-mm diameter. Cold draw 20% to 5.0-mm diameter. ◄

Example 5–6.2

The time–temperature relationship for the completed recrystallization of the aluminum in Fig. 5–6.6 follows an Arrhenius-type relationship. Establish an appropriate empirical equation.

Procedure. The Arrhenius relationship is

$$\ln t = C + B/T. \qquad\qquad (5\text{–}6.1)$$

We need to determine the constants C and B. Pick t and T for points near the two ends of the R_f curve in Fig. 5–6.6. Solve simultaneously for C and B.

Calculations

$$T = 250°C;\ t = 200\text{hr}: \qquad \ln 200 = C + B/523.$$
$$T = 327°C;\ t = 0.14 \text{ hr}: \qquad \ln 0.14 = C + B/600.$$
$$C = -52; \qquad B = 30,000 \text{ K};$$
$$\ln t = -52 + 30,000 \text{ K}/T.$$

Comment. Just as the various diffusion couples of Table 4–7.2 had their own values of D_0 and E, these values of C and B apply only to *this* commercially pure aluminum that had been cold-worked 75%. ◄

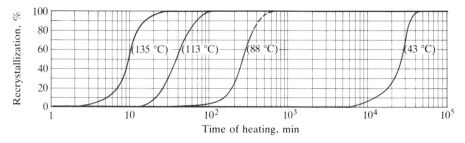

Figure 5–6.7 Isothermal recrystallization of (99.999% pure copper, cold-rolled 98%). Refer to the study problems. (Adapted from A. G. Guy and J. J. Hren, *Elements of Physical Metallurgy*, Addison-Wesley, after Decker and Harker.)

5–7

BEHAVIOR OF POLYCRYSTALLINE METALS

The solidified metal in a conventional casting (Fig. 5–7.1a) contains many grains, each an individual crystal. The same is true for rods, bars, sheets, tubes, etc. The size of the grains has an effect on the deformation of material. In brief, those metals with small grains are *stronger at low temperatures* and *weaker at elevated temperatures* than are metals with large grains. In each case, the effect is primarily a consequence of the amount of grain boundary area. Recall from Section 4–4 that there is an inverse relationship between the average grain dimension and the amount of boundary area.

Since the engineer has some control over the grain size, we will look first at grain growth, then at the effects of grain sizes on deformation both at low temperatures and at high temperatures.

Grain Growth in Metals

The average grain size of a single-phase metal increases with time if the temperature produces significant atom movements (Section 4–7). The driving force for *grain growth* is the energy released as an atom moves across the boundary from the grain with the convex surface to the grain with the concave surface. There,

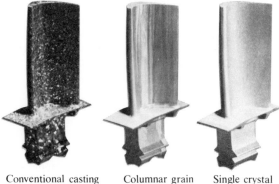

Conventional casting   Columnar grain   Single crystal
     (a)   (b)   (c)

Figure 5–7.1 Advances in turbine airfoil materials (cast turbine blades). (a) Since grain boundaries permit strain at high temperatures by creep, conventional processing specifies coarse grains (and reduced grain-boundary area). (b) A more sophisticated process that not only directionally solidifies the metal but also lowers the area of the boundaries further (and aligns them parallel to the centrifugal forces). (c) Advanced (and more difficult) procedures eliminate all grain boundaries by producing a blade out of a single crystal of metal. (Courtesy of M. Gell, D. H. Duhl, and A. F. Giamei, Pratt and Whitney Aircraft, *Superalloys 1980*, ASM.)

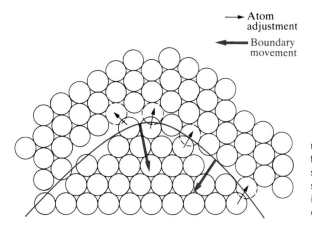

⟶ Atom
adjustment

⟵ Boundary
movement

Figure 5–7.2 Grain-boundary movement. The atoms move to the grain with the concave surface, where they are more stable. As a result the boundary is shifted toward the center of curvature.

the atom is, on the average, coordinated with a larger number of neighbors at equilibrium interatomic spacings (Fig. 5–7.2). As a result, the boundary moves toward the center of curvature. Since small grains tend to have surfaces of sharper convexity than large grains, they disappear because they feed the larger grains (Fig. 5–7.3). The net effect is grain growth (Fig. 5–7.4).

All crystalline materials, metals and nonmetals, exhibit this characteristic of grain growth. An interesting example of grain growth can be seen in the ice of a snow bank. Snowflakes start out as numerous small ice crystals, lose their identity with time, and are replaced by larger granular ice crystals. A few of the crystals grow at the expense of the many smaller crystals.

The growth rate depends heavily upon temperature. An increase in temperature increases the thermal vibrational energy, which in turn *accelerates* the net diffusion of atoms across the boundary from small to large grains. A subsequent decrease in temperature slows down the boundary movement, but *does not reverse it*. The only way to reduce (refine) the grain size in alloys that have only a single phase is to plastically deform the grains and start new grains by recrystallization (Section 5–6).

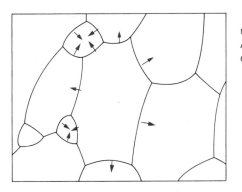

Figure 5–7.3 Grain growth. The boundaries move toward the center of curvature (arrows). As a result the small grains eventually disappear.

(a) 15 min at 580 °C   (b) 1 hr at 580 °C (1076 °F)

(c) 10 min at 700 °C   (d) 1 hr at 700 °C (1292 °F)

Figure 5–7.4 Grain growth (brass at ×40). (Courtesy of J. E. Burke, General Electric Co.)

Deformation at Low Temperatures

Figure 5–7.5 plots the hardness of an annealed 70–30 brass as a function of grain-boundary area. The boundaries interfere with slip because the dislocation movements must be redirected as they try to enter the new grain. The adjacent grain is invariably tilted or rotated with respect to the grain undergoing slip. (If the two were not misaligned, there would be no boundary.) Thus, it takes a greater force to continue the slip across a grain boundary. Commonly, there is a

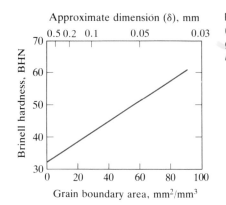

Figure 5–7.5 Hardness versus grain growth (annealed 70–30 brass). Larger grains with less grain boundary area are softer. (After F. Rhines, *Metal Progress*.)

Figure 5–7.6 Dislocation pile-up. A boundary or surface hinders continued dislocation movements at low temperatures. ⊥ = edge dislocation.

pile-up of dislocations that have the same effect as a traffic jam on an arterial highway (Fig. 5–7.6).

• Polycrystalline metals deform differently from single crystals of metals in another respect. The single crystal of Fig. 5–4.1 was not constrained by adjacent crystals. In contrast, observe Fig. 5–5.2 where it is evident that the large grain of copper in the center did not yield independently of its neighboring grains. The metallurgist can show that at least five slip systems (Table 5–4.1) must operate simultaneously within a grain if that grain is to deform in concert with its neighbors and not introduce cracks or gaps. Since not all slip planes are favorably oriented (Example 5–4.2b), we can appreciate why bcc and fcc metals, with a larger number of slip systems, are ductile metals; and hcp metals are less ductile (Table 5–4.1).

Deformation at High Temperatures

We just observed in Fig. 5–7.5 that fine-grained metals are harder and therefore stronger than coarse-grained metals—*at low temperatures*. At elevated temperatures, the situation is reversed. Above temperatures at which atoms start to move significantly, the grain boundary is a source of weakness to a material. We may understand this better with Fig. 5–7.7, which shows schematically

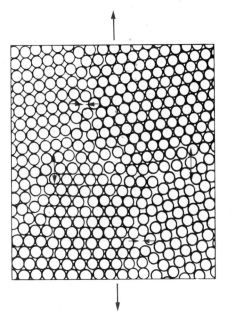

Figure 5–7.7 Grain boundaries and deformation (schematic). Under tension, vertical boundaries are crowded and lose atoms by diffusion; the horizontal boundaries receive relocated atoms because they develop extra space. The resulting creep becomes a significant factor in the use of materials at high temperatures, where the atoms diffuse readily. (Base sketch from Fig. 4–4.7.)

several grains loaded vertically in tension. With tension in one direction, there is a perpendicular contraction. (Cf. Poisson's ratio, Section 5–3.) Thus, atoms along vertically oriented boundaries are crowded; atoms along horizontally oriented boundaries have an increased amount of space. This induces diffusion from the vertical boundaries to the horizontal boundaries, with the net effect that there is a gradual change in the shape of the metal.*

The strain just described is one mechanism of *creep*. With smaller grains and, therefore, more grain boundary area, creep is more rapid. There are more "sinks" for atoms along horizontal boundaries, and more "sources" of atoms from vertical boundaries (Fig. 5–7.7). Equally important is the fact that the diffusion distances are shorter in fine-grained materials.

Of course, this mechanism of creep does not occur at low temperatures where there are negligible atom movements, but it does increase exponentially with the increases in the values of diffusivity, D, for self-diffusion (Section 4–7). As with recrystallization, the temperature for the reversal of these grain-size effects is a function of time, bond strength, impurities, etc. Grain size and shape control offers the engineer a means of minimizing creep (Fig. 5–7.1).

Review and Study

SUMMARY

1. *Single-phase alloys* are metallic solid solutions that contain only one crystal structure. They are stronger and harder than either component alone. Conversely, solid solution alloys have lower electrical conductivity (and thermal conductivity since the two are related). Microstructural variables within single-phase metals include grain *size*, grain *shape*, and crystal *orientation*.

• 2. Processing of single-phase metals includes *melting* and *casting*, and *mechanical working*. The latter involves plastic deformation.

3. Analysis of elastic deformation involves not only Young's modulus E (Section 1–5) but also a *shear modulus, G,* and the *bulk modulus, K,* for shear stresses and hydrostatic stresses respectively. We may relate the three by using *Poisson's ratio, ν.* The elastic moduli are anisotropic and usually are lower at elevated temperatures.

4. Plastic deformation occurs in crystals by *dislocation movements* along slip planes. *Schmid's law* lets us resolve an axial stress into a shear stress on a slip plane. The *slip systems* with the lowest *critical shear stresses* are in the most densely packed directions on the most densely packed planes. *Solution hardening* results because dislocations become anchored to solute atoms.

5. *Strain hardening* develops from the entangled dislocations that accumulate during *cold working*. Strength, hardness, and electrical resistivity increase; also, ductility

---

* Compressive forces induce similar, but opposite, shape changes.

and toughness are reduced. We index the amount of cold work by the percent change in cross-sectional area.

6. During *annealing,* a strain-hardened (cold-worked) metal *recrystallizes.* The *recrystallization temperature* typically lies between 0.3 and 0.6 of the absolute melting temperature, T_m, and depends upon: i) available time, ii) the amount strain hardening, and iii) the purity of the metal, each of which facilitates recrystallization. In a *hot-working* process, the temperature is high enough for recrystallization to be nearly simultaneous with deformation. *Cold-work–anneal* cycles are used in production to accomplish extensive reshaping and to avoid cracking from loss of ductility.

7. There is *grain growth* in all materials. This may be infinitely slow at ambient temperatures, but it becomes very significant at elevated temperatures. At low temperatures with low atom mobility, *fine-grained* metals are stronger and less ductile than coarse-grained metals, because the grain boundaries restrict slip. At elevated temperatures where the atoms can move to and from the grain boundaries, the fine-grained materials undergo *creep.* Thus, coarse-grained, or even single-crystal, parts are specified for critical high-temperature service.

TERMS AND CONCEPTS

Alloy   A metal containing two or more elements.

Anisotropic   Having different properties in different directions.

Annealing   Heating and cooling to produce softening.

Brass   A copper-based alloy containing zinc.

• **Casting**   The process of pouring a liquid or suspension into a mold, or the object produced by this process.

Cold work, %   Amount of cold working calculated from the change in cross-sectional area, $100(A_o - A_f)/A_o$.

Cold-working   Plastic deformation below the recrystallization temperature.

• **Creep**   Time-dependent strain occurring under stress (commonly at elevated temperatures).

• **Creep rate**   Creep strain per unit of time.

Critical shear stress   Minimum resolved stress to produce shear.

Deformation, elastic   Reversible deformation without permanent atomic (or molecular) displacements.

Deformation, plastic   Permanent deformation arising from the displacement of atoms (or molecules) to new surroundings.

Dislocation   See Chapter 4.

• **Drawing**   Mechanical forming by tension through or in a die, e.g., wire drawing (Fig. 5–2.3) and sheet drawing; usually carried out at temperatures below the recrystallization temperature.

• **Extrusion**   Shaping by pushing the material through a die (Fig. 5–2.2).

• **Forging**   Mechanical shaping by compression (Fig. 5–2.2).

Grain growth   Increase in average grain size by atoms diffusing across grain (or phase) boundaries.

Hot-working   Deformation that is performed above the recrystallization temperature, so that annealing occurs concurrently.

• **Ingot**   A large casting suitable for subsequent rolling or forging.

Microstructure Structure of grains and phases. Generally requires magnification for observation.

Modulus, bulk (K) Hydrostatic pressure per unit volume strain.

Modulus, shear (G) Shear stress per unit shear strain.

Poisson's ratio (ν) Absolute ratio of lateral to axial strain.

Polycrystalline Materials with more than one crystal, and therefore with grain boundaries.

Preferred orientation A nonrandom alignment of crystals or molecules.

Recovery Loss of resistivity, or related behaviors that originated during strain hardening, by annealing out point defects. (Cf. recrystallization.)

Recrystallization The formation of new annealed grains from previously strain-hardened grains.

Recrystallization temperature Temperature above which recrystallization is spontaneous.

Resolved shear stress Stress vector in slip plane.

• **Rolling** Mechanical working through the use of two rotating rolls (Fig. 5–2.2).

Schmid's law Relates axial stress to resolved shear stress (Eq. 5–4.1).

Shear strain (γ) Tangent of shear angle α developed from shear stress.

Shear stress (τ) Shear force per unit area.

Slip direction Crystal direction of the displacement vector in which translations of slip takes place.

Slip plane Crystal plane along which slip occurs.

Slip system Combination of slip directions on slip planes that have low critical shear stresses.

Slip vector Displacement distance of a dislocation. It is parallel to a screw dislocation and perpendicular to an edge dislocation.

Solution hardening Increased strength arising from the creation of solid solutions (from pinning of dislocations by solute atoms).

Spinning (metals) Mechanical working of sheet metal over a mandrel (Fig. 5–2.3).

Strain hardening Increased hardness (and strength) arising from plastic deformation below the recrystallization temperature.

FOR CLASS DISCUSSION

A_5 Distinguish between grain shape and grain orientation in single-phase materials.

B_5 Why do alloys of copper and nickel have lower conductivity than either metal alone?

• C_5 Compare the mechanical forming processes of Figs. 5–2.2 and 5–2.3. Which will more uniformly deform the metal? Least?

D_5 Distinguish between the three moduli of elasticity.

E_5 The engineer commonly defines the stress for yield strength after an 0.2% offset strain has occurred (Fig. 1–5.2b). Based on Fig. 5–3.4, why is this necessary?

F_5 Assume a material deforms elastically in tension with no change in volume, that is, $(1 + e_x)(1 + e_y)(1 + e_z) = 1$. Calculate Poisson's ratio.

G_5 Alloying elements affect elastic moduli only slightly if at all. (Cf. copper and brass in Appendix C.) However, the strengths are significantly affected (Fig. 5–1.1). Discuss reasons for this difference.

H_5 At what combination of angles λ and ϕ (Eq. 5–4.1) is the resolved shear stress the greatest?

I_5 Differentiate between the terms, resolved shear stress, and critical shear stress.

J_5 Alloys of zinc and of magnesium are more widely used as casting alloys than as wrought (plastically deformed) alloys. Suggest a valid reason.

K_5 Why are alloys of copper and nickel stronger than either metal alone?

L_5 Impurity atoms inhibit dislocation movements. Explain.

M_5 A rod is cold-drawn through a die to twice its original length. Why is its ductility decreased? How much is it cold worked? Will its percent reduction in area be increased or decreased? (This question was asked in order to review the difference between *lengthening* as a process and *elongation* as a property, and between decrease in area arising from *cold work* and *reduction-in-area* as a property.)

N_5 Use the data of Figs. 5–1.1(b), 5–5.5, and 5–5.7. Sketch a plausible curve of yield strength versus cold work for copper.

O_5 How will longer times (>1 hour) affect the curves of Fig. 5–6.3? How will the curves differ for an 85–15 brass?

P_5 As a rule of thumb, one may consider that the recrystallization temperature is one half of the melting temperature. Why must this rule be taken with caution?

Q_5 Distinguish between hot-working and cold-working of tin; of tungsten.

R_5 Explain why a single-crystal metal is softer and more plastic than a polycrystalline metal.

QUIZ SAMPLES

5A The conductivity ratios, k/σ, of metals generally ___+−___ a temperature increase.
(a) increases with   (b) remains unchanged as   (c) inverts with
(d) decreases with   (e) is proportional to

5B The mean free path of electron movements increases with ___**-er___ temperatures, ___**-er___ purity, and ___**-er___ dislocations.

• 5C Heating is required for ___**___ mechanical working, which includes rolling, ___**___, and ___**___ .

5D ___**___ ratio equals ___**___ strain per unit axial strain. Its maximum value of 0.5 implies constant ___**___ during axial loading.

5E Shear strain is reported as the ___**___ of the elastic angular displacement.

5F The bulk modulus of elasticity is the ratio of ___**___ to ___**___ .

5G Possible slip systems in fcc metals include (111)[uvw] where [uvw] may be ___+−___ .
(a) [111]   (b) [1$\bar{1}$0]   (c) [$\bar{1}$01]   (d) [011]   (e) [001]   (f) [01$\bar{1}$]
(g) [101]

5H A force of 3N is applied in the [102] direction of a cubic crystal. The resolved force in the [001] direction is ___**___ N.

5I A 1-mm wire contains a single crystal across its full diameter with the [111] direction parallel to the wire's length. The wire is stressed in loading to 10 MPa. The maximum resolved shear stress on a {100} plane is ___**___ MPa.

5J   Plastic deformation introduces ___**___ as imperfections. When these are present, there is a decrease in _+−_ .

(a) thermal conductivity   (b) yield strength   (c) hardness   (d) ductility
(e) electrical resistivity   (f) None of the above, rather _____

5K   A 1.1-mm diameter wire is cold drawn through a 1.0-mm die. It is cold worked _+−_ .

(a) 10%   (b) 0.1/1.1   (c) 21%   (d) None of the above, rather _____

5L   A plate of copper must have hardness >125 BHN and a ductility >20% El (50 mm). Its specification window for cold working is _+−_ .

(a) 20 to 30%   (b) 25% to 50%   (c) best at ≈40%   (d) None of these, rather _____   (e) Impossible to meet specifications by cold working.

5M   Other factors equal, recrystallization occurs in shorter times when the metal _+−_ .

(a) has higher purity   (b) is strain hardened less   (c) is heated slowly
(d) has a high melting temperature

5N   In order to have a linear plot of recrystallization time as a function of temperature, we must plot _+−_ .

(a) t vs. K   (b) ln t vs. K   (c) log t vs. K   (d) D vs. K^{-1}
(e) None of the above, rather _____

5O   Grain boundaries move toward their center of ___**___ . Therefore, ___**___ grains get larger.

5P   The average grain size of a metal **-creases rapidly at high temperatures; it **-creases at intermediate temperatures.

5Q   Higher engineering design stresses are possible at elevated temperatures, if the grain size is ___**___ . At ambient temperatures, the opposite is true because atoms cannot ___**___ .

STUDY PROBLEMS

5-1.1   Based on Fig. 5–1.1, what is the electrical resistivity of annealed 70–30 brass?
Answer: 60×10^{-9} ohm·m (or 60 ohm·nm)

5-1.2   Refer to Fig. 5–1.1. Can the same k/σ ratio be used for brasses as for pure metals?

5-1.3   Estimate the thermal conductivity of copper at 90°C if, as in Example 5–1.2, the k/σ ratio is proportional to temperature. ($\sigma_{90} = 22 \times 10^{-9}$ ohm·m.)
Answer: 380 $(W/m^2)/(°C/m)$ [or 0.38 $(W/mm^2)/(°C/mm)$]

5-1.4   A copper wire has a resistance of 0.5 ohm per 100 m. Consideration is being given to the use of a 75–25 brass wire instead of copper. What would be the resistance, if the brass wire were the same size?

5-1.5   A brass alloy is to be used in an application that will have ultimate strength of more than 275 MPa (40,000 psi) and an electrical resistivity of less than 50×10^{-9} ohm·m (resistivity of Cu = 17×10^{-9} ohm·m). What percent zinc should the brass have?

Answer: 15% to 23% zinc

5–1.6 A copper alloy (brass, or Cu–Ni) is required that has an ultimate strength of at least 245 MPa (35,000 psi) and a resistivity of less than 40×10^{-9} ohm·m. (a) Select a suitable alloy bearing in mind that $\$_{Ni} > \$_{Cu} > \$_{Zn}$. (b) What is the ductility of the alloy you chose?

5–1.7 A certain application requires a piece of metal having a yield strength greater than 100 MPa and a thermal conductivity greater than 0.04 (W/mm²)/(°C/mm). Specify either an annealed brass or an annealed Cu–Ni alloy that could be used.

Answer: Use 80 Cu–20 Ni

5–1.8 A motorboat requires a seat brace. Iron is excluded because it rusts. Select the most appropriate alloy from Figs. 5–1.1 and 5–1.2. The requirements include an ultimate strength of at least 310 MPa (45,000 psi); a ductility of at least 45% elongation (in 50 mm); and low cost. (*Note:* Zinc is cheaper than copper.)

5–1.9 A brass wire must carry a load of 45 N (10 lb) without yielding and have a resistance of less than 0.033 ohm/m (0.01 ohm per foot). (a) What is the smallest wire that can be used if it is made of 60–40 brass? (b) 80–20 brass? (c) 100% Cu?

Answer: (a) 1.8-mm (0.07-in.) dia. (b) 1.4-mm (0.055-in.) dia. (c) 1.2-mm (0.05-in.) dia.

5–2.1 Two 20-m (65-ft) aluminum rods are each 14.0 mm (0.551 in.) in diameter. One bar is drawn through a 12.7-mm (0.50-in.) die. (a) What are the new dimensions of that bar? (b) Assume identical test samples are made from each bar (deformed and undeformed) and marked with 50-mm gage lengths. Which one, if either, of the bars will have the greater ductility? The greater yield strength?

Answer: (a) 24.3 m (b) Review Section 1–5.

5–3.1 What is the volume change of a rod of brass when it is loaded axially by a force of 233 MPa (33,800 psi)? Its yield strength is 270 MPa (39,200 psi), and its Poisson's ratio is 0.3.

Answer: 0.085 v/o

5–3.2 What is the bulk modulus of the brass in the previous problem?

5–3.3 A test bar 12.83 mm (0.5051 in.) in dia. with a 50-mm gage length is loaded elastically with 156 kN (35,000 lb) and is stretched 0.356 mm (0.014 in.). Its diameter is 12.80 mm (0.5040 in.) under load. (a) What is the bulk modulus of the bar? (b) The shear modulus?

Answer: (a) $\sim 1.5 \times 10^5$ MPa ($\sim 20 \times 10^6$ psi) (b) 65×10^3 MPa (10^7 psi)

5–3.4 Copper that has a modulus of elasticity of 110,000 MPa (16,000,000 psi) and a Poisson's ratio of 0.3 is under a hydrostatic pressure of 83 MPa (12,000 psi). What are the dimensions of the unit cell?

5–3.5 If copper has an axial stress of 97 MPa (14,000 psi), what will be the highest local stress within a polycrystalline copper bar?

Answer: 170 MPa (24,500 psi)

5–3.6   A precisely ground cube (100.00 mm)^3 of an aluminum alloy is compressed in the x-direction 70 MPa. (a) What are the new dimensions if Poisson's ratio is 0.3? (b) A second compression of 70 MPa is concurrently applied in the y-direction. Now, what are the dimensions? (c) Derive Eq. (5–3.6). (*Hint:* Consider the cube under a stress $s_x = s_y = s_z = P_h$ and recall that $(1 + \Delta V/V) = (1 + \Delta l/l)^3 = 1 + 3e + \cdots .$)

5–3.7   When iron is compressed hydrostatically with 205 MPa (30,000 psi), its volume is changed by 0.13%. How much will its volume change when it is stressed axially with 415 MPa (60,000 psi)?

Answer: 0.09 v/o

5–4.1   A force of 660 N (150 lb$_f$) is applied in the [111] direction of a cubic crystal. What is the resolved force in the [110] direction?

Answer: 540 N (120 lb$_f$)

5–4.2   What force must be applied in the [112] direction of copper to produce a resolved force of 410 N (92 lb$_f$) in the [110] direction?

5–4.3   A force of 3800 N (850 lb$_f$) occurs in the [111] direction of tin. What is the resolved force in the [110] direction? (Tin is bct with $a = 0.5820$ nm and $c = 0.3175$ nm.)

Answer: 3550 N (795 lb$_f$)

5–4.4   An axial stress of 123 MPa (17,800 psi) exists in the [110] direction of bcc iron. What is the resolved shear stress in the [101] direction on the (010) plane?

Answer: 43.5 MPa (6300 psi)

5–4.5   An aluminum crystal slips on the (111) plane and in the [1$\bar{1}$0] direction with a 3.5 MPa (500 psi) stress that was applied in the [1$\bar{1}$1] direction. What is the critical resolved shear stress?

Answer: 0.95 MPa (138 psi)

5–4.6   A single crystal lies completely across a copper wire (1.0-mm dia) that is being pulled in tension. The loading direction is lengthwise of the wire and very close to the [$\bar{1}$01] direction. The load is 1 kg. (a) What is the axial stress on a plane that cuts directly across the wire? (b) What is the shear stress in the [$\bar{1}$10] direction on the ($\bar{1}$11) plane?

5–4.7   Refer to Example 3–7.3. (a) If there are 8.2×10^{12} atoms/mm^2 on the {100} planes of lead, how far apart are the adjacent planes? (b) If there are 9.4×10^{12} atoms/mm^2 on the {111} planes of lead, how far apart are the adjacent planes? Solve via atoms/mm^3 and without regard to unit-cell geometry. (c) On which set will slip be expected to occur more readily?

Answer: (a) 0.248 nm   (b) 0.285 nm

5–4.8   What are the twelve $\langle$111$\rangle${101} slip systems in bcc metals? The twelve $\langle \bar{1}$11$\rangle${211} slip systems?

• **5–4.9**   Refer to Fig. 5–4.8. The shortest slip vector on the {111} plane is $\mathbf{b}_{01\bar{1}}$ (or its permutations such as $\mathbf{b}_{10\bar{1}}$). The second shortest is $\mathbf{b}_{11\bar{2}}$ (or its permutations). (a)

Excluding the directions of these slip vectors, which direction contains the next shortest slip vector? (b) How would its relative energy requirements compare with $\mathbf{b}_{01\bar{1}}$?

Answer: (a) $\mathbf{b}_{\bar{3}21}$ (b) $\times 7$

5–4.10 A flat brass plate must conduct more than 2.7 cal/mm²·sec (or 11.3 W/mm²) when its two faces have a temperature difference of 125°C. It must also have an ultimate strength-thickness product of more than 0.55 MPa (m) that is, $(S_T)(t) > 0.55$ MPa (m), (>3140 lb$_f$/in.). Consider the 90–10, 80–20, and 70–30 brasses of Fig. 5–1.1, and select the alloy that will meet the requirements least expensively.

5–5.1 How much cold work was performed on the aluminum rod in Study Problem 5–2.1?

Answer: 18%

5–5.2 Change the metal in Study Problem 5–2.1 from aluminum to copper. What are the ultimate strength and the ductility for the deformed metal?

5–5.3 A copper wire 2.5 mm (0.10 in.) in diameter was annealed before cold-drawing it through a die 2.0 mm (0.08 in.) in diameter. What ultimate strength does the wire have after cold-drawing?

Answer: 330 MPa (48,000 psi)

5–5.4 A pure iron sheet 2.5-mm (0.10-in.) thick is annealed before cold-rolling. It is rolled to 2.0 mm (0.08 in.), with negligible change in width. What will be the ductility of the iron after cold-rolling?

5–5.5 Change the metal in Study Problem 5–2.1 to 80 Cu–20 Zn. What are the ductilities for the two rods?

Answer: Annealed—50% elongation; 18% CW—20% elongation

5–5.6 Copper is to be used as a wire with at least 290 MPa (42,000 psi) ultimate strength and at least 18% elongation (50 mm, or 2 in.). How much cold work should the copper receive?

5–5.7 Iron is to have a BHN of at least 125 and an elongation of at least 32% (50 mm, or 2 in.). How much cold work should the iron receive?

Answer: Use 25% cold work

5–5.8 How much zinc should be in a cold-worked brass to give a ductility of at least 20% elongation, *and* a hardness of at least 70R$_B$? (b) How much should it be cold-worked? (Your answer should not exceed 36% Zn because of processing complications.)

5–6.1 A copper wire must have a diameter of 0.7 mm and an ultimate strength of 345 MPa (50,000 psi). It is to be processed from a 10-mm copper rod. What should be the diameter for annealing prior to the final cold-draw?

Answer: 0.95 mm

5–6.2 A round bar of 85 Cu–15 Zn alloy, 5 mm in diameter, is to be cold-reduced to a rod 1.25 mm in dia. Suggest a procedure to be followed if a final ultimate strength of 415 MPa (or greater) is to be achieved along with a final ductility of at least 10% elongation (in 50-mm gage length).

5–6.3 A 70–30 brass (Fig. 5–5.6) wire with an ultimate strength of more than 415 MPa (60,000 psi), a hardness of less than $75R_B$, and an elongation (50 mm, or 2 in.) of more than 25% is to be made by cold-drawing. The diameter, as received, is 2.5 mm (0.1 in.). The diameter of the final product is to be 1.0 mm (0.04 in.). Prescribe a procedure for obtaining these specifications.

Answer: 15% to 19% cold work; therefore it should be annealed when the diameter is 1.1 mm (0.043 in.) before the final 17% cold work.

5–6.4 A round bar of brass (85% Cu, 15% Zn), 5 mm (0.20 in.) in diameter, is to be cold-drawn to wire that is 2.5 mm (0.10 in.) in diameter. Specify a procedure for the drawing process such that the wire will have a hardness less than $72R_B$, an ultimate strength greater than 415 MPa (60,000 psi), and a ductility of greater than 10% elongation in the standard gage length.

5–6.5 A rolled 66 Cu–34 Zn brass plate 12.7-mm (0.500-in.) thick had a ductility of 2% elongation (in 50-mm (2-in.) gage length) when it was received from the supplier. This plate is to be rolled to a sheet with a final thickness of 3.2 mm (0.125 in.). In this final form, it is to have an ultimate strength of at least 483 MPa (70,000 psi) and a ductility of at least 7% elongation (in 50-mm gage length). Assuming that the rolling process that reduces the plate to a sheet does not change the width, specify *all steps* (including temperature, times, thickness, etc.) that are required.

5–6.6 Aluminum sheet has been shaped into a cake pan by spinning (Fig. 5–2.3). Assume the data from Fig. 5–6.6 apply to this cold-worked aluminum. Will it start to recrystallize while in a 180°C (350°F) oven to bake a cake? Solve, both graphically and mathematically.

5–6.7 Assume annealing should be completed in 1 second to permit the hot-working of the aluminum of Fig. 5–6.6 and Example 5–6.2. What temperature is required?

Answer: 685 K (or 412°C)

QUIZ CHECKS

5A	(a), (e)		**5I**	4.7 MPa
5B	lower		**5J**	dislocations
	greater			(a), (d)
	fewer		**5K**	(d): 17%
5C	primary		**5L**	(e)
	forging and extrusion		**5M**	(a)
5D	Poisson's, lateral		**5N**	(e): log t vs. K^{-1}
	volume		**5O**	curvature
5E	tangent			big
5F	hydrostatic pressure-to-$(\Delta V/V)$		**5P**	increases, increases
5G	(b), (c), (f)		**5Q**	coarse
5H	2.7 N			diffuse

Chapter Six

Molecular Phases

PREVIEW

"Plastics" have rapidly come into technical importance. While they have the major advantage of being lightweight, their greatest value is their ability to be processed into intricate geometric shapes with desired properties, using the minimum of labor. They constitute one of the fastest-growing areas of materials development and applications. Since they may be processed easily, it is common to find them used by designers without full knowledge of their characteristics and limitations. The purpose of this chapter is to outline the structural nature of these nonmetallic materials and to relate these structures to their properties and service behavior.

The technical name for "plastics" is *polymers*. These are large molecules, *macromolecules*, made up of many repeating units, or *mers*. We will examine these mers and how they combine into polymeric materials. We will look first at the ideal molecular arrangements. However, as with crystals, we will also look at irregularities, because they have a pronounced influence on the behavior of the polymers.

Polymers do not crystallize readily because large molecules rather than individual atoms must be arranged into an ordered structure. In addition, the molecules have relatively weak *inter*molecular bonds.

Since polymers are commonly soft at elevated temperatures, they lend themselves to a variety of plastic molding processes, some of which are described.

STUDY OBJECTIVES

1. To become familiar with the molecules that are introductory to polymers.

2. To calculate molecular sizes, and to describe the bond changes in polymerization reactions.

3. To understand the mechanisms of polymer growth that lead to linear and network structures, plus the isomeric variations that modify these structures.

4. To utilize the concept of the glass temperature in understanding the behavior of amorphous polymers, and to contrast their properties with those of their more crystalline counterparts.

198

5. To interpret visoelastic properties in terms of molecular responses, above and below the glass temperature.

6. To communicate effectively with other engineers using the terms and concepts of this chapter.

6—1

GIANT MOLECULES

Our next principal category of materials following metals is molecular solids. These are nonmetallic compounds (Section 1–4) that commonly originate from organic raw materials. Organic substances have served as engineering materials from a time preceding the first engineer. Wood has long been a common construction material, and such natural organic substances as leather for gaskets, felt for packings, cork for insulation, fibers for binding, oils for lubrication, and resins for protective coatings are extensively used by engineers.

Early in the history of the use of organic materials, attempts were made to improve their engineering properties. For example, the properties of wood are highly directional; the strength parallel to the grain is several times greater than in the perpendicular direction. The development of plywood has helped to overcome this nonuniformity, and still better physical properties are obtained when the pores of the wood are impregnated with a thermosetting resin (Chapter 16).

The ingenuity of technologists in working with molecular materials* has not been limited to improving natural organic materials; many synthetic sub-

---

* Students who feel "rusty" on molecular chemistry are advised to review "polymeric molecules" in *Study Aids for Introductory Materials Courses*. This short review includes essentially all of the organic chemistry required as a prerequisite for this chapter.

stances have been developed as well. For example, *polymers,* or less technically, *plastics,** have given engineers an infinitely greater variety of materials for product applications. Great strides have been, and continue to be, made in the utilization of such materials.

Whether the engineer who uses polymers is working with natural organic materials or with artificial ones, he or she is concerned primarily with the nature and characteristics of *large molecules.* In natural materials, large molecules are "built in" by nature; in artificial materials, these large *macromolecules* are built by *deliberately joining small molecules.*

Macromolecules

Polymers are large molecules with many (*poly*) repeatable units (*mers*), just as crystals have numerous unit cells. These large molecules are also called *macromolecules* in contrast to the small molecules of Fig. 2–2.1. The macromolecules are illustrated by the *polyvinyls,* one of several important families of polymers. The structural composition of a polyvinyl is

$$\begin{bmatrix} \text{H} & \text{H} \\ \text{C--C} \\ \text{H} & | \\ & \mathbf{R} \end{bmatrix}_n \qquad (6\text{--}1.1a)$$

where *n* is the *degree of polymerization.* It is commonly several hundred or even thousands. The **R** in a vinyl C_2H_3R is different for each vinyl (Table 6–1.1). Thus,

$$\begin{bmatrix} \text{H} & \text{H} \\ \text{C--C} \\ \text{H} & \text{H} \end{bmatrix}_n , \qquad \begin{bmatrix} \text{H} & \text{H} \\ \text{C--C} \\ \text{H} & | \\ & \text{Cl} \end{bmatrix}_n , \qquad \text{and} \quad \begin{bmatrix} \text{H} & \text{H} \\ \text{C--C} \\ \text{H} \\ \end{bmatrix}_n \qquad (6\text{--}1.1b)$$

are *polyethylene* (PE), *polyvinyl chloride* (PVC), and polyvinyl benzene, respectively. The latter is more commonly called *polystyrene* (PS). Of course, there are many other polymers that do not belong to the polyvinyl family. We shall encounter a limited number of these later.

Molecular Weights

Properties vary for different size molecules. For example, butane (C_4H_{10}), octane (C_8H_{18}), and "35-ane" ($C_{35}H_{72}$) have melting temperatures of $-138°C$, $-56°C$, and $75°C$, respectively (Fig. 6–1.1). The same is true for macromolecules that possess thousands of atoms so that mechanical, chemical, and service stability of polymers are also dependent upon molecular size. Figure 6–1.2

---

* Strictly speaking, *plastic* is an adjective defining a permanently deformable material (Section 1–5), but by common usage "plastics" denote nonmetallic products, many of which have been shaped by plastic deformation.

TABLE 6–1.1

Structures of Selected Vinyl Polymers

$$\left[\begin{matrix} 1 & 3 \\ -C-C- \\ 2 & 4 \end{matrix}\right]_n$$

Polymer	1	2	3	4	International abbreviation
Polyvinyls (general)	H	H	H	$-R$	
Polyethylene	H	H	H	$-H$	PE
Polyvinyl chloride	H	H	H	$-Cl$	PVC
Polyvinyl alcohol	H	H	H	$-OH$	PVAl
Polystyrene	H	H	H	$-\langle\bigcirc\rangle$	PS
Polypropylene	H	H	H	$-CH_3$	PP
Polyvinyl acetate	H	H	H	$-\overset{\overset{\displaystyle O}{\|\|}}{O}CCH_3$	PVAc
Polyacrylonitrile	H	H	H	$-CN$	PAN
Polyvinylidenes (general)	H	H	R'	R''	
Polyvinylidene chloride	H	H	$-Cl$	$-Cl$	PVDC
Polymethyl methacrylate	H	H	$-CH_3$	$-\overset{\overset{\displaystyle O}{\|\|}}{C}OCH_3$	PMMA
Polyisobuytlene	H	H	$-CH_3$	$-CH_3$	PIB
Polytetrafluoroethylene	F	F	F	F	PTFE

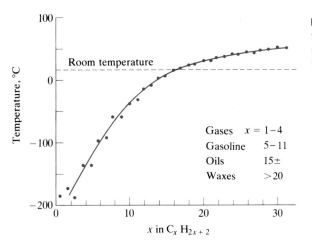

Figure 6–1.1 Melting temperatures versus molecule size in the C_xH_{2x+2} hydrocarbon series.

Room temperature

Gases	$x = 1-4$
Gasoline	$5-11$
Oils	$15\pm$
Waxes	>20

x in $C_x H_{2x+2}$

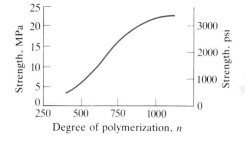

Figure 6–1.2 Strength versus molecular size (polyethylene). Larger molecules (with more mers/molecule, n) have increased mechanical and thermal properties, and thus are able to withstand more severe service requirements. (Adapted from Wyatt and Dew-Hughes, *Metals, Ceramics and Polymers*.)

shows the tensile strength of polyethylene as a function of the number of *repeat units* (mers) in the average molecule. It will be advantageous for us to quantify molecular size.

Molecular size can be reported in terms of the *degree of polymerization, n*. We do not determine the degree of polymerization by counting the repeat units (mers); rather the molecular weight, M, is determined. By knowing the mer weight, m, of our polymer,

$$n = \frac{M}{m} = \frac{\text{molecular weight}}{\text{mer weight}}.$$ (6–1.2a)

The dimensions for our calculation are

$$\frac{\text{amu/molecule}}{\text{amu/mer}} = \frac{\text{mers}}{\text{molecule}}.$$ (6–1.2b)

As an example, a polyvinyl chloride molecule may have a molecular weight of 35,000 amu (or 35,000 g/mol). Since the unit weight for C_2H_3Cl is 2(12 amu) + 3(1 amu) + 35.5 amu = 62.5 amu, the degree of polymerization is 35,000 amu/ 62.5 amu = 560 mers/molecule.

Molecular weights are obtained from a variety of laboratory tests.* Reported values are, of necessity, average values since any polymer product contains many molecules, few of which have grown identically. However, averages can be obtained by different calculations, depending on whether we use a weight basis or a number basis. Consider, for example, a syrup made with 18 g of sugar ($C_6H_{12}O_6$ with $M = 180$ amu) and 18 g of water (H_2O with $M = 18$ amu). Based on this 50/50 weight ratio, the *mass-average molecular weight*, $\overline{M}_m$, is

$$0.50(180 \text{ amu}) + 0.50(18) = 99 \text{ amu} \qquad \text{(or 99 g/mole)}.$$

Generalizing our calculation, we have

$$\overline{M}_m = \Sigma W_i M_i$$ (6–1.3)

---

* Principal among these are techniques that (1) involve osmosis, or (2) utilize light scattering. Molecular weights are also measured indirectly against calibrated standards by viscosity and gel-permeation chromatographic procedures. It will not be necessary to detail these for our requirements in this text.

where W_i is the weight fraction, and M_i is the molecular weight of each component, i.

The reader is aware that the 18 grams of water will involve ten times as many molecules as the 18 grams of sugar (because the molecular weights are 18 and 180 amu each). A *number-average molecular weight*, $\overline{M}_n$, based on a number fraction, X_i, will be

$$\overline{M}_n = \Sigma X_i M_i. \qquad (6\text{–}1.4)$$

Again, M_i is the molecular weight of each component. Since we have ten times as many water molecules as sugar molecules, the average molecular weight (on a number-fraction basis) is

$$(10/11)(18 \text{ amu}) + (1/11)(180 \text{ amu}) = 32.7 \text{ amu} \qquad (\text{or } 32.7 \text{ g/mol}).$$

Figure 6–1.3 illustrates this contrast between the two average molecular weights of a polymer where we have a range of various sizes. As indicated in the figure legend, the two presentations for the *same* material differ because there are more small molecules per gram than large molecules per gram. By inspecting Fig. 6–1.3 (or by making the calculation of Example 6–1.3), we observe that the number-average molecular weight is less than the mass-average. This is a universal situation, whether we consider a sugar-water syrup or any type of polymer. Only in a *monodisperse* (only one size of molecules) will the two figures be equal. In no case will the number-average be larger.

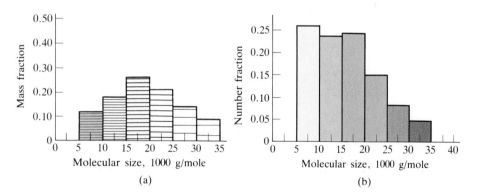

Figure 6–1.3 Polymer size distributions (a) based on mass, (b) based on numbers. These two presentations are for the same material. In (a), the data indicate that 12% of the total *mass* is accounted for by molecules between 5,000 amu and 10,000 amu (mid-value = 7,500 amu). Other percentages of the total mass are in other size intervals. In (b), we observe that 26% of the total *number* of molecules fall in the 5,000-amu to 10,000-amu range. These values of 26% for (b) versus 12% for (a) differ because it takes many of the small molecules to account for 12% of the mass. In contrast, 4% of the molecules account for 9% of the mass that is found in the 30,000–35,000 interval. (See Example 6–1.3.)

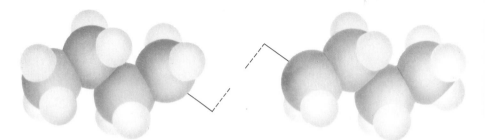

Figure 6–1.4 Linear polymer (polyethylene). The molecular length varies from molecule to molecule. Typically they contain hundreds of carbon atoms.

The ratio of the two average molecular weights is called the *polydispersivity index* (PDI); it is a measure of the variation of molecular sizes:

$$PDI = \overline{M}_m / \overline{M}_n. \qquad (6\text{–}1.5)$$

It ranges from PDI = 1 for laboratory monodisperses to PDI = 4 or 5 for many commercial products. A high value of the PDI indicates a large quantity of small molecules that adversely affect thermal stability in service.

Molecular Lengths

We could make a calculation from Table 2–2.1 that the average length of a polyethylene molecule with $n = 500$ [i.e., $+C_2H_4+_{500}$] is 154 nm, because each of the 1000 C–C bonds is 0.154 nm long. However, this calculation needs a correction because the bond angles across the carbon atoms are 109.5° (Fig. 6–1.4). Thus, the "sawtooth" length would be (154 nm) sin 54.75°, or 126 nm. Even this is overly simplified in many polymers where the single bonds of the carbon chain are free to rotate. This freedom is available at higher temperatures in amorphous (noncrystalline) polymers and in polymers dissolved in a liquid solvent.

With only three C–C bonds in butane (Fig. 6–1.5), the a–d distance is 0.4 nm while a–d' is less than 0.3 nm. The length varies randomly between these limits as the thermal agitation rotates the bond angles. The maximum limit for the end-to-end length of the polyethylene in Fig. 6–1.4 with $n = 500$ is the 126 nm value calculated above; the minimum end-to-end length would be a fraction of a namometer (if the molecule were twisted or kinked to bring the two ends into contact). The average, or *root-mean-square* length, $\overline{L}$, in noncrystalline polymers lies between these two extremes (Fig. 6–1.6) and can be calculated on a statistical basis as

$$\overline{L} = l\sqrt{x}, \qquad (6\text{–}1.6)^*$$

---

* The mathematics required to derive this equation is beyond the scope of this book. It utilizes the statistics of a "random-walk" process.

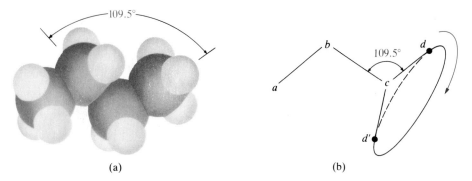

(a) (b)

Figure 6–1.5 Bond rotation (butane). Although there is a constant 109.5° bond angle, the end-to-end distance can vary from *a–d* to *a–d'*. Large molecules will have much more variation.

where l is the individual bond length and x is the number of bonds in the chain. (In vinyl polymers, l is the C–C bond length of 0.154 nm and $x = 2n$, since there are two C–C bonds per C_2H_4 mer. Using Eq. (6–1.6), the mean length of the PE with $n = 500$ is 4.9 nm.)

The rotations that introduce the twisting and coiling just described are important to properties. For example, a rubber has high strain values (low modulus of elasticity) because the kinked and coiled molecules can be straightened significantly with the application of a small stress. Furthermore, the molecules rekink when the stress is removed. The rubber is visibly elastic!

Example 6–1.1

The composition of Teflon is identical to the polyethylene of Fig. 6–1.4 except that fluorine atoms replace hydrogen atoms. Therefore, its repeat unit is $+C_2F_4+$. (a) What

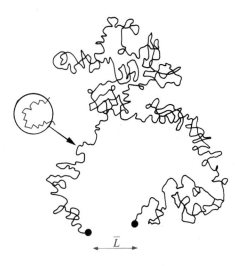

Figure 6–1.6 Kinked conformation (noncrystalline molecules). Since each C–C bond can rotate (Fig. 6–1.5), a long molecule is normally kinked and has a relatively short mean length, $\bar{L}$ (Eq. 6–1.6).

is the degree of polymerization if the mass of the molecule is 33,000 amu (or 33,000 g/mole)? (b) How many molecules per gram?

Procedure. (a) We want repeat units per molecule; therefore, we need the mer weight. As with ethylene in Example 2–2.1, we consider C_2F_4 to be the repeat unit rather than CF_2. The *monomer*, or starting molecules, must be C_2F_4, since CF_2 cannot be stable. (The carbon must have four bonds.) (b) There are 33,000 g per Avogadro's number of molecules.

Solution

a) The mer mass is $(4)(19) + (2)(12) = 100$ amu:

$$(33{,}000 \text{ amu/molecule})/(100 \text{ amu/mer}) = 330 \text{ mers molecule.}$$

b) $(0.6 \times 10^{24} \text{ molecules})/(33{,}000 \text{ g}) = 1.8 \times 10^{19}/\text{g}.$

Comment. Teflon is a trade name; PTFE (for polytetrafluoroethylene) is generic. ◀

Example 6–1.2

A solution contains 15 g of water, 4 g of ethanol (C_2H_5OH), and 1 g of sugar ($C_6H_{12}O_6$). (a) What is the mass fraction of each molecular component? (b) What is the number fraction of each molecular component?

Solution

a) Basis: 20 g

	W_i
H_2O	15 g/20 g = 0.75
C_2H_5OH	4 g/20 g = 0.20
$C_6H_{12}O_6$	1 g/20 g = 0.05.

b) Basis: 20 amu

	molecules		X_i
H_2O	15 amu/18 amu	= 0.833	= 0.900
C_2H_5OH	4 amu/46 amu	= 0.087	= 0.094
$C_6H_{12}O_6$	1 amu/180 amu	= 0.006	= 0.006.
		Total = 0.926	

◀

Example 6–1.3

It has been determined that a polyvinyl chloride (PVC) has the molecular weight distribution shown in Figs. 6–1.3(a) and (b)—for mass fraction and number fraction, respectively. (a) What is the "mass-average" molecular weight? (b) What is the "number-average" molecular weight? (c) What is the degree of polymerization n based on M_m? (d) What is the polydispersivity index?

Procedure. Consider the six size classes in Fig. 6–1.3 as six molecular types with molecular weights of 7,500 g/mol; 12,500 g/mol; etc. Then determine the averages as we did for the sugar-water syrup on p. 203.

Calculation

a) and (b)

Molecular size interval, amu	$(M)_i$, mid-value, amu	W_i, Mass fraction (Fig. 6–1.3a)	$(W_i)(M_i)$, amu	X_i, Number fraction (Fig. 6–1.3b)	$(X_i)(M_i)$, amu
5–10,000	7,500	0.12	900	0.26	1,950
10–15,000	12,500	0.18	2,250	0.23	2,875
15–20,000	17,500	0.26	4,550	0.24	4,200
20–25,000	22,500	0.21	4,725	0.15	3,375
25–30,000	27,500	0.14	3,850	0.08	2,200
30–35,000	32,500	0.09	2,925	0.04	1,300
			$\Sigma = 19,200$ amu/molecule (a) M_m		$\Sigma = 15,900$ amu/molecule (b) M_n

c) Amu/mer of PVC (Table 6–1.1):

$$(C_2H_3Cl) = 24 + 3 + 35.5 = 62.5 \text{ amu/mer.}$$

$$\text{Degree of polymerization} = \frac{(19,200 \text{ amu/molecule})}{(62.5 \text{ amu/mer})}$$

$$= 307 \text{ mers/molecule (based on } \overline{M}_m).$$

d) From Eq. (6–1.5), the polydispersivity index (PDI) equals

$$\overline{M}_m/\overline{M}_n = 19,200 \text{ amu}/15,900 \text{ amu} = 1.2.$$

Comments. Whenever there is a distribution of sizes, the "number-average" molecular size is always less than the "mass-average" value, because of the large number of smaller molecules per gram in the smaller size intervals. The two averages diverge more as the range of the size distribution increases. ◀

Example 6–1.4

A polyethylene $+C_2H_4+_n$ molecule (see Fig. 6–1.4), with a molecular mass of 22,400 amu, is dissolved in a liquid solvent. (a) What is the longest possible end-to-end distance of the polyethylene (without altering the 109.5° C–C–C bond angle)? (b) The shortest? (c) The most probable?

Procedure. We need to know the number of C–C bonds for both (a) and (c). Since there are 2 C–C bonds per mer in $(C_2H_4)_n$, we must determine n, the degree of polymerization. From Table 2–2.1, $l = 0.154$ nm.

Calculations

$$\text{Degree of polymerization} = (22,400 \text{ amu/molecule})/(28 \text{ amu/mer})$$

$$= 800 \text{ mers/molecule} \qquad \text{(or 1600 bonds).}$$

a) "Sawtooth length" = (1600)(0.154 nm)(sin 109.5°/2)

$$= 200 \text{ nm.}$$

b) Shortest distance = <1 nm (with ends in contact).

c) Eq. (6–1.6): $L = 0.154 \text{ nm } \sqrt{1600} = 6.2 \text{ nm.}$

Comments. Since the molecule is under continuous thermal agitation, the probability of its attaining a length of 200 nm would be extremely remote. A force would have to be used to stretch it out to that length. Furthermore, the required force would become greater at higher temperatures because the kinking becomes more persistent with increased thermal agitation. ◀

6–2

LINEAR POLYMERS

The polyvinyls are *linear* molecules. They possess a chain of carbon atoms that are covalently bonded. Although the "backbone" has "ribs" of hydrogen, chlorine, or various side radicals (Table 6–1.1), one chain has only weak attractions to the next chain.

There are linear polymers other than the polyvinyls. These include the *polyesters, polyurethanes,* and the *rubbers,* among others. In each case, the polymer product has improved properties if the molecular chain is lengthened, with a larger degree of polymerization, n (Section 6–1).

In order to produce long molecules out of molecules that are initially small and have only one unit (a *monomer*), each monomer must be *bifunctional*. That is, the monomer must be able to react with *two* neighbors.* The vinyl monomers can do this, as shown schematically by

$$n \begin{array}{c} \text{H} \ \text{H} \\ \text{C=C} \\ \text{H} \ | \\ \textbf{R} \end{array} \rightarrow \cdots \cdot \begin{array}{c} \text{H} \ \text{H} \ \text{H} \ \text{H} \ \text{H} \ \text{H} \\ -\text{C-C-C-C-C-C-} \\ \text{H} \ | \ \text{H} \ | \ \text{H} \ | \\ \textbf{R} \quad \textbf{R} \quad \textbf{R} \end{array} \cdots, \tag{6–2.1}$$

$$\text{or} \quad \begin{bmatrix} \text{H} \ \text{H} \\ -\text{C-C-} \\ \text{H} \ | \\ \textbf{R} \end{bmatrix}_n.$$

Most reactions that produce polymers are either a *chain mechanism* or a *step mechanism*. We will discuss these two. The reader is referred to a polymer text for other reaction mechanisms.

Chain-growth Mechanism (addition polymerization)

Chain-growth polymerization occurs by the sequential *addition* of mers to the molecule. However, simply placing monomers of ethylene (or some other vinyl) near other molecules does not lead to polymerization. The reaction,

$$n \ C_2H_4 \rightarrow (C_2H_4)_n, \tag{6–2.2}$$

---

* By analogy, a train can be lengthened to include "many units," because all of the freight cars have couplers at *both* ends. A highway semitrailer has a coupler at only the front; the tractor-trailer becomes a "bimer."

must be *initiated,* followed by *propagation*; finally, the resulting chain is *terminated.*

An *initiator* for chain-growth polymerization is commonly a free radical, such as a split H_2O_2 molecule,

$$HO-OH \rightarrow 2 \ HO\bullet \qquad (6-2.3a)$$

where the reactive dot • is an unfilled orbital of the HO pair.* Therefore it will react readily. For example, when it encounters a vinyl molecule,

$$\begin{array}{ccc} & H\ H & H\ H \\ HO\bullet\ + & C{=}C & \rightarrow HO{-}C{-}C\bullet \\ & H\ \ | & H\ \ | \\ & \textbf{R} & \textbf{R} \end{array} \qquad (6-2.5)$$

We still have a free radical (admittedly longer), so it can react with another;

$$\begin{array}{cccc} H\ H & H\ H & H\ H\ H\ H \\ HO{-}C{-}C\bullet\ + & C{=}C & \rightarrow HO{-}C{-}C{-}C{-}C\bullet \\ H\ \ | & H\ \ | & H\ \ |\ \ H\ \ | \\ \textbf{R} & \textbf{R} & \textbf{R}\ \ \ \ \textbf{R} \end{array} \qquad (6-2.6)$$

A chain-growth reaction continues to *propagate,* to produce molecules with many units, i.e., polymers.

The driving force for Eq. (6–2.5) and continuing reactions by addition is the change in energy. For example the left side of Eq. (6–2.6) contains one double C=C bond, whereas the right side contains an additional *pair* of single C–C bonds. All other bonds are unchanged. Table 2–2.1 indicates that the C=C bonds have 680 kilojoules per mole of 0.6×10^{24} bonds. Each C–C bond has 370 kJ per mole. By eliminating the double bond of Eq. (6–2.6), we must supply energy:

$$+680,000 \ J/(0.6 \times 10^{24}) = +1.1 \times 10^{-18} \text{ joules.}$$

By forming two single bonds,

$$-2(370,000 \ J)/(0.6 \times 10^{24}) = -1.2 \times 10^{-18} \text{ joules.}$$

Thus, 0.1×10^{-18} joules of energy are given off (−) by Eq. (6–2.6), or −60 kJ/mole. The resulting polymer compounds are more stable than the original monomers because they have less energy.†

---

* This may be illustrated as

$$H{:}\ddot{O}{:}\ddot{O}{:}H \rightarrow H{:}\ddot{O}\bullet \ + \ \bullet\ddot{O}{:}H \qquad (6-2.3b)$$

where neither •OH radical is charged. In contrast, an OH⁻ ion carries the charge of an excess electron:

$${:}\ddot{O}{:}H \qquad (6-2.4)$$

† Another factor, that of entropy, must also be considered; however, it does not alter the basic picture.

The propagation of Eqs. (6–2.5) and (6–2.6) *terminates* when (1) the supply of monomers is exhausted, or (2) the ends of the two growing chains encounter and join. A range of molecular weights is produced, since different molecules terminate their growth at different times (Fig. 6–1.3).

Step-growth Mechanism of Polymerization

This polymerization mechanism involves a reaction between two bifunctional molecules, *large* or *small,* to form still larger molecules. This is in contrast to chain-growth in which *only* monomers are added to the growing molecules.

From among several types of step-growth reactions, only the *condensation* mechanism will be presented. Our prototype will be the following, which produces Nylon 6/6:

$$\underset{\text{HO–C–(CH}_2)_4\text{–C–OH}}{\overset{\text{O}\qquad\quad\text{O}}{}} \quad + \quad \underset{\text{N–(CH}_2)_6\text{–NH}_2}{\overset{\text{H}}{}}$$

$$\rightarrow \text{HO–C–C–C–C–C–C–N–C–C–C–C–C–C–NH} \; + \; H_2O. \quad (6\text{–}2.7)$$

$$\underbrace{\qquad\qquad\qquad}_{6C^*} \underbrace{\qquad\qquad\qquad}_{6C^*}$$

This reaction is the very first step; later steps are identical, involving the ends of any two previously reacted molecules:

$$\cdots\text{–C–C–C–OH} \; + \; \text{N–C–C–C–}\cdots \; \cdots \rightarrow \cdots\text{–C–C–C–N–C–C–C–}\cdots \; + \; H_2O. \quad (6\text{–}2.8)^\dagger$$

Thus, the polymerization reaction is never completely terminated as it is with the chain-growth mechanism after the monomers are depleted or growing chains are encountered. This is a favorable feature where greater average molecular weights are sought.

---

* Note that the Nylon 6/6 gets its ID from the successive sequences of six carbons. There are other nylons. The starting molecules are, respectively, adipic acid and hexamethylenediamine. Do *not* memorize them.

† The product of Eq. (6–2.8) is a *polyamide,* since it contains a

$$\underset{\text{–C–N–}}{\overset{\text{O}}{}} \qquad\qquad (6\text{–}2.9\text{a})$$

linkage; comparably *polyesters* and *polyurethanes* possess

$$\underset{\text{–C–O–}}{\overset{\text{O}}{}} \quad \text{and} \quad \underset{\text{–O–C–N–}}{\overset{\text{O}}{}} \qquad (6\text{–}2.9\text{b,c})$$

linkages, respectively, along the polymer chain.

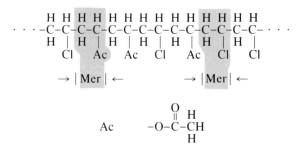

Figure 6–2.1 Copolymerization of vinyl chloride and vinyl acetate. When randomly added, a copolymer is comparable to a solid solution in metallic and ceramic crystals (Figs. 4–2.1 and 4–3.1).

Observe in Eqs. (6–2.7 and 6–2.8) that the condensation mechanism produces a water molecule as a *by-product*. Water is also released in producing various polyesters, etc. While a by-product is always formed, it can be other small molecules such as CH_3OH and HCl.

Copolymers

Each of the vinyl polymers previously considered contained only one type of mer. Polyethylene had only (C_2H_4) mers; polyvinyl chloride had only (C_2H_3Cl) mers, and polyvinyl alcohol had only (C_2H_3OH) mers. A marked advance in the technology of producing polymers occurred when scientists developed polymers containing mixtures of two or more different mers. They commonly have improved physical and mechanical properties.

It is possible, for example, to have a polymer chain composed of mers of vinyl chloride and vinyl acetate (Fig. 6–2.1). The resulting molecule is a *copolymer*. A copolymer may have properties quite different from those of either component member. Table 6–2.1 shows the variety of properties and applications of vinyl chloride–vinyl acetate mixtures with different degrees of copolymerization. The range is striking. It means that engineers may tailor their materials to a wide variety of requirements.

Thermoplasticity

As with a rope, linear molecules are inherently flexible, particularly at high temperatures where there is more thermal motion. Thus, linear polymers deform and flow with only small pressures as the temperature is increased. This heat-sensitive characteristic is called *thermoplasticity*. It is reversible, in that the plasticity is lost at low temperatures.

Thermoplasticity is used in processing. A polymer is heated to become soft for extrusion through a die or for injection into a mold (Section 6–8). After shaping by one or the other of these procedures, the product is cooled, after which it becomes rigid for use. Of course, a thermoplastic product has limitations on its service temperature.*

---

* As an example, a plastic spoon made specifically for use with an ice cream cup may not be fully satisfactory for stirring sugar into hot coffee.

TABLE 6–2.1

Vinyl Chloride-acetate Copolymers: Correlation between Composition, Molecular Weight, and Applications*

Item	w/o of vinyl chloride	No. of chloride mers per acetate mer	Range of average mol. wts.	Typical applications
Straight polyvinyl acetate	0	0	4,800–15,000	Limited chiefly to adhesives.
Chloride-acetate copolymers	85–87	8–9	8,500– 9,500	Lacquer for lining food cans; sufficiently soluble in ketone solvents for surface-coating purposes.
	85–87	8–9	9,500–10,500	Plastics of good strength and solvent resistance; molded by injection.
	88–90	10–13	16,000–23,000	Synthetic fibers made by dry spinning; excellent solvent and salt resistance.
	95	26	20,000–22,000	Substitute rubber for electrical-wire coating; must be plasticized; extrusion-molded.
Straight polyvinyl chloride	100	—	—	Limited, if any, commercial applications *per se*.

* Adapted from A. Schmidt and C. A. Marlies, *Principles of High Polymer Theory and Practice*. New York: McGraw-Hill.

Example 6–2.1

Polypropylene $+H_2C–CHCH_3+_n$ is a polyvinyl where **R** is $–CH_3$. During polymerization, how much energy is released per gram of product?

Procedure. Each
$$\begin{matrix} H & H \\ C{=}C & \\ H & | \\ & CH_3 \end{matrix}$$
added to the lengthening chain loses a double bond, but develops two single bonds,

$$\left[\begin{matrix} H & H \\ C{-}C \\ H & | \end{matrix}\right]_n \begin{matrix} H & H \\ C{-}C{-}\cdot \\ H & | \end{matrix}$$

Calculation. Basis: 1 mole = 2(12) + 3(1) + (12 + 3) = 42 g.

$$\text{Energy required} = +680{,}000 \text{ J}.$$
$$\text{Energy released} = -2(370{,}000 \text{ J}).$$
$$\text{Net energy/g} = (+680{,}000 \text{ J} - 740{,}000 \text{ J})/42 \text{ g}$$
$$= -1430 \text{ J/g}.$$

Comment. In general, vinyls release ~60 kJ of energy per mole during polymerization, because they involve the same change from one double bond to two single bonds. ◀

Example 6–2.2

A copolymer contains 92 w/o vinyl chloride and 8 w/o vinyl acetate. What is the mer fraction of VAc?

Procedure. Since a copolymer (Fig. 6–2.1) is the molecular equivalent to a crystalline solid solution, we will use a procedure comparable to that in Example 4–2.2. Choose a mass basis, and determine the numbers of each component.

Calculation. From Table 6–1.1,

$$VC = (C_2H_3Cl) = 24 + 3 + 35.5 = 62.5 \text{ amu/mer};$$

$$VAc = (C_2H_3OCOCH_3) = 24 + 3 + 59 = 86 \text{ amu/mer}.$$

Basis: 100,000 amu = 92,000 amu VC + 8,000 amu VAc.

$$VC \quad 92,000/62.5 = 1472 \text{ mers} = 0.941$$

$$VAc \quad 8,000/86 \quad = \underline{\quad 93} \text{ mers} = 0.059.$$

$$\text{Total} = 1565 \text{ mers}$$

Comment. This copolymer has ~16 PVC mers per acetate (PVAc) mer. Example uses are indicated in Table 6–2.1). ◀

Example 6–2.3

Nylon-6 is a condensation polymer of $HOCO(CH_2)_5NH_2$.

a) Sketch the structure of this molecule.

b) Show how condensation polymerization can occur.

c) What is the amount of energy released per mole of H_2O formed?

Answer

a) and (b)

$$\underset{\displaystyle \begin{matrix} & & H & H & H & H & H \\ HO-C-C-C-C-C-C-N- \end{matrix}}{O \atop \|} \text{H} + \text{HO} \underset{\displaystyle \begin{matrix} & H & H & H & H & H \\ -C-C-C-C-C-C-N-H \end{matrix}}{-\overset{O}{\underset{\|}{C}}} \quad (6\text{–}2.10)$$

c) A mole (0.6×10^{24}) of water requires elimination of that number of C–O and H–N bonds, and the formation of equal numbers of C–N and H–O bonds. From Table 2–2.1:

Bonds removed		Bonds formed	
C–O	+ 360 kJ/mole	C–N	− 305 kJ/mole
H–N	+ 430	H–O	− 500
	+ 790		− 805

The net energy difference is −15 kJ/mole, (or −3.5 kcal/mole). ◀

Example 6–2.4

A polyester labeled PET is formed by the condensation mechanism between the following two molecules. It yields methanol, CH_3OH, as a by-product.

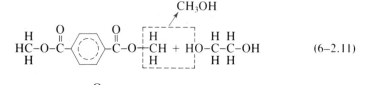

H_3COCO⟨ ⟩$COOCH_3$ and $HO(CH_2)_2OH$

a) Sketch the structure of these molecules.

b) Show the condensation reaction.

Answer

$$\underset{H}{\overset{H}{HC}}-O-\overset{O}{\overset{\|}{C}}-\!\!\left\langle\;\right\rangle\!\!-\overset{O}{\overset{\|}{C}}-O-\!\!\left[\!\begin{array}{c}H\\CH\\H\end{array}\!\right] + HO-\overset{H}{\underset{H}{C}}-\overset{H}{\underset{H}{C}}-OH \qquad (6\text{–}2.11)$$

CH_3OH

Comments. This polyester (with a $-\overset{O}{\overset{\|}{C}}-O-$ linkage) has various trade names—Terylene, Dacron (fiber), Mylar (film), Tergal, Tervina, etc., depending on the producing company. The chemical name is **polyethyleneterephthalate**; hence PET. ◄

6–3

THREE-DIMENSIONAL POLYMERS

The polymers discussed in the previous section were linear because they were bifunctional. Trifunctional (three possible reaction points) or polyfunctional molecules can lead to *network* (3-D) polymers. A vulcanizable rubber provides an example of a three-dimensional structure that forms by the addition mechanism. Phenol-formaldehyde (PF), which goes by Bakelite and other trade names, has a network structure that forms by the condensation mechanism. We will consider PF first and return to the rubbers later when we discuss cross-linking.

The atomic arrangements within formaldehyde (CH_2O) and phenol (C_6H_5OH) are shown in Fig. 6–3.1(a). At room temperature, formaldehyde is a gas; phenol is a low-melting solid. The polymerization that results from the reaction between these two compounds is shown schematically in Fig. 6–3.1(b). The formaldehyde has supplied a CH_2 "bridge" between the benzene rings of two phenols. In the process, a hydrogen is stripped from each of the two phenols, and an oxygen is removed from the formaldehyde. They form water, which can volatilize and escape. The reaction of Fig. 6–3.1 can occur at several points around the phenol molecule.* As a result of this polyfunctionality, a molecular network is formed, rather than a simple linear chain (Fig. 6–3.2).

---

* Three is the normal maximum, because there simply is not space to attach more than three CH_2 bridges. The number is limited by *stereohindrance*.

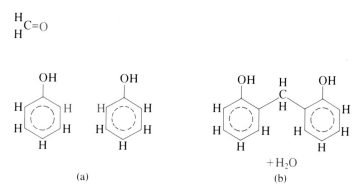

(a)   (b)

Figure 6–3.1 Phenol-formaldehyde reaction. The phenols (C_6H_5OH) contribute hydrogen and the formaldehyde (CH_2O) contributes oxygen to produce water as a by-product. The two rings are joined by a $-CH_2-$ bridge.

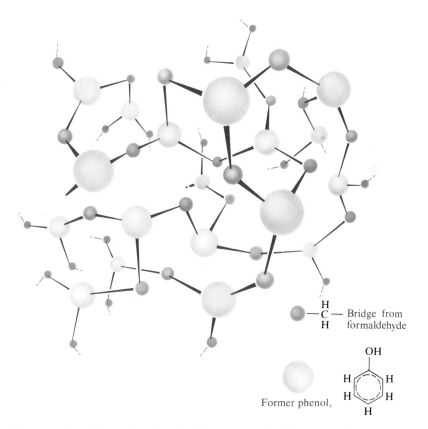

Figure 6–3.2 Network structure of polyfunctional units. Deformation does not occur as readily as in linear polymers, which are composed of bifunctional units.

Thermosetting and Thermoplastic Polymers

The polymer just described as phenol-formaldehyde (PF) was one of the first synthetic polymers. It has the attribute of not softening when heated because its three-dimensional structure keeps it rigid. Of course, this means that it cannot be molded in the same manner as linear polymers. In order to mold phenol-formaldehyde, it is necessary to start with a mixture that is only partially polymerized. At this stage, the combination is solid but averages less than three CH_2 bridges between adjacent phenols. Thus, it can be deformed under pressure. As it is held at 200°C–300°C, it completes its polymerization into a *more* rigid three-dimensional structure. It is thermal setting. The manufacturer calls polymers of this type *thermosets*. Once set, the product can be removed from the mold without waiting for cooling to occur.

The thermosetting polymers are in contrast to the previously discussed *thermoplastic polymers* that are linear and can be injected into a mold when warm because they soften at higher temperatures. The thermoplastic materials must be cooled before they can leave the mold, or they will lose their shape. There is no further polymerization in the molding process. The difference between thermosets and thermoplasts is mainly that the former has *three-dimensional* and the latter *linear* molecular structure. This difference has considerable technical implications to both processing and engineering applications because of effects on molding and on high-temperature uses.

Cross-linking

Some linear molecules, by virtue of their structure, can be tied together in three dimensions. Consider the molecule of Fig. 6–3.3(a) and its polymerized combination in Fig. 6–3.3(b). Intentional additives (divinyl benzene in this case, but we do not have to remember the name) tie together two chains of polystyrene. This causes restrictions with respect to the plastic deformation of polymers.

The *vulcanization* of rubber is a result of cross-linking by sulfur, as shown schematically in Fig. 6–3.4. The effect is pronounced. Without sulfur, rubber is a soft, even sticky, material that, when it is near room temperature, flows by viscous deformation. It could not be used in automobile tires because the

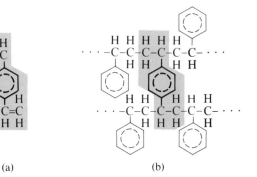

(a)   (b)

Figure 6–3.3 Cross-linking of polystyrene. The divinyl benzene (a) becomes part of two adjacent chains because it is tetrafunctional; i.e., it has four reaction points.

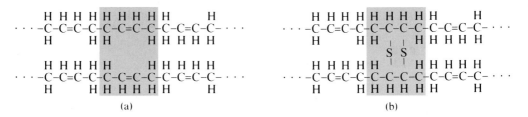

Figure 6–3.4 Vulcanization (butadiene-type rubbers). Sulfur atoms cross-link adjacent chains. In this representation of vulcanization, *two* sulfur atoms are required for each *pair* of mers. (Other cross-linking arrangements are also possible; for example, a pair of sulfur atoms in series, rather than a single sulfur between two carbons.)

service temperature would make it possible for molecules to slide by their neighbors, particularly at the pressures encountered. However, cross-linking by sulfur at about 10% of the possible sites gives the rubber mechanical stability under the above conditions, but still enables it to retain the flexibility that is obviously required. Hard rubber has a much larger percentage of sulfur and appreciably more cross-links. You can appreciate the effect of the addition of greater amounts of sulfur on the properties of rubber when you examine a hard-rubber product such as a pocket comb.

Example 6–3.1

How many grams of sulfur are needed per 100 g of final rubber product to completely cross-link a polybutadiene $(C_4H_6)_n$ rubber with sulfur according to the pattern started in Fig. 6–3.4?

Procedure. In this rubber, the mer is C_4H_6, or $-C-C=C-C-$ with $\begin{smallmatrix}H & H & H & H \\ & H & & H\end{smallmatrix}$. As shown in Fig. 6–3.4, complete cross-linking requires *two* sulfur atoms for a *pair* of C_4H_6 mers or an $S/(C_4H_6)$ ratio of 1 to 1.

Calculation

$$1 \text{ mole } C_4H_6 = 48 + 6 = 54 \text{ g}$$

$$1 \text{ atom} \quad S \quad\quad = 32 \text{ g}$$

$$\text{Product} = 86 \text{ g.}$$

$$\text{Fraction sulfur} = \frac{32 \text{ g}}{86 \text{ g}} = \frac{x}{100 \text{ g}};$$

$$x = 37 \text{ g.}$$

Comment. Alternate cross-linking structures can develop. ◀

Example 6–3.2

A plastics molding company buys a phenol-formaldehyde raw material that is only two thirds polymerized; that is, there is an average of only two $-CH_2-$ bridges joining each phenol rather than the maximum three. (See the footnote on p. 214.) (a) How many grams of additional formaldehyde are required per kg of the above raw material to complete the network formation? (To make the phenols fully trifunctional.) (b) How many grams of water will be formed in this thermosetting step?

Procedure. The raw material corresponds to a linear polymer with the following structure and bridging reaction.

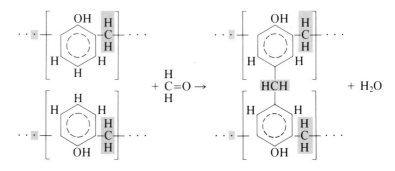

Calculation. (Using g/mole.)

Raw material   formaldehyde   product   water
$$2[6(12) + 3 + 16 + 1) + (12 + 2)] + (12 + 16 + 2) \rightarrow \quad 224 \quad + \quad 18$$

a) $x/1 \text{ kg} = 30/212; x = 0.142 \text{ kg}$ (or 142 g).

b) $y/1 \text{ kg} = 18/212; y = 85 \text{ g}.$

Comments. The heated raw material is thermoplastic during molding since it is essentially linear and lacks a network structure. While hot and still under pressure, the above "curing" reaction gives a "set" to the product. The material is thermosetting, because it develops a network structure. It will not soften on reheating. ◀

6–4

VARIATIONS IN MOLECULAR STRUCTURES

We have already seen that the structure of molecules affects the properties of the polymer product. For example, larger molecules have enhanced mechanical and thermal stabilities. Likewise, linear polymers are thermoplastic, while polymers with network structures do not soften significantly at higher temperature. Other structural features of molecules also have significant consequences.

Stereoisomers

The arrangement of the polypropylene mers show a high degree of regularity along the molecular chain of Fig. 6–4.1(a). Not only is there an addition sequence of $(C_2H_3CH_3)$ units that form the linear polymer, but there is also an identical orientation of each propylene unit. The isomer of Fig. 6–4.1(b) has the same chain of propylene mers, but the geometric (or stereo) arrangement of the $(C_2H_3CH_3)$ units differs. We speak of these two molecules as being *stereoisomers*; that in (a) being *isotactic*, that in (b) being *atactic*. (Literally, the same pattern, and without a pattern.) The properties of these two isomers are different. For example, the isotactic molecules can "mesh together" better and require less volume. We will see in a later section that isotactic molecules can

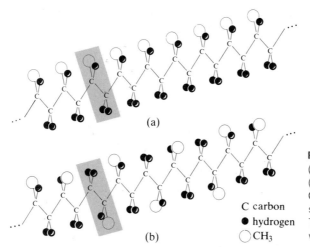

Figure 6–4.1 Stereoisomers (polypropylene). (a) Isotactic. (b) Atactic. In (a), each $C_2H_3CH_3$ mer has the same spatial (stereo) arrangement. The stereoisomer of (b) is without a spatial pattern.

C carbon
● hydrogen
○ CH₃

crystallize to a greater degree. The consequence is a slightly greater density and a measurably better resistance to adverse thermal and stress conditions in service.

A second example of geometric isomerism is found in rubbers that are made of butadiene-type molecules (Table 6–4.1). Natural rubber is polymerized isoprene with a mer of

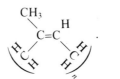

$$ (6\text{–}4.1) $$

The two double-bonded carbons have a CH_3 group and a hydrogen on the same side of the chain. This *cis* isomer has important consequences in the chain behavior, because it arcs the mer (Eq. 6–4.1). Another modification has the CH_3 group and the single hydrogen on opposite sides of the chain to give a *trans* isomer:

$$\left(\begin{array}{c} CH_3 \\ H \quad C \quad H \\ C = C \quad C \\ H \quad H \quad H \end{array} \right)_n \qquad (6\text{–}4.2)$$

Although identical in compositions, these two isomers of isoprene have different structures and therefore different properties. Natural rubber, with its cis-type structure, has a very highly kinked chain, as a result of the arc within the mer. The polymer of the trans-type called *gutta percha* has a bond-angle pattern that is more typical of linear polymers (Fig. 6–1.4). In effect, the unsaturated positions balance each other across the double bond. As a result, the properties differ markedly.

TABLE 6–4.1

Structures of Butadiene-type Rubbers

$$n\underset{\substack{H\ \\ }}{C}=\underset{\substack{R\ \\ }}{C}-\underset{\substack{H\ \\ }}{C}=\underset{\substack{H\ \\ }}{C} \quad \rightarrow \quad \left[\underset{\substack{H\ \\ }}{C}-\underset{\substack{R\ \\ }}{C}=\underset{\substack{H\ \\ }}{C}-\underset{\substack{H\ \\ }}{C}\right]_n \quad (6\text{–}4.3)$$

Name	R*	Equation
Butadiene	–H	
Chloroprene	–Cl	
Isoprene (cis)	–CH$_3$	(6–4.1)
Gutta percha (trans)	–CH$_3$	(6–4.2)

* Cf. Table 6–1.1 for vinyl polymers.

Branching

Ideally, linear polymer molecules such as we have studied to date are two-ended chains. There are cases, however, in which polymer chains branch. We can indicate this schematically as in Fig. 6–4.2. Although a branch is uncommon, once formed it is stable because each carbon atom has its complement of four bonds and each hydrogen atom has one bond. The significance of branching lies in the three-dimensional entanglements that can interfere with plastic deformation. Think of a pile of tree branches compared with a bundle of sticks; it is more difficult to move a branch with respect to its neighbors, than to move the individual sticks.

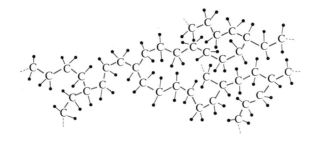

Figure 6–4.2 Branching (polyethylene). The linear molecule of Fig. 6–1.4 can be branched. (The hydrogen atoms are represented by dots.)

6–5

GLASS-TRANSITION TEMPERATURE

Unlike liquid aluminum, copper, and other metals, in which the atoms maneuver as individuals, a macromolecule involves many hundreds of atoms. Also, intramolecular bonds are highly specific in both polymeric liquids and solids. As a result, the molecules of a liquid do not readily organize themselves into a crystalline array. Crystallization from a melt is slow and tedious at best. Super-cooling below the true melting (and crystallization) temperature is common. Only structurally regular molecules can be coaxed into crystals. Atactic molecules, molecules with side ''lumps'' (such as the benzene ring of polystyrene), molecules that are branched, and molecules with arced mers (such as the isoprene of Eq. 6–4.1), all have low probabilities of crystallizing.

In any liquid there is a continual rearrangement of the atoms and molecules. This allows for flow in the presence of a shear stress. This mobility also lowers the packing factor by introducing ''free space'' around the molecules. As the temperature of a liquid decreases, the thermal agitation lessens, and there is a decrease in both the ''free space'' and the vibrational amplitude of the atoms. The resulting decrease in volume continues below the freezing point into the supercooled-liquid range (Fig. 4–5.3). However, the liquid structure is retained. As with liquids at higher temperatures, flow can occur; but the flow is more difficult because the viscosity increases as the temperature drops, and because there is a decrease in the amount of ''free space'' between molecules.

Those polymers that are cooled without crystallizing eventually reach a point at which the thermal agitation is not sufficient to allow for the rearrangement of the molecules into more efficient packing, nor can they respond readily to externally applied forces. Thus, the polymer becomes markedly more rigid and brittle. Likewise, its thermal expansion coefficient is limited to changes in the amplitude of thermal vibrations and does not reflect changes in the ''free space'' among the molecules. Thus, with this increased rigidity during cooling, there is a discontinuity in the slope of the volume-versus-temperature curve (Fig. 6–5.1). This point of change in the slope is called the *glass point*, or the *glass-transition temperature*, because this phenomenon is typical of all glasses. In fact, below this temperature, just as with normal silicate glasses, a noncrystalline polymer *is* a glass (although admittedly an organic one). Conversely, we shall observe in Chapter 7 that a normal glass is an inorganic polymer.

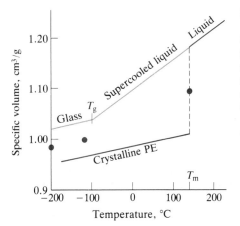

Figure 6–5.1   Glass-transition temperature T_g (polyethylene, PE). Supercooled liquids have not crystallized. Below the glass temperature T_g, the polymer becomes rigid, and various properties change significantly; e.g., the thermal expansion coefficient, dV/dT, and dielectric properties (Chapter 9). The glass retains the structure of the liquid; therefore, it is less dense than crystalline polyethylene.

TABLE 6–5.1

**Glass-transition Temperatures, T_g, of
Selected Linear Polymers**

Polymer	T_g, K (amorphous)	T_m, °C (if crystallized)
Polyethylene (HD)	−120	140
Polybutadiene	− 70±	—*
Polypropylene	− 15	175
Nylon 6/6	50	265
Polyvinyl chloride	85	210*
Polystyrene	100	240*

* Crystallizes with great difficulty.

The glass-transition temperature or, more briefly, the glass temperature, T_g, is as important to polymers as the melting (or freezing) temperature, T_m, is. The glass temperature of polystyrene is at approximately 100°C (Table 6–5.1); therefore it is glassy and brittle at room temperature. In contrast, a rubber whose T_g is at −73°C is flexible even in the most severe winter temperatures.

Example 6–5.1

Based on the data of Fig. 6–5.1, estimate the linear thermal expansion coefficient of (a) amorphous polyethylene at 20°C. (b) Of crystalline polyethylene.

Procedure. The slopes of the curves can provide the volume of coefficients, $\alpha_V = (\Delta V/V)/\Delta T$. Since $(1 + \Delta V/V) = (1 + \Delta L/L)^3$, $\Delta V/V \approx 3(\Delta L/L)$; therefore $\alpha_L \approx \alpha_V/3$.

Calculation. Basis: 1 g.

a) At −100°C, 1.04 cm³ per gram; and at 100°C, 1.16 cm³ per gram.

$$\frac{\Delta V/V}{\Delta T} = \frac{(0.12/1.1)}{200°C} = 540 \times 10^{-6}/°C;$$

$$\alpha_L = 180 \times 10^{-6}/°C.$$

b) At −100°C, 0.97 cm³; and at 100°C, 1.005 cm³.

$$\frac{\Delta V/V}{\Delta T} = \frac{0.035}{200°C} = 175 \times 10^{-6}/°C;$$

$$\alpha_L = 60 \times 10^{-6}/°C.$$

Comment. Below T_g, the expansion coefficients are approximately identical in both amorphous and crystalline PE because only thermal vibrations are involved. Above T_g, "free space" is added to the amorphous PE with increased temperatures because of molecular movements. ◀

6–6

**MOLECULAR
CRYSTALS**

Since each molecule contains more than one atom, we must look at a more complex packing arrangement in molecular crystals than in metal crystals. Methane is among the more easily visualized because the CH_4 molecules have a relatively symmetrical, almost spherelike shape (Fig. 6–6.1). Iodine molecules are not complex since the molecule contains only two atoms, I_2. However, that 2-atom molecule is not spherical. Therefore, crystalline iodine does not form a cubic unit cell. Rather, it is orthorhombic ($a \neq b \neq c$) as shown in Fig. 6–6.2.

More complex situations arise when polyethylene crystallizes (Fig. 6–6.3) because the molecules extend beyond the unit cell. However, from this figure we can determine that there are two mers per unit cell of polyethylene. In Example 6–6.1 we will calculate the density of an ideal polyethylene crystal and obtain an answer of 1.01 g/cm³. Densities measured from commercial

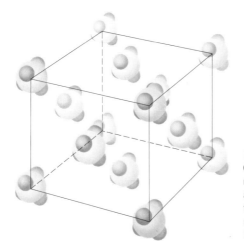

Figure 6–6.1 FCC compound (methane, CH_4). Each fcc lattice point contains a molecule of five atoms. Methane solidifies at −183°C (90 K). Between 20 K and 90 K, the molecules can rotate in their lattice sites. Below 20 K, the molecules have identical alignments as shown here.

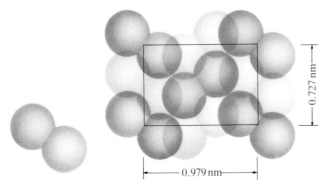

0.727 nm

0.979 nm

Figure 6–6.2 Molecular crystal (iodine). The molecule of I_2 acts as a unit in the repetitive crystal structure. This lattice is *simple orthorhombic* because $a \neq b \neq c$, and the face-centered positions are *not* identical to the corner positions. (The molecules are oriented differently.) The unit cell axes have 90° angles.

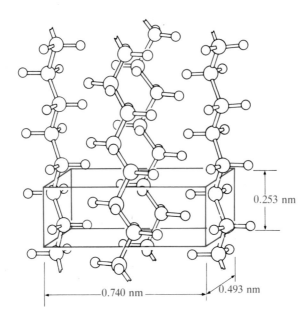

0.253 nm

0.740 nm   0.493 nm

Figure 6–6.3 Molecular crystal (polyethylene). The chains are aligned longitudinally. The unit cell is orthorhombic with 90° angles. (M. Gordon. *High Polymers*, Iliffe, and Addison-Wesley. After C. W. Bunn, *Chemical Crystallography*, Oxford.)

grades of polyethylene are normally 5% to 10% less than this value (Table 6–6.1), because the long polyethylene molecules seldom have the perfect "zipperlike" alignment shown in Fig. 6–6.3. If they are not perfectly aligned, their packing factor decreases to the extent that there may be as much as 10 percent additional "free space" among the atoms.

Crystals form with moderate difficulty in polyethylene because the molecules extend through hundreds of unit cells. Over these distances it is highly unlikely that adjacent chains will remain perfectly aligned, particularly when

TABLE 6–6.1		
Characteristics of Polyethylenes		
	Low-density polyethylene (LDPE)	**High-density polyethylene (HDPE)**
Density, Mg/m³ (= g/cm³)	0.92	0.96
Crystallinity, v/o	Near zero	~50
Thermal expansion, °C⁻¹	180×10^{-6}	120×10^{-6}
Thermal conductivity (watt/m²)(°C/m)	0.34	0.52
Tensile strength, MPa	5–15	20–40
Young's modulus, MPa	100–250	400–1200
Heat resistance for continuous use, °C	55–80	80–120
10-min. temp. exposure, °C	80–85	120–125

they are not of the same length and are subject to thermal vibration and folding. Furthermore, only weak intermolecular bonds hold them together. Diffraction analyses (Section 3–8) reveal negligible evidence of crystallinity in the low-density polyethylenes ($\rho_{LDPE} \sim 0.92$ g/cm^3). The higher-density polyethylenes ($\rho_{HDPE} \sim 0.96$ g/cm^3) have local regions that are well crystallized, but still have intervening amorphous areas. Neither full density nor complete crystallization is observed in practice.

Crystals form more readily in nylon than in polyethylene. This may be unexpected since much of the nylon chain contains (CH$_2$) segments (Eq. 6–2.7) as does polyethylene (Fig. 6–6.3). However, an examination of Fig. 6–6.3 reveals that there are negligible opportunities for bonding between adjacent chains of PE. In fact, the hydrogen atoms, which are only exposed protons on the ends of covalent bonds, have mutual repulsion with the hydrogen atoms of the adjacent chains. In a nylon (Fig. 6–6.4), the single hydrogen with the nitrogen, $\diagdown$N–H, forms a bond to the side oxygen, O=C$\diagdown$, in the adjacent molecule:

$$\diagup^{\diagdown}\hspace{-0.3em}N\text{---}H \cdots O=C^{\diagup}_{\diagdown} \tag{6–6.1}$$

The periodic presence of this *hydrogen bridge* helps align adjacent molecules into the straight and parallel rows found in crystals.

Figure 6–6.4 Molecular crystal (Nylon 6/6). The condensation mechanism of Eq. (6–2.7) produces the molecular chains. The polar C=O groups bond to the next chain through hydrogen bridges. This favors a matching of adjacent molecules and leads to more complete crystallinity than in polyethylene (Fig. 6–6.3). Within crystals, the molecules are not kinked and coiled as in Fig. 6–1.6 for amorphous polymers. (The nonbridging hydrogens are shown on only the upper molecular chain.)

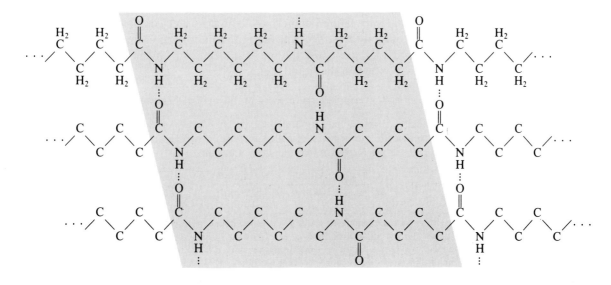

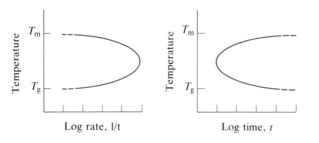

Figure 6–6.5 Crystallization versus temperature (schematic for 50% crystallized). (a) Crystallization rate. (b) Time for isothermal crystallization. The two are reciprocals. Crystallization is most rapid if the temperature is held constant at intermediate levels. Crystallization will occur more rapidly for isotactic polymers (Fig. 6–4.1), for polymers with polar groups (Fig. 6–6.4), and for linear polymers that are stretched to align the molecules (Fig. 6–7.3).

• **Rates of Crystallization**

No polymer of large molecules is 100 percent crystalline. As noted in Table 6–6.1, high-density polyethylene is ~50 percent crystalline. Within this product there are regions of crystalline order, separated by amorphous zones. Thus, even a "crystalline" polymer shows a glass-transition temperature T_g (as well as a melting temperature T_m). While exceptions abound, T_g is commonly between $\frac{1}{3}T_m$ and $\frac{2}{3}T_m$ (on the Kelvin scale).* The rate of crystallization is zero at the melting temperature and above, because any crystallites that develop are destroyed as fast as they form. The rate is also zero at T_g and below, because the molecules are not mobile for structural rearrangements. Between these limits there is an optimum temperature for crystallization, as shown schematically in Fig. 6–6.5(a).

Since time requirements are inversely related to the rates of crystallization, Fig. 6–6.5(b) illustrates the time–temperature relationships for crystallization at constant temperature (isothermal). The fastest crystallization occurs at intermediate temperatures. We will see in the next section that crystallization can be "forced" by stretching the linear molecules from the coiled conformation that they possess in the liquid state.

Example 6–6.1

From Fig. 6–6.3, calculate the density of fully crystalline polyethylene.

Solution. A $(C_2H_4)_n$ mer is parallel to the two ends of the rectangular cell, for an equivalent of one mer per unit cell. Likewise, a mer is parallel to the two sides. Together, there are a total of two mers per unit cell.

$$\rho = \frac{2(24 + 4 \text{ amu})/(0.602 \times 10^{24} \text{ amu/g})}{(0.253 \times 0.740 \times 0.493)(10^{-27} \text{ m}^3)} = 1.01 \times 10^6 \text{ g/m}^3 \qquad \text{(or 1.01 g/cm}^3\text{)}.$$

---

* This is partially analogous to the relationship between the recrystallization temperature T_R and T_m for metals (Section 5–6). Both T_R and T_g are the threshold temperatures for atomic and molecular rearrangements within the solid. Both vary with available time and other factors. However, the major differences in structures prevent us from carrying the analogy too far.

Comments. Densities normally lie in the 0.92 to 0.96 g/cm^3 range, depending on the degree of crystallinity (see Study Problem 6–6.3 and Table 6–6.1). As a result, there is additional space present. The chemist calls this "free space." ◀

6–7

DEFORMATION OF POLYMERIC MATERIALS

Polymers undergo both elastic and plastic deformation, just as metals do. As a class, polymers have lower elastic moduli than metals. (See Appendix C.*) Often we use this characteristic to advantage, because the low stress per unit strain is a design feature of rubber. Also, as a class, polymers can be plastically deformed more readily than the average metal. The everyday term *plastic* emphasizes this processing advantage of polymers. Our class comparisons are not to rate one higher than the other but to give us an index for making technical specifications. By knowing what makes a rubber visibly elastic, and what accounts for the viscous flow of polystyrene into a mold, we can better select polymers and know their limitations.

Elastomers

Polymers that have large elastic deformations are called elastomers. The basis of the large strain is the kinked conformation (discussed in Section 6–1). To illustrate, the molecule of Example 6–1.4 had a mean end-to-end length of 6.2 nm. It can be extended many nanometers when a stress is applied that unkinks the molecular chain from the conformation of Fig. 6–1.6. Of course, we never realize the full extension to 200 nm, even in nonvulcanized rubbers, but large percentages of elastic strain are common. The stretched conformation of rubber is not natural; therefore, the molecules quickly rekink by bond rotation after the stress is removed.

In order for this unkinking and rekinking to occur, the elastomer or rubber must be above its glass temperature, T_g. This permits the rearrangements between adjacent molecules. Below the T_g, the elastomer is brittle and hard.†

The elastic portion of the stress-strain curve for a rubber is unlike that for metals. The first strain occurs easily with little stress, because the molecules are simply unkinking. This results in a low elastic modulus. As the molecules become straightened and are aligned together, the stress requirements increase for each additional increment of strain (Fig. 6–7.1). Thus the modulus of elasticity increases. The data for rubbers in Appendix C are for the initial stress-strain ratios, since that is the range in which elastomers are usually used.

The molecules of rubber are stretched by restricting the bond rotation that causes kinking. Now, if a stretched rubber band is heated, the thermal agitation for rotation and the retractive forces increase; the band will *contract* if the

---

* There are, of course, exceptions. Lead and other low-melting metals have Young's moduli of less than 70,000 MPa (10 × 10^6 psi). Certain polymers with aliphatic rings in their backbones have moduli of higher figures. However, both of these examples are extremes.

† An interesting demonstration is to place an *unstretched* rubber band into liquid nitrogen (−196°C, or 77 K) and then break it like glass. The glass temperature T_g of common rubbers is about −75°C or 200 K.

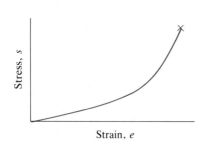

Figure 6–7.1 Stress-strain curve (rubber). The slope (elastic modulus) of the *s-e* curve increases as the molecules are unkinked. (Cf. Fig. 1–5.2 for metals.)

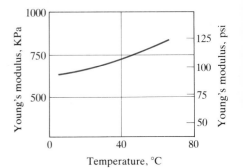

Figure 6–7.2 Young's modulus versus temperature (stretched rubber). The retractive forces of kinking are greater with more thermal agitation. Therefore, the elastic modulus increases with temperature. (Cf. Fig. 5–3.3.)

external load is not changed. Thus, a stretched rubber band has a *negative* coefficient of thermal expansion, and Young's modulus *increases* with temperature (Fig. 6–7.2).*

Finally, partial *crystallization* occurs more readily in any polymer that has been deformed so that its molecules are aligned. For example, a rubber is normally completely amorphous and noncrystalline. However, a tensile force will pull the molecules into a parallel orientation. X-ray diffraction (Fig. 6–7.3) reveals that this stretching permits the adjacent oriented molecules to mesh together into a crystalline array. The crystallographer can relate the diffraction spots in Fig. 6–7.3(b) to molecular alignment. When the stress is removed and the rubber contracts, the diffraction spots disappear.

Viscous Deformation   Shear stresses make a liquid flow. The rate of flow depends inversely upon the viscosity. Many polymers are supercooled liquids between their melting temperature T_m and their glass temperature T_g (Fig. 6–7.4b) and, therefore, are subject to viscous flow in addition to elastic deformation. The rate may be slow because the viscosity η is commonly high. However, even at a slow rate, this permits deformation when a polymer carries a long-term load. Likewise, it permits processing by deforming the polymer within a mold at high temperatures.

Figure 6–7.4(a) shows the shear stress τ required to produce one percent strain in a polymethyl methacrylate (PMMA†) as a function of temperature. It is obvious that a marked change occurs slightly above 100°C. This temperature corresponds to the glass temperature, T_g, of Fig. 6–7.4(b). Recall that the

---

* This sentence applies to *stretched* rubber only.

† Lucite is one trade name. The composition is given in Table 6–1.1

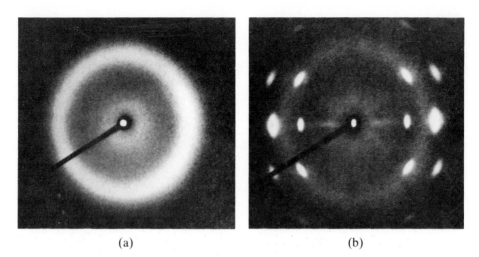

(a)   (b)

Figure 6–7.3 Deformation crystallization of natural rubber (polyisoprene) revealed by x-ray diffraction. (a) Unstretched. (b) Stretched. (S. D. Gehman, *Chemical Reviews*, vol. 26, page 203, by permission.)

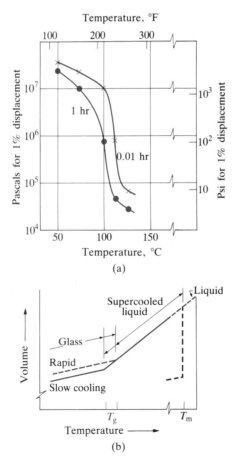

Figure 6–7.4 Viscoelastic deformation (ordinate) versus temperature (polymethyl methacrylate, PMMA). (a) With more time at a given temperature, less stress is required for unit deformation. There is a major decrease in the modulus, M_{ve}, at the glass temperature. (b) The glass temperature is lower with slower cooling, because more time permits molecular adjustments to the stress.

229

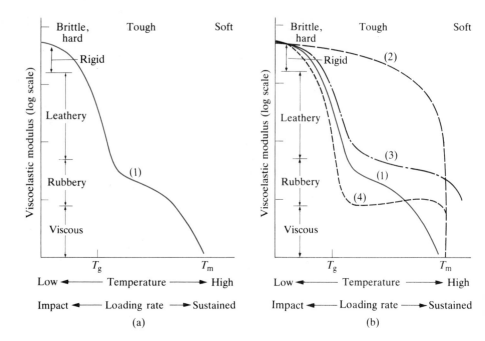

Figure 6–7.5 Viscoelastic modulus versus structure (schematic). (1) Amorphous linear polymer. (2) Crystalline (100%) polymer. (3) Cross-linked polymer. (4) Elastomer (rubber).

molecules have the freedom to kink and turn by thermal agitation above the glass temperature. Below that temperature, there is insufficient thermal agitation to permit rearrangements of molecules. Thus this represents a discontinuity in the thermal behavior of the material. Return to Fig. 6–7.4(a) and observe that the stress required for a deformation changes by more than two orders of magnitude at the glass temperature. Obviously the glass temperature is a temperature important for polymer behavior.

The two curves of Fig. 6–7.4(a) indicate that less stress is required when the time of stressing is increased from 36 sec (0.01 hr) to 1 hr. The two curves also indicate that the glass temperature drops ~10°C as the time is increased from 0.01 to 1.0 hr. This change is reflected in Fig. 6–7.4(b) as a drop in the glass transition temperature with slower cooling. With slower cooling rates, or longer times, the molecules can be rearranged at somewhat lower temperatures.

We may compare different molecular structures and their effect on deformation. In Fig. 6–7.5, the ordinate shows the *viscoelastic modulus* M_{ve} where

$$M_{ve} = s/(\gamma_e + \gamma_f). \tag{6–7.1}$$

As in Chapter 5, s is shear stress* and γ is shear deformation, γ_e being elastic

---

* In this section, shear stress is represented by s rather than by τ as in Eq. (5–3.3); τ will be used for relaxation time (Eq. 6–7.2 ff).

deformation and γ_f being displacement by viscous flow. The abscissa has been generalized. The right end includes higher temperatures and/or longer times, both of which introduce more deformation (and therefore lower values for M_{ve}). At the left end of Fig. 6–7.5(a) and below the glass temperature T_g, where only elastic deformation can occur, the material is comparatively *rigid;* a clear plastic tumbler used for soft drinks on air flights is an example. In the range of the glass temperature, the material is *leathery*; it can be deformed and even folded, but it does not spring back quickly to its original shape. In the *rubbery plateau,* polymers deform readily but quickly regain their previous shape if the stress is removed. A rubber ball and a polyethylene ''squeeze'' bottle serve as excellent examples of this behavior because they are soft and quickly elastic. At still higher temperatures, or under sustained loads, the polymer deforms extensively by *viscous flow*.

Figure 6–7.5(b) compares the deformation behavior for the different structural variants cited earlier with the amorphous polymer just described. A highly (100%) *crystalline* polymeric material (curve 2) does not have a glass temperature. Therefore it softens more gradually as the temperature increases until the melting temperature is approached, at which point fluid flow becomes significant. The higher-density polyethylenes (Table 6–6.1) lie between curves (1) and (2) of Fig. 6–7.5(b) because they possess approximately 50% crystallinity.

The behavior of *cross-linked* polymers is represented by curve (3) of Fig. 6–7.5. A vulcanized rubber, for example, is harder than a nonvulcanized one. Curve (3) is raised more and more as a larger fraction of the possible cross-links are connected. Note that the effects of cross-linking carry beyond the melting point into the true liquid. In this respect, a network polymer like phenol-formaldehyde (Fig. 6–3.2) may be considered as an extreme example of cross-linking, which gains its thermoset characteristics by the fact that the three-dimensional amorphous structure carries well beyond an imaginable melting temperature.

Once the glass temperature is exceeded, *elastomeric* molecules can be rotated and unkinked to produce considerable strain. If the stress is removed, the molecules quickly snap back to their kinked conformations (Fig. 6–1.6). This rekinking tendency increases with the greater thermal agitation at higher temperatures. Therefore the behavior curve (4) increases slightly to the right across the rubbery plateau (Fig. 6–7.5b). Of course, the elastomer finally reaches the temperature at which it becomes a true liquid, and then flow proceeds rapidly.

• **Stress Relaxation**   In the discussion of Fig. 6–7.5, we assumed constant stress and increasing strain. In some situations, strain is constant. Since viscous flow proceeds, the stress is reduced. The reader has undoubtedly observed such a phenomenon if he or she has removed a stretched rubber band from a book or bundle of papers after a period of time. The rubber band does not return to its original length. For this reason, it was not holding the papers as tightly. Some of the stress had disappeared.

The stress decreases in a viscoelastic material that is under constant strain because the molecules can gradually flow by one another. The rate of stress decrease ($-ds/dt$) is proportional to the stress level:

$$(-ds/dt)\tau = s. \tag{6-7.2}$$

Rearranging and integrating,

$$ds/s = -dt/\tau;$$

$$\ln s/s_0 = -t/\tau, \tag{6-7.3a}$$

or

$$s = s_0 e^{-t/\tau}. \tag{6-7.3b}$$

The stress ratio, s/s_0, is between the stress at time t and the original stress s_0 at t_0. The proportionality constant τ of Eq. (6-7.2) must have the units of time; it is called the *relaxation time*. When $t = \tau$, $s/s_0 = 1/e = 0.37$.

Stress relaxation is a result of molecular movements; therefore, we find that temperature affects stress relaxation in much the same manner as it affects diffusion. Since the relaxation time is the reciprocal of a rate,

$$1/\tau \propto e^{-E/kT}, \tag{6-7.4}$$

or, as in Eq. (4-7.4),

$$\ln 1/\tau = \ln 1/\tau_0 - E/(13.8 \times 10^{-24} \text{ J/K})(T). \tag{6-7.5}$$

In these equations, E, k, and T have the same meanings and units as in Eq. (4-7.4); $1/\tau$, like D, contains $\sec^{-1}$ in its units, since both involve rates.

Example 6-7.1

Detect evidence of orientation of rubber molecules by sensing a temperature change.

Procedure. For this simple experiment, use a heavy but easily deformable rubber band. Your lip can serve as a sensitive detector of temperature changes. Place the band in contact with your lower lip. Stretch it rapidly, then quickly (without snapping) return it to its original length; repeat this cycle several times. A little care will permit one to detect a temperature increase on stretching and a temperature decrease on release. These temperature changes occur because a heat of fusion is released from the rubber band to your lip during orientation. Energy is absorbed (as entropy) during deorientation when the stress is removed. ◀

• Example 6-7.2

A stress of 11 MPa (1600 psi) is required to stretch a 100-mm rubber band to 140 mm. After 42 days at 20°C in the same stretched position, the band exerts a stress of only 5.5 MPa (800 psi). (a) What is the relaxation time? (b) What stress would be exerted by the band in the same stretched position after 90 days?

Procedure. (a) The stress drops to 50% of its initial value. Therefore, $42/\tau = \ln 2.0$. (b) Here, $90/\tau = \ln(s_{90}/s_0)$, or we can use Eq. (6-7.3b).

Calculation

a) From Eq. (6–7.3a),

$$\ln \frac{5.5}{11} = -\frac{42}{\tau},$$

$$\tau = 61 \text{ days};$$

b) Eq. (6–7.3b),

$$s_{90} = 11e^{-90/61}$$

$$= 2.5 \text{ MPa} \qquad \text{(or 360 psi)}.$$

Alternative answer for (b), with 48 *additional* days:

$$s_{48} = 5.5e^{-48/61} = 2.5 \text{ MPa}. \blacktriangleleft$$

- **Example 6–7.3**

The relaxation time at 25°C is 50 days for the rubber band in Example 6–7.2. What will be the stress ratio, s/s_0, after 38 days at 30°C?

Procedure. Observe in Example 6–7.2 that $\tau_{20°} = 61$ days. With this and $\tau_{25°} = 50$ days, we use Eq. (6–7.5) to solve simultaneously for E and $\ln (1/\tau_0)$. With these we can calculate $\tau_{30°}$, and hence the remaining stress.

Calculation. At 20°C,

$$\ln 1/\tau_0 = \ln 1/61 + E/(13.8 \times 10^{-24} \text{ J/K})(293 \text{ K}).$$

At 25°C,

$$\ln 1/\tau_0 = \ln 1/50 + E/(13.8 \times 10^{-24} \text{ J/K})(298 \text{ K}).$$

Solving simultaneously,

$$E = 4.8 \times 10^{-20} \text{ J},$$

$$\ln 1/\tau_0 = 7.76.$$

At 30°C,

$$\ln 1/\tau = 7.76 - 4.8 \times 10^{-20} \text{ J}/(13.8 \times 10^{-24} \text{ J/K})(303 \text{ K});$$

$$\tau = 41 \text{ days},$$

$$s_{38} = s_0 e^{-38/41},$$

$$s_{38}/s_0 = 0.4.$$

Comments. The relaxation time shortens at higher temperatures.

A rubber is more subject to oxidation when it is under stress. When this occurs, the structure is modified, and we observe other changes such as hardening and/or cracking. ◀

• 6–8

**FORMING
PROCESSES**

The majority of polymer products are plastically formed into their final shapes. Hence, the term *plastics* has been applied both technically and generically. Forming processes include extrusion, injection molding, sheet molding, blow molding, and spinning. This list is far from exhaustive; however, it indicates common options that are open to the producer.

Extrusion

The extrusion process is sketched in Fig. 6–8.1(a). Starting materials are commonly granules of a thermoplastic polymer that had been sized for easy flow. An *auger* (or screw) feeds the granules into a heated zone where the thermoplastic pellets soften. However, the heating does not come entirely from external heaters. There is considerable energy that goes into the polymer melt from the work of the auger as it compresses the melt and forces it through the die. For example, the motor turning the auger uses as much as 50 kw (~70 Hp) to drive a 10-cm screw. This power is absorbed by the plastic in the form of heat.

In order to control the temperature as closely as possible, the heater is segmented. The temperature of each zone is measured by thermocouples and controlled separately so that the desired temperature profile is obtained along the barrel. This is important because various thermoplastics have different softening characteristics. For example, low-density polyethylene, which is amorphous, softens progressively as the temperature increases. In contrast, nylon softens rather abruptly at a higher temperature because it possesses high crystallinity. This leads to the need for modifications of the screw design of the auger depending on the product. As shown in Fig. 6–8.1, the feed into the heated zone and the compression to eliminate the pores is simultaneous and continuous for polyethylene (Fig. 6–8.1b). The compression is abrupt in Fig. 6–8.1(c) for nylon after the granules have passed well into the heated zones.

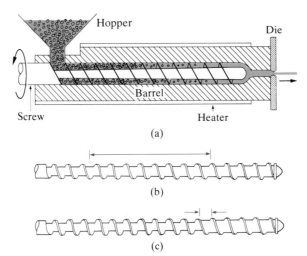

(a)

(b)

(c)

Figure 6–8.1 Extrusion (auger). (a) Schematic. Solid pellets are fed into the heated zone by an auger. The melted thermoplastic is extruded through an open-end die as bars, pipe, or sheet product. (b) Auger for polyethylene, which softens gradually. (c) Auger for nylon, which softens abruptly. The compression and metering section (arrows) is much shorter in the latter.

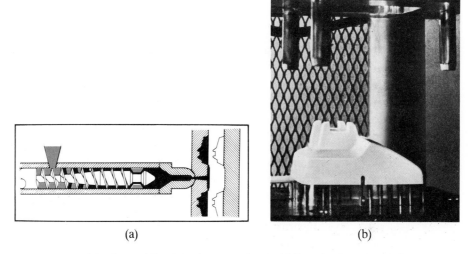

(a)　　　　　　　　　　　　　(b)

Figure 6–8.2　Injection molding (telephone receivers). (a) Extrusion into closed dies (schematic). The softened thermoplastic is forced through sprues to the cooled die cavity, where it hardens. (b) Formed product in the opened die (rotated 90° since the large auger in part (a) is operated horizontally). (Courtesy of Western Electric Co.)

The die of an extruder may have a variety of geometries. Of course, the cross section along the length of the product must remain constant. A circular orifice produces a solid rod, and a slit produces sheet or film products. It is also possible to manufacture longitudinal moldings of irregular cross sections. Tubing and pipe must be extruded by mounting a *mandrel* (or "torpedo") in the center of the hole so that extrusion is through the annulus. Alternatively, wire can be continuously fed into the center of the orifice to be coated with rubber insulation. Rates may approach a km/min. The extrusion process is very adaptable to large-scale production.

The plastic must flow as it emerges through the die but must harden immediately upon exit so as to retain its desired shape. The surface-to-volume ratio and the amount of mass to be cooled per second dictate the choice of air or water for cooling.

Injection Molding

The open die of the extrusion process may be replaced with a closed die as shown schematically in Fig. 6–8.2(a). The die is a split mold that contains the negative contours of the product to be made. The hot plastic is forced (injected) into the mold by either an auger or a hydraulic plunger. The injection process provides more flexibility in product geometry than does extrusion, because the cross section is not fixed longitudinally. Examples of injection-molded products are numerous and range from plastic ice cream spoons to telephone receivers (Fig. 6–8.2b).

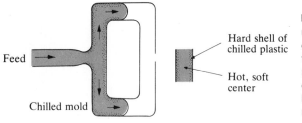

Figure 6–8.3 Flow in chilled mold (schematic). The mold is chilled to harden the plastic for stripping. However this produces a rigid shell that constricts the flow channels. High injection pressures are required.

The injection molding of *thermoplastic* polymers generally utilizes water-cooled molds for hardening the product. This facilitates production because the product becomes rigid almost immediately and may be removed so there can be a sequel injection. However, this presents complications since the viscous melt must enter through chilled sprues (feed channels) and mold cavities. This is illustrated in Fig. 6–8.3 where we see a rigid shell of chilled polymer that constricts the channel. Higher injection pressures are required in order to feed the far cavities and to avoid porosity from solidification shrinkage.

Thermosetting polymers may be injection molded more readily than they can be extruded, because the injection cycle may include time for *curing,* that is, for the completion of the polymerization reaction. It is not necessary to cool the mold inasmuch as the product "sets" rigidly while hot. Close control is required in the heating and feeding part of the cycle. The granular feed must be only partially polymerized with an average of approximately two bonds per mer (Example 6–3.2). This gives it thermoplastic characteristics for molding. Further polymerization reactions occur during curing in the mold to produce a rigid network or cross-linked structure. If that second stage polymerization were to start during the feeding step, the mold would not fill and the auger (or plunger) would bind.

Sheet Molding

In principle, sheet molding is the simplest forming process. An extruded sheet of thermoplastic material is clamped over the edges of a mold. The sheet is heated by infrared heaters. Gravity or a vacuum sags the sheet into the contour of the mold (Fig. 6–8.4). Variants of the process use air pressure or may even incorporate a mating mold on the opposite side.

Sheet molding is an inexpensive process for products with suitable geometries. Applications range from raised-letter signs to the housing for automobile instrument panels.

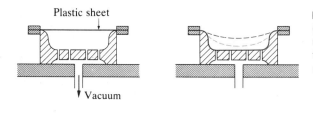

Figure 6–8.4 Sheet molding (vacuum molding). The molding pressure may also be from air pressure or from a mating mold in the upper side.

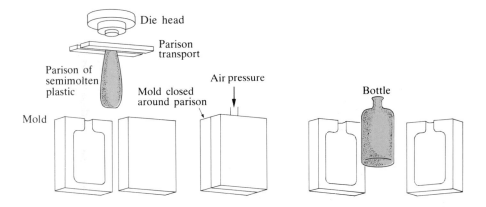

Figure 6–8.5 Blow molding. Air expands the viscous parison of plastic into the shape of the mold. Precise time and temperature controls are required in this process that originated in the glass industry.

Blow Molding

Bottles and related products having a constricted neck cannot be molded by any of the previous processes. There would be no way to remove the interior mold. In the blow-molding process, which is an adaptation from the container glass industry, a soft plastic tube is extruded and cut free from the die head (Fig. 6–8.5). The resulting ''hollow drop,'' or *parison,* of soft plastic is surrounded by a mold. Air pressure is used to expand the parison until it takes on the shape of the mold. Blow molding, more than any other plastic forming process, requires a knowledge of the relationships between viscosity, temperature, and viscoelastic behavior. And not to be ignored is the accumulated empirical experience of the operator.

Spinning

Fibers are made by forcing the plastic through a multiple orifice *spinnerette.* This contains as many as 50 or 100 holes, each less than 0.2 mm in diameter. In *melt spinning,* the thermoplastic polymers, such as nylon and polyesters, are heated to low viscosity. The slender, emerging fibers (Fig. 6–8.6a) cool quickly in a current of air before they travel over a take-up roll.

Rayon is made by a *dry-spinning* operation. It is dissolved in acetone to produce a thick solution that is extruded through the spinnerette. The acetone evaporates* to permit the fibers to dry before going over the take-up roll. In each of the above spinning operations, the filaments travel at about 15 m/s (3000 ft/min).

A third, slower process is *wet spinning.* It is used when the fibers must react with the bath to complete the polymerization reaction after they leave the spinnerette (Fig. 6–8.6b).

---

* It is collected and reused.

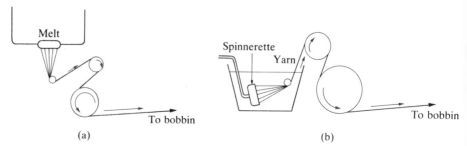

Figure 6–8.6 Fiber spinning. (a) Melt spinning (polyester fibers). The thermoplastic filaments solidify as they quickly cool. (b) Wet spinning (cellulose fibers). The extruded filaments complete their polymerization in the bath. By moving at different surface speeds (arrow lengths), filaments are stretched between the two take-up rolls to extend and orient the molecules.

Molecular Orientation

We previously noted that as linear molecules become uncoiled and straightened, they align themselves together, and the crystallinity increases (Fig. 6–7.3). Concurrently, the stress requirements increase for each additional increment of strain (Fig. 6–7.1). Thus, the elastic modulus increases. Also the strength, the heat resistance, and the moisture resistance increase. These facts are particularly useful to the processor of fiber products. Figure 6–8.6 illustrates an option.

The warm fibers emerge from the spinnerette as a viscous, but deformable, supercooled liquid. They have been wrapped 360+° around each roll; the second roll travels appreciably faster than the first roll. Therefore, the fibers are uniformly stretched several hundred percent, inducing crystallization. The slender fibers cool rapidly to ambient temperatures while still under tension, thus preserving the crystallization. Not only is the more fully crystalline product stronger and more stable in service, the property changes that accompany the glass-transition temperature are less abrupt, since only a minor fraction of the product is amorphous. Essentially all fibrous products that are made artificially are stretched in processing for molecular orientation and crystallization.

The molecules of film products can also be oriented; however, the process is more complex since the alignment must be bidirectional within the plane of the film, rather than uniaxially along the length of a fiber.

Review and Study

SUMMARY

1. Most molecular solids are composed of nonmetallic elements. The common feature of those molecular solids that have engineering applications is that their molecules are exceptionally large. We call them *macromolecules*, and also *polymers* since their molecular structures contain many *repeating units*, or *mers*.

The size of large molecules is reported (1) as the *degree of polymerization, n,* which is the number of mers per molecule, (2) by their *average molecular weight,* $\overline{M}$, and (3) by their *root-mean square length,* $\overline{L}$. The average molecular weight may be based on (a) the *weight* distribution or (b) the *number* distribution. The root-mean length takes into account the twisting and coiling of the molecular chains.

2. *Linear polymers* are formed through the combination of small *bifunctional* molecules into long extended molecules. Two of the principal reactions include a *chain-growth* mechanism (addition polymerization) and a *step-growth* mechanism. The latter mechanism commonly involves a condensation reaction that produces a by-product such as water or CH_3OH. *Copolymers* are polymers with more than one kind of mer among their repeating units. Linear polymers are *thermoplastic,* becoming soft and moldable at elevated temperatures, and rigid at low temperatures.

3. *Network,* or 3-dimensional, polymers must contain trifunctional mers. These restrict movements in the molecular solid. In fact, by controlling the progress of polymerization, the product may be *thermosetting,* with rigidity increased at elevated temperatures.

   Cross-linking is realized in linear polymers if adjacent chains are tied together with sulfur (vulcanization), oxygen, or appropriate tetrafunctional molecules. The properties may be adjusted through the number of cross-links.

4. The structures of isotactic *stereoisomers* are extremely regular compared with atactic polymers. This affects molecular packing, and hence mechanical and thermal properties. Rubbers of the *butadiene* type have elastic strains of over 100% because of the *cis* (rather than *trans*) structure in their mers. *Branching* also affects packing and properties.

5. While metals and small molecules crystallize readily at their melting temperature, T_m, polymeric liquids are readily *supercooled.* They become very viscous but retain the *amorphous* structure of the liquid.

   The *glass-transition temperature,* T_g, is at the lower limit of molecular rearrangements in amorphous materials. At this temperature there are major changes in those properties that relate to thermal agitation. These include the thermal expansion coefficient, rigidity, the viscoelastic modulus (Section 6–7), and the dielectric constant (Section 9–2).

6. *Crystalline* polymers generally are stronger and have better heat resistance than their amorphous counterparts. Crystallization is favored among the molecules with greater stereoregularity (e.g., isotactic polymers) and by those that have some hydrogen bonding (e.g., the nylons). Crystallization occurs most readily at intermediate temperatures between T_g and T_m. Even the most crystalline of the polymeric solids contain large amorphous fractions.

7. *Elastomers* are amorphous polymers with exceptionally high elastic strains (low Young's moduli). The molecules are uncoiled by small stresses, but they will recoil readily when the stress is removed. With elastomers, higher stresses are required at higher temperatures, thus producing positive values of dE/dT and negative values of dL/dT in the direction of stressing.

   The *visoelastic modulus,* M_{ve} is the ratio of shear stress s to total displacement—both viscous flow γ_f and elastic strain γ_e. Therefore, this property is both temperature and time dependent. There is a major decrease in the viscoelastic modulus at the T_g of amorphous solids. *Stress relaxation* and *creep* arise from slow viscous movements between adjacent molecules. These are proportional to the imposed shear stresses.

- 8. Various *forming* processes include extrusion, injection molding, sheet molding, blow molding, and spinning. Each involves plastic deformation, commonly at an elevated temperature. The properties of fiber products are greatly enhanced by molecular orientation, which leads to increased crystallinity.

TERMS AND CONCEPTS

Atactic   Lack of long-range repetition in a polymer (as contrasted to isotactic).

Bifunctional   Molecule with two reaction sites for joining with adjacent molecules.

- **Blow molding**   Processing by expanding a parison into a mold by air pressure.

Branching   Bifurcation of a polymer chain.

Butadiene-type compound   Prototype for several rubbers based on C=C–C=C. (See Table 6–4.1.)

Chain-growth   Mechanism for polymerization by sequential addition of monomers.

Cis   A prefix denoting unsaturated positions on the same side of the polymer chain. (Cf. *trans*.)

Copolymer   Polymers with more than one type of mer.

Cross-linking   The tying together of adjacent polymer chains.

Crystallinity (polymers)   Volume fraction of a solid that has a crystalline (as contrasted to an amorphous) structure.

Degree of polymerization, *n*   Mers per average molecule. Also, molecular mass/mer mass.

Elastomer   Polymer with a large elastic strain. This strain arises from the unkinking of the polymer chain.

- **Extrusion**   Mechanical forming by compression through open-ended dies. (Cf. Fig. 6–8.1.)

Free Space   Added volume in an amorphous polymer (as contrasted to a crystalline polymer).

Glass   An amorphous solid below its transition temperature. A glass lacks long-range crystalline order but normally has short-range order.

Glass-transition temperature, T_g   Transition temperature between a supercooled liquid and its glassy solid.

Hydrogen bridge   Secondary bonds in which the hydrogen atom (proton) is attracted to electrons of neighboring atoms.

- **Injection molding**   Process of molding a material in a closed die. For thermoplasts, the die is appropriately cooled. For thermosets, the die is maintained at the curing temperature for the plastic.

Isotactic   Long-range repetition in a polymer chain (in contrast to atactic).

Isothermal crystallization   Crystallization at constant temperature from a supercooled liquid.

Macromolecules   Molecules made up of hundreds to thousands of atoms.

Mer, ɟ ɟ   The smallest repetitive unit in a polymer.

Molecular crystal   Crystals with molecules as basic units (as contrasted to atoms).

Molecular length   End-to-end root-mean-length.

Molecular weight (mass-average), $\overline{M}_m$ Average molecular size based on the mass fraction.

Molecular weight (number-average), $\overline{M}_n$ Average molecular size based on the number fraction.

Monomer   A molecule with a single mer.

Plastics   Materials consisting predominantly of polymers.

Polar group   A local electric dipole within the polymer molecule.

Polydispersivity index, PDI   The $\overline{M}_m/\overline{M}_n$ ratio. (Used as a measure of molecular size distribution.)

Polymer   Nonmetallic material consisting of (large) macromolecules composed of many repeating units; the technical term for plastics.

Polymer, linear   Polymer of bifunctional mers. Chain polymer.

Polymer, network   Polymers containing polyfunctional mers that form a 3-dimensional structure.

Polymerization, addition   Polymerization by chain growth

Polymerization, condensation   The most common mechanism of step-growth polymerization.

• **Relaxation time, τ**   Time required to decay an exponentially dependent value to 37%, that is, $1/e$, of the original value.

Root-mean-square length, $\overline{L}$ Statistical end-to-end length of linear molecules.

• **Spinning (polymers)**   Fiber-making process by filament extrusion.

Step-growth   Mechanism for polymer growth that involves a direct reaction between two bifunctional molecules. (Free radicals are not required, and a small by-product molecule is common.)

Stereoisomers   Isomers differing in mer patterns along the molecular chain.

• **Stress relaxation**   Decay of stress at constant strain by molecular rearrangement.

Thermoplasts   Plastics that soften and are moldable due to the effect of heat. They are hardened by cooling, but they soften again during subsequent heating cycles.

Thermosets   Plastics that polymerize further on heating; therefore heat causes them to take on an additional set. They do not soften with subsequent heating.

Trans **(polymers)**   Prefix denoting unsaturated positions on the *opposite* side of the polymer chain. (Cf. *cis*.)

Trifunctional   Molecule with three reaction sites for joining with adjacent molecules.

Vinyl compounds, C_2H_3R   See Table 6–1.1.

Viscoelastic modulus, M_{ve}   Ratio of shear stress to the sum of elastic deformation, γ_e, and viscous flow, γ_f.

Vulcanization   Treatment of rubber with sulfur to cross-link the elastomer chains.

FOR CLASS DISCUSSION

A_6 Explain why the "number-average" molecular size is always smaller than the "mass-average" molecular size.

B_6 Why are polyfunctional mers necessary for network molecules?

C_6 Sketch the structure of three vinyl monomers. Repeat for the mers of the same three vinyl polymers.

D_6 Show how bonds are altered for the addition polymerization of propylene (CH_2CHCH_3).

E_6 Show how bonds are altered for the addition polymerization of butadiene $[CH_2(CH)_2CH_2]$.

F_6 Hydrogen peroxide (H_2O_2) promotes polymerization. "Twice as much is better yet." Discuss.

G_6 Heat removal requires attention by the engineers who design polymerization processes. Why?

H_6 Rewrite Eq. (6–2.8) so that NH_3 is a by-product, rather than H_2O. Suggest why H_2O is the normal by-product, and not NH_3.

I_6 By a sketch show why the $-\overset{\displaystyle Cl}{\underset{\displaystyle H}{\overset{|}{\underset{|}{C}}}}-$ portion of polyvinyl chloride is a polar group.

J_6 Why is it that metals are seldom glassy but amorphous polymers are common?

K_6 Explain why the thermal-expansion coefficient (slope of the curve in Fig. 6–5.1) is greater above T_g than below.

L_6 The exact value of T_g depends on the cooling rate. Why?

M_6 Why does polyvinyl chloride seldom crystallize?

N_6 A copolymer of two vinyls may be *random*, or they may be in *blocks* with the mer groups clustered along the chain (e.g., –BBAAAAAABBBAAAAABBBBBBB-A–). The latter crystallizes more than the random copolymer (–BAABABBBA-BAABAAABABBABAB–). Explain.

O_6 The glass temperature of a copolymer is lower than the average glass temperature of the two component polymers. Explain.

P_6 What is the by-product when phenol and formaldehyde are polymerized? When urea and formaldehyde are polymerized (Fig. 2–2.1)?

Q_6 Will scrap from thermoplasts or from thermosets be more amenable to recycling? Why?

R_6 It is suggested that butadiene (Table 6–4.1) be used to cross-link polyethylene. Discuss pros and cons.

S_6 Explain why rubber is brittle at liquid-nitrogen temperatures (77 K).

T_6 Play again with the "silly putty" you had as a child. Explain its behavior.

U_6 Explain the curves of Fig. 6–7.5(b) to someone who is not taking this course.

V_6 Discuss the relationships between the viscoelastic modulus, glass temperature, and dielectric constant (Section 9–2).

QUIZ SAMPLES

6A   Macromolecules contain thousands of atoms made up of repeating units called __**__ . A molecule with many such units is called a(n) __**__ .

6B   Vinyl compounds have the composition of $C_2H_3\mathbf{R}$. The $\mathbf{R}$ of vinyl chloride is __**__ ; in styrene, the $\mathbf{R}$ is __**__ ; in __**__ it is $-OH$; and in polypropylene it is __**__ .

6C   The __**__ is calculated as the molecular weight divided by the mer weight.

6D   The average molecular weight may be calculated on either a __**__ basis or a __**__ basis; the __**__ basis gives the higher value.

6E During polymerization of ethylene, the addition of each mer destroys *one* double carbon bond and introduces __**__ single carbon bonds. These numbers are __**__ double, and __**__ single bonds for polymerization of styrene.

6F Doubling the degree of polymerization of PVC increases the mean-square length of the molecule by __**__ %.

6G Common by-products from step-growth (condensation) polymerization include __+−__.

(a) NH_3 (b) CH_2O (c) $C_2H_3\mathbf{R}$ (d) CO_2
(e) None of the above, rather _____

6H A solid-solution alloy possesses two components but only one phase; likewise, a(n) __**__ possesses two types of mers within a single molecule.

6I A(n) __**__ stereoisomer has greater regularity than does a(n) __**__ stereoisomer.

6J Most of the polyvinyls typically possess __+−__ below their glass temperature.

(a) brittleness   (b) linear molecules   (c) thermoplasticity
(d) crystallinity   (e) None of the above, rather _____

6K Thermosetting polymers involve __+−__.

(a) network structure   (b) oven drying   (c) transition temperature
(d) polyfunctional mers   (e) None of the above, rather _____

6L Vulcanization requires the __+−__ of polymeric molecules.

(a) branching   (b) thermosetting   (c) oxidation   (d) degradation
(e) None of the above, rather _____

6M Several common rubbers have the composition of $C_4H_5\mathbf{R}$. The $\mathbf{R}$ of butadiene is __**__; in isoprene it is __**__; and in __**__ it is chlorine. $\mathbf{R}$ is $-CH_3$ in __**__.

6N A(n) __**__ solid is noncrystalline. It is called a __**__ below its transition temperature, and a __**__ liquid immediately above that temperature.

6O When compared to a crystalline solid, a glass has excess volume called __**__.

6P Since polymer molecules unkink under tension, the elastic modulus __**-creases__ as strain increases.

6Q The elastic modulus of an oriented polymer is __**-er__ than the elastic modulus of an amorphous polymer.

6R High-density polyethylene, HDPE, has higher values of __+−__ than LDPE.

(a) thermal expansion   (b) thermal conductivity   (c) molecular weights
(d) crystallinity   (e) heat resistance   (f) Young's modulus

• **6S** A long-term load above the glass-transition temperature of a polymer produces __**__; a fixed strain can, with time, lead to __**__.

6T There is a major drop in the viscoelastic modulus at the __**__ temperature. This temperature is slightly lower when __**__ time is available.

STUDY PROBLEMS   **6-1.1** A common polymer has $C_2H_2Cl_2$ as a mer (Table 6–1.1). It has an average mass of 60,000 amu per molecule. (a) What is its mer mass? (b) What is its degree of polymerization?

Answer: (a) 97 amu (or 97 g/mole)   (b) 620

6–1.2 There are 10^{20} molecules per gram of polyvinyl chloride. (a) What is the average molecular size? (b) What is the degree of polymerization?

6–1.3 The melting temperature, T_m, in K of the paraffins (C_xH_{2x+2}) of Fig. 6–1.1 is sometimes given by the empirical equation

$$T_m = [0.0024 + 0.017/x]^{-1}. \qquad (6-SP.1)$$

What melting point would polyethylene have with (a) a degree of polymerization n of 10? (b) of 100? (c) of 1000? (*Note:* Mer = C_2H_4.)

Answer: (a) 35°C (b) 130°C (c) 142°C

6–1.4 Two grams of dextrose ($C_6H_{12}O_6$) are dissolved in 14 grams of water. What is the number-fraction of each type of molecule?

6–1.5 (a) What is the "mass-average" molecular size of the molecules in Example 6–1.2? (b) The "number-average" molecular size?

Answer: (a) 31.7 amu (b) 21.6 amu

6–1.6 The following data were obtained in a determination of the average molecular size of a polymer:

Molecular size, interval midvalue	Mass
30,000 amu	3.0 g
20,000	5.0
10,000	2.5

Compute the "mass-average" molecular size.

6–1.7 (a) What is the number-fraction of molecules in each of the three size categories of Study Problem 6–1.6? (b) What is the "number-average" molecular size?

Answer: (a) 0.167, 0.417, 0.417 (b) 17,500 amu

6–1.8 Polyethylene contains equal numbers of molecules with 100 mers, 200 mers, 300 mers, 400 mers and 500 mers. What is the "mass-average" molecular size?

6–1.9 Polyvinyl chloride $\{C_2H_3Cl\}_n$ is dissolved in an organic solvent. (a) What is the mean square length of a molecule with a molecular mass of 28,500 g/mole? (b) What would be the molecular mass of a molecule with one half the mean square length as in part (a)?

Answer: (a) 4.65 nm (b) 7125 g/mole

6–1.10 Determine the degree of polymerization and the mean square length of the average molecule in Study Problem 6–1.7, if the polymer is polyvinyl alcohol (number–average).

6–1.11 A polymeric material contains polyvinyl chloride molecules that have an average of 900 mers/molecule. What is the theoretical maximum strain this polymer could undergo if every molecule could be unkinked into a straight molecule (except for the 109.5° bond angles)?

Answer: 3300% strain

6–2.1 Utilizing the data of Chapter 2, what is the net energy change as 7 grams of ethylene polymerize to polyethylene?

Answer: $-15,000$ J

6–2.2 A copolymer has equal masses of styrene and butadiene. What is the mer fraction of each?

6–2.3 A triple copolymer, ABS, has equal mass fractions of each. What is the mer fraction of each? (**A**, acrylonitrile; **B**, butadiene; **S**, styrene.)

Answer: acrylonitrile, 0.4; butadiene, 0.4; styrene, 0.2

6–2.4 How much energy is released when seven grams of the ABS copolymer of Study Problem 6–2.3 are polymerized?

6–2.5 Refer to Example 6–2.3. How much water is released per gram of product?

Answer: 0.16 g

6–2.6 Refer to Example 6–2.3. Assume the reaction formed NH_3 as a by-product (rather than H_2O). (a) Show which bonds would have to be broken. (b) What is the amount of energy change per mole of NH_3 formed? (c) Why is this reaction not expected?

6–2.7 H_2O_2 is added to 280 kg (616 lb) of ethylene prior to polymerization. The average degree of polymerization obtained is 1000. Assuming all the H_2O_2 was used to form terminal groups for polymer molecules, how many grams of H_2O_2 were added? (Mer $= C_2H_4$.)

Answer: 340 g (0.75 lb)

6–2.8 Two tenths of one percent by weight of H_2O_2 was added to ethylene prior to polymerization. What would the average degree of polymerization be if all the H_2O_2 were used as terminals for the molecules? (Mer $= C_2H_4$.)

6–3.1 What percent sulfur would be present if it were used as a cross-link at every possible point (a) in polyisoprene? (b) in polychloroprene?

Answer: (a) 32% S (b) 26.5% S

6–3.2 A rubber contains 91% polymerized chloroprene and 9% sulfur. What fraction of the possible cross-links are joined by vulcanization? (Assume that all the sulfur is used for cross-links of the type shown in Fig. 6–3.4.)

6–3.3 A rubber contains 54% butadiene, 34% isoprene, 9% sulfur, and 3% carbon black. What fraction of the possible cross-links are joined by vulcanization, assuming that all the sulfur is used in cross-linking?

Answer: 0.188

6–3.4 Rubber A (200 g) contains 168 g of butadiene $+C_4H_6\!+_n$ and 32 g sulfur. (a) What fraction of the cross-links are utilized if all of the sulfur forms those links?

Rubber B (217 g) contains 168 g of isoprene $+C_5H_8\!+_n$ and 49 g of selenium. Although the use of selenium has some major disadvantages, it can cross-link rubber. (It lies immediately below sulfur in the periodic table.) (b) Which rubber, A or B, is more highly cross-linked?

6–3.5 What is the percent weight loss if phenol polymerizes trifunctionally with formaldehyde and all of the by-product water evaporates?

Answer: 19%

6–3.6 Company X buys a partially polymerized resin of phenol and formaldehyde. A sample (10.3 g) was pressed into a hot mold. It was initially thermoplastic, but it set after a few minutes. After the sample was thoroughly dried, it weighed 9.6 g. What was the number of $-CH_2-$ connections that the average phenol had initially if we assume the final product has 3 $-CH_2-$ bridges?

6–3.7 One kg of divinyl benzene (Fig. 6–3.3) is added to 50 kg of styrene. What is the maximum number of cross-links per gram of product?

Answer: $9 \times 10^{19}/g$

6–3.8 How much sulfur must be added to 100 g of chloroprene rubber to cross-link 10% of the mers? (Assume all of the available sulfur is used, comparable to that in Fig. 6–3.4 for butadiene.)

6–4.1 Refer to Study Problem 6–4.2. How much energy is released by polymerizing 7 grams of butadiene?

Answer: -7800 J

6–4.2 (a) How many C=C bonds are eliminated per mer during the polymerization of butadiene? (b) How many additional C–C bonds are formed?

6–6.1 Solid methane has a density of 0.54 g/cm³ ($=0.54$ Mg/m³). (a) What is the lattice constant of its crystal that is fcc (Fig. 6–6.1)? (b) What is the "radius" of the CH_4 molecule?

Answer: (a) 0.582 nm   (b) 0.206 nm

6–6.2 Crystalline iodine has a density of 4.93 Mg/m³ ($=4.93$ g/cm³) and the structure of Fig. 6–6.2. What is the third dimension of the unit cell?

6–6.3 A polyethylene with no evidence of crystallinity has a density of 0.90 Mg/m³. Commercial grades of low-density polyethylene (LDPE) have 0.92 Mg/m³, while HDPE has a density of 0.96 Mg/m³. Estimate the volume fraction of crystallinity in each case.

Answer: $C_{LDPE} = 0.18$; $C_{HDPE} = 0.55$

6–6.4 From Example 6–6.1 and Study Problem 6–6.3, calculate the amount of "free space" in LDPE; in HDPE.

6–6.5 Sketch the three orthogonal views of the atoms in the polyethylene unit cell (Fig. 6–6.3).

● **6–7.1** A stress relaxes from 0.7 MPa to 0.5 MPa in 123 days. (a) What is the relaxation time? (b) How long would it take to relax to 0.3 MPa?

Answer: (a) 366 days   (b) an additional 187 days (total of 310 days)

● **6–7.2** An initial stress of 10.4 MPa (1500 psi) is required to strain a piece of rubber 50%. After the strain has been maintained constant for 40 days, the stress required is only 5.2 MPa (750 psi). What would be the stress required to maintain the strain after 80 days? Solve without a calculator.

● **6–7.3** The relaxation time for a polymer is known to be 45 days and the modulus of elasticity is 70 MPa (both at 100°C). The polymer is compressed 5% and held at 100°C. What is the stress (a) initially? (b) after 1 day? (c) after 1 month? (d) after 1 year?

Answer: (a) 3.5 MPa   (b) 3.4 MPa   (c) 1.8 MPa   (d) 1000 Pa

● **6–7.4** The relaxation time for a nylon thread is reduced from 4000 min to 3000 min if the temperature is increased from 20°C to 30°C. (a) Determine the activation energy for relaxation. (b) At what temperature is the relaxation time 2000 minutes?

6–7.5 Lucite (PMMA of Fig. 6–7.4) is loaded at 125°C for 1 hr. (a) How much would the stress have to be increased to give the same strain in 36 sec? (b) Repeat for 100°C.

Answer: (a) $s_{0.01}/s_1 = 2.5$ (b) $14(s_1)$

QUIZ CHECKS

6A	mers polymer		**6K**	(a), (d)
6B	chlorine benzene ring vinyl alcohol $-CH_3$		**6L** **6M**	(e): cross-linking H CH_3 chloroprene isoprene
6C	degree of polymerization		**6N**	amorphous
6D	mass or number mass			glass supercooled
6E	two one and two		**6O**	free space
6F	40%		**6P**	increases
6G	(e): H_2O (or CH_3OH)		**6Q**	higher
6H	copolymer		**6R**	(b), (d), (e), (f)
6I	isotactic atactic		**6S**	creep stress relaxation
6J	(a), (b)		**6T**	glass more

Chapter Seven

Ceramic Materials

PREVIEW

Ceramic materials contain phases that are *compounds of metallic and nonmetallic elements*. We can cite many of these compounds ranging from Al_2O_3 to inorganic glasses, to clay products, and on to sophisticated piezoelectrics such as $Pb(Zr,Ti)O_3$.

In general, we will find that ceramic compounds are more stable with respect to thermal and chemical environments than their components; for example, Al_2O_3 as a compound versus aluminum and oxygen separately. Since compounds inherently involve more complex atomic coordinations than their corresponding components, we will find that there is more resistance to slip, so that ceramics are generally harder, and always less ductile than their metallic or polymeric counterparts.

The dielectric, semiconductive, and magnetic characteristics of selected ceramics are especially valuable to the scientists and engineers who design or utilize devices for electronic circuits.

CONTENTS

STUDY OBJECTIVES

1. To visualize the atomic arrangements in close-packed AX structures where CN = 8, CN = 6, and CN = 4.

2. To calculate density of AX compounds as a means of understanding their structures.

3. To use the previous structures as a basis for understanding several additional structures that are widely encountered in engineering applications (CaF_2, Al_2O_3, $BaTiO_3$, and ferrites).

4. To use the background of polymers (Chapter 6) as a basis for understanding the nature and behavior of silica and the silicates.

5. To recognize (a) why ceramic materials (like other 3-D compounds) are strong in compression, weak in tension, and brittle, and (b) that these materials must utilize compressive stresses for optimum performance.

250

6. To associate atom movements and displacements with the processes of viscous forming, sintering, and tempering.

7. To incorporate terms and concepts pertinent to ceramics into one's technical communication capabilities.

7–1

CERAMIC PHASES

The term *ceramic* is most familiar as an adjective describing artware. For the engineer, however, ceramics include a wide variety of substances such as glass, brick, stone, concrete, abrasives, porcelain enamels, dielectric insulators, nonmetallic magnetic materials, high-temperature refractories, and many others (Fig. 7–1.1). The characteristic feature all these materials have in common is that they are *compounds of metals and nonmetals*. These compounds are held together with ionic and/or covalent bonds. Thus, their properties differ from metals. They are generally insulators, commonly transparent (or translucent), typically nondeformable, and unusually stable under severe environments (Fig. 7–1.1).

The compound MgO is representative of a simple ceramic material with a one-to-one ratio of metallic and nonmetallic atoms. It is used extensively as a refractory because it can withstand exceedingly high temperatures ($1500°C–2500°C$, or $3000°F–4500°F$) without dissociation or melting. Clay is also a common ceramic material but it is more complex than MgO. The simplest clay is $Al_2Si_2O_5(OH)_4$. Its crystal structure has the four different units: Al, Si, O, and the (OH) radical. Although ceramic materials are not as simple as metals, they

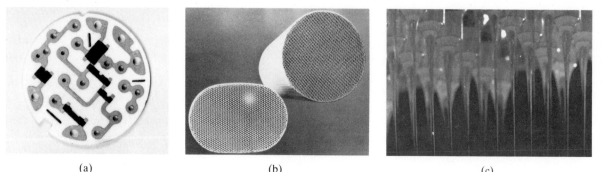

(a)   (b)   (c)

Figure 7–1.1   Examples of technical ceramic products. (a) Printed circuit. The substrate is Al_2O_3 because its strong ionic bonds preclude electrical transport. The printed components are powders of semiconducting oxides and metals that provide the functional circuitry. (Courtesy AC Spark Plug Division—General Motors Corp.) (b) Catalyst supports. These honeycombed, ceramic parts are placed in automotive exhaust systems. There, they support the catalysts that eliminate the unburned hydrocarbons through oxidation. Here again, Al_2O_3 is used, because it resists elevated temperatures and may be made strong enough to withstand the vibrational punishment in service. (Alumina ceramic parts by Corning Glass Works and photo by Aluminum Company of America.) (c) Glass fibers. These glass filaments are being drawn in the laboratory as continuous fibers through platinum orifices. The fiber dimensions are controlled by the orifice temperature, and by the velocity at which the glass is pulled away from the hot orifice. Commercial textile production involves hundreds of orifices and speeds of kilometers per minute to produce strands that contain fibers with diameters in the micrometer range. (Courtesy Owens-Corning Fiberglas Co.)

may also be classified and understood in terms of their internal structures. Furthermore, we may use the structure of the simpler metals as our point of departure.

Comparison of Ceramic and Nonceramic Phases

Most ceramic phases, like metals, are crystalline. Unlike metals, however, their structures do not contain large numbers of free electrons. Either the electrons are shared covalently with adjacent atoms or they are transferred from one atom to another to produce an ionic bond, in which case the atoms become ionized and possess a charge.

Ionic bonds give ceramic materials relatively high stability. As a class, they have a much higher melting temperature, on the average, than do metals or organic materials. Generally speaking, they are also harder and more resistant to chemical alteration. Like the organic materials, solid ceramic minerals are usually insulators. At elevated temperatures where they have more thermal energy, ceramics do conduct electricity, but poorly as compared with metals. Due to the absence of free electrons, most ceramic materials are transparent, at least in thin sections, and are poor thermal conductors.

Crystalline characteristics may be observed in many ceramic materials. Mica, for example, has cleavage planes that permit easy splitting. Plastic deformation, similar to slip in metals, may occur in some of the simpler crystals, such as MgO. Crystal outlines can form during growth, as exemplified by the cubic outline of ordinary table salt. In asbestos, the crystals have a marked tendency toward linearity; in micas and clays, the crystals form two-dimensional sheet structures. The stronger, more stable ceramic materials commonly possess three-dimensional network structures with bonding equally strong in all three directions.

As compared with those of metals, the crystal structures of ceramic materials are relatively complex. This complexity and the greater strength of the bonds holding the atoms together make ceramic reactions sluggish. For example, at normal cooling rates glass does not have time to rearrange itself into a complicated crystalline structure, and therefore at room temperature it remains as a supercooled liquid for a long time.

The structures and properties of compounds such as the refractory carbides and nitrides fall somewhere between those of ceramic and metallic materials. These include such compounds as TiC, SiC, BN, and ZrN, which contain semimetallic elements and whose structures comprise a combination of metallic and covalent bonds. Ferromagnetic spinels are another example. Because they lack free electrons, they are not good conductors of electricity; however, the atoms can be oriented within the crystal structure so as to possess the magnetic properties normally associated with iron and related metals.

The structure and properties of silicones fall somewhere between those of ceramic and organic materials. They are often called inorganic polymers. Finally, there are close structural relationships between amorphous polymers and commercial glasses, a fact we have already encountered through the examination of the glass temperature T_g in Chapter 6.

7-2
CERAMIC CRYSTALS (AX)

The simplest ceramic compounds possess equal numbers of metal and nonmetal atoms. These may be ionic like MgO, where two electrons have been transferred from the metallic to the nonmetallic atoms and have produced cations (Mg^{2+}) and anions (O^{2-}). AX compounds may also be covalent, with a large degree of sharing of the valence electrons. Zinc sulfide (ZnS) provides an example of this type of compound.

The characteristic feature of any AX compound is that the A atoms are coordinated with only X atoms as immediate neighbors, and the X atoms have only A atoms as *first-neighbors*. Thus the A and X atoms or ions are highly *ordered*. There are three principal ways in which AX compounds can form cubic crystals so that the two types of atoms are in equal numbers and possess the ordered coordination just described. The prototypes are

CsCl with CN = 8;

NaCl with CN = 6;

ZnS with CN = 4.

CsCl has a simple cubic array of atoms. NaCl and ZnS have an fcc array. Of course, there are also noncubic AX structures. However, we shall not consider them within our limited time and space.

AX Structures (CsCl-type)

Each A atom of Fig. 7–2.1 has eight X neighbors (and an extension of the structure would show each X atom has eight A neighbors). This structure is *not* bcc because the 0,0,0 and $\frac{1}{2},\frac{1}{2},\frac{1}{2}$ locations are not identical. Rather it is simple cubic with equivalent sites related only by integer translations of a. Since the locations occupied by the A atoms, or ions, in Fig. 7–2.1 are in sites surrounded by eight neighbors, we can speak of them as being in 8-*fold interstitial sites* (or more simply 8-f sites) within a simple cubic pattern.

This same structure was previously noted in Section 3–5 and Fig. 3–5.3 as that of β'-brass. However, in general it is not common among ceramics, because cations, stripped of electrons, tend to be considerably smaller than anions with excess electrons. Thus, few ionic compounds meet the r/R ratio of 0.73 or more, which is required for CN = 8 (Table 2–3.2).

The lattice constant a of a CsCl-type compound is directly related to the ionic radii because $2(r + R)$ is the body diagonal:

$$2(r + R)_{\mathrm{CsCl}} = a\sqrt{3}. \tag{7–2.1}$$

Of course, since CN = 8, an appropriate correction must be made to the ionic radii that are listed in Appendix B. Recall that $R_{\mathrm{CN}=8} \approx (R_{\mathrm{CN}=6})/0.97$.

AX Structures (NaCl-type)

The NaCl-type structure has an fcc array of anions with positive ions located in 6-*fold interstitial sites* (6-f sites). One such site is identified in Fig. 7–2.2. The NaCl structure was initially shown in Fig. 3–1.1. A composite of these two

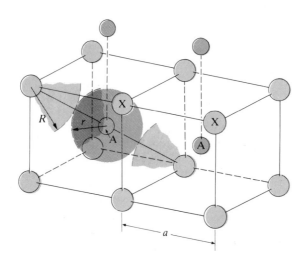

Figure 7–2.1 AX structure (CsCl-type). The A atom, or ion, sits in the interstitial site among eight X atoms (8-f sites). These $\frac{1}{2},\frac{1}{2},\frac{1}{2}$ locations are occupied in all unit cells. (Cf. Fig. 3–5.3.) Also note that X atoms sit among eight A atoms. From Eq. (7–2.1), the lattice constant a equals $2(r + R)/\sqrt{3}$.

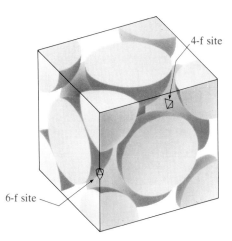

Figure 7–2.2 Interstitial sites (fcc). The 6-f site is among six neighboring atoms. There are four such sites per unit cell. The 4-f site is among four neighboring atoms. There are eight of these sites per unit cell.

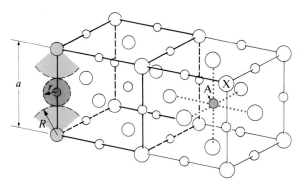

Figure 7–2.3 AX structure (NaCl-type). Compare this figure with Figs. 3–1.1 and 7–2.2. Each A atom is among six X atoms (and each X atom is among six A atoms). There are four of these 6-f interstitial sites per unit cell. All are occupied. The lattice constant a equals $(2r + 2R)$.

figures has been resketched in Fig. 7–2.3. The fcc sites are normally positioned at

$$0,0,0 \qquad \tfrac{1}{2},\tfrac{1}{2},0 \qquad \tfrac{1}{2},0,\tfrac{1}{2} \qquad 0,\tfrac{1}{2},\tfrac{1}{2};* \qquad (7\text{–}2.2)$$

and the 6-f sites are

$$\tfrac{1}{2},\tfrac{1}{2},\tfrac{1}{2} \qquad 0,0,\tfrac{1}{2} \qquad 0,\tfrac{1}{2},0 \qquad \tfrac{1}{2},0,0.* \qquad (7\text{–}2.3)$$

There are equal numbers of each; thus all 6-f sites are occupied in an AX compound. Here, as with the CsCl-type structure, it makes no difference whether all of the A ions are assigned to the fcc locations, or conversely, to the 6-f sites.

There are several hundred compounds with an NaCl-type structure; for example, MgO, NiO, FeO (Fig. 4–3.2), and MnS are among those more commonly encountered in technology. We have already observed in Eq. (3–2.4)

---

* Or at integer translations from these sites.

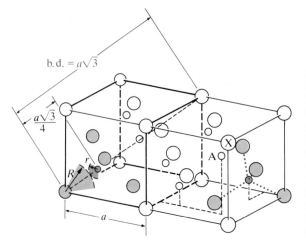

Figure 7–2.4 AX structure (ZnS-type). Compare this figure with Fig. 7–2.2 and 8–4.3(b). Each A atom is among four X atoms (and each X atom is among four A atoms). Only four of the eight 4-f sites are occupied in each unit cell. The lattice constant a equals $4(r + R)/\sqrt{3}$. The left-most A atom is at $\frac{3}{4},\frac{1}{4},\frac{1}{4}$; therefore, $r + R$ equals one fourth of the body diagonal.

that the lattice constant a is equal to twice the sum of the two radii, $2(r + R)$, since unlike ions "touch" and like ions do not. The minimum radius ratio, r/R, for this structure is 0.41 (Table 2–3.2). There is no maximum ratio; however, those ionic compounds with $r/R > 0.73$ commonly are more stable with the CsCl-type structures, where CN = 8.

AX Structures (ZnS-type)

The third structure of AX compounds has ZnS as the prototype. It is sketched in Fig. 7–2.4, as well as in Fig. 8–4.3(b) where it will receive attention for major semiconducting compounds. As in the NaCl structure, the X atoms may be located at

$$0,0,0 \qquad \tfrac{1}{2},\tfrac{1}{2},0 \qquad \tfrac{1}{2},0,\tfrac{1}{2} \qquad 0,\tfrac{1}{2},\tfrac{1}{2}; \tag{7–2.2}$$

or at their integer translations. However, the A atoms are located at

$$\tfrac{3}{4},\tfrac{3}{4},\tfrac{3}{4} \qquad \tfrac{1}{4},\tfrac{1}{4},\tfrac{3}{4} \qquad \tfrac{1}{4},\tfrac{3}{4},\tfrac{1}{4} \qquad \tfrac{3}{4},\tfrac{1}{4},\tfrac{1}{4}.* \tag{7–2.4a}$$

These are four of the eight 4-f sites in the structure. (One of the eight is identified in Fig. 7–2.2.) This structure, with CN = 4, is formed by AX compounds where the atoms must share four covalent bonds (Fig. 8–4.3). Relatively few fully ionic compounds are forced into this structure, because $r/R \leq$ 0.4 is rather uncommon.

The distance in Fig. 7–2.2 from an atom at 1,1,1 to the 4-f site at $\tfrac{3}{4},\tfrac{3}{4},\tfrac{3}{4}$ is one quarter of the body diagonal of the cubic cell. That distance is also $(r + R)$;

---

* Or alternatively at

$$\tfrac{1}{4},\tfrac{1}{4},\tfrac{1}{4} \qquad \tfrac{3}{4},\tfrac{3}{4},\tfrac{1}{4} \qquad \tfrac{3}{4},\tfrac{1}{4},\tfrac{3}{4} \qquad \tfrac{1}{4},\tfrac{3}{4},\tfrac{3}{4} \tag{7–2.4b}$$

There are eight 4-f sites per fcc unit cell. Thus, only half of these are occupied by an AX compound, because there are only four fcc positions per unit cell.

therefore,

$$\text{body diagonal} = 4(r + R) = a\sqrt{3};$$

or

$$a = 4(r + R)/\sqrt{3}. \tag{7-2.5}$$

Example 7-2.1

The compound CsBr has the same structure as CsCl. The centers of the two unlike ions are separated 0.37 nm. (a) What is the density of CsBr? (b) What is the radius of the Br^- ions in this structure?

Procedure. (a) It is necessary to obtain the lattice constant a to calculate the unit cell volume a^3 for density, $\rho = m/a^3$. Atomic mass of Br is from Fig. 2-1.1. (b) Since CN = 8, the radius of Cs^+ must be increased 3% from the value in Appendix B, where CN = 6.

Calculation

a)
$$a = 2(r + R)/\sqrt{3} = 2(0.37 \text{ nm})/\sqrt{3} = 0.427 \text{ nm}.$$

$$\rho = \frac{(132.9 + 79.9 \text{ g})/(0.602 \times 10^{24})}{(0.427 \times 10^{-9} \text{ m})^3} = 4.5 \times 10^6 \text{ g/m}^3$$

$$= 4.5 \text{ Mg/m}^3 \qquad\qquad (\text{or } 4.5 \text{ g/cm}^3).$$

b)
$$R_{Br^-} = 0.37 \text{ nm} - (0.167 \text{ nm})/0.97 = 0.20 \text{ nm}. \quad \blacktriangleleft$$

Example 7-2.2

Calculate the density of FeO which has an NaCl-type structure. (Assume equal numbers of Fe and O ions.)

Solution. From Table 2-4.1, or Appendix B,

$$r_{Fe^{2+}} = 0.074 \text{ nm} \qquad \text{and} \qquad R_{O^{2-}} = 0.140 \text{ nm}$$

when CN = 6. There are 4 Fe^{2+} and 4 O^{2-} per unit cell. (See Fig. 7-2.3.) Then

$$V = a^3 = [2(0.074 + 0.140) \times 10^{-9} \text{ m}]^3$$
$$= 78.4 \times 10^{-30} \text{ m}^3;$$

$$m = 4(55.8 + 16.0 \text{ amu})/(0.6 \times 10^{24} \text{ amu/g})$$
$$= 479 \times 10^{-24} \text{ g};$$

$$\rho = m/V = \frac{479 \times 10^{-24} \text{ g}}{78.4 \times 10^{-30} \text{ m}^3} = 6.1 \times 10^6 \text{ g/m}^3 \qquad (\text{or } 6.1 \text{ g/cm}^3).$$

Comments. The measured density is about 5.7 g/cm^3 because the structure contains cation vacancies (Fig. 4-3.2 and Example 4-3.1). $\quad \blacktriangleleft$

Example 7-2.3

The compound CdS has the same structure as ZnS. The centers of the two unlike ions are separated 0.25 nm. (a) What is the volume of the unit cell? (b) What is the density?

Procedure. Follow the procedure of Example 7-2.1, *except* for the fact that the body diagonal is $4(r + R) = a\sqrt{3}$ in the ZnS structure.

Calculation

a)
$$a = 4(r + R)/\sqrt{3}$$
$$= 4(0.25)/\sqrt{3} = 0.577 \text{ nm.}$$
$$a^3 = 0.192 \text{ nm}^3.$$

b) Since Cd = 112.4 amu, and S = 32.1 amu (Fig. 2–1.1),

$$\rho = 4(144.5 \text{ g}/0.6 \times 10^{24})/(0.577 \times 10^{-9} \text{ m})^3 = 5.0 \text{ Mg/m}^3 \qquad \text{(or 5.0 g/cm}^3\text{).}$$

Added information. Since CN = 4, $R_{S^{2-}}$ = 0.181 nm/1.1 from Appendix B, where $R_{S^{2-}}$ is given for CN = 6. Therefore, $r_{Cd^{2+}}$ = 0.25 nm − 0.181/1.1 = 0.085 nm (= 0.094 for CN = 6). ◀

Example 7–2.4

MnS has three polymorphs. Two of these are (a) the NaCl-type structure (Fig. 7–2.3), and (b) the ZnS-type structure (Fig. 7–2.4). What percent volume change occurs when the second type (ZnS) changes to the first type (NaCl)? (See Appendix B for radii when CN = 6, and the attached footnote when CN ≠ 6.)

Procedure. Each unit cell (Figs. 7–2.3 and 7–2.4) has 4 Mn^{2-} ions and 4 S^{2-} ions. Therefore, we can consider one unit cell of each. Also, $a_{NaCl} = 2(r + R)$; and $a_{ZnS}\sqrt{3} = 4(r + R)$.

Calculation

a)
$$V_{NaCl} = a^3 = [2(0.080 + 0.184 \text{ nm})]^3 = 0.147 \text{ nm}^3;$$

b)
$$V_{ZnS} = a^3 = [4(0.073 + 0.167 \text{ nm})/\sqrt{3}]^3 = 0.170 \text{ nm}^3;$$

$$(\Delta V/V)_{b \to a} = -14 \text{ v/o.}$$

Comment. Only the NaCl-type is stable; however, the other two polymorphs can be formed with appropriate starting materials, and crystallization procedures. ◀

7–3

CERAMIC CRYSTALS ($A_m X_p$)

Not all binary compounds have equal numbers of A and X atoms (or ions). We shall consider only two cases out of many to illustrate the point, the CaF_2 structure and the Al_2O_3 structure. Fluorite (CaF_2) is the basic structure for UO_2, which is used in nuclear fuel elements, and provides the pattern for one of the polymorphs of ZrO_2, which is a useful high-temperature oxide. Corundum, Al_2O_3 is one of the most widely used ceramics for technical purposes. We will encounter Al_2O_3 in a sparkplug (Fig. 7–7.1b). Other uses range from emery grinding wheels to acid pumps, to substrates for printed circuits (Fig. 7–1.1a) and on to high-temperature materials for catalyst supports in exhaust systems (Fig. 7–1.1b).

The *CaF₂ structure* has cations, Ca^{2+}, at fcc lattice points: $0,0,0$; $\frac{1}{2},\frac{1}{2},0$; $\frac{1}{2},0,\frac{1}{2}$; and $0,\frac{1}{2},\frac{1}{2}$. The anions, F^-, are in 4-f interstitial positions among the cations and include *both* sets listed in "Eqs." (7–2.4a and b). This 4-to-8 ratio of Ca^{2+} to F^-

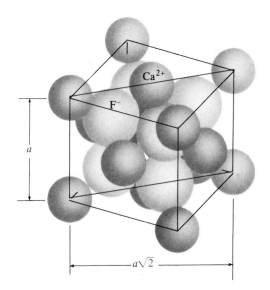

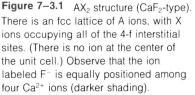

Figure 7–3.1 AX$_2$ structure (CaF$_2$-type). There is an fcc lattice of A ions, with X ions occupying all of the 4-f interstitial sites. (There is no ion at the center of the unit cell.) Observe that the ion labeled F$^-$ is equally positioned among four Ca^{2+} ions (darker shading).

accommodates the 1-to-2 ratio of *m*-to-*p* in A$_m$X$_p$. A sketch is shown in Fig. 7–3.1.

The structure just described is fcc with F$^-$ ions in 4-f sites among Ca^{2+} ions. In addition, the calcium ions are among eight fluorine ions. However, note that not all such 8-f sites are occupied; e.g., those at the center of unit cells (Fig. 7–3.1), and those at the mid-points of the edges of the unit cells. These vacant sites are important to the UO$_2$ of nuclear fuel elements, which has this crystal structure, because these holes provide space for fission products to reside.*

The structure of Al$_2$O$_3$ will be the final binary compound to receive consideration. It has an hcp array of O^{2-} ions. Two thirds of the 6-f interstitial sites are occupied by Al^{3+} ions.† With an interatomic distance of only 0.191 nm separating the 3$^+$ and 2$^-$ charges of the aluminum ion and its six oxygen neighbors, it is not surprising that the bonding energies are high.‡ This is reflected in the melting temperature (>2000°C), hardness (Mohs = 9), and the resistance of Al$_2$O$_3$ to a large number of chemicals. Furthermore, its combination of low electrical conductivity and a relatively high thermal conductivity (Appendix C) makes Al$_2$O$_3$ useful for many electrical applications.

---

* If uranium metal were used as a fuel element, it would expand excessively when atoms are split into two or more daughter atoms, because the metallic crystal does not have comparable vacant sites.

† This oversimplifies the structure, since it does not take into account a distortion that develops as a consequence of the vacant 6-f sites.

‡ Recall from Eq. (2–4.1) that the coulombic attractive forces are proportional to $-Z_1Z_2/a_{1-2}^2$ where Z_1 and Z_2 are the valences.

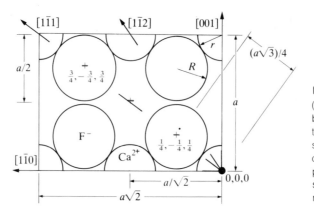

Figure 7–3.2 CaF_2 structure. (The (110) plane of Fig. 7–3.1, but with the origin shifted to the lower, right corner.) This sketch, matching the plane outlined in color on the previous figure is drawn to scale. Note that the F^- ions do not "touch."

Example 7–3.1

The lattice constant of CaF_2 is 0.547 nm. (a) Sketch the arrangement of ions on a {110} plane of CaF_2. (b) What is the sum of the two radii ($r_{Ca^{2+}} + R_{F^-}$)? •(c) What is the linear density of lattice points in the [1$\bar{1}$2] directions?

Procedure. The [1$\bar{1}$1] direction in Fig. 7–3.2 is the body diagonal (= $a\sqrt{3}$) of the unit cell in Fig. 7–3.1. Since the ion labeled F^- in Fig. 7–3.1 is equally positioned among four Ca^{2+} ions, it is at a quarter point, and $4(r + R) = a\sqrt{3}$.

Solution

a) Figure 7–3.2.

b)
$$4(r + R) = \sqrt{a^2 + a^2 + a^2}$$
$$r + R = (0.547 \text{ nm})\sqrt{3}/4 = 0.237 \text{ nm}.$$

• c) Since CaF_2 is fcc, the repetition distance t will be from 0,0,0 to $\frac{1}{2}, -\frac{1}{2}, 1$, or
$$\sqrt{a^2 + (-a/2)^2 + (a/2)^2}.$$

Therefore:
$$t = a\sqrt{1.5} = 0.547 \text{ nm } (1.225) = 0.67 \text{ nm},$$
and

$$\text{linear density} = 1.49/\text{nm} \qquad (\text{or } 1.5 \times 10^6/\text{mm}).$$

Comments. The F^- ions are separated by about 0.03 nm (Fig. 7–3.2).

The data from Appendix B would indicate $a = 0.515$ nm; however, this does not take into account the mutual repulsion of the 8 F^- ions around the vacant 8-f sites. This repulsion expands the structure.

This is the same structure as UO_2. The unoccupied 8-f sites provide space for fission products to reside within the crystal. ◀

• Example 7–3.2

Uranium dioxide has the CaF_2 structure. As such, it has a large interstitial site at the center of the unit cell. (See Fig. 7–3.2.) (a) How many such holes are there per unit cell? (b) Using adjusted data from Appendix B, what is the radius ratio, r/R, for UO_2?

Discussion

a) The unit cell is fcc, therefore, there will be four equivalent vacant sites. These are $\frac{1}{2},\frac{1}{2},\frac{1}{2}$ plus the permutations of $\pm\frac{a}{2},0,\pm\frac{a}{2}$ (Section 3–5). These include the center vacant site plus the midpoints of all 12 unit cell edges (which in turn are associated with 4 unit cells) to give $1 + \frac{12}{4} = 4$.

b) Since the coordination numbers are 8 and 4, respectively,

$$r_{U^{4+}} = 0.097 \text{ nm}/0.97 = 0.100 \text{ nm} \quad \text{and} \quad R_{O^{2-}} = 0.140 \text{ nm}/1.1 = 0.127 \text{ nm}.$$

$$r/R = 0.79.$$

A structure with CN = 6 (and therefore 0.097 nm/0.140 nm = 0.69) could be proposed for UO_2. However, that coordination does not permit a 1-to-2 ratio of the two atoms. ◀

7–4

MULTIPLE COMPOUNDS

$A_m B_n X_p$ **Compounds**

Although the presence of three types of atoms lends additional complexity, several $A_m B_n X_p$ compounds are of sufficient interest to warrant our attention. First among these is $BaTiO_3$, the prototype for the ceramic materials used in applications such as cartridges for record players. Above 120°C, $BaTiO_3$ has a cubic unit cell with Ba^{2+} ions at the corners, O^{2-} ions at the center of the faces, and a Ti^{4+} ion at the center of the cell (Fig. 7–4.1). This structure is altered very slightly below 120°C. With the change, it becomes a useful piezoelectric material (Section 9–3).

Nonmetallic magnets may also be $A_m B_n X_p$ compounds, the most common being a ferrospinel (often called a ferrite) with the composition of MFe_2O_4, where the M are divalent cations with radii of 0.075 ± 0.01 nm. These *spinel* structures have a close-packed (fcc) array of O^{2-} ions with cations located in

Figure 7–4.1 Cubic $BaTiO_3$. This structure is stable above 120°C, where it has a Ti^{4+} ion in the center of the cube, Ba^{2+} ions at the corners, and O^{2-} ions at the center of each face.

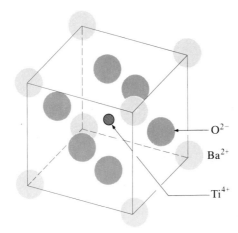

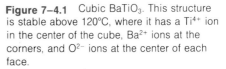

one half of the 6-f sites and in one eighth of the 4-f *interstitial sites* (Fig. 7–2.2). The magnetic characteristics of these materials are influenced by the cation occupants (Section 9–1).

• **Solid Solutions**   Brief mention was given in Section 4–3 to solid solutions in ionic compounds. Two chief requirements were cited as necessary for solid solutions to occur: (1) compatibility in size and (2) balance in charge. These limitations are not as rigid as might be surmised, because compensation may be made in charge. For example, Li^+ ions may replace Mg^{2+} ions in MgO *if* F^- is simultaneously present to replace O^{2-} ions. Conversely, MgO may be dissolved in LiF. We may also find Mg^{2+} dissolved in LiF without comparable O^{2-} ions; however, in this case cation vacancies must be included. As a result, 2 Li^+ are replaced by $(Mg^{2+} + \square)$.

Ceramists depend heavily on solid solutions in the previously mentioned magnetic spinels, because optimum magnetic characteristics exist when part of the divalent metal ions, R^{2+}, are zinc ($r = 0.074$ nm) and the balance of the divalent ions are ferromagnetic; for example, Ni^{2+} ($r = 0.069$ nm). In this case it is a simple matter of direct substitution; however, for certain applications it is desirable to replace $2R^{2+}$ with an Li^+Fe^{3+} pair, or to replace $2Fe^{3+}$ with an $Mg^{2+}Ti^{4+}$ pair.

Study Aid (ceramic compounds)   The coordination relationships of ceramic compounds are presented with visual sketches in *Study Aids for Introductory Materials Courses,* Topic IX. This is recommended for the reader who has difficulty in viewing the 3-D patterns of ceramic compounds.

Example 7–4.1   A ferrospinel has a lattice of 32 oxygen ions, 16 ferric ions, and 8 divalent ions. (The unit cell contains 8 times as many oxygen ions as MgO does, so that the repeating pattern can be developed.) If the divalent ions are Zn^{2+} and Ni^{2+} in a 3:5 ratio, what weight fraction of ZnO, NiO, and Fe_2O_3 must be mixed for processing?

Solution. Basis: 8 mol wt of Fe_2O_3.

$$5 \text{ NiO} + 3 \text{ ZnO} + 8 \text{ Fe}_2\text{O}_3 \rightarrow (Zn_3, Ni_5)Fe_{16}O_{32}.$$

Wt fraction

5 NiO = 5(58.71 + 16.00)	=	373.5	= 0.197
3 ZnO = 3(65.37 + 16.00)	=	244.1	= 0.129
8 Fe_2O_3 = 8[2(55.85) + 3(16.00)]	=	1277.6	= 0.674
		1895.2	1.000

◄

• **Example 7–4.2**   A solid solution of MgO and LiF contains 11 w/o oxygen, 52 fluorine, 20 Mg, and 17 Li. How many positive ion vacancies are there per 200 unit cells?

Procedure. The structure of MgO and/or LiF is that of NaCl (Fig. 7–2.3). Thus, there are 800 anions per 200 unit cells. Determine the numbers of negative (anions) and positive ions, and calculate the Δ per 800 anions.

Calculation. Basis: 100 amu.

$$\begin{array}{lll}
\text{Oxygen:} & 11/16 = 0.69 \\
\text{Fluorine:} & 52/19 = 2.74
\end{array} \right\} \; 3.43 \text{ anions}$$

$$\begin{array}{lll}
\text{Magnesium:} & 20/24.3 = 0.82 \\
\text{Lithium:} & 17/6.9 = 2.46
\end{array} \right\} \; 3.28 \text{ cations}$$

$$\Delta = \overline{0.15} \; \square$$

Basis: 200 unit cells.

$$800 \text{ anions } (0.15\square/3.43 \text{ anions}) = 35\square/200 \text{ unit cells.}$$

Comments. There is a charge balance:

$$0.82(2) + (2.46)(1) = 4.1 = 0.69(2) + 2.74(1).$$

Because of these vacancies, positive ions can diffuse much more rapidly than in either pure MgO or pure LiF. ◄

7–5
SILICATES

Many ceramic materials contain *silicates,* partly because silicates are plentiful and cheap and partly because they have certain distinct properties that are useful in engineering applications. Probably the most widely known silicate is portland cement, which has the very definite advantage of being able to bond rock aggregates into a monolithic material. Many other construction materials, such as brick, tile, glass (Fig. 7–1.1), and vitreous enamel, are also made of silicates. Other engineering applications of silicates include electrical insulators, chemical ware, and reinforcing glass fibers.

Silicate Tetrahedral Units

The primary structural unit of silicates is the "SiO_4" *tetrahedron* (Fig. 7–5.1a), in which one silicon atom is coordinated interstitially among four oxygen atoms. The forces holding these tetrahedra together involve both ionic and covalent bonds; consequently the tetrahedra are tightly bonded. However, with either the ionic or covalent bonding mechanism, each oxygen of the tetrahedra has only seven electrons rather than eight in its outer shell.

Two methods are available to overcome this deficiency of electrons for the oxygen ions: (1) An electron may be obtained from metal atoms. In this case SiO_4^{4-} ions and Metal$^+$ ions are developed. (2) Each oxygen may share an electron pair with a second silicon. In this case multiple tetrahedral coordination groups are formed (Fig. 7–5.1b). The shared oxygen is the *bridging* oxygen.

Network Silicates

With pure SiO_2 there are no metal ions, and every oxgyen atom is a *bridging* atom between two silicon atoms (and every silicon atom is among four oxygen atoms, as shown in Fig. 7–5.2). This gives a networklike structure. Silica (SiO_2)

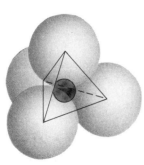

(a)

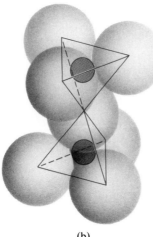

(b)

Figure 7–5.1 (a) Tetrahedral arrangement of SiO_4^{4-}. Compare with Fig. 2–3.3(b). This SiO_4^{4-} ion has obtained four electrons from other sources. (b) Double tetrahedral unit $(Si_2O_7)^{6-}$. The center oxygen atom is shared by each tetrahedral unit. Thus, it becomes a bridging oxygen.

can have several different crystal structures, just as carbon can be in the form of graphite or diamond. The structure shown in Fig. 7–5.2 is a high-temperature form. A more common structure of silica is quartz, the predominant material found in the sands of many beaches. Just as SiO_2 in Fig. 7–5.2 contains SiO_4 tetrahedra, so does quartz, but with a more complex lattice.

Another natural silicate is feldspar. The pink or tan mineral of granite is $KAlSi_3O_8$, which may be visualized as a network silicate with one of every four silicons replaced by an Al^{3+} ion. The latter, however, has only three charges as compared to four for silicon. Thus K^+ is present to balance out the charges. The K^+ ions are interstitial ions. However, these network structures are quite open, so that there is space for extra ions to be present (cf. Fig. 7–5.2).

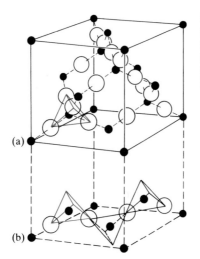

(a)

(b)

Figure 7–5.2 Network structure (silica). This SiO_2 polymorph (cristobalite) is fcc and stable at high temperatures. (a) Unit cell. Each silicon atom is surrounded by four oxygen atoms. Every oxygen atom joins two tetrahedra. (b) Part of the adjoining unit cell to show the bridging oxygens.

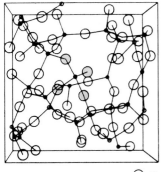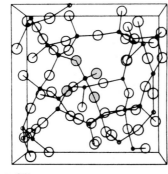

○ Oxygen • Silicon

Figure 7–5.3 A stereographic sketch of silica glass at 20°C. The figure should be viewed with a suitable stereo-viewer (or by controlled eye focus). As with the crystalline SiO_2 in Fig. 7–5.2, each silicon is among four oxygens, and each oxygen joins (bridges) two tetrahedral units. (By permission of T. F. Soules, General Electric, Cleveland. Published in *J. of Non-Crystalline Solids*, Vol. 49.)

Glasses

The principal commercial glasses are silicates. They have the SiO_4 tetrahedra described in the preceding section, plus some modifying ions. They are *amorphous*, i.e., noncrystalline. Figure 7–5.3 is a stereographic view of an SiO_2 glass, which may be contrasted with the crystalline SiO_2 of Fig. 7–5.2. Atoms in both have the same first neighbors, i.e., each oxygen atom is between two silicon atoms and *bridges* two tetrahedra. Each silicon is among four oxygens. However, in glass, differences arise in more distant neighbors because the glass, unlike the crystalline silicon, does not have a regular long-range pattern.

When only silica (SiO_2) is present, and every oxygen atom serves as a bridge, the glass is very rigid. Fused silica, for example, is extremely viscous even when it gets into the temperature range in which it is a true liquid. In polymer terms, its units are polyfunctional and its *network* structure is highly cross-linked. Fused silica is very useful in some applications because it has a low thermal expansion. However, its high viscosity makes it extremely difficult to shape.

Network Modifiers

Most silicate glasses contain *network modifiers*. These are oxides such as CaO and Na_2O that supply cations (positive ions) to the structure. As shown in Eq. (7–5.1a) and in Fig. 7–5.4, the addition of Na_2O to a glass introduces *two* Na^+ ions, and produces two *nonbridging oxygens,* each with a single negative charge. These oxygens are attached to only one silicon. Likewise, the introduction of *one* Ca^{2+} ion also leads to *two* nonbridging oxygens (Eq. 7–5.1b).

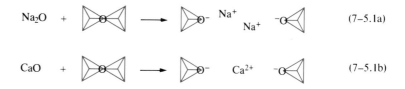

$$Na_2O \; + \quad\quad \longrightarrow \quad\quad Na^+ \quad Na^+ \quad\quad\quad\quad (7\text{–}5.1a)$$

$$CaO \; + \quad\quad \longrightarrow \quad\quad Ca^{2+} \quad\quad\quad\quad (7\text{–}5.1b)$$

The presence of these nonbridging oxygens reduces the activation energy required for the atom in movements that permit flow in molten glass. This is

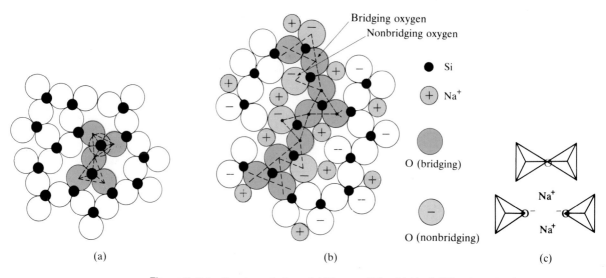

Figure 7–5.4 Structure of glass. (a) Vitreous SiO_2. (b) Na_2O-SiO_2 glass. (c) Oxygen bridging. (The fourth oxygen of each tetrahedron has been deleted for clarity.) In vitreous silica, all oxygen atoms are bridging. In a soda-silica glass, there is one nonbridging oxygen for each Na^+ ion. (In a lime-silica glass, there are two nonbridging oxygens for each Ca^{2+} ion.)

shown in Fig. 7–5.5 for glasses with increasing Na_2O content (and, therefore, an increasing fraction of nonbridging oxygens).

In commercial glasses that contain soda and lime, less than three fourths of the oxygen atoms serve as a bridge between SiO_4 tetrahedra. Thus, some of the units are directly connected to only two other tetrahedra. As a result, the glass is thermoplastic at elevated temperatures and can be shaped into products such as light bulbs, window glass, and fibers that become rigid during cooling.

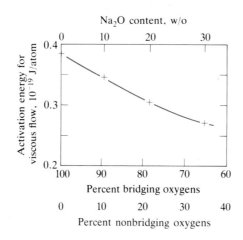

Figure 7–5.5 Energy for viscous flow versus bridging oxygens (Na_2O-SiO_2 glasses). The activation energy required for viscous flow decreases as a smaller percentage of the oxygen atoms share adjacent tetrahedra. (Adapted from Stevels, *Encyclopedia of Physics*.)

There are thousands of commercial glasses, because there are more than a dozen possible components that can be used in varying amounts to produce specific properties, such as index of refraction, color dispersion, or viscosity. The more prominent categories of glasses are shown in Table 7–5.1.

Example 7–5.1 Quartz (SiO_2) has a density of 2.65 Mg/m^3 (=2.65 g/cm^3). (a) How many silicon atoms (and oxygen atoms) are there per m^3? (b) What is the packing factor, given that the radii of silicon and oxygen are 0.038 nm and 0.114 nm, respectively?

Procedure. (a) There are 60.1 $g/(0.6 \times 10^{24}$ SiO_2). Thus, we can calculate the number of SiO_2 units in 2.65×10^6 g, or 1 m^3. That number n will also be the number of silicon atoms per m^3. (b) Calculate the volume of n silicon and $2n$ oxygen atoms. (Assume spheres).

Calculation

a)
$$SiO_2/m^3 = \frac{2.65 \times 10^6 \text{ g/m}^3}{(28.1 + 32.0) \text{ g}/(0.6 \times 10^{24} \text{ } SiO_2)}$$
$$= 2.645 \times 10^{28} \text{ } SiO_2/m^3$$
$$= 2.645 \times 10^{28} \text{ } Si/m^3 = 5.29 \times 10^{28} \text{ } O/m^3.$$

b)
$$V_{Si}/m^3 = (2.645 \times 10^{28}/m^3)(4\pi/3)(0.038 \times 10^{-9} \text{ m})^3 = 0.006 \text{ m}^3/m^3$$
$$V_O/m^3 = (5.29 \times 10^{28}/m^3)(4\pi/3)(0.114 \times 10^{-9} \text{ m})^3 = \underline{0.328} \text{ m}^3/m^3$$
$$\text{Packing factor} = 0.33.$$

Comments. Although there is considerable open space within this structure, most single atoms (except for helium) must diffuse through SiO_2 as ions. Thus their charges prohibit measurable movements at ambient temperatures. ◀

Example 7–5.2 A glass contains 80 w/o SiO_2 and 20 w/o Na_2O. What fraction of the oxygen atoms is nonbridging?

TABLE 7–5.1

Common Commercial Glass Types

Type	Major components, weight percent						Requirements
	SiO_2	Na_2O	CaO	Al_2O_3	B_2O_3	MgO	
Window	72	14	10	1		2	High durability
Plate (Arch.)	73	13	13	1			High durability
Container	74	15	5	1		4	Easy workability, chemical resistant
Lamp bulbs	74	16	5	1		2	Easy workability
Fiber (Elect.)	54		16	14	10	4	Low alkali
Pyrex	81	4		2	12		Low thermal expansion, low ion exchange
Fused silica	99						Very low thermal expansion

Procedure. Recalculate into mole percent so we may obtain the numbers of oxygen and sodium atoms. Each Na^+ accounts for one nonbridging oxygen.

Calculation. Basis: 100 g = 80 g SiO_2 + 20 g Na_2O.

$$80 \text{ g } SiO_2/[(28.1 + (2 \times 16.0)) \text{ g/mole}] = 1.33 \text{ mole } SiO_2 \quad or \quad 80.6 \text{ m/o } SiO_2;$$

$$20 \text{ g } Na_2O/[((2 \times 23.0) + 16.0) \text{ g/mole}] = 0.32 \text{ mole } Na_2O \quad or \quad 19.4 \text{ m/o } Na_2O.$$

New basis: 100 moles.

$$80.6 \text{ } SiO_2 = 80.6 \text{ Si} + 161.2 \text{ O}$$

$$19.4 \text{ } Na_2O = \frac{19.4 \text{ O} + 38.8 \text{ } Na^+}{80.6 \text{ Si} + 180.6 \text{ O} + 38.8 \text{ } Na^+}$$

Note (from Eq. 7–5.1a) that there is one nonbridging oxygen for each Na^+ added. Thus, there are 38.8 nonbridging oxygens (and 141.8 bridging oxygens):

$$\text{Fraction nonbridging oxygens} = 38.8/180.6 = 0.215.$$

Comments. A commercial soda-lime glass used for containers, etc. (Table 7–5.1), also has CaO additions. Each Ca^{2+} ion produces two nonbridging oxygens, both of which carry a single negative charge. (Cf. Eq. 7–5.1.) ◀

• **Example 7–5.3**   Refer to Fig. 7–5.2 and compare with Figs. 2–2.3(a) and 2–3.3(b). The O–Si–O bond angle is 109.5°. (a) What is the closest approach (center-to-center) of the oxygens? (b) Of the silicons? (With CN = 2, $R_{O^{2-}} \approx 0.114$ nm.)

Procedure. Sketch an isosceles triangle with a silicon atom at the 109.5° apex. The equal legs are $(r + R)$ in length. The base equals the center-to-center distance for nearest oxygen atoms (O-to-O). Silicon has $r = 0.042/1.1$, because CN = 4.

Calculation

a)
$$\frac{[\text{O-to-O}]/2}{r + R} = \sin(109.5°/2) = 0.82;$$

$$\text{O-to-O} = 2[(0.042 \text{ nm}/1.1) + 0.114 \text{ nm}][0.82] = 0.25 \text{ nm}.$$

b) Assume the Si–O–Si bond angle is 180° (Fig. 7–5.2).

$$\text{Si-to-Si} = 2[(0.042 \text{ nm}/1.1) + 0.114 \text{ nm}] = 0.30 \text{ nm}.$$

Comment. In some silica polymorphs, the Si–O–Si bond angle is less than 180° to accommodate the lone-pair electrons. ◀

7–6

MECHANICAL BEHAVIOR OF CERAMICS

With the exception of a few materials such as plasticized clay, ceramic materials are characterized by high shear strengths; thus, they are not easily deformed. We will see that this leads to high hardnesses and high compressive strengths, combined with notch sensitivity and low fracture strengths.

Compounds have ordered arrangements of dissimilar atoms (or ions). This was discussed for intermetallic compounds as an optional topic in Section 5–4.

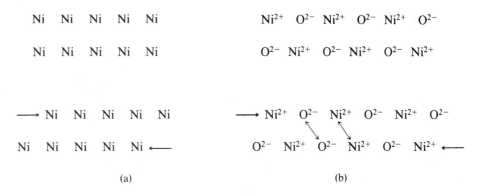

Figure 7–6.1 Comparison of slip processes (metallic nickel and nickel oxide). More force is required to displace the ions in NiO than the atoms in nickel. The strong repulsive forces between like ions become significant. Nickel also has more slip systems than does nickel oxide.

In ceramic phases, cations are coordinated with anions; and anions are coordinated with cations (Section 7–2). This is natural, since unlike ions attract and like ions repel. However this interferes with slip, as shown in Fig. 7–6.1(b). The slip process in the horizontal direction (as drawn) is precluded in NiO, MgO, and other NaCl-type crystals. The displaced arrangement would have to be achieved by first breaking strong positive-to-negative bonds of the Ni^{2+} and O^{2-} ions. Next, like charges would have to pass adjacently as the Ni^{2+} ion moves to join the next O^{2-} ions. This consideration does not exist in the metal.

In NiO, slip is possible in certain other directions without the above restriction, e.g., at a 45° direction. This would be slip in a $\langle 1\bar{1}0 \rangle$ direction on a $\{110\}$ plane; however, there are only six slip systems possible. This compares with twelve slip systems for $\{111\}\langle 1\bar{1}0 \rangle$ slip in fcc nickel (Table 5–4.1). Furthermore the slip vector $\mathbf{b}$ in NiO is $(r + R)\sqrt{2}$, or 0.3 nm, which is greater than the 0.25 nm in nickel. Everything considered—(1) coulombic repulsion, (2) fewer slip systems, and (3) longer displacement distances*—adds up to a much greater resistance to plastic deformation in crystalline ceramics than in their metallic counterparts. This difference is accentuated even further when more complex ceramic phases are considered, e.g., silicates, the spinel of Section 7–4, or portland cement.†

Hardness

Ceramic phases are hard because they generally cannot undergo plastic deformation. As a result, abrasive materials such as emery consist of Al_2O_3 (usually with some Fe_2O_3 and/or TiO_2 in solid solution). Silicon carbide (SiC) and TiC

---

* The energy for dislocation movements increases with the square of the slip vector, $\mathbf{b}$ (Eq. 5–4.2).

† Some chemists consider the chief hydrated phase of portland cement to be

$$Ca_2[SiO_2(OH)_2]_2 \cdot (CaO)_{y-1} \cdot xH_2O!!$$

(See Section 15–2.)

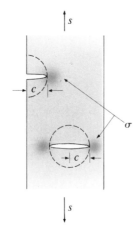

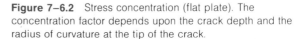

Figure 7–6.2 Stress concentration (flat plate). The concentration factor depends upon the crack depth and the radius of curvature at the tip of the crack.

are equally important for grinding and cutting metals, and similar manufacturing processes. One form of SiC has the ZnS-type structure (Fig. 7–2.4); TiC has the NaCl-type structure (Fig. 7–2.3).

Exceptions to the general rule that ceramics are relatively hard can be related directly to structures. Talc, clays, and mica are soft because they have sheet structures. They are strong within the sheets; however, the sheets are held together by weak secondary bonds.

Notch Sensitivity

A notch or crack is a *stress-raiser*. The effect is much greater than would be anticipated simply on the basis of the decreased cross-sectional area. The true stress, σ, at the tip of the notch or flaw (Fig. 7–6.2) exceeds the nominal tensile stress, s, by a factor that includes $\sqrt{c}$, where c is the crack length from the surface (or half of an internal crack's width). Furthermore, the stress concentration is greater when the tip of the notch is sharp, and less when it is rounded to a larger radius of curvature.

If σ exceeds the yield strength in a *ductile* material, the tip of the notch will deform to a larger radius, and the stress concentration will be reduced. If the notch is a crack in a *nonductile* material, the radius of curvature may be of atomic dimensions (say 0.1 nm). A crack that reaches only 0.1 μm to 10 μm (10^{-4} mm to 10^{-2} mm) into the surface may multiply the stress by a factor of 10^2 or 10^3. Deeper cracks are more severe. Even though ceramics are strong in shear, we find stress concentrations σ that exceed the bond strength between the atoms. Thus the crack may propagate. This increases the crack depth c, and accentuates the stress concentration still further, until catastrophic failure occurs.

Ceramic materials (and intermetallic compounds, Section 5–4) are generally weak in tension because of their resistance to shear at the crack tip. The same factor, shear strength, makes them strong in compression. A compressive load can be supported across a microcrack without extending the flaw.

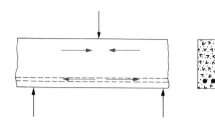

Figure 7–6.3 Reinforced concrete beam. This beam uses the nonductile material in the compressive positions.

Use of Nonductile Materials

This relationship of high compressive strength and lower tensile strength is important to the design engineer. Concrete, brick, and other ceramics are used primarily in compressive locations (Fig. 7–6.3). When it is necessary to subject such materials as glass to bending (and therefore tensile loading), it is usually necessary to increase dimensions. For example, the viewing glass of a television picture tube may be as much as 15 mm thick.

Since ceramic materials are stronger in compression than in tension, "tempered" glass is used for glass doors, rear windows of cars, and similar high-strength applications. To produce *tempered glass*, the glass plate is heated to a temperature high enough to permit adjustments to stresses among the atoms and is then quickly cooled by an air blast or oil quench (Fig. 7–6.4). The surface contracts because of the drop in temperature and becomes rigid, while the center is still hot and can adjust its dimensions to the surface contractions. When the center cools and contracts slightly later, compressive stresses are produced at the surface (and tensile stresses in the center). The stresses that remain in the cross section of the glass are diagrammed in Fig. 7–6.5. A consid-

Hot glass
No stresses

Surface cooled quickly
Surface contracts
Center adjusts
Only minor stresses

Center cools
Center contracts
Surface is compressed
Center in tension

Figure 7–6.4 Dimensional changes in "tempered" glass.

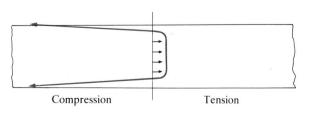

Compression   Tension

Figure 7–6.5 Surface compression of "tempered" glass. These compressive stresses must be overcome before the surface can be broken in tension.

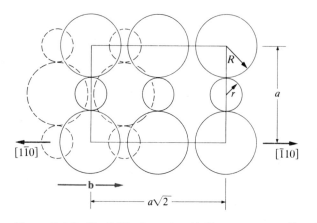

Figure 7–6.6 The (110) plane of an NaCl-type structure. The slip system is [1$\bar{1}$0] (110) and the slip vector **b** is equal to $a/\sqrt{2}$. (The dashed circles indicate the locations of atoms in the underlying (110) plane.) (See Example 7–6.1.)

erable deflection must be applied to the glass before tensile stresses can be developed in the surface of the glass where cracks start.* In effect, since the compressive stresses must be overcome first, the overall strength of the glass is greatly enhanced.

Example 7–6.1

(a) Identify the six $\langle 1\bar{1}0\rangle\{110\}$ slip systems for MnS (which has an NaCl structure).
• (b) How long is the slip vector in these slip systems?

Solution

a) Figure 7–6.6 shows the (110) plane of the NaCl-type structure. Each plane has only one set of slip directions, in this case [1$\bar{1}$0] or its negative [$\bar{1}$10]. Since there are six planes in the {110}, the six slip systems are

$$[1\bar{1}0](110); \quad [\bar{1}01](101); \quad [0\bar{1}1](011);$$
$$[110](1\bar{1}0); \quad [101](10\bar{1}); \quad [011](01\bar{1}).$$

b) Since $r_{Mn^{2+}} = 0.080$ nm and $R_{S^{2-}} = 0.184$ nm,

$$a = 0.528 \text{ nm.}$$
$$\mathbf{b} = (0.528 \text{ nm})\sqrt{2}/2$$
$$= 0.373 \text{ nm.} \quad \blacktriangleleft$$

---

* If a crack penetrates through the compressive skin (e.g., by scratching) into the tension zone shown in Fig. 7–6.5, the crack may become rapidly self-propagating. The aftermath of this effect can be observed in a broken rear window of a car, where the crack pattern is a mosaic, rather than spearlike shards.

7–7

PROCESSING OF CERAMIC MATERIALS

Attention will be directed in this section to three principal procedures for processing ceramic materials—(1) viscous forming, (2) sintering processes, and (3) single crystal preparation. Of course, these are not the only processes common to ceramic production. In deference to time and space, no attention will be given to raw material preparation, such as purification, particle sizing, and spray drying.

Viscous forming involves the melting and shaping of a viscous silicate liquid. The *sintering* process starts with finely divided particles that are agglomerated into the desired shape; this is generally followed by *firing* to produce a bond between the particles. *Single crystal preparation* of ceramic compounds almost always are grown by solidification of a melt upon a seed crystal and directional solidification.

Viscous Forming

Commercial glasses (Table 7–5.1) are manufactured by this process. When heated, the glass becomes sufficiently thermoplastic to be shaped into the final product. Prior to the final shaping, however, it is necessary to completely melt the component oxides in order to assume a homogeneous composition and to remove entrapped gases. This removal is particularly important because the soda and lime contents of glasses (Table 7–5.1) are obtained from Na_2CO_3 and $CaCO_3$, both of which release CO_2. A gas bubble that remains in the glass is, of course, a defect. The final shaping process may be pressing (for structural glass block), or sagging (for many car windows), or blowing (for light bulbs), or drawing (for glass fibers—Fig. 7–1.1c), etc.

Ceramic glasses have many of the characteristics of thermoplastic polymers. In fact, the initial recognition of the *glass temperature* T_g was for silicous glasses (Figs. 4–5.3 and 6–5.1). The *viscosity* η of a glass is temperature-sensitive:

$$\log_{10} \eta = C' + \frac{B'}{T},\qquad (7\text{–}7.1)$$

where C' and B' are constants* for a glass.

Viscosity control is important in glass manufacture. First, the glass must have a low viscosity (<30 Pa·s, or 300 poises) in the *melting range* so that

---

* The flow of glass requires the movement of atoms past other atoms and therefore depends on an activation energy E. A glass with a high *fluidity*, f (low viscosity, since $\eta = 1/f$) corresponds to a glass that has fast diffusion. Thus, as in Eqs. (4–6.6) and (4–7.4),

$$\ln f = \ln f_0 - E/kT = -\ln \eta.\qquad (7\text{–}7.2)$$

This leads to Eq. (7–7.1). Since $\ln \eta = 2.3 \log \eta$, the terms C' and B' are, respectively, $-\ln f_0/2.3$ and $E/2.3k$. Fluidity is a diffusivity (m^2/s) per unit force, that is, m^2/s·N. Thus, viscosity has dimensions of Pa·s. Former cgs units were "poises," where 10 poises = 1 Pa·s.

bubbles can escape and homogenization can occur as discussed above. The *working range* covers several orders of magnitude. Fast operations, such as the production of light bulbs, require a relatively fluid glass ($\approx 10^{2.5}$ Pa·s, or $10^{3.5}$ poises); the drawing operation for heavy sheet glass requires a more viscous product ($\approx 10^6$ Pa·s).

Two other viscosity points are important: (1) the *annealing point*, the temperature with $\eta = 10^{12}$ Pa·s (or 10^{13} poises), and (2) the *strain point*, the temperature with $\eta = 10^{13.5}$ Pa·s. At the temperature of the annealing point, atoms can move sufficiently so that residual thermal stresses may be relieved within about 15 minutes, a time compatible with production schedules. Below the strain point, the glass is sufficiently rigid so that it may be handled without the generation of new residual stresses. The strain point is below the glass transition temperature.

Sintering Processes

Most nonvitreous ceramic materials are made from very finely ground particles that are sintered (fired) into a monolithic product. The artist follows the prehistoric procedure of using finely divided clays that are *hydroplastic* when wet. Products such as brick can be shaped into the desired form, allowed to dry, and then fired so that the altered clay particles sinter together and form a strong, hard bond. Alternatively, it is possible to slip-cast ceramics by making a suspension, or *slip,* that can be poured into a porous plaster mold. The mold dewaters the slip adjacent to the mold wall. The balance of the slip is dumped; the product is then *dried* and *fired*. Considerable shrinkage accompanies this process because the packing factor for the solid particles in the original slip casting is low.

Technical ceramic products are commonly formed by a *pressing* process. A spark plug insulator, to which we have referred several times, is made by *isostatic molding* (Fig. 7–7.1). This has a major manufacturing advantage in that high pressures may be applied uniformly. Also, a negligible amount of plasticizing liquid is required that must be dried. Thus, the subsequent shrinkage is small and easily controlled. Of course, high temperatures are required for firing or sintering, since these insulators are predominantly Al_2O_3. Such oxides have strong bonds and therefore low diffusivities for atom movements.

Sintering requires heating in order to agglomerate small particles into bulk materials. Sintering without the formation of a liquid requires diffusion within the solid and therefore occurs most rapidly at very high temperatures (short of melting). Many powdered metal parts and various dielectric and magnetic ceramics are produced by *solid sintering*. These ceramic materials cannot be made by melting, and often no feasible crucible or mold is available.

The principle involved in solid sintering is illustrated in Fig. 7–7.2. As shown in part (a), there are two surfaces between any two particles before sintering. After sintering, there is a single grain boundary. The two surfaces are high-energy boundaries; the grain boundary has much less energy. Thus, this reaction occurs naturally if the temperature is high enough for a significant number of atoms to diffuse.

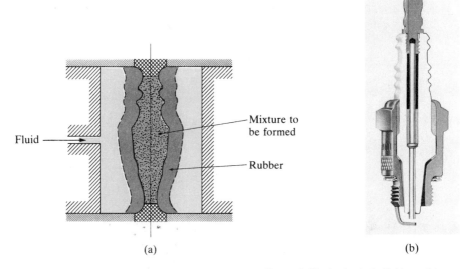

Fluid

Mixture to
be formed

Rubber

(a)

(b)

Figure 7–7.1 (a) Isostatic molding (sparkplug manufacture). The hydrostatic fluid provides radial pressure to the product. Maximum compaction is possible with this procedure. (Jeffery, U.S. Patent 1,863,854.) (b) Cut-away view of modern automotive sparkplug. The insulator is primarily Al_2O_3. (Courtesy of the Champion Spark Plug Co.)

The actual mechanism of sintering is shown in Fig. 7–7.3 in photographs taken through a scanning electron microscope. The points of contact between particles grow by a diffusion of atoms. The net diffusion produces a *shrinkage,* and an accompanying reduction of porosity. Figure 7–7.4 shows the progress of sintering for sodium fluoride.

Single-crystal Preparation

Metals are only rarely used as single crystals; single crystals of polymers are rare, small, and usually imperfect. However, there are numerous ceramic phases (and also semiconductors) that find technical usage as single crystals. Figure 7–7.5 shows natural and artificial quartz crystals. The latter are used as crystal oscillators to maintain frequency control at radio and microwave frequencies. Hydrothermal and fusion techniques will be described.

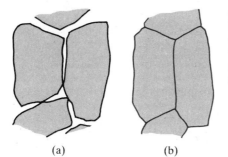

(a)

(b)

Figure 7–7.2 Solid sintering. (a) Particles before sintering have two adjacent surfaces. (b) Grains after sintering have one boundary. The driving force for the sintering is the reduction of surface area (and therefore of surface energy).

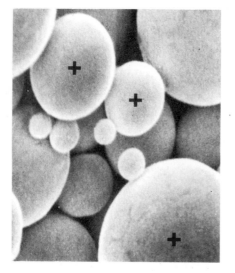

(a)

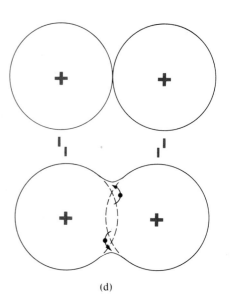

(d)

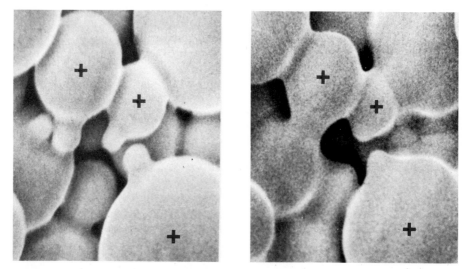

(b)

(c)

Figure 7–7.3 Sintering (nickel powder). The initial points of contact in (a) become areas of contact in (b) and (c) while heating to 1100°C. (d) The atoms diffuse from the contact points to enlarge the contact area. (Vacancies diffuse in the opposite direction.) The particles of powder move closer together, and the amount of particle surface is reduced. (R. M. Fulrath, dec., University of California, Berkeley.)

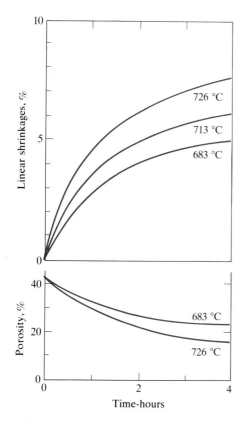

Figure 7-7.4 Sintering shrinkage (NaF). Powdered NaF (-330 mesh) loses porosity as it shrinks during sintering. (Adapted from Allison and Murray, *Acta Metallurgica*.)

Figure 7-7.5 Quartz crystals (SiO$_2$). (a) Natural crystals. (b) Artificial crystals being removed from the hydrothermal cell. They will be used for frequency controls of radio circuits. (Courtesy of Western Electric.)

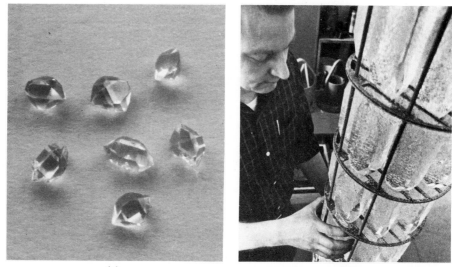

(a)   (b)

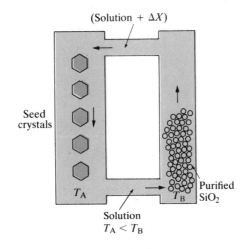

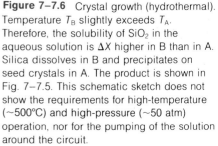

Figure 7–7.6   Crystal growth (hydrothermal). Temperature T_B slightly exceeds T_A. Therefore, the solubility of SiO_2 in the aqueous solution is ΔX higher in B than in A. Silica dissolves in B and precipitates on seed crystals in A. The product is shown in Fig. 7–7.5. This schematic sketch does not show the requirements for high-temperature ($\sim$500°C) and high-pressure ($\sim$50 atm) operation, nor for the pumping of the solution around the circuit.

The *hydrothermal* growth of quartz crystals is shown schematically in Fig. 7–7.6. Since the solubility of SiO_2 in an alkaline solution of NaOH and H_2O varies with temperature, quartz sand can be dissolved at T_B and SiO_2 precipitated onto seed crystals at T_A as sketched in Fig. 7–7.6. The temperatures that are used are as high as 500°C (930°F); therefore, the process must be performed at high pressures (or the water would volatilize). Close temperature control is required: (a) T_B should be as high as possible to accelerate solution; (b) greater differentials between T_B and T_A increase the chemical efficiency. However, excessive supercooling in A will cause spontaneous nucleation, which must be avoided if we wish to obtain single crystals.

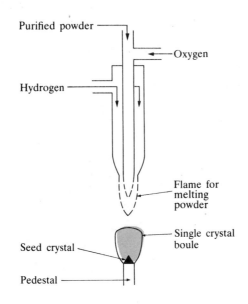

Figure 7–7.7   Crystal growth (flame fusion). Single crystals of sapphire and ruby (both Al_2O_3 plus solutes) are grown in this manner, because H_2-O_2 flames can exceed the 2050°C melting requirement. Powder enters through the torch, melts in the flame, and falls onto the boule, where it solidifies as part of the growing crystal.

Fusion techniques may be by $H_2 + \frac{1}{2}O_2$ flames as shown schematically in Fig. 7–7.7. In order to grow a single crystal, previously prepared powders of the correct composition are melted in the hydrogen-oxygen flame. The small droplets fall onto a preselected *seed* crystal that grows into a large single crystal boule. Alternatively, the Czochralski method (Fig. 7–7.8) starts with a seed crystal that is lowered against the surface of a melt and then pulled slowly upward, allowing growth to proceed.

Example 7–7.1

An electrical glass has a working range of 870°C ($\eta = 10^6$ Pa·s, or 10^7 poises) to 1300°C ($\eta = 10^{2.5}$ Pa·s, or $10^{3.5}$ poises). Estimate its annealing point ($\eta = 10^{12}$ Pa·s, or 10^{13} poises).

Procedure. Since the flow of glass involves the movements of atoms with respect to their neighbors, both viscosity, η, and fluidity, f, respond to thermal activation. Therefore, either Eq. (7–7.1) or Eq. (7–7.2) can be our basis for relating viscosity to temperature.

Calculation

$$6 = C' + B'/(870 + 273),$$

$$2.5 = C' + B'/(1300 + 273).$$

Solving simultaneously,

$$B' = 14{,}600 \quad \text{and} \quad C' = -6.8$$

For $\eta = 10^{12}$ Pa·s,

$$12 = -6.8 + 14{,}600/T,$$

$$T = 775 \text{ K } (\approx 500°C).$$

Comment. The annealing point corresponds approximately to the glass temperature, T_g. ◄

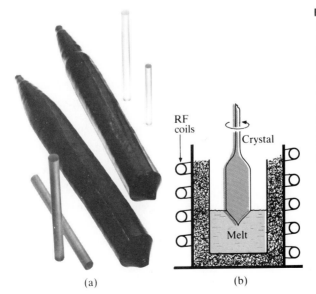

(a) (b)

Figure 7–7.8 Crystal growth (pulling). (a) Boules (up to 100-mm dia.) and laser rods. (b) Crystal growth by the Czochralski, or "pulling," method (schematic). A seed crystal is lowered to the surface of the melt and rotated as it is slowly raised, growing in the process. (Courtesy of Union Carbide, Electronics Division.)

RF coils

Crystal

Melt

Example 7–7.2   A ceramic magnet has a porosity of 28 v/o before sintering and a density of 5.03 g/cm³ after sintering. The true density is 5.14 g/cm³ (=5.14 Mg/m³). (a) What is the porosity after firing (sintering)? (b) If the final dimension should be 16.3 mm, what should be the die dimension?

Procedure. This problem may be set up various ways. However, the volumes of one gram are convenient because they let us compare volume changes directly.

Calculation

a)   True volume of 1 g = 1/5.14 = 0.1946 cm³.

                 Fired volume of 1 g = 1/5.03 = 0.1988 cm³.

                 Final porosity = (0.1988 − 0.1946)/0.1988 = 2%.

b)   True volume = $0.72V_0 = 0.98V_f$;

$$\frac{V_0}{V_f} = \frac{L_0^3}{L_f^3} = \frac{0.98}{0.72}.$$

$$L_0 = L_f \sqrt[3]{0.98/0.72}$$

$$= (16.3 \text{ mm})(1.108) = 18.1 \text{ mm}.$$

Comment. Note that if the die is set up for a 10% shrinkage (18.1 mm to 16.3 mm), then processing procedures must be consistent so that there is *always* a 28% porosity in the pressed (presintered) stage. Otherwise, the dimensional specifications will be missed. Processing variables must be closely controlled. ◀

Example 7–7.3   A ceramic magnet has a true density of 5.41 Mg/m³ (=5.41 g/cm³). A poorly sintered sample weighs 3.79 g dry, and 3.84 g when saturated with water. The saturated sample weighs 3.08 g when suspended in water. (a) What is its *true volume*? (b) What is its *bulk volume* (total volume)? (c) What is its *apparent (open) porosity*? (d) What is its *total porosity*?

Background. There are three measurements of volumes:

True volume = total (bulk) volume − total pore volume;

Apparent volume = total volume − open pore volume;

= true volume + closed pore volume;

Bulk (total) volume = true volume + total pore volume.

Since $\rho = m/V$, there are also three densities.

Recall from Archimedes principle that "an object is buoyed up by the weight of the displaced fluid."

Calculation

a)   True volume = m/ρ_{tr} = 3.79 g/(5.41 g/cm³) = 0.70 cm³.

b) From Archimedes principle:

Water displaced by bulk sample = buoyancy = 3.84 g − 3.08 g = 0.76 g;

$$\therefore \text{total volume} = 0.76 \text{ g/(1 g/cm}^3) = 0.76 \text{ cm}^3.$$

This includes material plus all pore space.

c)   Apparent porosity = (open pore volume)/(total volume)

$$= [(3.84 - 3.79 \text{ g})/(1 \text{ g/cm}^3)]/0.76 \text{ cm}^3$$

$$= 0.066 \qquad \text{(or 6.6 v/o, bulk basis)}.$$

d)   Total porosity = $(V_{total} - V_{true})/V_{total}$

$$= (0.76 - 0.70 \text{ cm}^3)/0.76 \text{ cm}^3$$

$$= 0.079 \qquad \text{(or 7.9 v/o bulk basis)}.$$

Comment. The *closed porosity* is 1.3 v/o. The *bulk density* is 3.79 g/0.76 cm³ = 5.0 g/cm³. We may also speak of an *apparent density*:

mass/(bulk volume − open pore volume).

This density is (3.79 g)/(0.76 cm³)(1 − 0.066) = 5.34 g/cm³   (or 5.34 Mg/m³). ◄

Review and Study

SUMMARY

1. Since ceramic phases are *compounds of metallic and nonmetallic elements*, we find that it is not as easy to generalize about structure and properties as for metals or nonmetals alone. Metals, for example, are always electrical and thermal conductors; most ceramics are insulators but some possess semiconductivity that has technical uses. Organic polymers always transmit light in thin sections; ceramics may have the transparency of optical glass or may be opaque in magnetic spinels. Structural ceramics are very strong in compression but must be considered weak in tension. Fiberglass may have a tensile strength greater than steel and therefore can be used as a reinforcement; glass is also recognized as a brittle, friable material, so that special care must be taken when it is shipped or handled.

   The examples above are just a few of those we could cite to indicate that ceramic materials have a variety of characteristics. Many of these are indispensable for current-day technology and societal needs. However, they are more complex than other materials and therefore require greater technical familiarity and understanding.

2. The simplest ceramics are AX compounds with cubic structures, since these binary compounds have equal numbers of positive and negative ions. Among these, the *CsCl-type,* the *NaCl-type,* and the *ZnS-type structures* have coordination numbers of 8, 6, and 4, respectively. Nature generally prefers larger CN's, if the *r/R* ratios permit. The CsCl structure is simple cubic; the latter two are fcc.

3. The binary compound CaF_2 was used as an A_mX_p example. The coordination numbers for Ca^{2+} and F^- are different because m/p does not equal unity. Unoccupied interstitial sites lead to lattice distortion.

4. Two $A_mB_nX_p$ *compounds*, $BaTiO_3$ and a ferrospinel, were presented to describe somewhat more complicated structures. Their use will be encountered in Chapter 9. •*Solid solutions* in ceramic compounds require (1) compatibility in size and (2) balance in charges. The latter is sometimes achieved through a defect structure of either ion vacancies or interstitial atoms.

5. The structural units in silicates are the SiO_4 *tetrahedra,* in which one silicon atom is surrounded by four oxygen atoms. All of the oxygen atoms are *bridging* in silica, SiO_2, since each is shared by two adjacent tetrahedra. This produces a very rigid *network* structure. Silicate glasses have the same short-range order as found in crystalline silicates, but lack the long-range order. *Network modifiers* include Na_2O and CaO, which introduce *nonbridging* oxygens. The soda-lime-silicate glasses are therefore more readily shaped than silica glass into commercial products for container, fiber, and window applications.

6. Ceramics, like other compounds, have a *high shear strength,* because slip disorders the coordination of unlike neighbors. Consequently, ceramics can be hard and strong in compression, but *nonductile*. Being nonductile, they are sensitive to notches and flaws during tensile loading. Thus, design principles dictate that ceramics should be used under *compression*. Examples include the use of concrete in reinforced beams, and brick in buildings. It is also possible to introduce surface compression in glass and other materials by *tempering*. This can strengthen a glass significantly.

• 7. *Viscous forming* involves the melting and shaping of a viscous silicate liquid. Viscosity and temperature are related by an Arrhenius equation (7–7.1). The melting range, working range, annealing point, and strain point are all defined in terms of the temperature for a given viscosity—10 to $10^{1.5}$, $10^{2.5}$ to 10^6, 10^{12}, and $10^{13.5}$ Pa·s, respectively.

The *sintering* process bonds powders into an integral mass, by heating to promote atomic diffusion and thereby pore elimination. The driving force is the reduction of surface energy.

Hydrothermal and fusion procedures were described for *single-crystal preparation*. The former proceeds by solution and precipitation under a controlled temperature gradient. Fusion processes melt the material, from which a seed crystal initiates growth in a temperature gradient. Other methods of crystal growth were not described.

KEY TERMS AND CONCEPTS

Abrasive   Hard, mechanically resistant material used for grinding or cutting; commonly made of a ceramic material.

Amorphous   Noncrystalline and without long-range order.

$A_mB_nX_p$ **compounds**   Ternary compounds. In this text X is generally oxygen; A and B are commonly metal atoms.

Annealing point   Stress-relief temperature (approx. T_g). This temperature provides a viscosity of $\sim 10^{12}$ Pa·s ($\sim 10^{13}$ poises).

AX compounds Binary compounds with a 1-to-1 ratio of the two elements; commonly ionic.

AX structure (CsCl) Structure of a binary compound with CN = 8. Simple cubic.

AX structure (NaCl) Structure of a binary compound with CN = 6. Face-centered cubic.

AX structure (ZnS) Structure of a binary compound with CN = 4. Face-centered cubic.

A_mX_p compounds Binary compounds with A ≠ X. (See Section 7–3.)

Bridging oxygens Oxygen atoms shared by two adjacent silica tetrahedra.

Ceramics Materials consisting of compounds of metallic and nonmetallic elements.

Clay Fine soil particles (<0.1 mm). In ceramics, clays are specifically sheetlike alumino-silicates.

Compound A phase composed of two or more elements in a given ratio.

Coulombic forces Forces between charged particles, particularly ions.

Ferrite (ceramics) Compounds containing trivalent iron; commonly magnetic.

Firing Bonding by thermal means.

Fracture, nonductile Failure by crack propagation in the absence of significant ductility.

Glass An amorphous solid below its transition temperature. A glass lacks long-range crystalline order but normally has short-range order.

Glass-transition temperature (T_g) Transition temperature between a supercooled liquid and its glassy solid.

Interstitial site (n-fold) An interstice with n (4, 6, or 8) immediate atomic (or ionic) neighbors).

Ion vacancy Unoccupied ion site within a crystal structure. The charge of the missing ion must be appropriately compensated.

Isostatic molding Compression of powder into the desired shape, utilizing a rubber die and a hydraulic fluid (Fig. 7–7.1).

Long-range order A repetitive pattern over many atomic distances.

Network modifier A component that introduces nonbridging oxygen into a silicate glass.

Nonbridging oxygens Oxygens attached to one SiO_4 tetrahedron only, thus not tying tetrahedra together.

Pores, closed Pores without access to the surrounding environment.

Pores, open Interconnecting pores. Permeable pores.

Quartz The most common phase of SiO_2.

Radius ratio (r/R) Ratio of ionic radii of coordinated cations, r, and anions, R.

Residual stresses Stresses induced as a result of differences in temperature or volume.

Short-range order Specific first-neighbor arrangement of atoms, but random long-range arrangements.

Silicate structures Materials containing SiO_4 tetrahedra.

Silicates, network Three-dimensional silicate structure with tetrahedral oxygens shared.

Sintering Bonding by thermal means.

Sintering, solid Agglomeration by solid diffusion.

SiO_4 tetrahedron Coordination unit of four oxygens surrounding a silicon atom.

Soda-lime-silica glass The most widely produced glass; it serves as a basis for window and container glass. Contains $Na_2O/CaO/SiO_2$ in approximately a 1/1/6 ratio.

Strain point (glass)   Temperature at which the viscosity, η, of a glass is $10^{13.5}$ Pa·s ($10^{14.5}$ poises).

Tempered glass   Glass with surface compressive stresses induced by heat treatment.

Viscosity (η)   The ratio of shear stress to velocity gradient.

Viscous forming   Thermoplastic shaping of glass or polymeric products.

Vitreous   Glassy or glasslike.

Volume, apparent   True volume plus closed-pore volume (or total volume minus open-pore volume).

Volume, total (bulk)   True volume plus pore volumes (both closed and open).

Volume, true   Volume inherent to the material (exclusive of pore volumes).

Working range (glass)   Temperature range lying between viscosities of $10^{2.5}$ Pa·s to 10^6 Pa·s.

**FOR CLASS
DISCUSSION**

A_7 Make a list of six ceramic materials that are used in technical applications and that have not been discussed in this text.

B_7 Distinguish between CsCl-type, NaCl-type, and ZnS-type structures with regard to coordination number, minimum ion–size ratio, and the number of ions per unit cell.

C_7 Both KCl and MnSe have the same structure (NaCl-type). Although the K^+ is lighter than Mn^{2+} (39 versus 55 amu) and Cl^- is lighter than Se^{2-} (35 versus 79 amu), KCl has a larger lattice constant ($a > 0.6$ nm) than MnSe ($a < 0.6$ nm). Explain why.

D_7 Sketch the atom arrangements on the ($1\bar{1}0$) plane of ZnS. (Cf. Fig. 7–3.2 for CaF_2.) Repeat for the (110) plane.

E_7 As shown in Fig. 7–4.1, cubic $BaTiO_3$ has an oxygen atom at the center of each face. However, it is not face-centered cubic. Why?

F_7 Figure 4–3.1 with Mg^{2+}, Fe^{2+}, and O^{2-} ions, and Fig. 7–4.1 with Ba^{2+}, Ti^{4+}, and O^{2-} ions, both have three types of ions. We call one a solid solution and the other an $A_m B_n X_p$ compound. Why the difference?

G_7 Both MnO and MnS have NaCl-type structures. A limited amount (~1 a/o) of solid solution occurs. Suggest the mechanism. Why is the extent of solid solution limited?

H_7 Use electron dots (cf. Fig. 2–3.1a) to show covalent bonding in an SiO_4^{4-} ion; show ionic bonding.

I_7 Compare and contrast the structure of cristobalite (Fig. 7–5.2) with that of ZnS (Fig. 7–2.4), and with silicon, which has the structure of diamond (Fig. 2–3.1b).

J_7 Cristobalite (Fig. 7–5.2) has almost no expansion when it melts to fused silica. Suggest why.

K_7 Distinguish between bridging oxygens and nonbridging oxygens, and their effect on glass deformation above the glass temperature.

L_7 B_2O_3 is a good glass-former. Compare its structure and viscous behavior with fused silica (SiO_2).

M₇ What characteristics are required of an oxide for it to be a network modifier in the structure of glass? Suggest six oxides that will fill this role.

N₇ A ceramic (Al_2O_3) cutting tool has to be mounted in a lathe differently from a tool made of "high-speed" steel. Why?

O₇ A glaze for dinnerware is chosen so that it has a slightly lower thermal expansion coefficient than the underlying porcelain. Explain why this makes the product stronger.

QUIZ SAMPLES

7A Ceramic materials contain __**__ of metallic and nonmetallic __**__.

7B Of the two, a magnesium atom in a metal, and a magnesium ion in its oxide, the radius of the __**__ is larger; the radius of the __**__ can be measured more precisely.

7C For an AX compound to have CN = 6, the r/R ratio must be at least __**__. If r/R is 0.6, the CN may be __**__.

7D An AX compound is cubic and has anions located at the origin and other fcc positions. Among these anions are 6-f sites at __+−__.
(a) $\frac{1}{2},0,\frac{1}{2}$ (b) $0,\frac{1}{2},0$ (c) $\frac{1}{4},\frac{1}{4},\frac{1}{4}$ (d) $\frac{1}{2},\frac{1}{2},\frac{1}{2}$ (e) $\frac{3}{4},\frac{1}{4},\frac{3}{4}$
(f) None of the above, rather _____

7E An AX compound is cubic and has anions located at the origin and other fcc positions. Among these are 4-f sites at __+−__.
(a) $0,\frac{1}{2},\frac{1}{2}$ (b) $\frac{1}{4},\frac{3}{4},\frac{3}{4}$ (c) $\frac{3}{4},\frac{3}{4},\frac{3}{4}$ (d) $\frac{1}{2},\frac{1}{2},\frac{1}{2}$ (e) $\frac{1}{2},\frac{1}{4},\frac{3}{4}$
(f) None of the above, rather _____

7F The ionic radii (CN = 6) of Mn^{2+} and S^{2-} are 0.074 nm and 0.184 nm. The lattice constant for the NaCl version of MnS is ~__**__ nm; that for the ZnS version is ~__**__ nm.

7G The compound ZrO_2 may possess the CaF_2 structure. One hundred unit cells contain __+−__ O^{2-} ions; __+−__ Zr^{4+} ions, __+−__ unoccupied 8-f sites among the O^{2-} ions and __+−__ unoccupied 4-f sites among the Zr^{4+} ions.
(a) 100   (b) 200   (c) 400   (d) 800
(e) None of the above, rather _____

7H In general, higher __**__ favor polymorphs with lower density and greater symmetry; higher __**__ favor polymorphs with greater density.

7I Ferrous oxide contains a significant number of ferric, Fe^{3+} ions. For every 100 Fe^{3+} ions present, there must be 50 __**__ in order to balance charges. A __**__ structure results, and the oxide is __**__-metric.

7J In cubic $BaTiO_3$, the CN of Ti^{4+} is __**__; the CN of Ba^{2+} is __**__.
(a) 2   (b) 3   (c) 4   (d) 6   (e) 8   (f) 12

7K The __**__ tetrahedron is a very stable form of the two most plentiful elements, __**__ and __**__.

7L In silicate glasses, a __**__ oxygen is shared by two adjacent tetrahedra.

7M A network __**__ reduces the numbers of bridging oxygens and, therefore, makes the glass less __**__ at high temperature.

7N Ceramic compounds, like all compounds with 3-D structures, characteristically have low __+−__ .

(a) hardness   (b) tensile strength   (c) ductility   (d) shear strength
(e) compressive strength   (f) yield strength

7O As a rule, ceramic materials have a very high __**__ strength, since they are compounds with a preference for __**__ neighbors.

7P Since ceramics have a high shear resistance, they support large __**__ stresses but are sensitive to flaws and imperfections when loaded in __**__ .

7Q The principal chemical component of commercial glass products is __**__ .

7R The viscosity of glass at its __**__ point (temperature) is such that residual stresses can be removed in ~15 min.; from its __**__ point, the glass can be cooled rapidly without the introduction of new residual stresses. The __**__ temperature lies between these two temperatures.

7S The residual surface stresses of tempered glass are in __**__ .

7T Solid sintering of powders replaces __**__ with a single __**__ boundary.

7U The __**__ growth of single crystals depends on the variation of solubility with temperature. A common __**__ technique of single crystal production uses a seed crystal.

STUDY PROBLEMS

7–2.1 Calculate the density of CsCl from the data in the Appendix. (Remember that CN = 8.)

Answer: 3.94 Mg/m³

7–2.2 CsCl has the simple cubic structure of Cl^- ions with Cs^+ ions in the 8-f sites. (a) The radii are 0.187 nm and 0.172 nm, respectively, for CN = 8; what is the packing factor? (b) What would this factor be if $r/R = 0.73$?

7–2.3 The intermetallic compound AlNi has the CsCl-type structure with $a = 0.2881$ nm. Calculate its density.

Answer: 6.0 Mg/m³

7–2.4 X-ray data show that the unit-cell dimensions of cubic MgO are 0.42 nm. It has a density of 3.6 Mg/m³. How many Mg^{2+} ions and O^{2-} ions are there per unit cell?

7–2.5 Periclase (MgO) has an fcc structure of O^{2-} ions with Mg^{2+} ions in all the 6-f sites. (a) The radii are 0.140 nm and 0.066 nm, respectively; what is the packing factor? (b) What would this factor be if $r/R = 0.41$?

Answer: (a) 0.73   (b) 0.80

7–2.6 Nickel oxide is cubic and has a density of 6.87 Mg/m³ (=6.87 g/cm³). Its lattice constant is 0.417 nm. How many atoms are there per unit cell? (The nickel ion is Ni^{2+}.)

7–2.7 Lithium fluoride has a density of 2.6 Mg/m³ and the NaCl structure. Use these data to calculate the unit cell size, a, and compare it with the value you get from the ionic radii.

Answer: 0.405 nm versus 0.402 nm

7–2.8   ZnS (Fig. 7–2.4) has a density of 4.1 Mg/m^3. Based on this, what is the spacing between the centers of the two ions?

7–2.9   Estimate the densities of the two polymorphs of MnS in Example 7–2.4.

Answer: (a) 3.9 Mg/m^3 (b) 3.4 Mg/m^3

7–2.10   The density of 1.5 g/cm^3 (or 1.5 Mg/m^3) is given for NH_4Cl in the Chemical Handbook. X-ray files state that there are two polymorphs for NH_4Cl; one has an NaCl-type structure with $a = 0.726$ nm, the other has a CsCl-type structure with $a = 0.387$ nm. The density value is for which polymorph? (The NH_4^+ ion occupies the crystal lattice as a unit.)

7–2.11   It is hypothesized that at high pressures NaCl can be forced into a CsCl-type structure. What would be the percent volume change? (See the footnote of Appendix B for differences in radii with CN = 6 and CN = 8.)

Answer: -16 v/o

7–3.1   With CN = 4, $R_{O^{2-}} = 0.127$ nm (Table 2–4.1). With CN = 8, $r_{Zr^{4+}} \approx 0.085$ nm. Estimate the size of the unit cell of cubic ZrO_2, which has the CaF_2-type structure.

Answer: 0.49 nm (versus 0.507 nm by experiment)

7–3.2   Using the radii from the answer of Example 7–3.2, will the estimated unit cell size be larger than the experimentally measured value of $a = 0.547$ nm? (See the comment with Example 7–3.1 for an explanation.)

• **7–3.3**   The unit cell size a of UO_2 is 0.547 nm. (a) Based on this information and Fig. 7–3.1, what is the linear density of U^{4+} ions in the $\langle 110 \rangle$ directions? Of the O^{2-} ions? (b) What is the linear density of U^{4+} ions in the $\langle 100 \rangle$ directions? Of the O^{2-} ions? (c) What is the linear density of U^{4+} ions in the $\langle 111 \rangle$ directions? Of the O^{2-} ions?

Answer: (a) 2.6×10^6/mm, 2.6×10^6/mm   (b) 1.8×10^6/mm, 3.7×10^6/mm (c) 1.1×10^6/mm, 2.1×10^6/mm

7–3.4   Can MgF_2 have the same structure as CaF_2? Explain.

7–4.1 $BaTiO_3$ is made by calcining $BaCO_3$ to BaO and mixing it with TiO_2. (a) How much BaO and TiO_2 are required per 100 g of product? (b) How much $BaCO_3$ is required per gram of product? (Ba = 137.34 amu.)

Answer: (a) 65.7 g BaO and 34.3 g TiO_2 (b) 0.846 g $BaCO_3$

7–4.2   Calcium titanite ($CaTiO_3$) has the same structure as $BaTiO_3$. It is cubic and has a density of 4 Mg/m^3. Calculate its lattice constant.

7–4.3   The density of cubic $PbTiO_3$ is 7.5 Mg/m^3. What is the lattice dimension of the unit cell?

Answer: 0.406 nm (versus 0.397 nm by experiment)

7–4.4   Calculate the oxygen content (a) of Fe_3O_4; (b) of $NiFe_2O_4$.

• **7–4.5**   A solid solution contains 30 mole percent MgO and 70 mole percent LiF. (a) What are the weight percents of Li^+, Mg^{2+}, F^-, O^{2-}? (b) What is the density?

Answer: (a) Li^+, 16 w/o; Mg^{2+}, 24 w/o; F^-, 44 w/o; O^{2-}, 16 w/o;   (b) 3.0 g/cm^3 (=3.0 Mg/m^3)

• **7–4.6** (a) What type of vacancies, anion or cation, must be introduced with MgF_2 in order for it to dissolve in LiF? (b) What type must be introduced with LiF for it to dissolve in MgF_2?

7–4.7 A cubic form of ZrO_2 is possible when one Ca^{2+} ion is added in a solid solution for every six Zr^{4+} ions present. Thus the cations form an fcc structure, and O^{2-} ions are located in the 4-f sites. (a) How many O^{2-} ions are there for every 100 cations? (b) What fraction of the 4-f sites is occupied?

Answer: (a) 185.7 O^{2-} ions   (b) 92.9%

7–4.8 (a) What is the coordination number of Ba^{2+} in $BaTiO_3$? (b) Of Ti^{4+}? (c) The O^{2-} ion cannot be assumed to be spherical. Why?

7–4.9 The unit cell size a of the ferrospinel in Example 7–4.1 is 0.84 nm. (a) Calculate its density. (b) What is the distance of closest approach of O^{2-} ions (center-to-center)?

Answer: (a) 5.3 Mg/m^3 (b) 0.297 nm

7–4.10 The lattice constant a of cubic $BaTiO_3$ is 0.40 nm. Consider $r_{Ti^{4+}} = 0.068$ nm (App. B), and $r_{Ba^{2+}} = 0.148$ nm when CN = 12. (a) What is the radius R of the O^{2-} ion in the $\langle 100 \rangle$ directions? (b) In the $\langle 110 \rangle$ directions?

7–4.11 One-tenth weight percent Fe_2O_3 is in solid solution with NiO. As such, 3 Ni^{2+} are replaced by (2 Fe^{3+} + ☐) to maintain a charge balance. How many cation vacancies are there per m^3?

7–5.1 Refer to Fig. 7–5.2, where $a = 0.693$ nm. How many silicon and oxygen atoms are there per m^3?

Answer: 2.40×10^{28} Si/m^3, 4.80×10^{28} O/m^3

7–5.2 What is the volume expansion as quartz (Example 7–5.1) changes to the SiO_2 polymorph shown in Fig. 7–5.2? (Use data from Study Problem 7–5.1 as required.)

7–5.3 What is the packing factor of SiO_2 in cristobalite (Fig. 7–5.2)? (If needed, use information from Study Problem 7–5.1 and Example 7–5.3.)

Answer: 0.29

7–5.4 From the information in Example 7–5.3, determine the unit cell size a of the SiO_2 in Fig. 7–5.2.

7–5.5 Coesite, the high-pressure polymorph of SiO_2, has a density of 2.9 Mg/m^3. What is the packing factor? ($r_{Si} = 0.038$ nm; $R_O = 0.114$ nm.)

Answer: 0.36

7–5.6 A glass contains only Na_2O and SiO_2 and has a structure in which 72% of the oxygen atoms serve as bridges between adjacent silicons. How many sodium and silicon atoms are there per 100 oxygen atoms?

7–5.7 A glass contains 75 w/o SiO_2 and 25 w/o CaO. What fraction of the oxygens serve as bridges between pairs of silicon atoms?

Answer: 70 a/o

7–5.8 A window glass is a soda-lime glass with a weight ratio of 14 Na_2O–14 CaO–72 SiO_2. What fraction of the oxygen atoms are nonbridging?

7–6.1 Under certain conditions, the $\langle 011 \rangle \{100\}$ slip systems can operate in NaCl-type structures in addition to those listed in Example 7–6.1. (a) How many slip systems belong to this set? List them. (b) What is the slip vector in PbS, which has the NaCl-type structure and the above slip system?

Answer: (a) 6   (b) 0.43 nm

7–6.2 The theoretical strengths of amorphous materials are calculated to be between $G/6$ and $G/4$, where G is the shear modulus. Estimate the theoretical strength of glass based on its elastic properties, including $\nu = 0.25$.

7–6.3 A 1-mm round rod with the composition of plate glass (Appendix C) is coated with 0.1 mm of borosilicate glass, so that the rod is now 1.2 mm in diameter. Assuming no initial stresses at 200°C, what longitudinal stress is developed when the composite rod is cooled to 20°C?

Answer: $\sigma_{boro} = -55$ MPa (−8000 psi), $\sigma_{pl} = +25$ MPa (3500 psi)

7–6.4 Repeat the previous problem, but interchange the locations of the two glasses. (a) Which will have the higher stresses, plate or borosilicate? (b) Comment on the strength of this composite glass rod.

7–7.1 Assume that the flow rate of a glass is inversely proportional to the viscosity η. How much faster will a molten electrical glass (Example 7–7.1) flow from a furnace at 920°C than at 900°C?

Answer: $\eta_{920}/\eta_{900} = 0.618$; $\dot{F}_{920} = 1.6\,\dot{F}_{900}$

7–7.2 The viscosity of window glass drops from 10^6 Pa·s at 680°C to 10^3 Pa·s at 1035°C. What is its viscosity at 900°C?

7–7.3 A ceramic insulator will have 1 v/o porosity after sintering and should have a length of 13.7 mm. During manufacturing, the powders can be pressed to contain 24 v/o porosity. What should the die dimensions be?

Answer: 15.0 mm

7–7.4 A magnetic ferrite for an oscilloscope component is to have a final dimension of 15.8 mm (0.621 in.). Its volume shrinkage during sintering is 33.1% (unsintered basis). What initial dimension should the powdered compact have?

7–7.5 A powdered metal part has a porosity of 23% after compacting and before sintering. What linear shrinkage allowance should be made if the total porosity after sintering is expected to be 2%?

Answer: 8% (sintered basis)

7–7.6 A brick weighs 3.3 kg when dry, 3.45 kg when saturated with water, and 1.9 kg when suspended in water. (a) What is the apparent porosity? (b) Its bulk density? (c) Its apparent density?

7–7.7 A piece of ceramic wall tile 5 mm × 200 mm × 200 mm absorbs 2.5 g of water. What is the apparent porosity of the tile?

Answer: 1.25 v/o

7–7.8   An insulating brick weighs 1.77 kg (3.9 lb) dry, 2.25 kg (4.95 lb) when saturated with water, and 1.05 kg (2.3 lb) when suspended in water. (a) What is the apparent porosity? (b) Its bulk density? (c) Its apparent density?

7–7.9   An electrical porcelain product has only 1 v/o porosity as sold. It had 27 v/o porosity (total) after pressing and drying. How much linear shrinkage occurred during sintering?

Answer: 9.7 l/o

7–7.10   Refer to Fig. 7–7.3. (a) How much linear shrinkage occurred during the sintering of this powder—from part (a) to part (c)? (b) How much volume shrinkage?

QUIZ CHECKS

7A	compounds, elements	7L	bridging
7B	atom, atom	7M	modifier, viscous
7C	0.41	7N	(b), (c)
	4 or 6	7O	shear
7D	(b), (d)		unlike
7E	(b), (c)	7P	compressive
7F	~0.52 nm		tension
	~0.54 nm	7Q	SiO_2
7G	(d)	7R	annealing
	(c)		strain
	(c)		glass-transition
	(e): 0	7S	compression
7H	temperatures	7T	two surfaces
	pressures		grain
7I	vacancies	7U	hydrothermal
	defect,		fusion
	nonstoichiometric		
7J	6, 12		
7K	SiO_4		
	Si and O		

Chapter Eight

Conducting
Materials

PREVIEW

Atoms and their arrangements have received the bulk of our attention to date. In this chapter we will focus attention on electrons and their freedom to move among atoms.

Metals, with their weak hold on valence electrons, are good conductors of both electricity and heat. This conductivity occurs because very little energy is required to activate delocalized electrons into conduction levels. In contrast, electrons must be raised across a large energy gap in an insulator. Semiconductors have small energy gaps so that a useful number can jump to the conduction band to transport charge. This leaves electron holes in the lower bands that serve as positive carriers of charge.

Simplified concepts of many devices are possible with an introductory understanding of energy gaps and junctions. The basic principles of crystal growing and device preparation are presented.

CONTENTS

1. To handle simple calculations for conductivity based on the concept of the electron as a negative charge carrier. Conversely, a missing electron, called an electron hole, is a positive charge carrier.

2. To relate resistivity (and conductivity) changes to the effects of impurities, temperature changes, and cold work through the concept of the mean free path.

3. To associate energy bands and energy gaps qualitatively with the conductivity of metals, semiconductors, and insulators.

4. To solve introductory problems (a) on the conductivity of intrinsic semiconductors based on their charge mobility, and the size of their energy gap, (b) on the conductivity of extrinsic semiconductors based on the doping concentration, and (c) on the progress of recombination based on relaxation times.

• 5. To understand the bases for simple semiconductor devices, such as the thermistor, photoconductor, L.E.D., and junction rectifier.

• 6. To become acquainted with the basic principles of crystal growth and device preparation.

7. To develop sufficient familiarity with terms used with electronic materials to communicate as required with design engineers.

8–1

CHARGE CARRIERS Various materials that are available to the engineer or scientist exhibit a wide range of conductivities (or resistivities, since $\sigma = 1/\rho$). As shown in Fig. 8–1.1, we commonly divide materials into three categories, *conductors, semiconductors*, and *insulators*. Metals fall in the first category, since they have delocalized electrons that are free to move throughout the structure (Section 2–5). Insulators include those ceramics and polymeric materials with strongly held

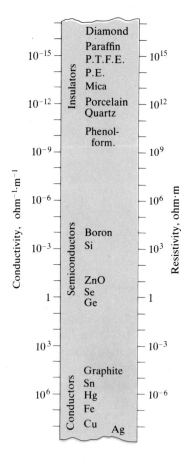

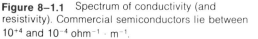

Figure 8–1.1 Spectrum of conductivity (and resistivity). Commercial semiconductors lie between 10^{+4} and 10^{-4} ohm$^{-1} \cdot$ m^{-1}.

electrons and nondiffusing ions. Their function is to isolate neighboring conductors. It was not very long ago that only the two ends of the spectrum of Fig. 8–1.1 were considered to be useful. Today, however, the middle, semiconducting category has become exceedingly important and will, in fact, be the chief subject of this chapter.

In those types of materials that conduct electricity, the charge is carried in modules of 0.16×10^{-18} coul, this being the charge on an individual electron. In metals, it is the individual electron that moves. In ionic materials, charge can be carried by diffusing ions. However, their charge is simply an integer number of electron charges ($-$ or $+$, for anions and cations, respectively). Thus, an SO_4^{2-} ion carries 0.32×10^{-18} coul of charge within a storage battery, and all Pb^{2+} ions have an absence of two electron charges as they move in the opposite direction.

Electrons and anions are *negative charge carriers*. In contrast, a cation such as Pb^{2+} is a *positive charge carrier* because, as we have just seen, it has an *absence* of electrons. There is another positive charge carrier that becomes important in semiconductors, namely, an *electron hole*. It is an absence of an electron within the energy band for the delocalized electrons discussed in Section 2–5. We will come back to these in Section 8–4.

Conductivity σ and *resistivity* ρ ($= 1/\sigma$) values for a material depend upon the *number, n*, of charge carriers, the *charge, q*, on each, and their *mobility, μ*:

$$\sigma = 1/\rho = nq\mu. \qquad (8\text{–}1.1)$$

In this chapter much of our attention will be directed at the number, *n*, of carriers per m^3 that are available for conduction. The value of *q* is fixed at 0.16×10^{-18} C (or 0.16×10^{-18} A·s) in electronic conductors. The mobility μ relates the drift velocity $\bar{v}$ to the electric field $\mathcal{E}$ (as indicated in Section 2–5):

$$\mu = \bar{v}/\mathcal{E}. \qquad (8\text{–}1.2)$$

Thus, the units for mobility are (m/s) per (V/m), or m²/V·s. As a result,

$$\sigma = (m^{-3})(A{\cdot}s)(m^2/V{\cdot}s) = ohm^{-1}{\cdot}m^{-1}, \qquad (8\text{–}1.3a)$$

and

$$\rho = ohm{\cdot}m. \qquad (8\text{–}1.3b)$$

Example 8–1.1

A semiconductor with 10^{21} charge carriers/m³ has a resistivity of 0.1 ohm·m at 20°C. What is the drift velocity of the electrons if one ampere of current is carried across a gradient of 0.15 volt/mm?

Procedure. We have resistivity (and thus conductivity) plus the number of carriers and, of course, the charge *q* on each. We can calculate the mobility. The electric field is also available, so we can calculate the drift velocity.

Calculation

$$\mu = \sigma/qn = (1/0.1 \text{ ohm·m})/(0.16 \times 10^{-18} \text{ A·s})(10^{21}/m^3)$$
$$= 0.0625 \text{ m}^2/V{\cdot}s.$$
$$\bar{v} = \mu\mathcal{E} = (0.0625 \text{ m}^2/V{\cdot}s)(0.15 \text{ V}/0.001 \text{ m})$$
$$= 9 \text{ m/s.}$$

Comment. Of course, the electrons move much faster than 9 m/s. This is the *drift* or *net velocity* across the semiconductor. ◀

8–2

METALLIC CONDUCTIVITY

The metallic bond is described in terms of *delocalized electrons* (Section 2–5). Specifically, the valence electrons are able to move throughout the metal as standing waves. Thus, there is no net charge transport in the absence of an electric field.

If the metal is placed in an electrical circuit, the electrons moving toward the positive electrode *acquire more energy* and gain velocity in that direction.

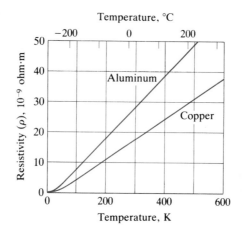

Figure 8–2.1   Resistivity versus temperature (metals). The resistivity of metals is linear with temperature in the 100 K to 500 K range (−200°C to 200°C). Except in superconductors, the "toes" of the curves are finite.

Conversely, those electrons moving toward the negative electrode *reduce their energy* and velocity. As a result, a *drift velocity* is developed. (See Example 8–1.1.)

Mean Free Path

In a review of Section 2–5, we recall that waves move through periodic structures without interruption. A well-ordered crystal (Chapter 3) provides one of the most regular of the periodic structures available. Thus a metallic crystal lattice provides an excellent medium for electron movements. However, any irregularity in the repetitive structures through which a wave travels may deflect the wave. Thus, if an electron had been traveling toward the positive electrode and was then deflected, it would no longer continue to gain velocity in that direction. The net effect is to *reduce the drift velocity* just cited, even though we have not altered the electric field. In brief, irregularities in the lattice *decrease* the mobility of Eq. (8–1.2); therefore, they *decrease* the conductivity and *increase* the resistivity (Eq. 8–1.1).

The average distance that an electron can travel in its wavelike pattern without deflection is called the *mean free path*. We will want to identify irregularities that deflect electron movements, because that will help us understand why resistivities of metals are not all the same. We can identify two effects on the basis of Chapter 4.

Resistivity versus Temperature

The resistivity of a metal increases with temperature (Fig. 8–2.1). To a first approximation it is linear (except near absolute zero). We have no basis on which to conclude that the *n* of Eq. (8–1.1) decreases significantly with increased temperature in a metal;* rather we must look at the mobility μ. Ther-

---

* The number of charge carriers increases with increased temperature in a semiconductor (Section 8–4).

TABLE 8–2.1

Temperature Resistivity Coefficients

Metal		Resistivity at 0°C*, ohm·nm	Temperature resistivity coefficient, y_T, °C^{-1}
Aluminum		27	0.0039
Copper		16	0.0039
Gold		23	0.0034
Iron		90	0.0045
Lead		190	0.0039
Magnesium		42	0.004
Nickel		69	0.006
Silver		15	0.0038
Tungsten		50	0.0045
Zinc		53	0.0037
Brass	(Cu–Zn)	~60	0.002
Bronze	(Cu–Sn)	~100	0.001
Constantan	(Cu–Ni)	~500	0.00001
Monel	(Ni–Cu)	~450	0.002
Nichrome	(Ni–Cr)	~1000	0.0004

* These values will not agree with those in Appendix C, since they are based on different reference temperatures.

mal agitation (Section 4–6) increases in intensity in proportion to increased temperature (except at very low temperatures). This increased agitation decreases the mean free path of the electrons by decreasing the regularity of the crystal and therefore decreases the mobility of electrons in a metal. The consequent change in resistivity is important to the engineer who is designing electrical equipment. In some cases, compensation must be introduced into a circuit to avoid an unwanted temperature sensitivity. In other cases, this temperature sensitivity provides a useful "brake." Example 8–2.3 will point this out in a familiar application (the toaster).

We may determine the ρ-versus-T relationship with a *temperature resistivity coefficient* y_T as follows:

$$\rho_T = \rho_{0°C}(1 + y_T \,\Delta T), \qquad (8–2.1)*$$

where $\rho_{0°C}$ is the resistivity at 0°C, and ΔT is $(T - 0°C)$. This value of this coefficient is approximately 0.004/°C for pure metals (Table 8–2.1). This suggests that the mean free path of electrons is reduced by a factor of two between 0°C and 250°C.

---

* The decreases in resistivity (increase in conductivity) we are describing are not related to *superconductivity* that appears near absolute zero. That involves another phenomenon (quantum mechanical), which is beyond the scope of this book.

Resistivity in Solid Solutions

Another factor that can reduce the mean free path of electrons in a metal is the presence of solute atoms. A solid-solution alloy always has a higher resistivity than do its pure component metals.*

The reason for this generalization is that an electron encounters an irregularity in the local potential of the crystal lattice when it approaches an impurity atom. In the first place, the lattice is slightly distorted in an alloy such as brass, because the atomic radii differ a few percent; in addition, a zinc atom has 30 protons rather than the 29 in copper. This also alters the local field. Although these differences seem small, they deflect additional electrons and reduce the mean free path. Since brass (70 Cu–30 Zn) has a resistivity 3 or 4 times as great as that of pure copper, we can assume that the mean free path for electrons is only 25%–30% as long in brass as in pure copper. If high conductivity is paramount in a design, the engineer will turn to pure metals (Section 5–1).

Energy Bands

Recall from Fig. 2–5.3 that electrons of isolated atoms occupy only specific orbitals or energy levels, and that forbidden-energy gaps exist between these levels. In effect, the electrons establish standing waves around an individual atom. This pattern is also found in the inner or subvalence electrons of metals; however, the outer or valence electrons are delocalized when we have a large number of coordinated atoms. As a result, the valence orbitals form a band (Fig. 2–5.3b), and the standing wave is influenced by every atom that is involved. A consequence of this fact is that a band possesses as many standing wave forms and discrete energy *levels* as there are atoms in the system. Since the number is exceedingly great, and the energy bands are usually only a few electron volts wide, it follows that the energy levels within a band are so infinitesimally separated that we may pretend the band forms a continuum.

A physical principle† states that only two electrons may occupy the same level (and these two must be of opposite magnetic spin). Thus with its multitude of levels, *a band may contain twice as many electrons as there are atoms.* As a result, a monovalent metal such as sodium has its first valence band only half-filled (Fig. 8–2.2a). Since aluminum has three valence electrons per atom, its first valence band is filled and its second band is half full (Fig. 8–2.2c). Naturally the lower energy levels of the band fill first.

Definition of a Metallic Conductor

We characterized metals in Section 1–4 by their "ability to give up valence electrons" and pointed out that they thus were conductors. We now have a better definition of metallic conductors, namely, they have *unfilled valence bands.* Figures 8–2.2(a) and (c) show this schematically for sodium and aluminum.

The empty energy levels within a band are important for conduction because they permit an electron to rise to a higher energy level when it moves

---

* Examine the data for metals and alloys in either Table 8–2.1 or Appendix C.

† Pauli exclusion principle.

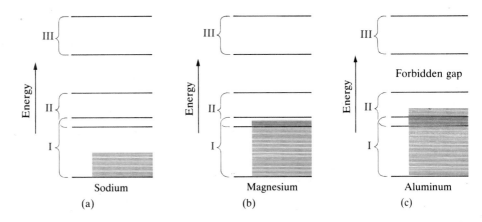

Figure 8–2.2 Energy bands. (a) Sodium. Since it has only one valence electron per atom, its first valence band (**I**) is only half filled. (b) Magnesium. Its first band would be full, except that its second band (**II**) overlaps, to contain a few electrons. (c) Aluminum. With three valance electrons, its first band is filled, and its second band is half full. All of these metals have empty levels in the upper part of their valence bands.

toward the positive electrode. This would not be possible if the energy band were completely filled and an overlying forbidden-energy gap were present.

Magnesium, with its two valence electrons per atom, is expected to fill the first valence band. It so happens, however, that the first and second bands overlap (Fig. 8–2.2b). Thus some of the $2N$ electrons (where N is the number of atoms) spill over into the second band, where there are plenty of vacant levels to receive the accelerating electrons. As a result, magnesium is metallic.

Silicon, however, presents another story because its four valence electrons per atom completely fill the first two valence bands (Fig. 8–2.3). Furthermore,

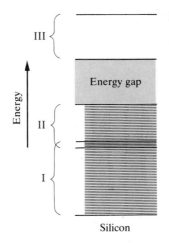

Figure 8–2.3 Energy gap (silicon). The four valence electrons per atom of silicon fill the first two energy bands. There is a forbidden-energy gap between the second and third bands.

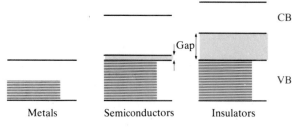

CB

Gap

VB

Metals   Semiconductors   Insulators

Figure 8–2.4 Metals, semiconductors, and insulators. Metals have unfilled energy bands. Semiconductors have a narrow forbidden gap above the top filled valence band (VB). A few electrons can "jump the gap" to the conduction band (CB). Insulators have a wide energy gap, which is a barrier to the electrons.

there is a forbidden-energy gap above the second band. Thus, electrons cannot be energized within these valence bands, and there is a large gap below the energy levels of the third band (III). Silicon is not a metallic conductor (with pure materials, at 20°C, $\rho_{Si} = 2 \times 10^3$ ohm·m; $\rho_{Cu} = 17 \times 10^{-9}$ ohm·m—a ratio of about 10^{11}). We shall see, in the next two sections, that the difference between insulators and semiconductors is related to the size of the *energy gap* E_g that overlies the filled *valence band* (Fig. 8–2.4). Silicon is a semiconductor (Fig. 8–1.1) because its energy gap is of such size that some highly energized electrons are able to "jump the gap" into the *conduction band* (Section 8–4).

Study Aids (metallic conductivity)

Energy bands, mean free path, and factors that affect metallic resistivity are presented by a series of nine sketches in the paperback *Study Aids for Introductory Materials Courses*. These are recommended for the student who finds difficulty with the visual concepts. In addition, Fermi energy and energy distributions are introduced for a more detailed understanding.

Example 8–2.1

Calculate the resistivity of silver at −40°C.

Procedure. Table 8–2.1 shows the resistivity at 0°C, and the temperature resistivity coefficient, i.e., the change per degree. Thus, this calculation is like one using a thermal expansion coefficient.

Calculation. From Table 8–2.1, $\rho = 15$ ohm·nm at 0°C, and $y_T = 0.0038/°C$.

$$\rho_{-40°C} = 15 \text{ ohm·nm} [1 + (0.0038/°C)(-40°C)]$$
$$= 13 \text{ ohm·nm}. \quad \blacktriangleleft$$

Example 8–2.2

Each atom percent of tin increases the resistivity of copper 30 ohm·nm regardless of temperature. What is the resistivity at 100°C of an alloy containing 99Cu–1Sn (weight basis)?

Procedure. First it is necessary to change 99Cu–1Sn to atom percent to be able to calculate the *solution resistivity*, ρ_x. It is also necessary to determine the resistivity of copper at 100°C by using the temperature resistivity coefficient, y_T. The answer for the problem will be the sum of the two:

$$\rho = \rho_x + \rho_{0°C} (1 + y_T \, \Delta T). \tag{8–2.2}$$

Calculation. Basis: 100 amu.

Sn: 1 amu/(118.7 amu/atom) = 0.0084 = 0.54 a/o.

Cu: 99 amu/(63.54)   = 1.5581

Total atoms = 1.5665.

ρ = (0.54 a/o)(30 ohm·nm per a/o) + 16 ohm·nm [1 + (0.0039/°C)(100°C)]

= 38 ohm·nm   (or 38 × 10⁻⁹ ohm·m).

Comment. The increase in resistivity is nearly linear with composition at low solute concentrations. ◄

Example 8–2.3

A toaster uses 300 watts when it is in operation and the nichrome element is at 870°C. It operates off a 110-V line. (a) How many amperes does it draw when it is hot? (b) When the switch is first snapped on?

Procedure. Calculate the 870°C amperage and resistance from the basic relationships of $P = EI$ and $E = IR$. Since dimension changes are very minor (and partially compensating),

$$R_{20°C}/R_{870°C} = \rho_{20°C}/\rho_{870°C}.$$

We can calculate $R_{870°C}$ and can calculate $\rho_{20°C}/\rho_{870°C}$ from Eq. (8–2.1). This lets us determine the amperage at 20°C; again by $E = IR$.

Calculation

a) I = 300 W/110 V = 2.7 amp.

b) $R_{870°C}$ = 110 V/2.7 A = 40 ohm.

$$\frac{R_{20}}{R_{870}} = \frac{\rho_0(1 + y_T 20°C)}{\rho_0(1 + y_T 870°C)}$$

R_{20} = 40 ohm [1 + 0.0004(20)]/[1 + 0.0004(870)]

= 40 ohm [1.008/1.35] = 30 ohm.

I_{20} = 110 V/30 Ω = 3.7 amp.

Comments. Had the element continued to draw 3.7 amperes, the temperature would continue to rise beyond 870°C, subjecting it to faster oxidation and related service deterioration, even melting.

Note that the temperature coefficients of resistivity of alloys are less than for pure metals. This is due in part to the fact that the mean free path for the electron is already short and the resistivity is initially higher. ◄

8–3

INSULATORS

In terms of energy bands, an insulator is a material with a large energy gap between the highest filled valence band and the next empty band (Fig. 8–2.4). The gap is so large that for all intents and purposes we can state that electrons are trapped in the lower band. Their number n in the conduction band for Eq. (8–1.1) is insignificantly low.

We commonly describe the valence electrons as being bound within the negative ions (or in a covalent bond). Approximately 7 eV ($=1.1 \times 10^{-18}$ J) of energy would be required to break an electron loose from the Cl⁻ ions in NaCl, and about 6 eV of energy to separate an electron from the covalent bond of diamond. These activation energies of 7 eV and 6 eV are also the dimensions of the energy gaps and may be compared to 1.1 eV and 0.7 eV for silicon and germanium, respectively. The physicist considers an energy gap of about 4 eV ($=0.64 \times 10^{-18}$ J) as an arbitrary distinction between semiconductors and insulators. Thus, NaCl and diamond are electronic insulators, and silicon and germanium are labeled semiconductors* (Fig. 8–1.1).

Example 8–3.1

An 0.1-mm film of polyethylene (PE) is used as a dielectric to separate two electrodes at 110 V. Based on Fig. 8–1.1, what is the electron flux through the film?

Procedure. Calculate R from $R = \rho L/A$; and the amperage from $E = I ?$. Finally, the charge per electron (el) is 0.16×10^{-18} A·s.

Calculation. Basis = 1 m² and 1 s.

$$R = \frac{(10^{14}\ \text{ohm·m})(10^{-4}\ \text{m})}{(1\ \text{m}^2)} = 10^{10}\ \text{ohm}.$$

$$I = \frac{110\ \text{V}}{10^{10}\ \text{ohm}} = 11 \times 10^{-9}\ \text{amp}.$$

$$\frac{(11 \times 10^{-9}\ \text{A/m}^2)}{(0.16 \times 10^{-18}\ \text{A·s/el})} = 7 \times 10^{10}\ \text{electrons/s·m}^2.$$

Comments. This nanoampere current is small, but measurable. Very commonly, extraneous factors, such as impurities, pinhole porosity, surface leakage, must be considered when making measurements, because they will have larger effects. ◀

8–4

INTRINSIC SEMICONDUCTORS

Energy Gaps

Semiconductors and insulators are differentiated on the basis of the size of their forbidden-energy gaps (see Fig. 8–2.4). In a semiconductor, the energy gap is such that usable numbers of electrons are able to jump the gap from the filled valence band to the empty conduction band (Fig. 8–4.1). Those energized electrons can now carry a charge toward the positive electrode; furthermore, the resulting electron holes in the valence band become available for conduction because electrons deeper in the band can move up into those vacated levels.

Figure 8–4.2 shows the energy gap schematically for C(diamond), Si, Ge, and Sn(gray). The gap is too large in diamond to provide a usable number of charge carriers, so diamond is categorized as an insulator (Table 8–4.1). The number of carriers increases as we move down through Group IV of the peri-

---

* Diamond can be an electronic semiconductor if impurities are present; and NaCl can be an ionic semiconductor if conditions are favorable for sodium *ion* diffusion.

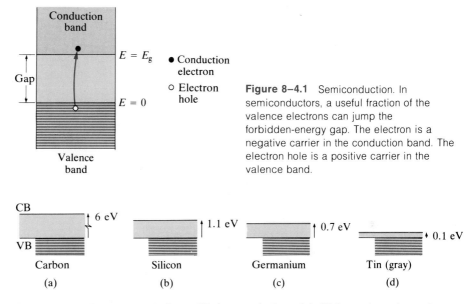

Figure 8-4.1 Semiconduction. In semiconductors, a useful fraction of the valence electrons can jump the forbidden-energy gap. The electron is a negative carrier in the conduction band. The electron hole is a positive carrier in the valence band.

Figure 8-4.2 Energy gaps in Group IV elements (schematic). All these elements can have the same structure, and they all have filled bands. Because tin has the smallest energy gap, it has the most electrons in the conduction band (CB) at normal temperatures, and therefore the highest conductivity. (Cf. Table 8-4.1.)

odic table to silicon, germanium, and tin; as a result, the conductivity increases, as shown in the accompanying table. This conductivity is an inherent property of these materials and does not arise from impurities. Therefore, it is called *intrinsic semiconductivity*.

The crystal structure of diamond is repeated (from Chapters 2 and 3) in Fig. 8-4.3(a). Each carbon atom has a coordination number of 4, and each neighboring pair of atoms shares a pair of electrons (Section 2-3). Silicon, germanium,

TABLE 8-4.1

Energy Gaps in Semiconducting Elements

| | Energy gap E_g | | At 20°C (68°F) | |
| | | | Fraction of valence electrons with | |
Element	10^{-18} J	eV	energy $> E_g$	Conductivity σ, ohm^{-1}·m^{-1}
C(diamond)	0.96	≈6	≈1/30 × 10^{21}	<10^{-16}
Si	0.176	1.1	≈1/10^{13}	5 × 10^{-4}
Ge	0.112	0.7	≈1/10^{10}	2
Sn(gray)	0.016	0.1	≈1/5000	10^6

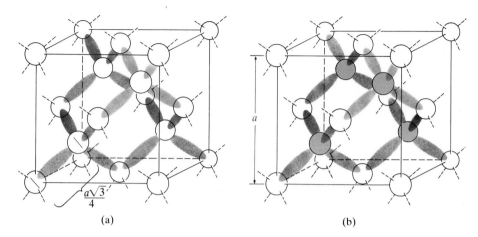

Figure 8–4.3 Crystal structures of familiar semiconductors. (a) Diamond, silicon, germanium, gray tin. (b) ZnS, GaP, GaAs, InP, etc. (Cf. Fig. 7–2.4.) The two structures are similar, except that two types of atoms are in alternate positions in the semiconducting compounds. All atoms have CN = 4; each material has an average of four valence electrons per atom, and two electrons per bond.

and gray tin have the same structure.* Figure 8–4.4 uses germanium to represent schematically the mechanism of semiconductivity in these elements.

The above four Group IV elements are the only elements that obtain semiconductivity from the structure of Fig. 8–4.3(a). However, a number of III-V *compounds* are based on the same structure (Fig. 8–4.3b), which has ZnS as its prototype (Fig. 7–2.4). Atoms of elements from Group III of the periodic table (B, Al, Ga, In) alternate with atoms of elements from Group V of the periodic table (N, P, As, Sb). Most of the 16 III-V compounds that can form from these elements are semiconductors because every atom has four neighbors, and the average number of shared valence electrons is *four*. This matches exactly the situation for silicon and germanium, our predominant semiconductors.

The materials and electrical engineers also design electronic devices with *alloys of compounds*; for example, a solid solution of InAs and GaAs to give (In,Ga)As, or a solid solution of InP and InAs to give In(As,P). Furthermore, 4-component compounds of (In,Ga)(As,P) are possible. All of these possess the same structure as the individual III-V compounds (Fig. 8–4.3b). The energy gaps of these III-V alloys are intermediate (but not necessarily linear) between the energy gaps of the end members (Fig. 8–4.5). Likewise, the lattice constants are intermediate. The relationships of Fig. 8–4.5 are useful to the design-

---

* White tin is the more familiar polymorph. It is stable above 13°C (but may be supercooled to lower temperatures). White tin (bct) is denser than gray tin ($\rho_w = 7.3$ Mg/m^3 = 7.3 g/cm^3, while $\rho_g = 5.7$ Mg/m^3); therefore, the energy bands of white tin overlap, and this phase is a metallic conductor (cf. Fig. 8–2.2b).

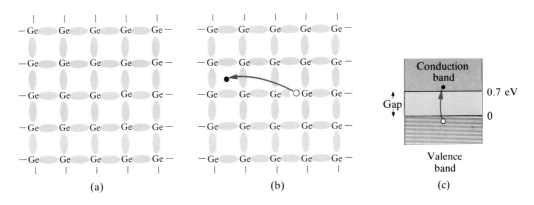

Figure 8–4.4 Intrinsic semiconductor (germanium). (a) Schematic presentation showing electrons in their covalent bonds (and their valence bands). (b) Electron-hole pair. (Positive electrode at the left.) (c) Energy gap, across which an electron must be raised to provide conduction. For each conduction electron, there is a hole produced among the valence electrons.

ers of complex devices where it is necessary to match the lattice size with specific energy-gap values.

Charge Mobility Our introductory equation on conductivity (Eq. 8–1.1) must now be modified to match Fig. 8–4.1, since an intrinsic semiconductor has both negative and positive carriers.

The *electrons* that jump into the conduction band are the *negative*-type carriers. The conductivity they produce depends on their mobility μ_n through

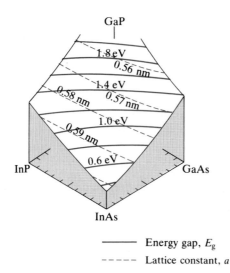

Figure 8–4.5 Alloys of III–V compounds. By adjusting the In–Ga ratio and/or the P–As ratio, alloys may be selected with desired combinations of energy gaps and lattice constants. (Adapted from C. J. Nuese, *J. Ed. Mod. Mat'ls Sci. & Engr.*)

——— Energy gap, E_g

- - - - - Lattice constant, a

TABLE 8–4.2

Properties of Common Semiconductors (20°C)*

Material	Energy gap E_g		Mobilities, m²/volt·sec		Intrinsic conductivity, ohm^{-1}·m^{-1}	Lattice constant, a, nm
	10^{-18} J	eV	Electron, μ_n	Hole, μ_p		
Elements						
C(diamond)	0.96	≈6	0.17	0.12	$<10^{-16}$	0.357
Silicon	0.176	1.1	0.19	0.0425	5×10^{-4}	0.543
Germanium	0.112	0.7	0.36	0.23	2	0.566
Tin (gray)	0.016	0.1	0.20	0.10	10^6	0.649
Compounds						
AlSb	0.26	1.6	0.02	—	—	0.613
GaP	0.37	2.3	0.019	0.012	—	0.545
GaAs	0.22	1.4	0.88	0.04	10^{-6}	0.565
GaSb	0.11	0.7	0.60	0.08	—	0.612
InP	0.21	1.3	0.47	0.015	500	0.587
InAs	0.058	0.36	2.26	0.026	10^4	0.604
InSb	0.029	0.18	8.2	0.17	—	0.648
ZnS	0.59	3.7	0.014	0.0005	—	—
SiC (hex)	0.48	3	0.01	0.002	—	—

* Revised data collected by B. Mattes.

the conduction band of the semiconductor. The *electron holes* that are formed in the valence band are the *positive*-type carriers.* The conductivity they produce depends on their mobility μ_p through the valence band of the semiconductor. The total conductivity in an intrinsic semiconductor arises from both contributors

$$\sigma = n_n q \mu_n + n_p q \mu_p. \qquad (8\text{–}4.1)$$

Of course, both the hole and the electron carry the same basic charge unit of 0.16×10^{-18} coul. In an intrinsic semiconductor, where there is a one-for-one formation of conduction electrons and electron holes, $n_n = n_p$; thus we could simplify Eq. (8–4.1). However, let's leave it in its present form because n_n does not equal n_p for *extrinsic* semiconductors (Section 8–5).

Table 8–4.2 summarizes the properties of a number of semiconductors. Note that we can make two generalizations.

1. The size of the energy gap commonly decreases as we move down in the periodic table (C → Si → Ge → Sn), or (GaP → GaAs → GaSb), or (AlSb → GaSb → InSb).

---

* Comparably, anions, with extra electrons, are negative-type, and cations, which are deficient in electrons, are positive-type.

2. The mobility of electrons within a given semiconductor is greater than the mobility of electron holes in the same semiconductor.* The latter difference will be important when considering the use of *n*-type semiconductors in contrast to *p*-type semiconductors.

Semiconductivity (intrinsic) versus Temperature

Unlike metals, which have increased resistivity and decreased conductivity at higher temperatures (Fig. 8–2.1), the conductivity of intrinsic semiconductors *increases* at higher temperatures. The explanation is straightforward when one considers that the number of charge carriers, *n*, increases directly with the number of electrons that jump the gap (Fig. 8–4.1), and both increase exponentially with temperature. At 0 K, *no* electron would have the necessary energy to do this; however, as the temperature rises, the electrons receive energy, just as the atoms do. At 20°C, a useful fraction of the valence electrons in silicon, germanium, and tin have energy in excess of E_g, the energy gap (Table 8–4.1). The same is true for compound semiconductors.

By analogy with Eq. (4–6.3a) the distribution of these thermally energetic electrons is

$$n_i \propto e^{-(E-\overline{E})/kT}, \tag{8–4.2a}†$$

where n_i is the number of electrons per m^3 in the conduction band (and the number of holes per m^3 in the valence band). Within the forbidden-energy gap of an intrinsic semiconductor, the average energy $\overline{E}$ is at the middle of the energy gap, or $E_g/2$. Therefore,

$$n_i \propto e^{-E_g/2kT}. \tag{8–4.2b}$$

As in Chapter 4, T is the absolute temperature (K), and k is Boltzmann's constant, often expressed as 86.1×10^{-6} eV/K rather than 13.8×10^{-24} J/K.

According to Eq. (8–1.1), conductivity σ is directly proportional to the number of carriers n; therefore,

$$\sigma = \sigma_0 e^{-E_g/2kT}, \tag{8–4.3a}$$

where σ_0 is the proportionality constant that includes, among other factors, both q and μ of Eq. (8–1.1). Admittedly, the mobility μ varies some with temperature; however, its variation within the normal working range of most semiconductors is small compared with the exponential variation of the number of carriers n. Thus, we can rewrite this last equation in an Arrhenius form:

$$\ln \sigma = \ln \sigma_0 - E_g/2kT. \tag{8–4.3b}$$

---

* This relationship exists for all of the semiconductors of Table 8–4.2, with the possible exception of AlSb, where the mobility data have not been accurately determined.

† Equations (8–4.2a) and (4–6.3a) are analogous at the upper end of the energy range only. In that range, the Pauli exclusion principle of two electrons per quantum state is not restrictive, since the probabilities of occupancy are very low. Of course, our interest is in this upper end of the range and in the initial electrons to jump the gap.

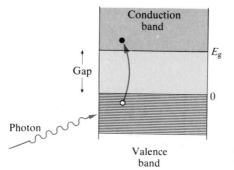

Figure 8–4.6 Photoconduction. A photon (i.e., light energy) raises the electron across the energy gap, producing a "conduction electron + valence hole" pair, forming charge carriers. Recombination (Eq. 8–4.4b) occurs when the electron drops back to the valence band.

If we measure the conductivity (or resistivity) of a semiconductor in the laboratory and plot $\ln \sigma$ versus $1/T$, we can calculate E_g from the slope of the curve, i.e., slope $= -E_g/2k$. Conversely, from E_g and one known value of σ, we may calculate σ at a second temperature (Example 8–4.4).

Photoconduction

There is only a small probability that an electron in the valence band of silicon may be raised across the energy gap by thermal activation into the conduction band (~ 1 out of 10^{13}, according to Table 8–4.1). In contrast, a photon of red light (wavelength = 660 nm) has 1.9 eV of energy, which is more than enough to cause an electron to jump the 1.1 eV energy gap in silicon (Fig. 8–4.6). Thus the conductivity of silicon increases markedly by photoactivation when it is exposed to light.

Recombination

The reaction that produces an *electron–hole pair*, as shown in Fig. 8–4.6 may be written as

$$E \rightarrow n + p \tag{8–4.4a}$$

where E is energy, n is the conduction electron, and p is the hole in the valence band. In this case the energy came from light, but it could have come from other energy sources such as heat or fast-moving electrons.

Since all materials are more stable when they reduce their energies, electron–hole pairs recombine sooner or later:

$$n + p \rightarrow E. \tag{8–4.4b}$$

In effect, the electron drops from the conduction band back to the valence band, just the reverse of Fig. 8–4.4(c). Were it not for the fact that heat, light, or some other energy source continually produces additional electron–hole pairs, the conduction band would soon become depleted.

The time required for recombination varies from material to material. However, it follows a regular pattern because, within a specific material, every conduction electron has the same probability of recombining within the next

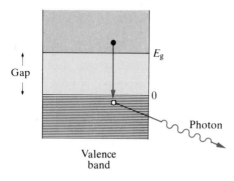

Figure 8–4.7   Luminescence. Each millisecond, a fraction of the electrons energized to the conduction band return to the valence band. As an electron drops across the gap, the energy may be released as a photon of light.

second (or minute). This leads to the relationship

$$N = N_0 e^{-t/\tau}, \tag{8–4.5a}$$

which we usually rearrange to

$$\ln (N_0/N) = t/\tau. \tag{8–4.5b}*$$

In these equations, N_0 is the number of electrons in the conduction band at a particular moment of time (say, when the source of light is turned off). After an additional time t, the number of remaining conduction electrons is N. The term τ is called the *relaxation*, or *recombination, time*, and is characteristic of the material.

• **Luminescence**

The energy released in Eq. (8–4.4b) may appear as heat. It may also appear as light. When it does, we speak of *luminescence* (Fig. 8–4.7). Sometimes we subdivide luminescence into several categories. *Photoluminescence* is the light emitted after electrons have been activated to the conduction band by light photons. *Chemoluminescence* is the word used when the initial activation is due to chemical reactions. Probably *electroluminescence* is best known, because this is what occurs in a TV tube, where a stream of electrons (cathode rays) scans the screen, activating the electrons in the phosphor to their conduction band. Almost immediately, however, the electrons and holes recombine, emitting energy as visible light.

Since the recombination rate is proportional to the number of activated electrons, the intensity I of luminescence also follows Eq. (8–4.5b):

$$\ln (I_0/I) = t/\tau. \tag{8–4.7}$$

---

* Equation (8–4.5) can be derived through calculus (by those who wish to do so) from the information stated above:

$$dN/dt = -N/\tau. \tag{8–4.6}$$

Rearranging,

$$dN/N = -dt/\tau;$$

then integrating we get,

$$\ln (N/N_0) = -t/\tau. \tag{8–4.5c}$$

For a TV tube, the engineer chooses a phosphor with a relaxation time such that light continues to be emitted as the next scan comes across. Thus our eyes do not see a light–dark flickering. However, the light intensity from the previous trace should be weak enough so that it does not compete with the new scan that follows one-thirtieth of a second later. (See Example 8–4.5.)

Study Aids (intrinsic semiconductors)

The paperback *Study Aids for Introductory Materials Courses* has a section on "intrinsic semiconduction" that reviews energy bands; contrasts conductors, semiconductors, and insulators; then relates conductivity to gap size and temperature. Optional sketches describe several intrinsic semiconductor devices.

Example 8–4.1

(a) What fraction of the charge in intrinsic silicon is carried by electrons? (b) By electron holes?

Solution. Recall that in an intrinsic semiconductor,

$$\sigma = \sigma_n + \sigma_p = (nq\mu)_n + (nq\mu)_p,$$

and with $n_n = n_p$,

$$\sigma_n/\sigma = \mu_n/(\mu_n + \mu_p)$$
$$= (0.19 \text{ m}^2/\text{V·s})/(0.19 + 0.0425 \text{ m}^2/\text{V·s}) = 0.82.$$

$$\sigma_p/\sigma = 0.18.$$

Comment. Typically, conductivity is higher in the conduction band than from the electron holes in the valence band because $\mu_n > \mu_p$. (See Table 8–4.2.) ◀

Example 8–4.2

From Table 8–4.2, the compound gallium arsenide has an intrinsic conductivity of 10^{-6} ohm^{-1}·m^{-1} at 20°C. How many electrons have jumped the energy gap?

Procedure. Again we use $\sigma = (nq\mu)_n + (nq\mu)_p = nq(\mu_n + \mu_p)$, and data from Table 8–4.2.

Calculation

$$n = (10^{-6} \text{ ohm}^{-1}\text{·m}^{-1})/(0.16 \times 10^{-18} \text{ A·s})(0.88 + 0.04 \text{ m}^2/\text{V·s})$$
$$= 6.8 \times 10^{12}/\text{m}^3.$$

Comment. There will be 1.36×10^{13} carriers/m^3 since an electron hole remains for each electron activated across the energy gap. ◀

Example 8–4.3

Each gray tin atom has four valence electrons. The unit cell (uc) size (Fig. 8–4.3a) is 0.649 nm. Separate calculations indicate that there are 2×10^{25} conduction electrons per m^3. What fraction of the electrons have been activated to the conduction band?

Procedure. The unit cell of tin has eight atoms (Fig. 8–4.3a); therefore 32 valence electrons per a^3.

Calculation

$$\text{Valence electrons/m}^3 = \frac{(8 \text{ atoms/uc})(4 \text{ el/atom})}{(0.649 \times 10^{-9} \text{ m})^3/\text{uc}} = 1.17 \times 10^{29}/\text{m}^3.$$

$$\text{Fraction activated} = \frac{2 \times 10^{25}}{1.17 \times 10^{29}} \approx 0.0002. \quad \blacktriangleleft$$

Example 8–4.4

The resistivity of germanium at 20°C (68°F) is 0.5 ohm·m. What is its resistivity at 40°C (104°F)?

Procedure. Two procedures are possible. The first is to solve for the $\rho_{20°C}/\rho_{40°C}$ ratio from $\rho_{20°C}$ and E_g. The second is to solve graphically (Fig. 8–4.8) from $\sigma_{20°C}$ and the slope, $E_g/2k$.

Solution. Based on Eq. (8–4.3) and an energy gap of 0.7 eV (Table 8–4.2):

$$\frac{\rho_1}{\rho_2} = \frac{\sigma_2}{\sigma_1} = \frac{\sigma_0 e^{-E_g/2kT_2}}{\sigma_0 e^{-E_g/2kT_1}},$$

$$\ln \frac{\rho_1}{\rho_2} = \frac{E_g}{2k}\left[\frac{1}{T_1} - \frac{1}{T_2}\right]. \tag{8–4.8}$$

$$\ln \frac{\rho_{20}}{\rho_{40}} = \frac{0.7 \text{ eV}}{2(86.1 \times 10^{-6} \text{ eV/K})}\left[\frac{1}{293 \text{ K}} - \frac{1}{313 \text{ K}}\right] = 0.9.$$

$$\rho_{20}/\rho_{40} = {\sim}2.5.$$

Thus, since $\rho_{20°} = 0.5$ ohm·m, $\rho_{40°C} = 0.2$ ohm·m.

Alternate solution. From the slope (Fig. 8–4.8):

$$\text{Slope} = -0.7 \text{ eV}/2(86.1 \times 10^{-6} \text{ eV/K}) = -4060 \text{ K};$$

or

$$\text{slope} = -(0.112 \times 10^{-18} \text{ J})/2(13.8 \times 10^{-24} \text{ J/K}) = -4060 \text{ K}.$$

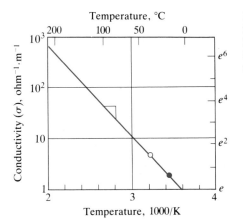

Figure 8–4.8 Semiconduction versus temperature (intrinsic germanium). The slope is $-E_g/2k$ of Eq. (8–4.3) when the ordinate is $\ln \sigma$ and the abscissa is reciprocal temperature, K^{-1}. See Example 8–4.4.

At 20°C (●),

$$\ln \sigma = \ln (1/0.5 \text{ ohm·m}) = 0.693 \qquad \text{(or } \log_{10} = 0.301\text{)};$$

$$1/T = 1/293 \text{ K} = 0.00341/\text{K} \qquad \text{(or } 1000/T = 3.41\text{)}.$$

At 40°C (○),

$$1/T = 1/313 \text{ K} = 0.00319/\text{K} \qquad \text{(or } 1000/T = 3.19\text{)}.$$

$$\text{Slope} = -4060 \text{ K} = \frac{0.693 - \ln \sigma_{40°C}}{(0.00341 - 0.00319)/\text{K}};$$

$$\ln \sigma_{40°C} = 1.6 \qquad \text{(or } \log \sigma_{40°C} = 0.7\text{)};$$

$$\sigma_{40°C} = 5 \text{ ohm}^{-1}\text{·m}^{-1} \qquad \text{(or } \rho = 0.2 \text{ ohm·m)}.$$

Comments. It is possible to measure resistance changes (and therefore resistivity changes) of <0.1%. Therefore one can measure temperature changes of a small fraction of a degree. (See Study Problem 8–6.1.) ◀

● Example 8–4.5

The scanning beam of a television tube covers the screen with 30 frames per second. What must the relaxation time for the activated electrons of the phosphor be if only 20% of the intensity is to remain when the following frame is scanned?

Procedure. Use Eq. (8–4.7). The time t in question is $\frac{1}{30}$ s.

Calculation

$$\ln (1.00/0.20) = (0.033 \text{ s})/\tau,$$

$$\tau = 0.02 \text{ s}.$$

Comments. We use the term *fluorescence* when the relaxation time is short compared to the time of our visual perception (~25 ms). If the luminescence has a noticeable afterglow, we use the term *phosphorescence*. ◀

8–5

EXTRINSIC SEMICONDUCTORS

N-type Semiconductors

Impurities alter the semiconducting characteristics of materials by introducing excess electrons or excess electron holes. Consider, for example, some silicon containing an atom of phosphorus. Phosphorus has five valence electrons rather than the four that are found with silicon. In Fig. 8–5.1(a), the extra electron is present independently of the electron pairs that serve as bonds between neighboring atoms. By supplying it with a small amount of energy, this electron can be pulled away from the phosphorus atom and carry a charge toward the positive electrode (Fig. 8–5.1b). Alternatively, in Fig. 8–5.1(c), the extra electron—which cannot reside in the valence band because that is already full—is located near the top of the energy gap. From this position—called a *donor* level E_d—the extra electron can easily be donated to the conduction band. Regardless of which model is used, Fig. 8–5.1(b) or 8–5.1(c), we can see

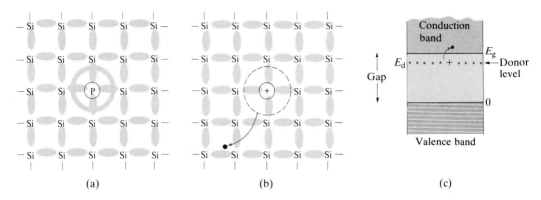

Figure 8–5.1 Extrinsic semiconductors (*n*-type). A Group V atom has an extra valence electron beyond the average of four sketched in Fig. 8–4.3. This fifth electron can be pulled away from its parent atom with very little added energy and "donated" to the conduction band, to become a charge carrier. We observe the donor energy level, E_d, as being just below the top of the energy gap. (a) An *n*-type impurity, such as phosphorous. (b) Ionized phosphorus atom. (Positive electrode at left.) (c) Band model.

that atoms from Group V (N, P, As, and Sb) of the periodic table (Fig. 2–1.1) can supply negative, or *n*-type, charge carriers to semiconductors.

***P*-type Semiconductors**

Group III elements (B, Al, Ga, and In) have only three valence electrons. Therefore, when such elements are added to silicon as impurities, electron holes come into being. As shown in Fig. 8–5.2(a) and (b), each aluminum atom can accept one electron. In the process a positive charge moves toward the negative electrode. Using the band model (Fig. 8–5.2c), we note that the energy

Figure 8–5.2 Extrinsic semiconductors (*p*-type). A Group III atom has one less valence electron than the average of four sketched in Fig. 8–4.3. This atom can accept an electron from the valence band, thus leaving an electron hole as a charge carrier. The acceptor energy level, E_a, is just above the bottom of the energy gap. (a) A *p*-type impurity such as aluminum. (b) Ionized aluminum atom. (Negative electrode at right.) (c) Band model.

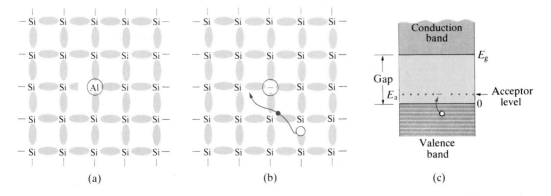

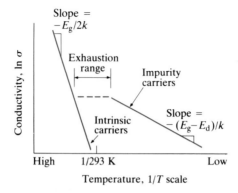

Figure 8–5.3 Donor exhaustion. Intrinsic (left-hand curve) and extrinsic (right-hand curve) conductivities require energies of E_g and $(E_g - E_d)$, respectively, to raise electrons into the conduction band (Fig. 8–5.1c). At lower temperatures, donor electrons provide most of the conductivity. Exhaustion occurs when all of the donor electrons have entered the conduction band, and before the temperature is raised high enough for valence electrons to jump the energy gap. The conductivity is nearly constant in this temperature range.

difference for electrons to move from the valence band to the *acceptor level*, E_a, is much less than the full energy gap. Therefore, electrons are more readily activated into the acceptor sites than into the conduction band. The electron holes remaining in the valence band are available as positive carriers for *p*-type semiconduction. Therefore, with $(n_p)_{ex} = n_{III}$,

$$\sigma_{ex} = n_{III} q \mu_p, \qquad (8\text{–}5.1)$$

where n_{III} is the number of acceptors per m³ (Gr. III atoms/m³).

Donor Exhaustion (and acceptor saturation)

Since donated electrons have only a small jump to the conduction band, they initiate *extrinsic* conductivity at relatively low temperatures. As the temperature is increased, the slope of the Arrhenius curve is $-(E_g - E_d)/k$ as shown at the right in Fig. 8–5.3.

If the donor impurities are limited in number (e.g., 10^{21} P/m³ within silicon), essentially all of the donated electrons have moved into the conduction band at temperatures below that of normal usage. This supply has been exhausted. In the above example of 10^{21}/m³, the extrinsic conductivity is

$$\sigma_{ex} = (10^{21}/\text{m}^3)(0.16 \times 10^{-18} \text{ A·s})(0.19 \text{ m}^2/\text{V·s})$$
$$= 30 \text{ ohm}^{-1}\text{·m}^{-1}.$$

The extrinsic conductivity will not continue to rise with further temperature increases and there is a conductivity plateau.*

In the meantime, the *intrinsic* conductivity is very low in a semiconductor such as silicon (5×10^{-4} ohm⁻¹·m⁻¹ at 20°C according to Table 8–4.1). Its Arrhenius curve is at the left in Fig. 8–5.3 with the intrinsic slope of $-E_g/2k$, that is, -1.1 eV/$2k$ = 6400 K. Only at elevated temperatures does the total conductivity increase above the exhaustion plateau. (Cf. Example 8–5.3.)

---

* With constant n, experiments can detect a slight decrease in μ that results from the shorter mean free path that accompanies increased temperatures (Section 8–2).

Donor exhaustion of *n*-type semiconductors has its parallel in *acceptor saturation* of *p*-type semiconductors. (The reader is asked to paraphrase the previous paragraphs for the saturation analog.) Donor exhaustion and acceptor saturation are important to materials and electrical engineers, since these situations provide a region of essentially constant conductivity. This means that it is less necessary to compensate for temperature changes in electrical circuits than it would be if the log σ-versus-$1/T$ characteristics followed an ever-ascending line.

Defect Semiconductors

The iron oxide of Fig. 4–3.2 possessed Fe^{3+} ions in addition to the regular Fe^{2+} ions. A similar situation occurs in Fig. 8–5.4(a) when NiO is oxidized to give some Ni^{3+} ions and, in fact, is relatively common among transition-metal oxides that have multiple valences. In nickel oxide, three Ni^{2+} are replaced by 2 Ni^{3+} and a vacancy, $\square$. This maintains the charge balance; it also permits easier diffusion and therefore some ionic conductivity. More important, however, is the fact that electrons can hop from an Ni^{2+} ion into acceptor sites in Ni^{3+} ions. Conversely, an electron hole moves from one nickel ion to another as it migrates toward the negative electrode. Nickel oxide and other oxides with $M_{1-x}O$ defect structures are *p*-type semiconductors.

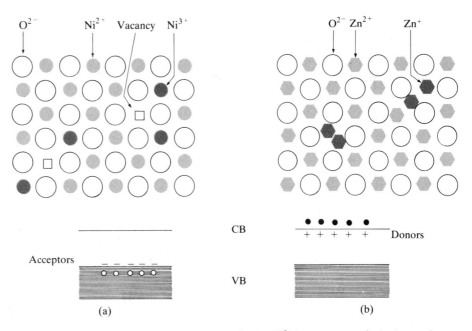

Figure 8–5.4 Defect semiconductors. (a) $Ni_{1-x}O$. The Ni^{3+} ions serve as electron acceptors, so that holes, $\bigcirc$, form in the valence band. (b) $Zn_{1+y}O$ (schematic). The Zn^+ ions are donors of electrons, $\bullet$, to the conduction band for *n*-type semiconduction.

There are also *n*-type oxides. Zinc oxide, when exposed to zinc vapors, produces $Zn_{1+y}O$. A zinc ion moves into an interstitial position (Fig. 8–5.4b). The Zn^+ ions that balance the charge have one electron more than the bulk of the Zn^{2+} ions. These can donate electrons to the conduction band for *n*-type semiconductivity.

Study Aids (extrinsic semiconductors)

A series of eleven sketches is available as an alternate approach to extrinsic semiconduction in *Study Aids for Introductory Materials Courses*. By using this alternate approach, it is possible to be more specific about junction devices, such as the Zener diode and transistors.

Example 8–5.1

Silicon, according to Table 8–4.2, has a conductivity of 5×10^{-4} ohm$^{-1}\cdot$m^{-1} when pure. Specifications call for a conductivity of 200 ohm$^{-1}\cdot$m^{-1} when it contains aluminum as an impurity. How many aluminum atoms are required per m^3?

Procedure. The total conductivity, σ, is the sum of $\sigma_{in} + \sigma_{ex}$. However, from Table 8–4.2, $\sigma_{in} = 5 \times 10^{-4}$ ohm$^{-1}\cdot$m$^{-1} \ll 200$ ohm$^{-1}\cdot$m^{-1}, as required here. Therefore, Eq. (8–5.1) can be used directly.

Calculation

$$n_{III} = n_p = (200 \text{ ohm}^{-1}\cdot\text{m}^{-1})/(0.16 \times 10^{-18} \text{ A}\cdot\text{s})(0.0425 \text{ m}^2/\text{V}\cdot\text{s})$$
$$= 3 \times 10^{22}/\text{m}^3.$$

Comments. Each aluminum atom contributes one acceptor site and hence one electron hole. Therefore 3×10^{22} aluminum atoms are required per m^3. This is, of course, a large number; however, it is still small (0.6 ppm) when compared with the number of silicon atoms per m^3. (See Study Problem 8–5.1a.) ◀

Example 8–5.2

Early transistors used germanium with an extrinsic resistivity of 200 ohm·m and a conduction electron concentration of 0.87×10^{17}/m^3. (a) What is the mobility of the electrons in the germanium? (b) What impurity elements could be added to the germanium to donate the conduction electrons?

Procedure. We can write an equation for *n*-type semiconduction analogous to Eq. (8–5.1):

$$\sigma_{ex} = n_V q \mu_n = 1/\rho_{ex} \qquad (8\text{–}5.2)$$

since the extrinsic conduction arises from electron donors.

Calculation

a) $$\mu_n = 1/(200 \text{ ohm·m})(0.87 \times 10^{17}/\text{m}^3)(0.16 \times 10^{-18} \text{ A}\cdot\text{s})$$
$$= 0.36 \text{ m}^2/\text{V}\cdot\text{s}.$$

b) Group V elements: N, P, As, Sb.

Comments. Note that the electron mobility does not depend on which of these Group V elements is added, since the electron, once in the conduction band, moves through the germanium structure independent of its donor.

Group VI elements could also be added. Since they have a second additional electron (beyond the four necessary for bonding), it would take only $0.4 \times 10^{17}/m^3$ of these atoms to supply $0.8 \times 10^{17}/m^3$ conduction electrons. ◀

Example 8–5.3

The residual phosphorus content of purified silicon is 0.1 part per billion (by weight). Will the resulting conductivity exceed the intrinsic conductivity of silicon?

Solution. Based on density (Appendix B): 1 m^3 silicon = 2.33×10^6 g Si = 2.33×10^{-4} g P.

$$(2.33 \times 10^{-4} \text{ g/m}^3)/(30.97 \text{ g}/0.6 \times 10^{24}) = 4.5 \times 10^{18}/m^3.$$

$$\sigma_{ex} = (4.5 \times 10^{18}/m^3)(0.16 \times 10^{-18} \text{ A·s})(0.19 \text{ m}^2/\text{V·s})$$

$$= 0.14 \text{ ohm}^{-1} \cdot m^{-1};$$

versus

$$\sigma_{in} = 5 \times 10^{-4} \text{ ohm}^{-1} \cdot m^{-1} \qquad \text{(from Table 8–4.2).}$$

Comment. It was necessary to develop entirely new processing procedures in order to achieve the low impurity levels required for semiconduction production. ◀

Example 8–5.4

There are 10^{22} Al/m^3 in silicon to produce a p-type semiconductor. At what temperature will the intrinsic conductivity of silicon equal the maximum extrinsic conductivity?

Procedure. Saturation is required for maximum extrinsic conductivity; therefore, $n_p = n_{III}$. Of course, μ_p (only) for silicon must be used for this extrinsic conductivity.

From Table 8–4.2, $\sigma_{in} = 5 \times 10^{-4}$ ohm^{-1}·m^{-1} at 20°C and $E_g = 1.1$ eV. Calculate the temperature at which $\sigma_{in} = \sigma_{ex}$.

Calculation

$$\sigma_{ex} = (10^{22}/m^3)(0.16 \times 10^{-18} \text{ A·s})(0.0425 \text{ m}^2/\text{V·s})$$

$$= 68 \text{ ohm}^{-1} \cdot m^{-1}.$$

$$\frac{\sigma_{20°}}{\sigma_T} = \frac{5 \times 10^{-4} \text{ ohm}^{-1} \cdot m^{-1}}{68 \text{ ohm}^{-1} \cdot m^{-1}} = \frac{\sigma_0 e^{-1.1/2k(293 \text{ K})}}{\sigma_0 e^{-1.1/2kT}};$$

$$\ln \frac{5 \times 10^{-4}}{68} = -11.8 = \frac{-1.1 \text{ eV}}{2(86.1 \times 10^{-6} \text{ eV/K})}\left[\frac{1}{293 \text{ K}} - \frac{1}{T}\right];$$

$$T = 640 \text{ K} \qquad \text{(or 367°C).}$$

Comments. The most general equation for conductivity in semiconductors is

$$\sigma = \sigma_{in} + (\sigma_n)_{ex} + (\sigma_p)_{ex}$$

$$= (n_{in}q)(\mu_n + \mu_p) + (n_n q\mu_n)_{ex} + (n_p q\mu_p)_{ex}. \qquad (8–5.1)$$

Typically, only one of the three terms is significant at a time, and the others can be dropped. In this example, only $(\sigma_p)_{ex}$ is significant at 20°C. ◀

• Example 8–5.5

Nickel oxide (Fig. 8–5.4a) normally has only a limited number of Ni^{3+} ions ($Ni^{3+}/Ni^{2+} < 10^{-4}$). When Li_2O is added to NiO, more Ni^{2+} ions oxidize to Ni^{3+}, and an Li^+Ni^{3+} pair replaces two previous Ni^{2+} ions. With 1 wt% Li_2O, how many charge carriers are there per m^3? (NiO has the same structure as NaCl.)

Procedure. Calculate the Li^+-to-O^{2-} ratio. Since there is one Ni^{3+} per Li^+, one *p*-type carrier per Ni^{3+}, and four O^{2-} per unit cell, we can calculate the carrier concentrations. The unit cell volume is based on the NaCl structure and the ionic radii of Appendix B. ($r = {\sim}0.069$ nm, since Ni^{2+} is predominant.)

Calculation. Based on 100 amu = 1 amu Li_2O + 99 amu NiO_2.

$$LiO_2: \quad 1/(13.88 + 16) = 0.0335 \ Li_2O = 0.067 \ Li^+ + 0.0335 \ O^{2-}$$

$$NiO: \quad 99/(58.71 + 16) = 1.3251 \ NiO = \underline{\qquad\qquad 1.3251 \ O^{2-}}$$

$$\text{Total } O^{2-} = 1.36.$$

$$\frac{Li^+}{O^{2-}} = \frac{0.067 \ Li^+}{1.36 \ O^{2-}} = \frac{0.2 \ Li^+}{4 \ O^{2-}} = \frac{0.2 \ Ni^{3+}}{\text{unit cell}} = \frac{0.2 \ \text{carriers}}{\text{unit cell}};$$

$$\frac{0.2 \ \text{carriers}}{a^3} = \frac{0.2 \ \text{carriers}}{[2({\sim}0.069 + 0.140)10^{-9} \ m]^3} = 2.7 \times 10^{27}/m^3.$$

Comment. One may visualize the conductivity of NiO occurring by electrons *hopping* from one nickel ion to another. Thus, as an electron hole moves from an Ni^{3+} to an Ni^{2+}, the latter atom becomes Ni^{3+} and the former Ni^{2+}. ◀

• 8–6

SEMICONDUCTING DEVICES

There are many electronic devices that use semiconductors. We shall consider but a few.

Conduction and Resistance Devices

We have already seen that the conductivity of a *photoconductor* will vary directly with the amount of incident light. This capability leads to *light sensing* devices. The radiation does not have to be visible—it may also be ultraviolet or infrared, providing the photons have energy comparable to or greater than the energy gap.

A second device is a *thermistor*. It is simply a semiconductor that has had its resistance calibrated against temperature. If the energy gap is large, so that the ln σ versus $1/T$ curve is steep, it is possible to design a thermistor that will detect temperature changes of $10^{-4}°C$.*

Because many semiconducting materials have low packing factors, they have a high compressibility. Experiments show that as the volume is compressed, the size of the energy gap is measurably reduced; this, of course, increases the number of electrons that can jump the energy gap. Thus pressure can be calibrated against resistance for *pressure gages*.

---

* In technical practice, thermistors are better than other types of thermometers for measuring small temperature *changes*. However, thermocouples, etc., are more convenient for measuring the temperature itself. The measurement of temperature changes is important in microcalorimetric studies involving chemical or biological reactions.

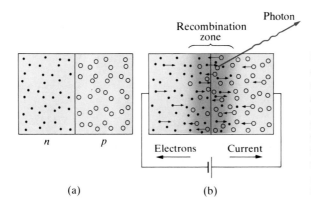

Figure 8–6.1 Light-emitting diode (schematic). (a) An LED is a junction device between *n*-type and *p*-type semiconductors. (b) When a forward bias is placed across the junction, carriers of both types cross the junction where they recombine, emitting a photon (Eq. (8–6.1) and Fig. 8–4.7).

A *photomultiplier* device makes use of electron activation, first by photons *and* then by the electrons themselves. Assume, for example, that a very weak light source, even just one photon, were to hit a valence electron. Our eye would not be able to detect it. However, if that electron is raised to the conduction band, and simultaneously the semiconductor is within a very strong electric field, that electron is accelerated to high velocities and high energies. In turn, it can activate one or more additional electrons that would also respond to the very strong field. The multiplying effect may be used to advantage. A very weak light signal may be amplified. With appropriate focusing, the image in nearly complete darkness may be brought into visible display.

Junction Devices (diodes)

A number of devices utilize junctions between *n*-type and *p*-type semiconductors. The most familiar of these is the *light-emitting diode* (LED). We see it used in the digital displays (red) that are placed in many hand calculators. An LED operates on the principle shown schematically in Fig. 8–6.1. The charge carriers on the *n*-side and *p*-side of the junction are electrons and holes, respectively. If a current is passed through the device in the direction shown, the holes of the valence band move through the junction, into the *n*-type material; conversely, the electrons of the conduction band cross into the *p*-type material. Adjacent to the junction, there are excess carriers that recombine and produce luminescence:

$$n + p \rightarrow \text{photon.} \tag{8–6.1}$$

When GaAs is used, the photons emitted in the recombination zone are red; a phosphorus substitution to give Ga(As,P) produces green photons.

The junction of Fig. 8–6.1 can also serve as a *rectifier*, that is, it is an electrical "check valve" that lets current pass one way and not the other. With the *forward bias* of Fig. 8–6.2(a), current can pass because carriers—both electrons and holes—move through the junction. With a *reverse bias* of Fig. 8–6.2(b), the carriers are pulled away from each side of the junction to leave a

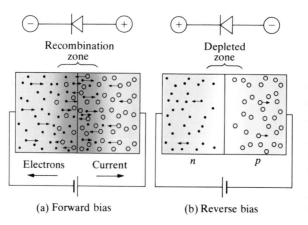

Recombination zone

Depleted zone

Electrons   Current

(a) Forward bias

n   p

(b) Reverse bias

Figure 8–6.2 Rectifier (schematic). (a) Current flows with a forward bias because charge carriers pass the junction. (b) With a reverse bias, charge carriers are depleted from the junction region. Extrinsic conductivity disappears from the junction region, and only a small amount of intrinsic conductivity remains.

carrier-depleted "insulating zone" at the junction. If a greater voltage is applied, this *depletion zone* is simply widened. Usable current passes only with the forward bias.

The last statement holds over a wide range of reverse voltages. There is a point, however, where a surge of current is allowed to pass because there is an electrical breakdown in the depleted zone. Specifically, those very few carriers that are in the depleted zone are accelerated to high velocities by the steep potential drop. As with the photomultiplier device that was previously described, these energetic electrons can knock other electrons loose. An *avalanche* develops that leads to a high current. In effect, we have a "safety valve" that opens at a definite voltage.

Diodes, based on the above principle, may be designed for breakdown voltages ranging from 1 or 2 volts to several hundred volts and with current ratings from milliamperes to a large number of amperes. Called *Zener diodes*, these devices can serve advantageously as filters, gates, and constant voltage controls.

Transistors

Transistors are junction devices that amplify weak signals into stronger usable outputs.

The simplest type of transistor* makes use of a carrier-depleted zone to amplify the output of a circuit. This is called the *field-effect transistor* (FET). In Fig. 8–6.3, the *p–n* junction is reverse-biased by the input signal. As the signal varies, the carrier-depleted zone varies in size and thus alters the resistivity between the *source* and the *drain*. In turn, the current going to the output varies in a controlled manner. A small signal produces a significant current fluctuation.

* The term *transistor* is a contraction of "transfer resistor," since early rationale for its performance utilized such a concept of operation.

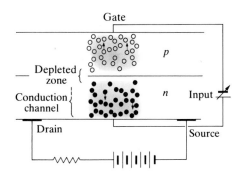

Figure 8–6.3 Transistor (field-effect). A single n–p junction is used. A voltage change across the gate, which is reverse-biased, changes the width of the depleted zone and, hence, the cross section of the conduction channel between the source and the drain. A small variation in the input voltage produces a major change in the resulting current through the conduction channel.

A more common transistor has two junctions in series. They may be either p–n–p or n–p–n and are called *junction transistors*. The former has been used somewhat more in the past; however, we shall consider the n–p–n transistor, since it is easier to visualize the movements of electrons, rather than the movements of holes. However, the principles behind each type are the same.

Before describing the makeup of the transistor, let us recall that as holes move across the junction with a forward bias (Fig. 8–6.2a), they recombine with the electrons in the n-type material according to Eq. (8–4.4b). Likewise, the electrons combine with the holes as the electrons move beyond the junction and into the p-type material. However, the reaction of Eq. (8–4.4b) does not occur immediately. In fact, an excess number of positive and negative carriers may move considerable distances beyond the junction. The number of excess, unrecombined carriers are an exponential function of the applied voltage and become important in transistor operation.

A transistor consists of an *emitter*, a *base*, and a *collector* (Fig. 8–6.4). For the moment, consider only the *emitter junction*, which is biased so that electrons move into the base (and toward the collector). As discussed a moment ago, the number of electrons that cross this junction and move into the p-type

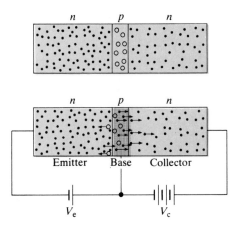

Figure 8–6.4 Transistor (n–p–n). The number of electrons crossing from the emitter-base junction is highly sensitive to the emitter voltage. If the base is narrow, these carriers move to the base–collector junction, and beyond, before recombination. The total current flux, emitter to base, is highly magnified, or amplified, by fluctuations in the voltage of the emitter.

material is an exponential function of the emitter voltage, V_e. Of course, these electrons at once start to combine with the holes in the base; however, if the base is narrow, or if the recombination time is long (τ of Eq. 8–4.5), the electrons keep on moving through the thickness of the base. Once they are at the second junction, or *collector junction*, the electrons have free sailing, because the collector is an *n*-type semiconductor. The total current that moves through the collector is controlled by the emitter voltage, V_e. As the emitter voltage fluctuates, the collector current, I_c, changes exponentially. Written logarithmically,

$$\ln I_c \approx \ln I_0 + V_e/B, \tag{8–6.2a}$$

or

$$I_c = I_0 e^{V_e/B}, \tag{8–6.2b}$$

where I_0 and B are constants for any given temperature. Thus, if the voltage in the emitter is increased even slightly, the amount of current is increased markedly. It is because of these relationships that a transistor serves as an amplifier.

Example 8–6.1
Zinc sulfide is used as a thermistor. To what fraction sensitivity, δ, must the resistance be measured to detect 0.001°C change at 20°C?

Procedure. We want $\delta(=\Delta R/R)$ for ±0.001°C. The shape of the thermistor is fixed; therefore, $R_2/R_1 = \rho_2/\rho_1 = \sigma_1/\sigma_2$.

Calculation

$$\delta = \frac{R_2 - R_1}{R_1} = \frac{\rho_2 - \rho_1}{\rho_1} = \frac{\rho_2}{\rho_1} - 1 = \frac{\sigma_1}{\sigma_2} - 1;$$

$$1 + \delta = \frac{\sigma_1}{\sigma_2} = \left[\frac{\sigma_0 e^{-3.7eV/2kT}}{\sigma_0 e^{-3.7eV/2k(T\pm0.001)}}\right];$$

$$\ln(1 + \delta) = \frac{-3.7 \text{ eV}}{2(86.1 \times 10^{-6} \text{ eV/K})}\left[\frac{1}{T} - \frac{1}{T \pm 10^{-3}}\right].$$

Since

$$\frac{1}{T} - \frac{1}{T \pm 10^{-3}} = \frac{(T \pm 10^{-3}) - T}{T(T \pm 10^{-3})} \approx \pm\frac{10^{-3}}{T^2},$$

$$\ln(1 + \delta) = (-21{,}500 \text{ K})(\pm10^{-3} \text{ K}/[293 \text{ K}]^2)$$
$$= \pm2.5 \times 10^{-4}.$$

$$\delta = \pm0.00025 \qquad \text{(or } \pm0.025\%).$$

Comment. A bridge-type instrument would be required. ◀

Example 8–6.2
A transistor has a collector current of 4.7 milliamperes when the emitter voltage is 17 millivolts. At 28 millivolts, the current is 27.5 milliamperes. Given that the emitter voltage is 39 millivolts, estimate the current.

Solution. Based on Eq. (8–6.2).

$$\ln 4.7 \approx \ln I_0 + 17/B = 1.55,$$

$$\ln 27.5 \approx \ln I_0 + 28/B = 3.31.$$

Solving simultaneously, using *milli*units, we have

$$\ln I_0 \approx -1.17, \quad \text{and} \quad B \approx 6.25.$$

At 39 millivolts,

$$\ln I_c \approx -1.17 + 39/6.25 \approx 5.07,$$

$$I_c \approx 160 \text{ milliamp.}$$

Comments. The electrical engineer modifies Eq. (8–6.2) to take care of added current effects. These, however, do not change the basic relationship: The variation of the collector current is much greater than the variation of the signal voltage. ◀

• 8–7

SEMICONDUCTOR PROCESSING

The composition of a semiconductor is very critical. Some impurities introduce donors and negative (*n*-type) carriers; others introduce acceptors and positive (*p*-type) carriers. Even when these *dopants* are desired, their amounts must be closely controlled to parts-per-million (ppm) levels, or less. Therefore, it is common practice to purify the silicon (or other semiconductors) to the highest level possible, and then make precise additions of the required dopant.

Crystal Growing

Single crystals are required for the large majority of semiconductor applications, since grain boundaries reduce carrier mobility and decrease the recombination time of excess carriers. The reduction of recombination time affects the performance of many junction devices. Single-crystal growth generally utilizes one of two techniques in semiconductor technology—crystal-pulling and floating-zone methods (Fig. 8–7.1).

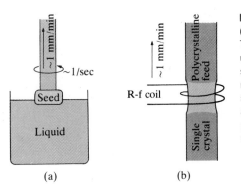

(a) (b)

Figure 8–7.1 Single-crystal growth (semiconductors). (a) Crystal-pulling method. The seed—a single crystal—is slowly pulled upward. The liquid crystallizes on its lower surface. (b) Floating-zone procedure. The molten zone is raised along the semiconductor bar solidifying the lower edge as a single crystal. The liquid is held in position by surface tension and does not come into contact with a container.

The *crystal-pulling* method previously received brief attention in Fig. 7–7.8. The semiconducting material is first melted; then a single-crystal seed crystal is touched to the surface and slowly pulled away (~1 mm/min) as it is rotated (~1/s). If the liquid is only slightly above its melting temperature, it will solidify on the seed crystal as the seed is pulled upward. The solidifying atoms continue the crystal structure of the seed. Dopants of Group III or Group V compounds may be added to the liquids in the amounts (~10^{-6} a/o) required to make *p*- and *n*-type products.

The above technique (Fig. 8–7.1a) is satisfactory for germanium and other materials that melt below 1000°C. However, it is not satisfactory for silicon for a couple of reasons. Silicon melts above 1400°C and, therefore, more readily picks up contaminants from the container and furnace walls. Also, the dopants more readily vaporize so that compositional control becomes more difficult. Therefore, a floating-zone process is used.

The *floating-zone* method starts with a rod (>5 cm dia.) of purified polycrystalline silicon sitting on a disc of a previously prepared single crystal. The two are melted where they are in contact by r-f heating. The r-f coil is then slowly raised (Fig. 8–7.1b) to move the molten zone upward. The polycrystalline solid melts with the upper movement and feeds the molten zone. The initial single crystal grows upward with the movement by crystallization at the lower side of the molten zone. As with the crystal-pulling process, the upward movement of the molten zone advances ~1 mm/min.

Thin (~0.25 mm) wafers are cut across the bar. These are polished and chemically cleaned; then dopants are added in an *epitaxial* layer. This is a layer that grows onto the surface as a continuation of the underlying crystal. The growth comes from a gaseous mixture whose composition is adjusted to give either an *n*-type or a *p*-type layer.* The diameters of commercially prepared single crystals have been increased almost without limits, since a larger wafer leads to more efficient device preparation (Fig. 8–7.2).

Device Preparation

Figure 8–7.3 shows a processed wafer that holds dozens of chips, each with literally thousands of circuit elements. These circuits have a *planar* configuration, that is, the preparation and contacts are on one flat surface (Fig. 8–7.4) rather than at the ends and sides as in Fig. 8–6.4. The thicker substrate, an *n*-type silicon base that was crystallized in the floating-zone process, is doped to have low resistivity for easy conduction. The overlying (epitaxial) layer was described in the previous paragraph and has higher, but controlled, resistivity. Following a masking procedure, boron is *diffused* from the surrounding gas into the epitaxial silicon, changing the selected regions from *n*-type to *p*-type. The latter becomes the base of the transistor. (Cf. Fig. 8–6.4.) A second masking and diffusion step introduces a Group V element to produce the *n*-type emitter. Final maskings, etching, and coating establish a passive silica (SiO_2) surface and contacts for circuit leads.

---

* $SiCl_4$, H_2, and PH_3 for *n*-type silicon, and $SiCl_4$, H_2, and B_2H_6 for *p*-type silicon.

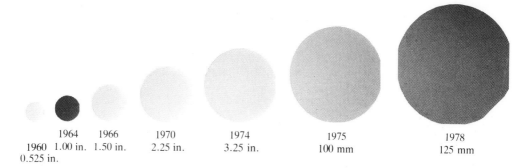

1960	1964	1966	1970	1974	1975	1978
0.525 in.	1.00 in.	1.50 in.	2.25 in.	3.25 in.	100 mm	125 mm

Figure 8–7.2 Polished silicon slices (Monsanto). The wafers are cross cuts from single-crystal silicon boules. Hundreds of duplicated semiconductor devices are vapor-deposited simultaneously onto these wafers before dicing them into chips. The commercial 1982 chip is 150 mm (6 in.) in diameter. The diameters of developmental 1984 wafers are up to 200 mm (8 in.), more than doubling the 1978 number of chips that can be processed simultaneously.

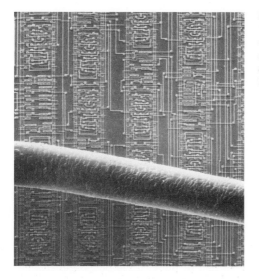

Figure 8–7.3 Processed wafer (silicon "chip" and human hair). The circuits of the chip are in planar configuration, made with sequential maskings and overlays. See Fig. 8–7.4 for cross sections. (Courtesy General Electric Co.)

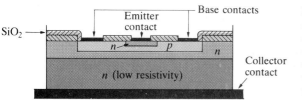

Figure 8–7.4 Planar transistor (junction-type). The substrate is a low-resistivity, n-type, single-crystal wafer of silicon. It is covered with an epitaxial layer of higher resistivity silicon that serves as the collector. Boron is diffused through the openings of an SiO_2 mask to produce the p-type base. A second masking admits phosphorus for the n-type emitter. Aluminum contacts are added.

The thickness of the base in Fig. 8–7.4 may be only 0.5 μm. With this short distance between the two junctions, any recombination can be minimized as described in Section 8–6. The dimensions and dopant concentrations are controlled by diffusion time and temperature, and by gas composition. This has been developed into a highly reproducible production process.

Review and Study

SUMMARY

This chapter examines the development of energy bands and energy gaps for valence electrons in solids. These concepts are used to understand charge transport and simple electron devices.

1. The electron (0.16×10^{-18} A·s) is the basis of charge transport within solid materials. *Electrons* are negative carriers, as are *anions,* because of their excess electron(s). Positive ions (*cations*) carry a positive charge by virtue of their missing electron(s). Likewise, a missing electron in the valence band, called an *electron hole,* can transfer a positive charge.

2. *Metals* have partially filled valence bands. These provide a large number of delocalized electrons for conductivity. Higher temperatures, impurities, and other disorder within a metal reduce the mean free path of the electrons and increase the resistivity. In the ambient range, resistivity is linear with temperature.

3. An *insulator* has a sufficiently large energy gap between a filled valence band and the conduction band to preclude significant conductivity.

4. A *semiconductor* has a sufficiently small energy gap between the valence band and the conduction band for a usable number of electrons to be activated by heat, light, or other electrons. *Intrinsic semiconductors* have equal numbers of negative carriers (electrons) and positive carriers (electron holes). The III-V compounds have the ZnS structure, which is similar to the structure of diamond, silicon, germanium, and gray tin. They all have filled valence bands as a result of their covalent bands. The numbers of electrons to "jump the gap," and therefore introduce conductivity into semiconductors, follows the Arrhenius relationship with temperature, that is, *ln n and ln σ are linear with 1/T.* In general, the *mobility* of the conduction electrons is greater than that of the electron holes. *Recombination* of electrons and holes obeys an exponential decay pattern.

5. *Extrinsic semiconduction* originates from impurities. *Donors* provide electrons to the conduction band. *Acceptors* receive electrons from the valence band. Thus, they produce negative and positive carriers, respectively. *Defect semiconduction* exists in nonstoichiometric compounds of those elements that have more than one valence, e.g., Fe^{2+} and Fe^{3+}.

• 6. There are two main categories of semiconducting devices. The first includes thermistors, photoconductors, pressure gages, and photomultipliers, all of which depend on the *conduction* (or resistance) variations of the semiconducting material. The second category involves *junctions* between *n*-type and *p*-type semiconductors. This category includes light-emitting diodes (LED's), rectifiers, Zener diodes, and transistors.

• **7.** The major steps of semiconduction processing includes materials purification, single-crystal growth, and device preparation. *Crystal pulling* (Czochralski) and *floating-zone* procedures are common methods of crystal growth. The latter has the advantage of avoiding contact with containers, a source of impurities. Device preparation involves *masking, etching,* and controlled *coatings* of planar circuit configurations.

TERMS AND CONCEPTS

Acceptor   Impurities that accept electrons from the valence band and, therefore, produce an electron hole in the valence band.

Acceptor saturation   Filling of acceptor sites. As a consequence, additional thermal activation does not increase the number of extrinsic carriers.

Band, conduction (CB)   Band above the energy gap. Electrons become carriers when they are activated to this band.

Band, valence (VB)   Filled energy band below the energy gap. Conduction in this band requires electron holes.

Charge carriers   Electrons in the conduction band provide *n*-type (negative) carriers. Electron holes in the valence band provide *p*-type (positive) carriers.

Compounds, III-V   Compounds of Group III and Group V elements with semiconducting properties.

Conductivity (electrical)   Product of (carrier density) (carrier charge) (carrier mobility). Reciprocal of resistivity.

• **Crystal pulling**   Method of growing single crystals by slowly pulling a seed crystal away from a molten pool.

• **Defect semiconductor**   Nonstoichiometric compounds of elements that have more than one valence.

• **Depleted zone**   The region adjacent to an *n–p* junction that lacks charge carriers. (A reverse bias increases the width of the depleted zone.)

Donor   Impurities that donate charge carriers to the conduction band.

Donor exhaustion   Depletion of donor electrons. Because of this, additional thermal activation does not increase the number of extrinsic carriers.

Drift velocity ($\bar{v}$)   Net velocity of electrons in an electric field.

Electric field ($\mathcal{E}$)   Voltage gradient, volts/m.

Electron charge (q)   The charge of 0.16×10^{-18} coul (or 0.16×10^{-18} amp·sec) carried by each electron.

Electron hole (p)   Electron vacancy in the valence band that serves as a positive charge carrier.

Electron-hole pair   A conduction electron in the conduction band and an accompanying electron hold in the valence band, which result when an electron jumps the gap in an intrinsic semiconductor.

Energy band   Permissible energy levels for valence electrons.

Energy gap (E_g)   Forbidden energies between the valence band and the conduction band.

• **Floating zone**   Method of growing single crystals, which melts an isolated zone within a bar. The material solidifies as a single crystal on the bottom side of the rising molten zone.

- **Fluorescence** Luminescence that occurs almost immediately after excitation.

  Forward bias Potential direction that moves the charge carriers across the n–p junction.

  Insulator Material with filled valence bands and a large energy gap.

  Junction (n–p) Interface between n-type and p-type semiconductors.

- **Light-emitting diode (LED)** A p–n junction device designed to produce photons by recombination.

- **Luminescence** Light emitted by the energy released as conduction electrons recombine with electron holes.

  Metal Material with partially filled valence bands.

  Mobility (μ) The drift velocity $\bar{v}$ of an electric charge per unit electric field, $\mathscr{E}$, (m/s)/(volt/m). Alternatively, the diffusion coefficient of a charge per volt, (m²/s)/volt.

- **Phosphorescence** Luminescence that is delayed by extended relaxation times.

  Photoconduction Conduction arising from activation of electrons across the energy gap by means of light.

- **Photomultiplier** Device that uses a photon to trigger an electron avalanche in a semiconductor. From this, weak light signals can be amplified.

  Photon A quantum of light.

- **Photovoltaic** Production of an electrical potential from incident light.

  Recombination Annihilation of electron-hole pairs.

- **Rectifier** Electric "valve" that permits forward current and prevents reverse current.

  Relaxation time (τ) Time required for electron-hole pairs to recombine and decrease the remaining numbers of 1/e of the original levels.

  Resistivity (ρ) Reciprocal of conductivity (usually expressed in ohm·m).

  Resistivity coefficient, thermal (y_T) Coefficient of resistivity versus temperature.

  Semiconductor A material with controllable conductivities, intermediate between insulators and conductors.

  Semiconductor, extrinsic Semiconduction from impurity sources.

  Semiconductor, intrinsic Semiconduction of a pure material. The electrons are excited across the energy gap.

  Semiconductor, n-type Impurities provide donor electrons to the conduction band. These electrons are the majority charge carriers.

  Semiconductor, p-type Impurities provide acceptor sites for electrons from the valence band. Electron holes are the majority charge carriers.

- **Thermistor** Semiconductor device with a high resistance dependence on temperature. It may be calibrated as a thermometer.

- **Transistor** Semiconductor device for the amplification of current. There are two principal types, field-effect and junction.

- **Zener diode** A p–n junction with controlled breakdown voltage under reverse bias.

FOR CLASS DISCUSSION

A_8 Cite and compare the various types of charge carriers.

B_8 Discuss various factors that affect the drift velocity of electrons in solids.

C_8 The conductivity mechanism is sometimes associated with diffusivity because we are concerned with transport processes. Based on their units, cite the similarity and a difference between mobility and diffusivity.

D_8 Why does brass have a higher resistivity than copper?

E_8 Why does the conductivity of metals decrease at elevated temperatures?

F_8 Why does the conductivity of semiconductors increase at elevated temperatures?

G_8 Explain to a classmate why there are energies that are forbidden to electrons that are associated with individual atoms.

H_8 Differentiate between metallic conductors, insulators, and semiconductors.

I_8 Differentiate between intrinsic and extrinsic semiconductors.

J_8 Predict whether boron nitride, with the structure of Fig. 8–4.3(b), will be an insulator or semiconductor.

K_8 What fraction of electrons remain in the conduction band after $t = \tau$? After $t = 2\tau$?

• L_8 Distinguish between fluorescence and phosphorescence.

M_8 ZnS may have the structure shown in Fig. 8–4.3(b). Assume some phosphorous atoms replace sulfur atoms. Will the compound be n-type or p-type?

N_8 Differentiate between acceptor and donor impurities.

O_8 Why is the slope for intrinsic carriers in Fig. 8–5.3 steeper than for impurity carriers?

P_8 Detail the mechanism of acceptor saturation in terms of Fig. 8–5.3.

Q_8 There are 10^{11} magnesium atoms, which replace the same number of silicon atoms, in a mm^3 of silicon. How will this affect the conductivity compared with an impurity of 10^{11} aluminum atoms?

• R_8 Cu_2O contains predominantly Cu^+ ions, but some Cu^{2+} ions. Will it be n-type or p-type?

• S_8 Describe the principle of the light-emitting diode.

• T_8 Explain how a silicon rectifier operates.

• U_8 Describe the principle of the photomultiplier tube.

QUIZ SAMPLES

8A A copper wire is to be 1 meter long and have a resistance of 2 milliohms. Its conductivity is 60×10^6 ohm$^{-1} \cdot$m^{-1}. Its cross-sectional area should be __+−__ .

(a) $\pi(3.25 \text{ mm})^2/4$ (b) $(0.002)(1)/(60 \times 10^6)$ (c) None of the above, rather _____   (d) Insufficient data to calculate.

8B Refer to Study Problem 1–5.6. Before loading, the wire was 3.07 m. Its resistance is __**__ . With 1.5 volts across the wire length, there will be __**__ electrons entering the wire per millisecond.

8C A copper wire (1.1-mm dia) was chosen to meet resistance specifications. How much heavier (+%) or lighter (−%) will a pure aluminum wire be that has the same resistance? __+−__

(a) −70%   (b) +41%   (c) −48%   (d) −52%   (e) None of the above, rather _____%   (f) Insufficient data to calculate.

8D The conductivity of copper at 0°C is 63×10^6 ohm$^{-1} \cdot$m^{-1}; at 100°C is 43×10^6 ohm$^{-1} \cdot$m^{-1}; and at 200°C (473K) is __+−__ ohm$^{-1} \cdot$m^{-1}. The resistivity of copper at 0°C is 16×10^{-9} ohm·m; at 100°C is 23×10^{-9} ohm·m; and at 200°C is __+−__ ohm·m.

(a) 30×10^{-9} (b) $>30 \times 10^{-9}$ (c) 23×10^6 (d) $<23 \times 10^6$
(e) None of the above, rather _____   (f) Insufficient data to calculate.

8E   The mean free path for electron drift increases with __+−__ .

(a) purity   (b) strain hardening   (c) recrystallization
(d) elastic modulus   (e) None of the above, rather _____

8F   The energy band of a __**__ is only partially filled, while for both __**__ and __**__ it is filled.

8G   In a semiconductor, a useful number of electrons can be raised across the energy __**__ ; very few possess that much energy in a(n) __**__ .

8H   Two Group IV elements have the same diamond cubic structure. The one with the larger __+−__ is expected to have the smaller energy gap.

(a) packing factor   (b) coordination number   (c) number of valence electrons   (d) None of the above, rather _____

8I   The conductivity of a semiconductor is the sum of the conductivity in the __**__ band by negative carriers and the conductivity in the __**__ band by positive carriers.

8J   The simplest compound semiconductors contain elements of Groups __**__ and __**__ with an average of __**__ valence electrons per atom. Those compounds with __**__ elements have narrower energy gaps.

8K   The numbers of electrons jumping the energy gap depend upon the reciprocal of the __**__ __**__ . In order to make a linear plot, the numbers must be expressed in terms of __**__ .

8L   Electrons can be raised across the energy gap, E_g, by __**__ , by __**__ , or by energetic __**__ . In the first case, the numbers are proportional to $e^{-E_g/2kT}$; in the second, the __**__-conduction depends upon the number of __**__ hitting the semiconductor.

8M   When an electron jumps the gap, an electron-__**__ pair is formed. This is eliminated by the __**__ of the electron and hole.

8N   The term __**__ applies to those semiconductors that possess semiconductivities by virtue of impurities; a semiconductor without impurities is called __**__ .

• **8O**   A semiconductor device for a __**__ works on the principle that the energy gap is __**__ in size with an increase in density, so that more electrons can enter the conduction band.

• **8P**   A __**__ device depends simply on the fact that more electrons enter the conduction band at higher __**__ .

• **8Q**   Current passes through the junction of a __**__ when a __**__ bias is applied.

STUDY PROBLEMS   **8–1.1**   Seventy millivolts are placed across the 0.5-mm dimension of a semiconductor with a carrier mobility of 0.23 m^2/V·s. What drift velocity develops?

Answer: 32 m/s

8–1.2   A semiconductor, which has a resistivity of 0.0313 ohm·m, develops a drift velocity of 6.7 m/s when 1.2 volts are applied across a 9-mm piece. How many carriers are there per m^3?

8–1.3 Laboratory measurements indicate that the drift velocity of electrons in a semiconductor is 149 m/s when the voltage gradient is 15 V/mm. The resistivity is 0.07 ohm·m. What is the carrier concentration?

Answer: $9 \times 10^{21}/m^3$

8–1.4 A flashlight bulb has a resistance of 5 ohms when it is used in a 2-cell flashlight battery. Assume 3 volts. How many electrons move through the filament per minute?

8–2.1 Determine the temperature at which the resistivity of silver is 10 ohm·nm.

Answer: $-88°C$

8–2.2 At what temperature does silver have the same resistivity as gold does at 50°C?

8–2.3 The resistivity of copper doubles between 20°C and 300°C. At what temperature does the resistivity of aluminum equal the higher value for copper?

Answer: 70°C

8–2.4 Based on the data of Table 8–2.1 and Appendix C, estimate the effect of copper on the resistivity of silver (ohm·nm per a/o).

8–2.5 Based on the data of Example 8–2.2, estimate the resistivity of a 95 Cu–5 Sn (weight basis) bronze at 0°C.

Answer: 97×10^{-9} ohm·m (=97 ohm·nm)

8–2.6 A 6% variation (maximum) is permitted in resistance between 0° and 25°C. Which metals of Table 8–2.1 meet the specification?

8–2.7 An iron wire (1.07 m long and 0.53 mm in diameter) is connected to 25 volts at 20°C. (a) What is the initial amperage? The amperage at 910°C, just below the bcc $\Leftrightarrow$ fcc transition temperature? (b) What is the approximate change in length between 20° and 910°C? Between 20° and 915°C? (Review Example 3–4.1. The thermal expansion for bcc iron will be directly proportional to the change in atomic radius.)

Answer: (a) 53A, 11A (b) 16 mm, 11 mm (Use this Study Problem for Discussion Topic T_{10}.)

8–4.1 A silicon chip is 1 mm × 1 mm × 0.1 mm. How fast do electrons drift through its short dimension if 37 millivolts are applied?

Answer: 70 m/s

8–4.2 Pure germanium has a conductivity of 2 ohm^{-1}·m^{-1} with equal numbers of negative carriers, n_n, and positive carriers, n_p. What fraction of the conductivity is due to electrons and what fraction is due to electron holes?

8–4.3 What fraction of the charge is carried by electron holes in intrinsic gallium arsenide (GaAs)?

Answer: 0.043 (or 4.3%)

8–4.4 How many electron carriers (and electron holes) does intrinsic silicon require to provide a conductivity of 1.1 ohm^{-1}·m^{-1}?

8–4.5 The resistivity of a semiconductor that possesses 10^{21} negative carriers/m^3 (and few positive carriers) is 0.016 ohm·m. (a) What is the conductivity? (b) What is the electron mobility? (c) What is the drift velocity when the potential gradient is 5 mV/mm? 0.5 V/m?

Answer: (a) 62.5 ohm^{-1}·m^{-1} (b) 0.39 m^2/V·s (c) 1.95 m/s, 0.195 m/s

8–4.6 Based on the data in Table 8–4.2, which is larger, (1) the conductivity from electrons in intrinsic InP, or (2) the conductivity from holes in intrinsic InAs?

8–4.7 Pure silicon has 32 valence electrons per unit cell (8 atoms with 4 electrons each). Its resistivity is 2×10^3 ohm·m. What fraction of the valence electrons are conductors?

Answer: 1 out of 1.5×10^{13}

8–4.8 At an elevated temperature, 1 of every 10^9 valence electrons in germanium is in the conduction band. What is the conductivity? (Germanium has the same structure as silicon, but with $a = 0.566$ nm.)

8–4.9 The mobility of electrons in silicon is 0.19 m^2/V·s. (a) What voltage is required across a 2-mm chip of Si to produce a drift velocity of the electrons of 0.7 m/s? (b) What electron concentration must be in the conduction band to produce a conductivity from negative carriers of 20 ohm^{-1}·m^{-1}? (c) What would the total conductivity for this silicon be if no impurities are present?

Answer: (a) 7.4 mV (b) 6.6×10^{20}/m^3 (c) 24.5 ohm^{-1}·m^{-1}

8–4.10 The conductivity of silicon is 5×10^{-4} ohm^{-1}·m^{-1} at 20°C (68°F). Estimate the conductivity at 30°C.

8–4.11 To what temperature must germanium be cooled in order for its conductivity to be reduced by a factor of two below its 20°C (68°F) value?

Answer: 6°C

8–4.12 A semiconductor has a conductivity of 7 ohm^{-1}·m^{-1} at 20°C and 20 ohm^{-1}·m^{-1} at 60°C. What is the size of the energy gap?

8–4.13 An intrinsic semiconductor has a conductivity of 111 ohm^{-1}·m^{-1} at 10°C and 172 ohm^{-1}·m^{-1} at 17°C. (a) What is the energy gap? (b) What is the conductivity at 13.5°C?

Answer: (a) 0.88 eV (b) 138 ohm^{-1}·m^{-1}

8–4.14 An intrinsic semiconductor has a conductivity of 390 ohm^{-1}·m^{-1} at 5°C and 1010 ohm^{-1}·m^{-1} at 25°C. (a) What is the size of the energy gap? (b) What is the conductivity at 15°C?

• **8–4.15** Refer to Example 8–4.5. Assume a phosphor is used with a recombination time of 0.04 s. What fraction of the light intensity will remain when the next scan is made?

Answer: 0.44

• **8–4.16** A phosphorescent material is exposed to ultraviolet light. The intensity of emitted light decreased 20% in the first 37 minutes after the ultraviolet light was removed. (a) How long will it be after the uv light has been removed before the emitted light has only 20% of the original intensity (a decrease of 80%)? (b) Only 1%?

8–4.17 A phosphorescent material must have an intensity of 50 (arbitrary units) after 24 hours and 20 after 48 hours. Based on these figures, what initial intensity is required? (Solve *without* a calculator.)

Answer: 125

8–4.18 With a calculator, determine the recombination time for the material in the previous problem.

8–5.1 Silicon has a density of 2.33 Mg/m^3 (=2.33 g/cm^3). (a) What is the concentration of silicon atoms per m^3? (b) Phosphorus is added to silicon to make it an *n*-type semiconductor with a conductivity of 100 mho/m and an electron mobility of 0.19 $m^2/V \cdot s$. What is the concentration of donor electrons per m^3?

Answer: (a) 5×10^{28} Si/m^3 (b) $3.3 \times 10^{21}/m^3$

8–5.2 How many silicon atoms are there for each aluminum atom in Example 8–5.1?

8–5.3 Extrinsic germanium is formed by melting 3.22×10^{-6} g of antimony (Sb) with 100 g of germanium. (a) Will the semiconductor be *n*-type or *p*-type? (b) Calculate the concentration of antimony (in atoms/cm^3) in germanium. (The density of germanium is 5.35 Mg/m^3, or 5.35 g/cm^3.)

Answer: (a) *n*-type   (b) $8.5 \times 10^{20}/m^3$

8–5.4 Gallium arsenide is made extrinsic by adding 0.000001 a/o phosphorus (and keeping a stoichiometric Ga/As ratio). Calculate the extrinsic conductivity at exhaustion.

8–5.5 Aluminum is a critical impurity in making silicon for semiconductors. Assume only 10 ppb (0.000001 a/o) remain. Will the resulting extrinsic conductivity be greater or less than the intrinsic conductivity of silicon at 20°C?

Answer: Greater (3.4 versus 5×10^{-4} $ohm^{-1} \cdot m^{-1}$)

8–5.6 Refer to Study Problem 8–5.5. At what temperature will 1% of the conductivity be intrinsic?

8–5.7 Three grams of *n*-type silicon that had been doped with phosphorus to produce a conductivity of 600 $ohm^{-1} \cdot m^{-1}$ are melted with three grams of *p*-type silicon that had been doped with aluminum to produce a conductivity of 600 $ohm^{-1} \cdot m^{-1}$. (a) What is the resulting conductivity? (b) Will it be *p*-type or *n*-type?

Answer: (a) 230 $ohm^{-1} \cdot m^{-1}$ (b) *p*-type

8–5.8 Silicon ($10^{21}/m^3$) is in solid solution in GaAs. Assume it replaces equal numbers of Ga and As atoms. (a) What is the anticipated conductivity? (b) Assume the 10^{21} Si/m^3 replaced only arsenic atoms. What is the anticipated conductivity?

8–5.9 Refer to Example 4–3.1. (a) Is the oxide *n*-type or *p*-type? (b) How many charge carriers are there per mm^3?

Answer: (a) *p*-type   (b) $7.8 \times 10^{18}/mm^3$

- **8–5.10**   Experiments indicate that 99% of the charge is carried by electrons in $Fe_{<1}O$. (The rest is carried by ions.) If the oxide in Study Problem 8–5.9 has a conductivity of 93 $ohm^{-1} \cdot m^{-1}$, what is the mobility of the electron holes?

- **8–6.1**   The resistance of a certain silicon wafer is 1031 ohms at 25.1°C. With no change in measurement procedure, the resistance decreases to 1029 ohms. What is the temperature change? The energy gap of silicon is 1.1 eV.

  Answer: +0.03°C.

- **8–6.2**   To what temperature must you raise InP in order to make its resistivity half what it is at 0°C? Its energy gap is 1.3 eV.

- **8–6.3**   A transistor operates between 10 millivolts and 100 millivolts across the emitter. At the lower voltage, the collector current is 6 milliamps; at the higher voltage, 600 milliamps. Estimate the current when the emitter voltage is 50 mV.

  Answer: 46 mA

- **8–6.4**   Refer to Example 8–6.2. If the emitter voltage is doubled from 17 to 34 millivolts, by what factor is the collector current increased?

QUIZ CHECKS		
8A	(a)	
8B	0.16 ohm, 6×10^{16} el/ms	
8C	(c)	
8D	(e): 33×10^6 $ohm^{-1} \cdot m^{-1}$	
	(a)	
8E	(a), (c)	
8F	metal	
	semiconductors and insulators	
8G	gap, insulator	
8H	(d): atomic weight, or atomic number	
8I	conduction	
	valence	
8J	III and V, 4	
	heavier	

8K	absolute temperature
	log n
8L	heat, light (radiation), electrons
	photo-, photons
8M	-hole
	recombination
8N	extrinsic
	intrinsic
8O	pressure gage
	reduced
8P	thermistor
	temperature
8Q	rectifier (LED), forward

Chapter Nine

Magnetic, Dielectric, and Optical Materials

PREVIEW

Electronic circuits require materials for purposes other than conductivity. These include *magnets, insulators, dielectrics, piezoelectrics,* and *optical* components.

Magnetic properties originate with the spin of each and every individual electron (magnetic moment = 9.27×10^{-24} A·m^2). Only in exceptional cases, however, are the effects not canceled by opposing spins. It will be our purpose to analyze those useful situations as they are related to structure.

Although nonconducting insulators, dielectrics are more than isolators. Dielectrics become functional in an electrical circuit through their *polarization*. This is associated with atomic and molecular structures and their displacements.

Optical materials are now being used for signal transmission as well as in visual paths. This is one of the existing frontiers of present-day technology.

STUDY OBJECTIVES

1. To distinguish between soft and hard magnets in terms of both the hysteresis loop and the controlling internal structures. This means identifying the factors that interfere with domain wall movements.

2. To be able to calculate the magnetization of the principal magnetic spinels.

3. To associate polarization with atomic, molecular, and crystal structures, so that the dielectric constant may be related to temperature and frequency.

4. To understand the principle of piezoelectric transducers and their relationship to structural changes in crystals.

5. To relate the light transmission characteristics of materials with their structure and composition.

6. To know the technical terms and concepts that are encountered in technical discussions.

MAGNETIC MATERIALS

The best known magnets contain metallic iron. However, several other elements can also show magnetism; furthermore, not all magnets are metallic. Modern technology uses both metallic and ceramic magnets. It also makes use of other elements to enhance the magnetic capabilities to meet requirements.

Magnetic Hysteresis

Magnets are commonly categorized as "soft" or "hard." A *hard magnet* attracts other magnetized materials to it. It retains its magnetism more or less permanently. A *soft magnet* can become magnetized and be drawn to another magnet; however, it has obvious magnetism only when it is in magnetic field. It is not permanently magnetized.

The differences between permanent, or hard, magnets and soft magnets, are best shown with the *hysteresis loop* that most readers have encountered in physics courses (Fig. 9–1.1). We will need to explore the reasons for these differences in more detail than commonly accompanies introductory physics.

When a magnetic material is inserted into a *magnetic field, H,* the adjacent "lines of force" are collected into the material to increase the *flux density.* In more technical terms, increased *magnetic induction, B,* results. Of course, the amount of induction depends on the magnetic field and on the nature of the material. In the example of Fig. 9–1.1, the B/H ratio is not linear, rather the induction jumps to a high level, then remains nearly constant in stronger fields.

In a *soft magnet,* there is a near perfect backtrack as the magnetic field is removed. A reversed magnetic field provides a symmetrical curve in the third quadrant.

The hysteresis curve of a *permanent magnet* differs significantly. When the magnetic field is removed, most of the induction is retained to give a *remanent induction, B_r.* A reversed field, called the *coercive field, $-H_c$,* is necessary before the induction drops to zero. In common with a soft magnet, the completed loop of a permanent magnet possesses 180° symmetry.

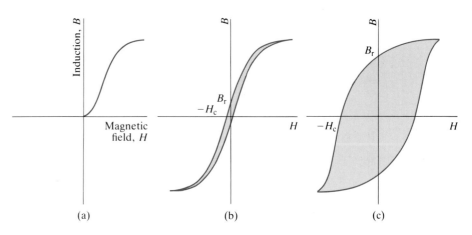

Figure 9–1.1 Magnetization curves. (a) Initial induction B versus magnetic field H. (b) Hysteresis loop (soft magnet). (c) Hysteresis loop (hard magnet). Both the remanent induction (flux density) and coercive field, B_r and $-H_c$, respectively, are large for hard magnets. The BH product is a measure of energy for demagnetization.

Since the product of the magnetic field (A/m) and the induction ($V \cdot s/m^2$) is energy per unit volume, the integrated area within the hysteresis loop is the energy required to complete one magnetization cycle from 0 to $+H$ to $-H$ to 0. Soft magnets require negligible energy; hard magnets require sufficient energy so that under ambient conditions demagnetization may be precluded. The magnetization is permanent.

The magnetic permanence can be indexed by the coercive field that is necessary to remove the induction (Table 9–1.1). A better measure is the BH product. The maximum instantaneous BH product is more commonly used because it represents the critical energy barrier that must be exceeded. The advantage of the BH product is apparent in Fig. 9–1.2, where the second-quadrant profiles for several permanent magnets are shown. $BaFe_{12}O_{19}$ has an extremely large $-H_c$ value, but only a moderately high BH_{max}, because its flux

TABLE 9–1.1			
Properties of Selected Hard Magnets (various sources)			
Magnetic material	**Remanence, B_r $V \cdot s/m^2$**	**Coercive field, $-H_c$ A/m**	**Maximum demagnetizing product, BH_{max}, J/m^3**
Carbon steel	1.0	0.4×10^4	0.1×10^4
Alnico V	1.2	5.5×10^4	3.4×10^4
Ferroxdur ($BaFe_{12}O_{19}$)	0.4	15.0×10^4	2.0×10^4

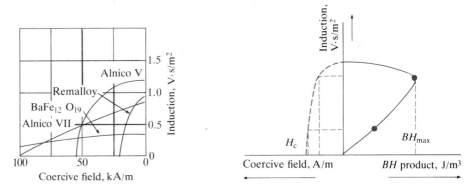

Figure 9–1.2 Demagnetization curves (second quadrant). (a) Hard magnets. (b) BH products. The maximum value is commonly used as an index of magnetic permanency.

density is lower than other permanent magnetic materials. Some readers may have experienced a sense of the magnetic "strength" of a steel magnet. Table 9–1.1 will let them anticipate the improvements that have been made possible by the scientists and engineers who have developed the Alnico (metallic) and $BaFe_{12}O_{19}$ (ceramic) magnets.

Soft magnets are the obvious choice for ac or high-frequency applications, since they must be magnetized and demagnetized many times per second. Among the more critical specifications for soft magnets are saturation induction (high), coercive field (low), and maximum permeability (high). Selected data are shown in Table 9–1.2 and may be contrasted with the data of Table 9–1.1 for permanent magnets. The *permeability* is the B/H ratio. However, since the BH behavior is nonlinear, the engineer commonly specifies the maximum B/H ratio. It is the secant of the curve as shown in Fig. 9–1.3. A high value of the

TABLE 9–1.2			
Properties of Selected Soft Magnets (various sources)			
Magnetic material	**Saturation induction, B_s V·s/m²**	**Coercive field, $-H_c$ A/m**	**Maximum relative permeability, μ_r(max)**
Pure iron (bcc)	2.2	80	5,000
Silicon ferrite transformer sheet (oriented)	2.0	40	15,000
Permalloy, Ni–Fe	1.6	10	2,000
Superpermalloy, Ni–Fe–Mo	0.2	0.2	100,000
Ferroxcube A, (Mn, Zn)Fe_2O_4	0.4	30	1,200
Ferroxcube B, (Ni, Zn)Fe_2O_4	0.3	30	700

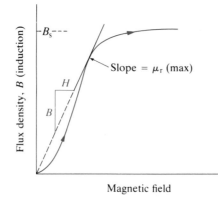

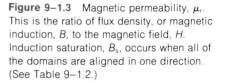

Figure 9–1.3   Magnetic permeability, μ_r. This is the ratio of flux density, or magnetic induction, B, to the magnetic field, H. Induction saturation, B_s, occurs when all of the domains are aligned in one direction. (See Table 9–1.2.)

B/H ratio implies easy magnetization because only a small magnetic field produces high flux densities (induction).

Magnetic Moments

A better understanding of the above magnetic characteristics is obtained by looking at the internal structures of materials.

Each electron of an atom possesses a *magnetic moment, p_m*. For this reason the physicist speaks of the *spin* of an electron. The value of this magnetic moment, called a Bohr *magneton*, is 9.27×10^{-27} A·m². Electrons normally pair off in their orbitals with as many spins "up" as "down." Thus, any external evidence of these moments is masked. An atom will appear magnetic only when there is an unbalance in the spins of the electrons. For the most part, this rules out half of the elements from having magnetic characteristics, since 50 percent of the elements have an even number of valence electrons. It also rules out the majority of ions since an ion has either shed or accepted electrons to achieve completed shells of eight electrons. Finally, covalent bonds always involve pairs of electrons (of opposite spins). The consequence is that few elements have atoms with unbalanced electron spins to achieve a net magnetic moment. Those that qualify are those transition elements with unfilled subvalent shells (Fig. 9–1.4). Iron, cobalt, and nickel are most familiar among these; in addition, gadolinium of the rare-earth series (unfilled 4-f shells) has pronounced magnetism.

	K	Ca	Sc	Ti	V	Cr	Mn	Fe	Co	Ni	Cu	Zn
$4s$	↑	↕	↕	↕	↕	↑	↕	↕	↕	↕	↑	↕
$3d$ { ↑	—	—	1	2	3	5	5	5	5	5	5	5
↓	—	—	—	—	—	—	—	1	2	3	5	5
Net ($3d$)	—	—	1	2	3	5	5	4	3	2	0	0

Figure 9–1.4   Unbalanced magnetic spins (isolated atoms). The 4s orbital fills before the 3d orbitals when each additional proton is introduced to the nucleus.

Domains

Each iron atom possesses four Bohr magnetons, 4β. That is, there is a mismatch of 4 spins per atom (Fig. 9–1.4). At elevated temperatures, thermal agitation prevents any magnetic coupling between atoms. With each "atomic" magnet acting independently, the net magnetic moment per unit volume $(=\Sigma\, p_m/V$, and called *magnetization, M*) is zero. The material is *paramagnetic;* i.e., no cooperative magnetic action exists among the atoms. The material does not have magnetic attractions.

At ambient temperatures, however, the magnetic moments of the adjacent atoms of iron become coupled. The coordinated orientation that is developed provides a net magnetization. The part of the crystal that has a common magnetic orientation is called a *domain*. Typically, a grain (individual crystal) will contain many domains with adjacent parts of the crystal having different magnetic orientations (Fig. 9–1.5). Here again, domains of opposing orientations cancel their magnetizations. Thus, if there are equal domain volumes aligned in the various crystal directions, we observe no net magnetization. This is the

Figure 9–1.5 Domains. Adjacent unit cells develop identical magnetic alignments to produce a domain. (a) Schematic. Adjacent domains cancel the external effects, unless a magnetic field is encountered that moves the boundaries, enlarging favored domains and restricting unfavorably oriented domains. (b) Domains of a permanent magnet retain this magnetization after the external field is removed.

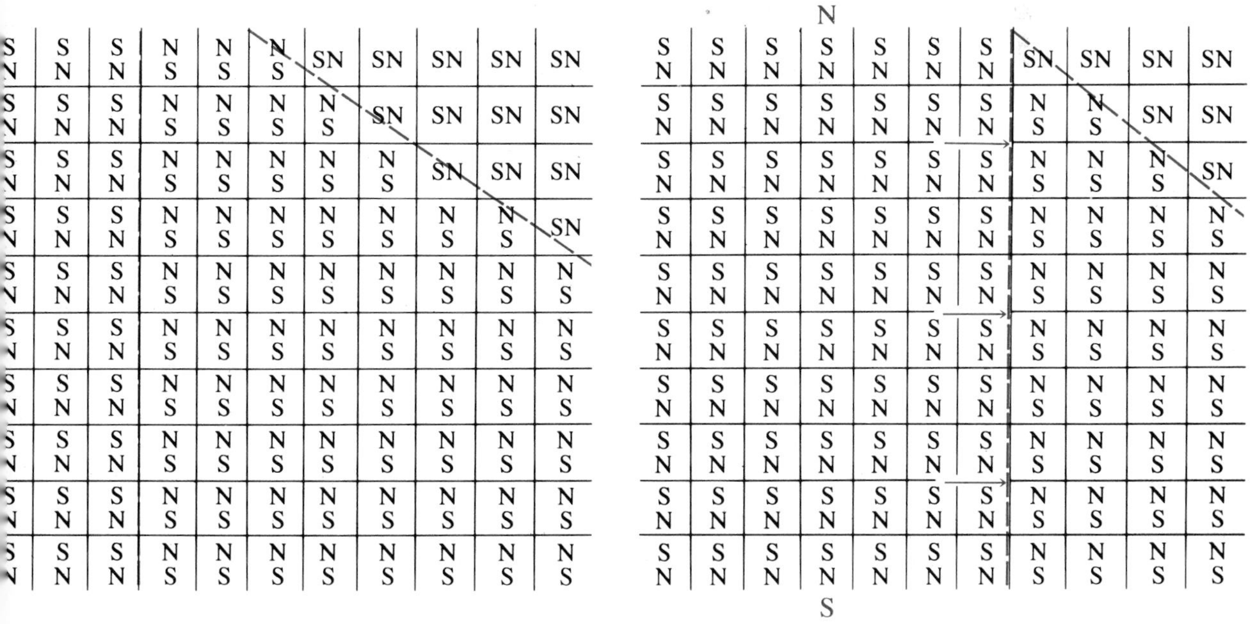

(a)

(b)

situation when a magnetic material is cooled from a paramagnetic temperature (above the *Curie temperature, T_C*) into the magnetic temperature range (below T_C) in the absence of an externally applied magnetic field.

Domain Growth and Domain Boundary Movements

If the magnetic material just described is placed in a magnetic field, those domains that are aligned with the external field possess lower energies than those that are not. Thus, the nonaligned domains are less stable. The atoms at their boundaries reorient into the more stable direction (Fig. 9–1.5b). The nonaligned domains shrink in size; the favorably oriented domains grow in size. The net magnetization M has increased because a balance no longer exists. In the context of Fig. 9–1.1, the external magnetic field has increased the flux density (induction) from zero up the *BH* curve. The growth of the domains occurs as a sideways movement of the domain boundaries.*

In a soft magnet, the removal of the external field is accompanied by a reversal of the domain boundary movements and a return to the initial situation of zero net magnetization; B has returned to nearly zero at $H = 0$ (Fig. 9–1.1b).

In a permanent magnet, the removal of the external field is not accompanied by a reversal of the domain boundary movements. The induction does not return to zero but remains at B_r, the remanent induction when $H = 0$ (Fig. 9–1.1c).

The difference between the behavior of soft and hard magnets arises from the mobility of the domain boundaries. In a soft magnet, very little energy is required for their movement; in a hard magnet, the domain boundary becomes "snagged" within the crystal and a large reverse field is required to supply the necessary energy.

The factors that arrest domain boundary movements have been of interest to the materials and electrical engineers, because these factors provide information for the design of permanent magnets with high *BH* energy products (Table 9–1.1). Conversely, the avoidance of these factors makes a magnet softer—an important consideration in power and high-frequency circuits.

i) Dislocations interfere with domain boundary movements. There is a strain field around each dislocation (Fig. 4–4.3). The domain boundary becomes stabilized in these regions and higher energies are required to force continued boundary movements. Because of this interaction, a work-hardened steel is also a magnetically hard steel. Conversely, a soft-annealed steel is magnetically soft.†

ii) Grain boundaries interfere with domain boundary movements. For this reason, specifications for soft magnetic materials may call for grain coars-

---

* These boundaries are sometimes called the Bloch wall. Actually, the reorientation of the magnetic moments of the atoms is not abrupt; rather, rotation occurs gradually over many layers of atoms.

† This is not a coincidence. It has long been known that cold work hardens a steel. It was an early observation that these hard steels were permanent magnets; hence the term, hard magnets.

(a)

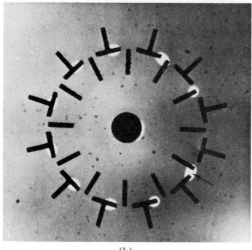

(b)

Figure 9-1.6 Magnetic materials. (a) Grain-oriented silicon steel (stator core of a large electrical generator). The individual crystals are aligned to give easy magnetization. (Courtesy of General Electric Co.) (b) Magnetic bubbles (ceramic garnets). A thin layer of magnetic garnet is deposited epitaxially on a nonmagnetic garnet single crystal. Opposite polarity (arrow) stores binary information. (Courtesy of A. H. Bobeck, Bell Telephone Laboratories.)

ening or even single crystals as shown in Fig. 9-1.6 for a stator core of a large generator and for a memory garnet for a microprocessor.

iii) The boundary between phases precludes domain boundary movements. All permanent metallic magnets contain more than one phase (to be discussed in Part II). Examples include not only the complex Alnico magnets but also plain-carbon steels, which contain a mixture of bcc iron and iron carbide.

iv) Domain boundaries have more restriction in their movements in noncubic materials than in cubic crystals. As an example, hexagonal $BaFe_{12}O_{19}$ is a permanent magnet (Table 9-1.1), while cubic $(Ni,Zn)Fe_2O_4$ is a soft magnet (Section 7-4 and Table 9-1.2).

Ceramic Magnets

Ceramic magnets are ionic compounds, for example, the ferrites. Thus, any iron is present as Fe^{2+} or Fe^{3+}. A ferrous ion has lost two electrons. These are the two $4s$ electrons; the *six* $3d$ electrons remain to give four unpaired electrons (Fig. 9–1.7). A ferric ion has lost the two $4s$ electrons and *one* $3d$ electron; thus *five* unpaired electrons remain.

Nature's original magnet is the mineral magnetite (Fe_3O_4), formerly called lodestone. Magnetite is a natural ceramic phase with an fcc lattice of O^{2-} ions. The iron ions are in both 4-fold and 6-fold interstitial sites (Fig. 7–2.2). More specifically, the Fe^{2+} ions are in 6-f sites; and the Fe^{3+} ions are equally divided between 6-f and 4-f sites. The resulting structure was previously described as the $NiFe_2O_4$-type structure in Section 7–4 as a *spinel*.

The unit cell of this structure is magnetic because the magnetic moments of the ions in the 6-f sites are all aligned in the same direction and those in the 4-f sites are all aligned in the opposite direction. We can make an accounting in the compound Fe_3O_4, or $[Fe^{2+}Fe_2^{3+}O_4]_8$, since there are eight formula weights per unit cell. For every 32 oxygen atoms, which are not magnetic, there are 8 Fe^{2+} ions and 8 Fe^{3+} ions with alignment in the ↑ direction. There are 8 Fe^{3+} ions with alignment in the ↓ direction. From this, plus the information in the previous two paragraphs:

Interstitial site	Spin alignment	Fe^{2+} (4β)	Fe^{3+} (5β)	β	Magnetic moment
6-f	↑	8		+32	$+8(4)(9.27 \times 10^{-24}$ A·m$^2)$
6-f	↑		8	+40	$+8(5)(9.27 \times 10^{-24}$ A·m$^2)$
4-f	↓		8	−40	$-8(5)(9.27 \times 10^{-24}$ A·m$^2)$
	Net:			+32	$+32(9.27 \times 10^{-24}$ A·m$^2)$

Magnetization, M, which may be described as the internal magnetic field of a material, is the total magnetic moment, p_m, per unit volume, V;

$$M = \Sigma p_m / V. \qquad (9–1.1)$$

A unit cell of $[Fe^{2+}Fe_2^{3+}O_4]_8$ is cubic and has a lattice constant, a, of 0.837 nm. Therefore, the *saturation* (maximum) *magnetization, M_s*, for magnetite is

$$32(9.27 \times 10^{-24} \text{ A·m}^2)/(0.837 \times 10^{-9} \text{ m})^3 = 0.5 \times 10^6 \text{ A/m}.$$

This compares with 0.53×10^6 A/m determined in laboratory experiments.

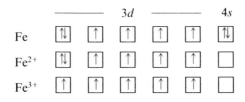

Figure 9–1.7 Unbalanced magnetic spins (iron). An Fe^{2+} ion has a mismatch of four electron moments ($\beta = 4$); for Fe^{3+} ions, $\beta = 5$. Thus the magnetic moment of an Fe^{3+} ion is $5(9.27 \times 10^{-24}$ A · m$^2)$. (The 4s electrons are lost first with ionization, since protons are not removed.)

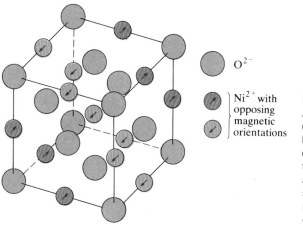

Figure 9–1.8

Antiferromagnetism (NiO). Magnetism exists; however, there is a balance of magnetic moments in the two opposing directions on alternate planes. Therefore, special laboratory procedures are required to detect its presence.

Labels on figure:
O^{2-}

Ni^{2+} with opposing magnetic orientations

Nickel ferrite $[Ni^{2+}Fe_2^{3+}O_4]_8$ has the same structure as magnetite; however, Ni^{2+} ions have only two unpaired electrons. (Compare Fig. 9–1.4 with Fig. 9–1.7.) Therefore, it has a magnetic moment of only $(+8)(2)(9.27 \times 10^{-24}$ A·m$^2)$ per unit cell. Its unit-cell dimensions are close to that of magnetite; thus, its saturation magnetization is about half that of Fe_3O_4.*

The ceramic magnets just described are *ferrimagnetic*. That is, they possess an opposing, but unbalanced, alignment of the magnetic atoms; thus, they have a net magnetization. Some ceramic compounds, for example, manganous oxide (MnO) and nickel oxide (NiO) are *antiferromagnetic*. In NiO, which has the NaCl structure (Fig. 9–1.8), alternate planes of Ni^{2+} ions have antiparallel (opposite) orientations. Thus, while each Ni^{2+} ion has two unpaired electrons, the alternate planes lead to zero net magnetization.

Metallic Magnets

Bcc iron is the most common metallic magnetic material. There are other metallic magnets as indicated in Tables 9–1.1 and 9–1.2. In particular, note the very high maximum permeability, μ_{max}, of superpermalloy, and the large coercive field, $-H_c$ of Alnico V. Rare earth–cobalt alloys are a recent development in magnetic materials.

Many tons of magnetic iron products are made and used each year as sheets for transformer cores and motor parts (Fig. 9–1.6a). In these applications, the magnet must be soft in order to respond to the 60-Hz power sources. Typically, these steels contain 1 to 4 percent silicon in solid solution to increase their permeability; their carbon content is near zero to avoid the presence of iron carbide that interferes with domain boundary movements during each half

---

* These two magnetic compounds, $[Fe_3O_4]_8$ and $[NiFe_2O_4]_8$, are called *ferrites* on the basis of their chemistry. They possess 32 O^{2-}, 16 Fe^{3+}, and 8 Fe^{2+} (or 8 Ni^{2+}) per unit cell, because the lattice constant must be twice that shown in Fig. 7–2.2 in order to obtain a matching of occupied interstices with full unit-cell translations.

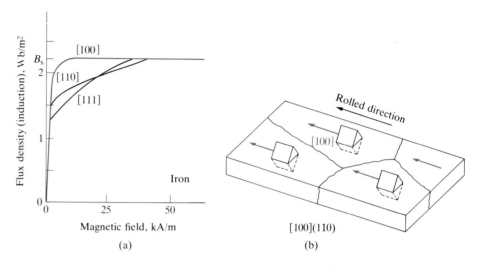

Figure 9–1.9   Magnetic anisotropy (bcc iron). (a) Permeability, (*B/H*) versus crystal direction. (b) Processed to give cube-on-edge. This processing permits the [100] direction to be predominant for easier magnetization of the transformer sheet. An alternate cube-on-face orientation is slightly more efficient in a transformer, but more difficult to process.

cycle. Likewise, the steels are annealed to eliminate dislocations; and the grain size is coarsened.

Steels for power transformers are commonly processed to provide a preferred orientation of the grains. As shown in Fig. 9–1.9(a), the ⟨100⟩ directions have greater permeability than other directions. Therefore, by orienting the grains to near common ⟨100⟩ orientation, a greater transformer efficiency is realized. The increase is from ~97% to 98%. At first, this 1% increase appears minor; however, one must consider that the distribution of gigawatts of power requires a step-up to a high voltage for cross-country transmission, then several step-down stages to the 120V–220V level for homes and industry. A 1% savings on *each* step totals giga-dollars of energy savings over a year!

Metallic magnets possess a major disadvantage in high-frequency circuits because the rapidly changing magnetic field induces current flow and I^2R losses in the core (E^2/R losses). This is the reason that thin laminates are used in transformers (to increase the resistance). At radio frequencies, however, the heating effect is excessive, so that it is necessary to switch to ceramic magnets, which have appreciably higher resistivities. (See the final two magnetic materials of Table 9–1.2.)

Example 9–1.1

Estimate the maximum energy product, BH_{max}, for Alnico VII in Fig. 9–1.2.

Procedure. At $H = 0$ and at $H = -H_c$, the *BH* product is zero. The maximum is somewhere in the mid-range. It will be quicker to take several pairs of data from the graph than to formulate and to differentiate an equation for the curve.

Estimate

At $H = $ 50,000 A/m	$B \approx$ 0.5 V·s/m²	$BH = $ **25,000** J/m³
75,000	$\approx$ 0.3	22,500
55,000	$\approx$ 0.45	24,750
45,000	$\approx$ 0.55	24,750

$$BH_{max} \approx 25,000 \text{ J/m}^2$$

Comparison. The maximum energy product for pure iron, a soft magnetic material, is $\approx$50 J/m². ◄

Example 9–1.2

Calculate the theoretical magnetization of $Mn^{2+}Fe_2O_4$, which has the same structure as $Ni^{2+}Fe_2O_4$ and magnetite. There are 8 $MnFe_2O_4$ formula weights per unit cell (cubic) that has a lattice constant of 0.85 nm.

Procedure. This problem is nearly identical to the calculation for the magnetization of magnetite (just prior to Eq. 9–1.1). Since Mn^{2+} has lost its two $4s$ electrons by ionization, it possesses five Bohr magnetons (net of Fig. 9–1.4). The magnetization is $\Sigma\, p_m/V$.

Calculation

$$\uparrow: \quad (8Mn^{2+})(5\beta/Mn^{2+})(9.27 \times 10^{-24} \text{ A·m}^2)$$

$$\uparrow: \quad (8Fe^{3+})(5\beta/Fe^{3+})(9.27 \times 10^{-24} \text{ A·m}^2) \quad \Big\} \text{net} = 3.7 \times 10^{-22} \text{ A·m}^2$$

$$\downarrow: \quad -(8Fe^{3+})(5\beta/Fe^{3+})(9.27 \times 10^{-24} \text{ A·m}^2)$$

$$M = 3.7 \times 10^{-22} \text{ A·m}^2/(0.85 \times 10^{-9} \text{ m})^3 = 600,000 \text{ A/m}$$

Comment. This compares with 570,000 A/m by experiment. ◄

9–2

DIELECTRIC MATERIALS

Dielectric materials separate two electrical conductors without a passage of current. Thus, metals cannot be dielectrics; but many (not all) ceramics and polymers fall in this category. In the simplest situation, a dielectric is an *insulator,* playing an inert role in the electrical circuit. The prime property for an insulator is *dielectric "strength."* This is the potential gradient value, V/mm, that the designer can use and expect to avoid an electrical breakdown. Table 9–2.1 lists values of the dielectric strengths of various insulators, together with their resistivities. A good insulator has high values of each; however, no correlation exists between the two because eventual electrical breakdown is generally a consequence of impurities, cracks, flaws, and other imperfections, rather than the inherent electrical characteristics of the material. This also accounts for the fact that the dielectric strength is a function of thickness.

Electrical Polarization

Dielectric materials do not conduct electricity. They are not, however, inert to an electrical field as suggested above for insulators. The electrons and the proton-containing atomic nuclei will shift their positions in response to the field. For example, the mean position of the electrons will be on the side of the

TABLE 9–2.1		
Typical Properties of Electrical Insulators		
Material	**Resistivity (20°C) ohm·m**	**Dielectric strength,* V/mm**
Ceramic Materials		
Soda-lime glass	10^{13}	10,000
Pyrex glass	10^{14}	14,000
Fused silica	10^{17}	10,000
Mica	10^{11}	40,000
Steatite porcelain	10^{12}	12,000
Mullite porcelain	10^{11}	12,000
Polymeric Materials		
Polyethylene	10^{13}–10^{16}	20,000
Polystyrene	10^{16}	20,000
Polyvinyl chloride	10^{14}	40,000
Natural rubber	—	16,000–24,000
Polybutadiene	—	16,000–24,000
Phenol-formaldehyde	10^{10}	12,000

* Not constant with thickness

atom nearer the positive electrode, while the atomic nucleus itself, which contains the protons, will shift slightly toward the negative electrode (Fig. 9–2.1). We call this *polarization*. If an ac field is applied, these internal charges "dance" forth and back in step with the frequency of the alternating field.

Polarization can be categorized into several different types in terms of the displaced units. *Electronic polarization* was just described. Being small, the electrons have a very high natural frequency ($\approx 10^{16}$ Hz) as they form their standing waves around the atoms (Section 2–1). Thus, this polarization can

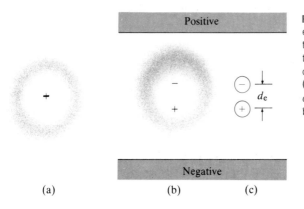

(a)   (b)   (c)

Figure 9–2.1 Electronic polarization (schematic). (a) No external field. The electrons are equally distributed around the nucleus. (b) External field. The positive nucleus shifts toward the negative electrode. The electrons shift in the other direction to separate the center of negative charge ⊖ from the center of positive charge ⊕ by the distance d_e. (c) The dipole moment, p_e, is the product, Qd_e (Q being the charge).

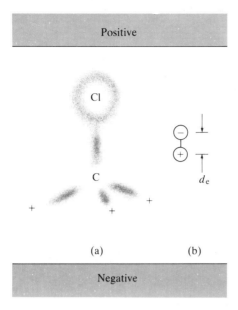

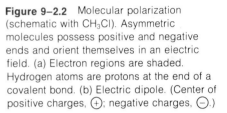

Figure 9–2.2 Molecular polarization (schematic with CH_3Cl). Asymmetric molecules possess positive and negative ends and orient themselves in an electric field. (a) Electron regions are shaded. Hydrogen atoms are protons at the end of a covalent bond. (b) Electric dipole. (Center of positive charges, ⊕; negative charges, ⊖.)

occur not only in 60 Hz circuits and at radio frequencies but also in response to light frequencies ($\approx 10^{15}$ Hz).

Ionic polarization is the displacement of negative and positive ions toward the positive and negative electrodes, respectively. Like electronic polarization, ionic polarization is *induced*, since the relative displacement of positive and negative ions occurs only when an external electrical field is present.

Being more massive than electrons, the ions cannot become polarized as rapidly. Ionic polymerization is limited to a maximum frequency of $\sim 10^{13}$ Hz. This is below the frequency of visible light. Therefore, incident light can not produce ionic polarization as it does for electronic polarization.

Molecular polarization occurs when polar molecules are in an electric field. In polar molecules, the "center-of-gravity" for the positive charges and for the negative charges are not coincident. A small *dipole* is present. An example is methyl chloride (CH_3Cl, Fig. 2–2.3b). The chlorine atom has a complement of 17 electrons while each of the three hydrogen atoms is an exposed proton at the far end of a covalent bond (Fig. 9–2.2). As shown schematically, the centers of positive and negative charges are separated by a distance d_e.

This polarization is permanent since it is inherent in the molecular structure. These dipoles can be oriented with the field. Furthermore, the molecule 'flips' each half cycle of an ac field. Since the masses involved depend upon the size of the molecule, the maximum frequency of response varies significantly from material to material. It is, however, always less than for electronic and ionic polarization. Furthermore, it is highly sensitive to temperature; e.g., polarization of polyvinyl chloride $(C_2H_3Cl)_n$ does not reverse below the glass temperature T_g, because molecular movements are restricted.

A *space charge* (also called *interfacial polarization*) develops when there is local conduction within a dielectric. For example, if an Al_2O_3 product, a non-conductor, contained small particles of aluminum, the conduction electrons in the latter will be shifted toward the positive electrode in an ac field. However, they are contained inside the metallic particles (Fig. 9–2.3). This example of metallic particles is for illustrative purposes and is not typical of engineering materials; however, it is not uncommon for some ceramic phases to contain particles of semiconducting oxides, e.g., Ti_2O_3 in TiO_2. Generally, this is to be avoided because it leads to dielectric losses in high-frequency circuits.

Polarization Calculations

The *dipole moment* p_e of a polarized molecule (or atom or unit cell) is the product of the charge Q and the distance d_e between the centers of positive and negative charges:

$$p_e = Qd_e. \tag{9–2.1}$$

This is illustrated in Fig. 9–2.2, where CH_3Cl has 26 electrons, each with 0.16×10^{-18} C. If the electric field produces an average displacement $\bar{d}_e$ of 0.01 nm between the center of positive charge and the center of negative charge, the dipole moment per molecule is $26(0.16 \times 10^{-18}C)(10^{-11} m) = 40 \times 10^{-30}$ C·m.

Polarization is measured as the sum of the dipole moment per unit volume:

$$\mathscr{P} = \Sigma_{p_e}/V = \Sigma Qd_e/V. \tag{9–2.2}$$

Thus, if there are 10^{20} methyl chloride molecules per cubic meter, the polarization of the CH_3Cl is

$$10^{20} (40 \times 10^{-30} \text{ C·m})/(1 \text{ m}^3) = 40 \times 10^{-10} \text{ C/m}^2.$$

Dielectric Constant

An electric field can induce electronic and ionic polarization and can orient permanently polarized molecules. Conversely, polarization leads to an increased charge density on a capacitor. We can show this by separating two

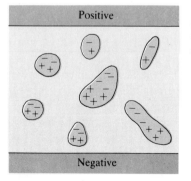

Figure 9–2.3 Space charge. When conducting particles are present within a dielectric, polarization can occur. Each particle develops a positive side and a negative side in an electric field.

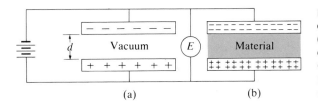

Figure 9–2.4 Charge density. The charge density (coulombs/m²) on the electrodes depends on the electric field, $\mathcal{E}$, and the polarization of material that is present.

capacitor plates by a distance d and applying a voltage E between them (Fig. 9–2.4). The *electric field* $\mathcal{E}$ is the voltage gradient:

$$\mathcal{E} = E/d. \tag{9–2.3}$$

Under these conditions, when there is nothing between the plates (Fig. 9–2.4a), the *charge density* $\mathcal{D}_0$ on each plate is proportional to field $\mathcal{E}$, with a proportionality constant, ε_0, of 8.85×10^{-12} C/V·m:

$$\mathcal{D}_0 = \varepsilon_0\mathcal{E} = (8.85 \times 10^{-12} \text{ C/V·m})\mathcal{E}. \tag{9–2.4*}$$

Thus, if the voltage gradient is 1 V/m, there will be 8.85×10^{-12} coulombs /m² on the electrodes. With each electron carrying a charge of 0.16×10^{-18} C, the electron density on the electrodes will be 55×10^6/m² (or 55/mm²). The voltage gradients are normally much higher in electrical circuits, and thus the charge densities are significantly greater.

If any material is placed between the capacitor plates in Fig. 9–2.4(b), the charge density on the plates is increased because of the polarization within the material. In effect, the upward shift of the negative charges in the material (and the downward shift of the positive charges) increases the value of $\mathcal{D}$ in Eq. (9–2.4) by a constant factor, κ:

$$\mathcal{D}_m = \kappa\varepsilon_0\mathcal{E}. \tag{9–2.5}$$

The factor κ, called the relative *dielectric constant*, is the ratio $\mathcal{D}_m/\mathcal{D}_0$, the charge density with and without the spacer material of Fig. 9–2.4. The relative dielectric constant is a property of the material that is used as the dielectric. Table 9–2.2 lists the relative dielectric constants for a number of common dielectrics. Note that they may be sensitive to temperature and to frequency.

Polarization may also be viewed as the added charge density that originates from the dielectric,

$$\mathcal{P} = \mathcal{D}_m - \mathcal{D}_0; \tag{9–2.6a}$$

and from Eqs. (9–2.4) and (9–2.5),

$$\mathcal{P} = (\kappa - 1)\varepsilon_0\mathcal{E}. \tag{9–2.6b}$$

This is illustrated schematically in Fig. 9–2.5.

---

* ε_0 is the *permittivity* of free space.

TABLE 9–2.2

Relative Dielectric Constants (20°C, unless stated otherwise)

	At 60 Hz	At 10^6 Hz
Soda-lime glass	7	7
Pyrex glass	4.3	4
"E" glass	4.2	4
Fused silica	4	3.8
Porcelain	6	5
Alumina	9	9
Nylon 6/6	4	3.5
Polyethylene $+C_2H_4+_n$	2.3	2.3
Teflon, $+C_2F_4+_n$	2.1	2.1
Polystyrene, PS	2.5	2.5
Polyvinyl chloride, PVC		
plasticized ($T_g \approx 0°C$)	7.0	3.4
rigid ($T_g = 85°C$)	3.4	3.4
Rubber (12 w/o S, $T_g \approx 0°C$)		
−25°C	2.6	2.6
+25°C	4.0	2.7
+50°C	3.8	3.2

Polymeric Dielectrics   The dielectric constant of polymers originates from both electronic polarization and molecular orientation. As revealed in Example 9–2.2, the dielectric constant is less than 3.0 above 10^8 Hz with *only* electronic polarization present. Molecular orientation may become a contributing factor below ~10^8 Hz. The protons of the side hydrogen atoms and the polar groups such as $\mathord{>}C{=}O$, $-C{\equiv}N$, and $-Cl$ respond to the electric field. Of course this is possible only above the glass temperature T_g.

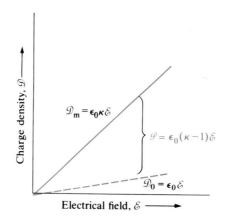

Figure 9–2.5 Charge density versus electric field. Polarization, $\mathscr{P}$, is that part of the charge density resulting from the presence of the material.

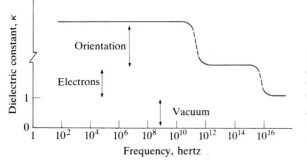

Figure 9–2.6 Relative dielectric constant versus frequency (schematic). Electrons respond to the alternating electric field below ~10^{15} Hz. Molecular dipoles can respond below about 10^{10} Hz.

Figure 9–2.6 shows schematically the effect of polarization $\mathscr{P}$ on the relative dielectric constant κ as a function of frequency. Below T_g, molecular movements cannot contribute to the dielectric constant. As soon as T_g is exceeded, the dielectric constant increases markedly as shown in Fig. 9–2.7. However, at still higher temperatures, we see a drop in the dielectric constant, because the more intense thermal agitation destroys the orientation of the molecular dipoles.

The curves of Fig. 9–2.7 also illustrate another point. The glass temperature T_g varies with frequency. Just as in Fig. 6–7.4, longer times (in this case, milliseconds for 60 Hz versus microseconds for 10^6 Hz) permit molecular movements at a somewhat lower temperature. A dc field has a still lower T_g because extended time is available for molecular orientation.

Example 9–2.1
Polystyrene has a relative dielectric constant of 2.5 when subjected to a dc field. What is the polarization within the PS when a 0.5-mm sheet separates 100 volts?

Solution. The field is 2×10^5 V/m. From Eq. (9–2.6b),

$$\mathscr{P} = (2.5 - 1)(8.85 \times 10^{-12}\ \text{C/V·m})(2 \times 10^5\ \text{V/m})$$
$$= 2.7 \times 10^{-6}\ \text{C/m}^2.$$

Comment. Since the electron charge is 0.16×10^{-18} C, this polarization is equivalent to 1.7×10^7 el/mm². ◄

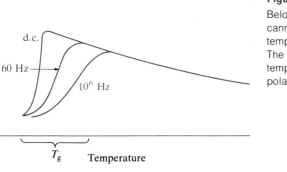

Figure 9–2.7 Dielectric constant versus temperature. Below the glass temperature, T_g, the molecular dipoles cannot respond to the alternating electric fields. (The glass temperature is lower with low frequencies—Fig. 6–7.4b.) The dielectric constant is greatest just above T_g. At higher temperatures, more vigorous thermal agitation destroys the polarization and therefore reduces the dielectric constant κ.

Example 9–2.2 The relative dielectric constants for polyvinyl chloride (PVC) and polytetrafluoroethylene (PTFE) are as follows:

Frequency, hertz	PVC	PTFE
10^2	6.5	2.1
10^3	5.6	2.1
10^4	4.7	2.1
10^5	3.9	2.1
10^6	3.3	2.1
10^7	2.9	2.1
10^8	2.8	2.1
10^9	2.6	2.1
10^{10}	2.6	2.1

a) Plot the capacitance-versus-frequency curves for three capacitors with 3.1 cm × 102 cm effective area separated by 0.025 mm of (1) vacuum, (2) PVC, and (3) PTFE.

b) Account for the decrease in the relative dielectric constant of PVC with increased frequency, and the constancy in the relative dielectric constant of PTFE.

Procedure. From physics, capacitance is in farads, F, or coulombs per volt, A·s/V. Rearranging the units of Eqs. (9–2.3) and (9–2.4):

$$A \cdot s/m^2 = \kappa\varepsilon_0(V/m) \qquad \text{to} \qquad A \cdot s/V = \kappa\varepsilon_0(m^2/m);$$

$$C = \kappa\varepsilon_0 A/d = \kappa(8.85 \times 10^{-12} \text{ A·s/V·m})A/d. \qquad (9\text{–}2.7)$$

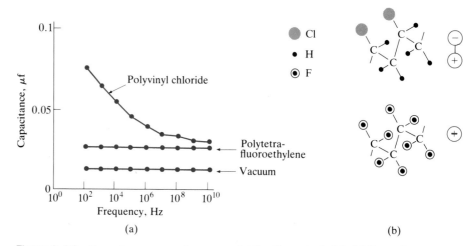

Figure 9–2.8 Capacitance versus frequency. (a) See Example 9–2.2. (b) Symmetry comparison of polyvinyl chloride and polytetrafluorethylene mers.

Sample calculation

a) At 10^2 cycles per second:

$C_{vac} = (1)(8.85 \times 10^{-12} \text{ A·s/V·m})(0.031 \text{ m})(1.02 \text{ m})/(25 \times 10^{-6} \text{ m}) = 0.0112 \text{ } \mu\text{F};$

$C_{pvc} = (6.5)(0.0112 \text{ } \mu\text{F}) = 0.073 \text{ } \mu\text{F}.$

See Fig. 9–2.8(a) for the remainder of the results.

b) The relative dielectric constant of PVC is high at low frequencies because PVC has an asymmetric mer with a large dipole moment (Fig. 9–2.8b). At high frequencies the dipoles cannot maintain alignment with the alternating field. On the other hand, PTFE has a symmetric mer, and therefore its polarization is only electronic. Although the dipoles in PTFE are weaker, they can be oscillated at the frequencies of Fig. 9–2.8. ◀

• 9–3

**PIEZOELECTRIC
MATERIALS**

We saw in the last section that certain molecules possess a permanent dipole; they are polar molecules. Likewise, certain crystals possess a permanent electrical dipole because their centers of positive and negative charges are not at the centers of the unit cells. These unit cells are polarized. This is illustrated by $BaTiO_3$. Recall from Fig. 7–4.1 that $BaTiO_3$ is cubic—above 120°C. Below that temperature, called the *Curie point*, there is a slight but important shift in the ions. The central Ti^{4+} ion shifts about 0.006 nm with respect to the corner Ba^{2+} ions. The O^{2-} ions shift in the opposite direction, as indicated in Fig. 9–3.1.* The center of positive charge and the center of negative charge are separated by the dipole length d.

A material such as $BaTiO_3$ changes its dimensions in an electric field because the negative charges are pulled toward the positive electrode, and the positive charges are pulled toward the negative electrode, thus increasing the dipole length, d. This also increases the dipole moment Qd, and the polarization $\mathcal{P}$, since the latter is the total of the dipole moments ΣQd per unit volume V (Eq. 9–2.2).

The above sequence of effects provides a means of changing mechanical energy into electrical energy and vice versa. To understand this, refer to Fig. 9–3.2(a). The cooperative alignment of the dipole moments of the many unit cells gives a polarization that collects positive charges at one end of the crystal and negative charges at the other end. Now consider two alternatives as shown in parts (b) and (d). (1) Compress (or pull) the crystal with a stress s. There is a strain e, which is dictated by the elastic modulus. This strain changes the dipole length d and directly affects the polarization $(=\Sigma Qd/V)$ since Q and V remain essentially constant. With a smaller polarization (from compression), there is excess of charge density on the two ends of the crystal. If the two ends are isolated, a voltage differential develops (Fig. 9–3.2b). If in electrical contact,

---

* The shift of the Ti^{4+} in Fig. 9–3.1 is upward as drawn. Actually, the shift could be in any one of the six coordinate directions. In any event, the O^{2-} ions shift in the opposite direction.

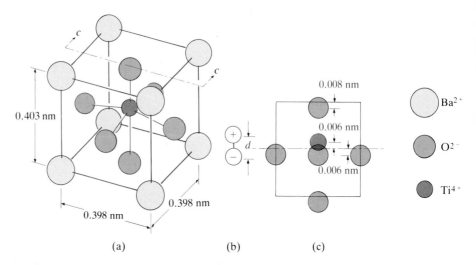

Figure 9–3.1   Tetragonal $BaTiO_3$. Above 120°C (250°F), $BaTiO_3$ is cubic (Fig. 7–4.1). Below that temperature, the ions shift with respect to the corner Ba^{2+} ions. Since the Ti^{4+} and the O^{2-} ions shift in opposite directions, the centers of positive and negative charges are not identical. The unit cell becomes noncubic.

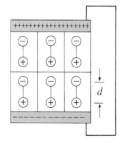

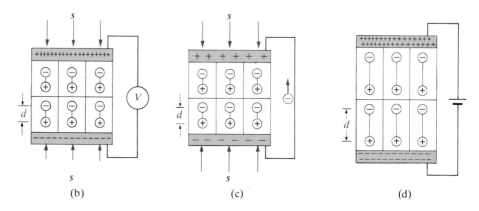

Figure 9–3.2   Piezoelectric material (schematic). (a) No external field: centers of positive and charges are not coincident. (b) Pressure applied: compression leads to voltage differential. (c) Pressure applied and electrical path present: compression leads to charge transfer. (d) External potential applied: dipole lengthening produces dimensional change.

electrons will flow from one end to the other (Fig. 9–3.2c). (2) No pressure is applied in Fig. 9–3.2(d); rather a voltage is applied, which increases the charge density at the two ends. The negative charges within the $BaTiO_3$ are pulled one way and the positive charges the opposite, thus changing not only the dipole length d, but also that dimension of the crystal.

These two situations indicate how mechanical forces and dimensions can be interchanged with electrical charges and/or voltages. Devices with these capabilities are called *transducers*. Materials with the above characteristics are called *piezoelectric*, i.e., pressure-electric. Crystals of the $BaTiO_3$ type* are used for pressure gages, for phonograph cartridges, and for high-frequency sound generators. Let us examine a ceramic phonograph cartridge as an example. The stylus, or needle, follows the groove on the record. A small transducer is in contact with the stylus, which detects the vibrational pattern recorded in the groove. Both the frequency and the amplitude can be sensed as voltage changes (Fig. 9–3.2b). Although the voltage signal is small, it can be amplified through electronic circuitry until it is capable of driving a speaker.

Quartz crystals (SiO_2) are also piezoelectric. They are produced (Fig. 7–7.5) for special applications in circuits requiring frequency control. Once cut to a selected geometry, the elastic vibrational frequency is constant to one part in 10^8! Thus, in resonance, the crystal can control the frequency of an electronic circuit to that same accuracy. Fine watch circuits and control circuits for radio broadcasting are but two applications that utilize quartz crystals for their piezoelectric characteristics.

Example 9–3.1

A piezoelectric material has a Young's modulus of 72,000 MPa (10,400,000 psi). What stress is required to change its polarization from 640 C·m/m³ to 645 C·m/m³?

Procedure. The increase in polarization must originate from a strain that increases d of the dipole moment. The stress must be positive (tension).

Calculation. $\Delta\mathscr{P}/\mathscr{P} = (645 - 640)/640 = +0.0078 = (\Delta Qd/V)/(Qd/V)$. However, Q and V do not change; therefore,

$$\Delta d/d = +0.0078 = e = s/E.$$

$$s = +0.0078 \ (72{,}000 \text{ MPa})$$

$$= +560 \text{ MPa};$$

or

$$s = +81{,}000 \text{ psi} \qquad \text{(tension)}.$$

Added information. The charge density changes 5 C/m² with this pressure. Therefore, if the two electrodes of Fig. 9–4.2(c) are connected, there will be a transfer of

$$\text{electrons} = 5 \text{ C/m}^2/(0.16 \times 10^{-18} \text{ C/el})$$

$$= 3 \times 10^{19} \text{ el/m}^2 \qquad \text{(or } 3 \times 10^{13} \text{ el/mm}^2\text{).} \blacktriangleleft$$

Example 9–3.2

Calculate the polarization of $\mathscr{P}$ of $BaTiO_3$, based on Fig. 9–3.1 and the information of this section.

Procedure. Calculate the dipole moment for each atom of the unit cell. Use the middle of the unit cell as the fulcrum for the moment arm.

---

* The most widely used piezoelectric ceramics are $PbZrO_3$–$PbTiO_3$ solid solutions called PZT's. They have the structure of $BaTiO_3$ in Fig. 9–3.1. However, their Curie points are higher than that of $BaTiO_3$, and they have a greater range of design applications.

Calculation

Ion		Q, coul	d, m	Qd, coul·m
Ba^{2+}		$+2(0.16 \times 10^{-18})$	0	0
Ti^{4+}		$+4(0.16 \times 10^{-18})$	$+0.006(10^{-9})$	3.84×10^{-30}
$2\ O^{2-}$	(side of cell)	$-4(0.16 \times 10^{-18})$	$-0.006(10^{-9})$	3.84×10^{-30}
O^{2-}	(top and bottom)	$-2(0.16 \times 10^{-18})$	$-0.008(10^{-9})$	2.56×10^{-30}

$$\Sigma = 10.24 \times 10^{-30}$$

From Eq. (9–2.2)

$$\mathscr{P} = \Sigma Qd/V$$
$$= (10.24 \times 10^{-30}\ \text{C·m})/(0.403 \times 0.398^2 \times 10^{-27}\ \text{m}^3)$$
$$= 0.16\ \text{coul/m}^2.$$

Added information. This means that polarized $BaTiO_3$ can possess a charge density of 0.16 coul/m², equivalent to 10^{12} electrons/mm². ◄

9–4

**OPTICAL
MATERIALS**

The near unanimous majority of materials that are used for optical purposes are assigned to light transmission roles. Filtering action (selective transmission) is sometimes specified. In the most familiar applications, such as architectural and automotive windows, the optical requirement is transmission without distortion. This calls for flat, parallel surfaces with the absence of internal flaws. In selected applications, infrared or ultraviolet radiation must be filtered out.

Optical *lenses* have the added purpose of refracting the light along an optical path (Fig. 9–4.1). In lenses for eyesight, the refraction is generally

Figure 9–4.1 Optical glass (prisms, lenses, and mirrors). Stringent requirements for composition and homogeneity are mandatory to assure clarity and precise optical paths for high resolution. (Courtesy of American Optical.)

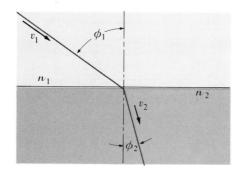

Figure 9–4.2 Refraction. The index of refraction is inversely proportional to the velocity of light and is measured by the refraction angles (Eq. 9–4.1). Materials with higher electronic polarization, $\Sigma\, p_e/V$, have higher indices.

controlled by grinding for surface curvature. The index of refraction, which is a property of the material, is a second factor in lens specification and must be considered in all optical systems.

Optical fibers for communication channels are a recent innovation into technical capabilities. Obviously, the fibers must have near zero light absorption. Somewhat surprisingly, the index of refraction is also a factor.

Laser optics is another high-technology application of light that has evolved sophisticated materials developments (Fig. III of Foreword).

Index of Refraction

The index of refraction n of a material is the ratio of light velocity, v, in a vacuum to the velocity in that material. The index is calculated by Snell's law, which relates the angle of the incident ray to that of the refracted ray (Fig. 9–4.2):

$$\frac{v_1}{v_2} = \frac{n_2}{n_1} = \frac{\sin \phi_1}{\sin \phi_2}. \tag{9–4.1}$$

The light ray may travel in either direction. If, however, ϕ_2 exceeds a critical value (at which ϕ_1 equals 90°), the ray is reflected internally.

• The index of refraction varies somewhat with frequency. The spread in index between blue and red is called *dispersion*. The shorter wavelengths have the higher indices. This becomes important in the design of lens systems where it is mandatory that all light colors have the same plane of focus. Table 9–4.1 lists the indices of refraction and dispersion for a few selected materials.

• The index of refraction is related to the polarization discussed in Section 9–2. Since only electronic polarization is operative at light frequencies ($\sim 10^{15}$ Hz), the relationship is between the index n and the electronic dielectric constant κ_e:

$$n = \sqrt{\kappa_e}. \tag{9–4.2}$$

Within glasses and cubic materials, the electronic polarization is uniform in all directions. In noncubic materials, however, the electronic polarizability varies

TABLE 9–4.1		
Selected Indices of Refraction		
	Index (20°)	
Material	$\lambda = 589.3$ nm	**656.3 nm**
Air	1.00028	
Water	1.333	
Optical glass (heavy flint)	1.650	1.644
Optical glass (crown)	1.517	1.514
Fused silica	1.458	1.457
Quartz and	1.544 1.553	
Calcite and	1.658 1.486	

with the direction of light vibrations. Thus, the index of refraction is anisotropic. For example, calcite ($CaCO_3$) has its $(CO_3)^{2-}$ ions oriented so that the
four atoms lie in the same (0001) plane of the hexagonal crystal (Fig. 9–4.3).
The electronic polarizability is 24% greater in this plane than perpendicular to
it. This leads to two indices of refraction—1.66 and 1.49, respectively. This
difference is called *birefringence* and appears as double refraction.

Absorption

The visible part of the electromagnetic spectrum lies near 5×10^{14} Hz. This is
below the limit of normal electronic polarization and well above the limit of
ionic polarization. Therefore, simple ionic and covalent materials are transparent, because the energy losses associated with phase lags are avoided.

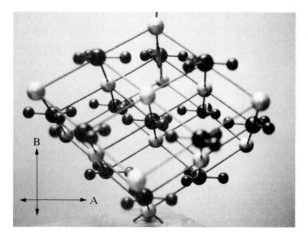

Figure 9–4.3 Optical anisotropy ($CaCO_3$). Light vibration
A perpendicular to the vertical axis—parallel to (0001).
Light vibration B parallel to the vertical
axis—perpendicular to (0001). The electronic polarization
from the transmitted light is 24% greater in A than in B;
therefore, the index of refraction is greater for the vibration
direction of A. This produces birefringence, giving two
indices of refraction.

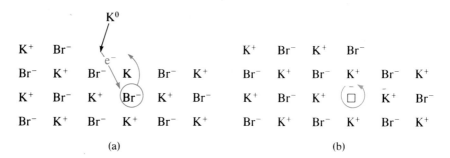

Figure 9–4.4   Color center (KBr). (a) An atom of potassium vapor ionizes to K^+ as it joins the surface, losing an electron. A Br^- ion migrates to the surface, leaving an anion vacancy. (b) Since the anion vacancy carries a positive charge, it traps the electron. The electron absorbs light that possesses its vibrational frequency within the vacancy. In KBr, the electron in this *color center* has a frequency of 4.85×10^{14} Hz, which corresponds to orange light; thus the remaining transmitted light is blue-green.

As a result, any attenuation that occurs is a result of imperfections within the material.

•   Colored materials are the consequence of partial absorption of white light and the resulting selective transmission of the balance of the spectrum. We will consider one of several mechanisms of light absorption in otherwise transparent materials—*color centers*. If KBr, which is normally transparent, is heated in the presence of potassium vapor, excess potassium atoms can deposit on the surface as shown in Fig. 9–4.4(a). This increases the probability that an interior Br^- ion will migrate to the surface, leaving a vacancy inside. This vacancy has a missing negative charge, which can attract an electron (e.g., from the ionizable potassium atom at the surface). The entrapped electron will respond to specific frequencies in its small "cage" (Fig. 9–4.4b). In this particular case, the resonant period of the electron is 2.06×10^{-15}s, for a frequency of 4.85×10^{14} Hz, the frequency of orange light ($\lambda = 618$ nm). Thus, the wavelengths in this vicinity are absorbed. The transmitted light is the complement, which is a blue-green. The structural defect just described is called an *F-center* from the original German label as a Farbe-point, or color-point. Other types of defects produce other absorptions.

The color of window glass is commonly a blue-green.* This color arises from the presence of iron ions within the glass. Ferrous ions have absorptions that introduce a blue color, while ferric ions bring about a yellow transmission.† Other ions introduce other colors, for example, Co^{2+} for blue.

---

* Look into the edge of a pane of glass so you observe a thickness of >10 cm.

† Glass melted under oxidizing conditions changes the Fe^{3+}/Fe^{2+} ratio, eliminating the blue color and accentuating the yellow. In days when glass containers were used for milk, specifications called for an oxidized glass for color control. The purchaser preferred a yellow, rather than green, tint when selecting milk products.

A transparent material becomes *translucent* if light is scattered by internal reflections or is refracted at grain and phase boundaries. Thus, maximum transparency dictates the use of glasses (Fig. 9–4.5), amorphous plastics, and single crystals.

Optic Fibers

The new multichannel, modulated-light communications systems require optical fibers that will transmit light greater distances. Thus, special attention must be given to the compositions, structure, and processing of glass fibers. The fibers must be free of all oxides that absorb light—in particular, all transition metal oxides such as FeO. We depend on total internal reflections to keep the light within the optical wave guide. In principle, this is simple, since the critical angle can be calculated from the index of refraction (Eq. 9–4.1). However, sharp curvatures of the fiber can lead to "leakages." More importantly, surface flaws, no matter how minute, can lead to the scattering of light, and, of course, any handling of a fiber will introduce flaws. The effects of these flaws can be avoided if the light-transmission glass is coated with a second glass of lower index. The light is totally reflected within the core glass and does not enter the outer glass coating that is exposed to the degrading environment. Such a dual fiber is made by jacketing the high-index glass rod as a core within a lower

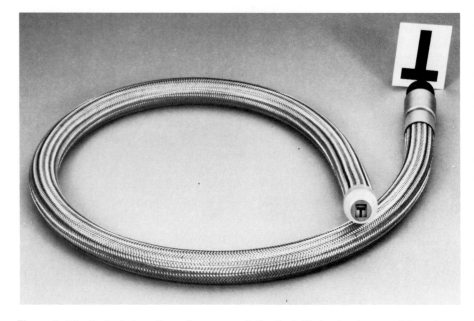

Figure 9–4.5 Optical glass fibers (imagescope). For flexibility to view inaccessible and hazardous locations, the 27,000 glass fibers in the bundle are only 60 μm in diameter. For transparency through their 900-mm length, the glass fiber must be free of any light-absorbing ions. Optical fibers for communication systems require kilometers of transparency, plus coherent light signals. (Courtesy of Galileo Electro-Optics Corp.)

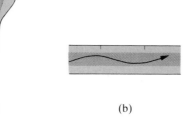

Figure 9–4.6 Optical wave guide (dual fiber). (a) Drawing. A low-index glass (outer) and a normal-index glass (center) are hot drawn together. (b) By internal reflection, the low-index glass keeps the light path away from the surface and its flaws.

(a)   (b)

index glass tube, and simultaneously drawing the hot glass into a composite filament (Fig. 9–4.6).*

• **Lasers**

A laser (**L**ight **A**mplification by **S**timulated **E**mission of **R**adiation) provides a coherent light source; that is, all of the light is in phase. As such, it must be monochromatic.

The original, and still used, laser material involves a single crystal rod of ruby, which is Al_2O_3 containing ~0.5 percent Cr^{3+} ions in solid solution. Numerous types of materials are currently used, including gases, liquids, solids, glasses, and semiconductors. In each case, electrons of the atoms are excited into a higher, or upper, energy state. In the ruby laser, the subvalence electrons of the Cr^{3+} ions are energized. The *pumping* of the electrons to the upper energy state is commonly achieved by photons from a flash lamp.

As with luminescence (Section 8–4), the activated electrons will return to their ground state in a random, statistical manner and release energy as a photon of light. The rate of relaxation is normally defined by the decay equation (8–4.5). However, this return to the lower energy state may be forced to occur prematurely if the electron is *stimulated* by a photon whose energy equals the release energy. Thus, one photon can release another (and another) in a chain reaction as long as the photons remain in the material. Furthermore, the succeeding photons are released in phase with triggering photon.

If none of the photons escaped, there would be an almost instantaneous cascade of all energized electrons, and a "burst" of light would result. The procedure to retain photons for repeated use is to polish and silver the ends of the laser to provide reflection (Fig. 9–4.7). This permits the pulse of energy to

* The product just described is overly simplified. Current practice calls for a jacket with a graded index and no sharp boundary. This lets the ray gain some velocity as it travels in the outer part of the fiber (lower index) and remain in phase with the other rays that travel along the fiber axis. This is necessary to retain coherency for modulated signals.

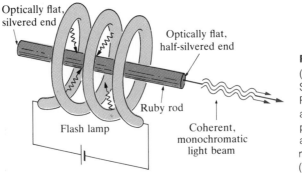

Figure 9–4.7 LASER device (Light Amplification by Stimulated Emission of Radiation). Photon energy is absorbed by the ruby (Al_2O_3 plus Cr_2O_3) rod and remitted as an intense, coherent, and monochromatic light beam. (Cf. Fig. III of Foreword.)

be released in nanoseconds, and power levels (J/s) in the watts to megawatt range because of the short time.

As a device (Fig. 9–4.7), a laser has a light pump, which is commonly a gas discharge tube, surrounded by a reflector to concentrate the activating energy into the ruby rod (or other laser material). The emerging beam is monochromatic, coherent, and parallel, so that it can be fed into an optical fiber of a communication system, or focused onto a "pinpoint." This latter attribute permits diverse applications ranging from eye surgery for detached retinas to surface hardening of steel.

Example 9–4.1

Water has an index of refraction of 1.331. (a) What is the critical angle for internal reflection of light at an air-water surface? (b) Will the index for ice be higher or lower than 1.331?

Procedure. (a) The refracted angle in air will be 90°, which can be used in Eq. (9–4.1). Consider that the index of refraction for air is 1.0. (b) The molecules of ice and water are the same kind, H_2O, but not present in equal numbers per unit volume. Which will have the greater polarization? (Compare densities, ρ.)

Answers

a) $1.331/1.0 = \sin 90°/\sin \phi_{H_2O}$:

$$\phi_{H_2O} = 48.7°.$$

b) With $\rho_{ice} < \rho_{water}$, $(\Sigma\ Qd/V)_{ice} < \mathscr{P}_{water}$. Therefore, κ and n are less for ice than for water.

Comment. The indices of air and of ice are 1.0003 and 1.31, respectively. ◄

Example 9–4.2

Sodium light has a frequency of 5.09×10^{14} Hz. (a) What is its wavelength in outer space? (b) In glass with $n = 1.56$?

Procedure. The velocity changes with the index; but the frequency ν must remain constant. From physics, $\lambda = c/\nu$.

Calculation

a)
$$\lambda = (3 \times 10^8 \text{ m/s})/(5.09 \times 10^{14}/\text{s})$$
$$= 589 \times 10^{-9} \text{ m} \qquad \text{(or 589 nm)}.$$

b)
$$\lambda = [(3 \times 10^8 \text{ m/s})/1.56]/(5.09 \times 10^{14}/\text{s})$$
$$= 378 \times 10^{-9} \text{ m} \qquad \text{(or 378 nm)}. \blacktriangleleft$$

Example 9–4.3

The attenuation of light in a glass fiber is purported to be -2.3 db per km. (a) What fraction of the initial intensity remains after 1 km? (b) After 8 km?

Procedure. A "bel" is the exponent (base 10) of I_2/I_1;

$$\therefore 1 \text{ db} = 0.1 \text{ bel} = 0.1 \log_{10} (I_2/I_1).$$

Calculation

a) -2.3 db $= -0.23$ b $= \log_{10}(I_2/I_1)$;

$$I_2/I_1 = 10^{-0.23} = 0.59.$$

b) $(-2.3 \text{ db/km})(8 \text{ km}) = -18.4 \text{ db} = -1.84 \text{ b}$;

$$I_3/I_1 = 10^{-1.84} = 0.014.$$

Comment. Ordinary window glass has an attenuation in excess of -2.3 db/m. $\blacktriangleleft$

Review and Study

SUMMARY

Nonconducting materials can have useful functions in an electrical circuit. These include magnetic, dielectric, piezoelectric, and optical materials.

1. The difference between *soft* and *permanent* (hard) magnets is the amount of *remanent induction*, B_r, after a magnetic field is removed. It takes an opposing *coercive field*, $-H_c$, to eliminate the induction (also called the flux density).

   The *BH product* is a measure of the energy required for demagnetization and therefore of the permanence of the magnet. Energy is required for domain boundary movements. In hard magnets, *domain boundaries* become anchored to (i) dislocations, (ii) grain boundaries, and (iii) phase boundaries; also (iv) they have less mobility in anisotropic materials. These factors must be taken into account in selecting magnetic materials.

   The *magnetization* of ceramic materials may be calculated from the magnetic moments of the ions. Metallic magnets cannot be used in high-frequency circuits because their low resistivity permits heating from induced currents. High efficiencies have been achieved for transformer sheets through grain orientation and other microstructural controls.

2. *Dielectric "strength"* is the resistance of an insulator to electrical breakdown.

Induced polarization involves both electrical and ionic displacements by an electric field without charge transfer. Such a field may also orient the permanent dipole of polar molecules to produce *polarization $\mathcal{P}$*. This is an integration of the dipole moments per unit volume. It is also equal to the excess *charge density*, $\mathcal{D}_m - \mathcal{D}_0$, developed in a capacitor. The relative *dielectric constant* κ is the ratio of the charge densities on a capacitor with and without a material present.

Polymer dielectrics have only electronic polarization below the glass temperature, T_g, and at high frequencies ($>10^{10}$ Hz). Molecular polarization becomes operative above T_g, so that the dielectric constant takes a big jump, particularly for dc and low-frequency ac circuits. At still higher temperatures, thermal agitation destroys the molecular orientation with the electric field, so the dielectric constant decreases.

3. Crystalline materials without a center of symmetry are *piezoelectric*. As such, they can be used for *transducers,* which can be made to convert electric signals to mechanical signals, and vice versa.

4. The *index of refraction* is the inverse measure of light velocity and wavelengths within materials. It depends upon the electronic polarization and varies slightly with wavelength. *Optical fibers* for communication transmission must have exceptionally low light absorption. Use is made of lower index (and graded) coatings to keep the light path away from the fiber surfaces, which can be damaged to permit light leakages.

The first photons that are released in *lasers* when activated electrons return to their lower energy levels stimulate additional photon emission. A chain reaction of released photons produces a power burst of coherent (in phase), monochromatic light.

TERMS AND CONCEPTS

Attenuation   Reduction of radiation with distance.

BH product   Energy required for demagnetization. The maximum product of the second quadrant is usually used.

• **Birefringence**   Difference in index of refraction with orientation.

Charge density   Charge per unit area, e.g., on an electrode.

Coercive field (magnetic, $-H_c$)   Magnetic field required to remove residual magnetic induction.

• **Color centers**   Points of structural imperfection that absorb light.

Curie point (electric)   Transition temperature between a symmetric crystal and polar crystal.

Curie temperature (magnetic)   Transition temperature for magnetic domain formation.

Dielectric   An insulator. A material that can be placed between two electrodes without conduction.

Dielectric constant, relative (κ)   Ratio of charge density arising from an electric field (1) with and (2) without the material present.

• **Dielectric strength**   Electrical breakdown potential of an insulator per unit thickness.

Dipole moment (p)   Product of electric charge and charge separation distance.

- **Dispersion**   Difference in index of refraction with wavelength.

  Domain   Microstructural area of coordinated magnetic alignments.

  Domain boundary   Transition zone between domains; moves during magnetization and demagnetization.

  Electric field ($\mathcal{E}$)   Potential gradient, V/m.

- **Ferrimagnetism**   Net magnetism arising from unbalanced alignment of magnetic ions within a crystal.

  Ferrites   Compounds containing trivalent iron; commonly magnetic.

  Index of refraction (n)   Ratio of light velocity in a vacuum to that in the material.

  Induction (magnetic, B)   Flux density in a magnetic field.

  Induction (remanent, B_r)   Flux density remaining after the external magnetic field has been removed.

- **Laser**   *Light Amplifications Stimulated by Emission of Radiation.*

  Magnet, permanent (hard)   Magnet with a large $(-BH)$ energy product, so that it maintains domain alignment.

  Magnet, soft   Magnet that requires negligible energy for domain randomization.

  Magnetization (M)   Magnetic moment density, that is, magnetic moment per unit volume.

  Magneton, Bohr (β)   Magnetic moment of an electron (9.27×10^{-24} A·m^2).

  Permeability (magnetic, μ)   Ratio of induction to magnetic field, B/H.

- **Piezoelectric**   Dielectric materials with structures that are asymmetric, so that their centers of positive and negative charges are not coincident. As a result the polarity is sensitive to pressures that change the dipole distance, and the polarization.

  Polar molecules   Molecules with a permanent electric dipole, i.e., with a positive and negative end.

  Polarization (electric, $\mathcal{P}$)   Electric dipole density, that is, dipole moment per unit volume.

- **Transducer**   A material or device that converts energy from one form to another, specifically electrical energy to or from mechanical energy.

FOR CLASS DISCUSSION

A_9 Why are strain-hardened steels also harder magnetically, and annealed steels softer magnetically?

B_9 Small magnets have been used for memory devices. A "square loop" is preferred. What is meant by this and why should it be preferred?

C_9 Soft ceramic magnets commonly have the spinel structure described on p. 344. Relate the size of that unit cell with the sketch of Fig. 7–2.2.

D_9 (a) Discuss two ways in which cracks affect an electrical insulator. (b) Why are insulators glazed before use on high-voltage lines?

E_9 Distinguish between dielectric constant and dielectric strength.

F_9 By a sketch show why the $-\overset{\text{Cl}}{\underset{\text{H}}{\text{C}}}-$ portion of polyvinyl chloride is a polar group.

G$_9$ Is it possible to have a relative dielectric constant of less than one? Why?

H$_9$ Dielectric constants of polymers are greatest at intermediate temperatures. Why? How does the dielectric constant differ between ac and dc?

I$_9$ Discuss the relationships among the viscoelastic modulus, glass temperature, and dielectric constant.

• J$_9$ Explain to someone who is not taking this course how a ceramic cartridge on a phonograph provides a signal that can then be amplified.

• K$_9$ Suggest a concept for a piezoelectric "spark plug."

L$_9$ Cite two reasons for having a lower index glass in the outer portions of an optical fiber.

• M$_9$ How does the stimulated emission of a laser lead to high power densities?

QUIZ SAMPLES

9A Significant magnetic induction, B, remains in a __**__ magnet after the external magnetic field is removed. A reversed field, called the __**__, is required to drop the induction to zero.

9B Fe^{2+} and Fe^{3+} ions have six and five 3-d electrons, respectively. Therefore, they have __**__ and __**__ Bohr magnetons each. Vanadium, with three 3-d electrons, has __**__ Bohr magnetons.

9C Magnetic domains expand by the sideway movements of domain __**__: __+−__ interfere with these movements.

(a) dislocations   (b) precipitates   (c) grain boundaries   (d) impurities
(e) None of these, rather _____

9D To make a steel magnetically softer, processing includes __+−__.

(a) annealing   (b) grain growth   (c) decarburization   (d) quenching
(e) None of these, rather _____

9E In magnetite, a ferrite found in nature, the iron ions are in __**__ sites and __**__ sites. The iron ions in the two sites have __**__ magnetic orientation.

9F Since the net magnetic moment of a unit cell of $[NiFe_2O_4]_8$ is $\sim 150 \times 10^{-24}$ A·m^2, there are __**__ unpaired electrons per unit cell. Since the magnetization of this compound is 260,000 A/m, we can back-calculate $a =$ __**__ nm for the size of the cubic unit cell.

9G The permeability of a magnet is the ratio of __**__ and __**__. In selecting hard and soft magnets, permeability receives more attention in __**__ magnets.

9H Dielectric strength can be reduced by __**__, by __**__, and by __**__.

9I Polarization is a measure of __+−__.

(a) dielectric constant per unit volume   (b) voltage gradient to produce electrical breakdown   (c) product of charge and distance
(d) None of the above, rather _____

9J The dielectric constant is increased by __**__, __**__, and/or __**__ polarization.

9K   Relative dielectric constants are always greater than ___**___ ; in general, they have ___**___ values at higher frequencies.

9L   With dc, the highest dielectric constant for a polymer is just above the ___**___ . At low temperatures, ___**___ polarization is the principal contributor to the dielectric constant.

• **9M**   The polarization of a ___**___ material can be changed by a stress that introduces an elastic ___**___ .

• **9N**   The interrelationship between polarization and strain provides a basis for ___**___ devices that convert mechanical energy to electrical, and vice versa.

9O   A lower index of refraction accompanies ___+−___ .

(a) slower light velocity   (b) high electronic polarization
(c) greater light absorption   (d) None of the above, rather _____

9P   Optical fibers for communication are produced to have a glass with a low ___**___ at the surface.

STUDY PROBLEMS

9–1.1   A ceramic magnetic material *nickel ferrite* has eight $[NiFe_2O_4]$'s per unit cell, which is cubic with $a = 0.834$ nm. Assume all the unit cells have the same magnetic orientation; what is the saturation magnetization? (Each Ni^{2+} ion has two Bohr magnetons.)

Answer: 250,000 amp/m

9–1.2   A magnetic ferrite has a unit cell with 32 oxygen ions, 16 ferric ions, and 8 divalent ions. If the divalent ions are Zn^{2+} and Ni^{2+} in a $3:5$ ratio, what weight fraction of ZnO, NiO, and Fe_2O_3 must be mixed together for the product?

9–1.3   Refer to Study Problem 9–1.2. In this particular magnetic ferrite, the Zn^{2+} ions preferentially choose 4-f sites and force a corresponding number of Fe^{3+} ions into 6-f sites. How many unbalanced electron spins does this produce per unit cell of $(Zn_3Ni_5)Fe_{16}O_{32}$?

Answer: Net 40 ↑

9–1.4   Refer to Example 9–1.2. Assume that domains have their N–S polarity reversed 180° in 40% of the unit cells after the external field has been decreased. How much magnetization remains?

9–2.1   Two capacitor plates (30 mm × 20 mm each) are parallel and 2.2 mm apart, with nothing between them. What voltage is required to produce a charge of 0.24×10^{-10} coul on the electrodes?

Answer: 10 V

9–2.2   What is the electron density on the electrodes of Study Problem 9–2.1?

9–2.3   The space between the plates of Study Problem 9–2.1 is filled with a 2.2-mm thick sheet of polystyrene. With 10 volts, the charge is increased from 0.24×10^{-10}

coul to 0.6×10^{-10} coul. What is the relative dielectric constant of the polystyrene?

Answer: 2.5

9–2.4 The dielectric constant of a glass ribbon is 6.1. Would a capacitor using such a glass ribbon 0.25 mm thick have greater or less capacitance than another capacitor having the same area, but using a 100-μm plastic with a dielectric constant of 2.1?

9–2.5 The relative dielectric constant κ of an isoprene rubber at 0°C is 3.0. What voltage difference (dc) is required to develop a charge density of 10^{-5} C/m^2 in a metal coating covering a 0.1 mm film?

Answer: 37.7 V

9–2.6 The polarization of the rubber in Study Problem 9–2.5 contributed how many electrons per mm^2 in the metal coating?

9–2.7 (a) What is the polarization in the polyvinyl chloride of Example 9–2.2 at 100 Hz when 100 volts is applied to the capacitor? (b) What will the electron density be on the capacitor plate?

Answer: (a) 0.0002 C/m^2 (b) 1.4×10^{15} electrons/m^2

9–2.8 A plate capacitor must have a capacitance of 0.25 μf. What should its area be if the 0.0005-in. (0.013-mm) mylar film that is used as a spacer has a dielectric constant of 3.0? (Cf. procedure in Example 9–2.2.)

• **9–3.1** (a) What dimensional change (%) is realized in a piezoelectric quartz crystal (E = 300,000 MPa) when an electric field changes the polarization from 35.1 C/m^2 to 35.4 C/m^2? (b) What stress would be required to produce the same strain?

Answer: (a) 0.85% (b) 2600 MPa

• **9–3.2** A piezoelectric crystal has a Young's modulus of 130,000 MPa (19,000,000 psi). What stress must be applied to reduce its polarization from 560 to 557 C/m^2?

• **9–3.3** Refer to Example 9–3.2. A 10-mm cube of $BaTiO_3$ is compressed 1%. The two ends receiving pressure are connected electrically. How many electrons travel from the negative end to the positive end?

Answer: 10^{12} electrons

• **9–3.4** Refer to Example 9–3.2. What is the distance between the centers of positive and negative charges of each unit cell?

• **9–3.5** Potassium niobate ($KNbO_3$) has the same structure as $BaTiO_3$, but with K^+ replacing Ba^{2+}, and Nb^{5+} replacing Ti^{4+}. Assume that the dimensions of the unit cell of $KNbO_3$ are approximately the same as those given for $BaTiO_3$ in Fig. 9–3.1. (a) What is the distance between the center of positive charges and the center of negative charges? (b) What is the dipole moment, $p = Qd$, of the unit cell? (c) What polarization is possible?

Answer: (a) 0.0117 nm (b) 11.2×10^{-30} C/m^2 (c) 0.175 C/m^2

QUIZ CHECKS

9A hard
    coersive field

9B 4, 5
    3

9C boundary, (a), (b), (c), (d)

9D (a), (b), (c)

9E 4-f, 6-f
    opposite

9F 16
    0.83 nm

9G induction, magnetic field
    soft

9H cracks, impurities, pores

9I (d): total dipole moment per
    volume, or excess charge density

9J electronic, molecular, ionic (also
    interfacial)

9K one
    lower

9L glass temperature
    electronic

9M piezoelectric, strain

9N transducer

9O (d): greater velocity
    • or less electronic polarization
    • or longer wavelength

9P index of refraction

Part Two

Structure, Properties, and Their Control in Multiphase Solids

The preceding chapters were concerned with the structure and properties of single-phase materials. Each had only a single basic structure to be related to its nature and behavior. In Part II, the next four chapters shall consider multiphase materials—materials that contain mixtures of two or more phases. This adds another dimension to the engineer's ability to control structure, since the phases that are present must perform in concert.

Chapter 10 introduces the equilibrium relationships between two phases. These are presented in the form of phase diagrams, from which we can determine (1) what phases exist, (2) what is the composition of each phase, and (3) what are the relative amounts of two phases. The answers can be very specific if we have allowed time for all reactions to be completed.

In Chapter 11, attention is given first to the microstructural geometry of phase mixtures, and then to the mechanics and kinetics of reactions within solids. These serve as bases for commercial heat treating processes, in which materials can be engineered to provide a variety of desired properties for service.

Chapter 12 introduces composites, which contain two or more materials that are unified into one. Thus, composites are multiphase. These engineered materials are receiving expanded attention to meet design requirements for new technical products.

The final chapter (13) of Part II focuses upon the performance of materials in service. Attention is given in turn (a) to fracture under mechanical loading, (b) to deterioration at elevated temperatures, (c) to radiation damage, and (d) to metallic corrosion.

Chapter Ten

Phase Diagrams

PREVIEW

A *phase diagram* is a collection of curves showing *solubility limits*. On one side of the curve is a single-phase unsaturated solution (liquid or solid). Beyond the curve, the solubility limit is exceeded. Therefore, a second phase must be present and a mixture of two phases exists.

We can learn to use the phase diagrams (1) to predict *what phases* are in equilibrium for selected alloy compositions at desired temperatures; (2) to determine the *chemical composition* of each phase; and (3) to calculate the *quantity* of each phase that is present. Phase diagrams are powerful tools in the hands of scientists and engineers who design materials for specific applications, as well as others who must anticipate the stability of specific materials when designing products for service environments.

Throughout this chapter, we shall assume that *equilibrium* is attained; that is, no further reaction is possible between the phases that are present.

CONTENTS

STUDY OBJECTIVES

1. To view a phase diagram as curves for solubility limits, and thus be able to conclude what phase(s) will be present for a specified alloy composition and at specified temperatures.

2. To determine the chemical composition of a phase at its limit of solubility.

3. To calculate by interpolation the relative amounts of two phases when the solubility limits are exceeded.

4. To take commercial alloys, such as brass, bronze, solder, sterling silver, and aluminum, under equilibrium conditions and predict phase relationships.

5. To be familiar with the structure and compositional characteristics of ferrite (α), austenite (γ), and carbide (Fe_3C).

6. To know the Fe–Fe_3C phase diagram sufficiently so that you can quickly visualize it in later discussions regarding the heat treatment of steels. This involves (a) the eutectoid temperature, composition, and reaction, and (b) the solubility limits of carbon in α and γ. (<1000°C and <1%C).

7. To handle quantitatively the phase changes encountered during austenite decomposition in either (a) plain-carbon steels (Fe–C) or (b) simple low-alloy steels (Fe–C–X). This requires a familiarity with the AISI–SAE nomenclature pattern.

10–1

QUALITATIVE PHASE RELATIONSHIPS

The three preceding chapters have been concerned with single-phase materials, and with the dependence of properties on their structure. Many useful engineering materials consist predominantly of one phase;* however, a greater number are *mixtures of phases*; for example, steel, solder, portland cement, grinding wheels, paints, and glass-reinforced plastics. The mixture of two or more phases in one material permits interaction among the phases, and the resulting properties are usually different from the properties of individual phases. It is possible, also, to modify these properties by changing either the shape or the distribution of the phases (see Chapter 11).

---

* See the end of Section 4–5 for our definition of phases.

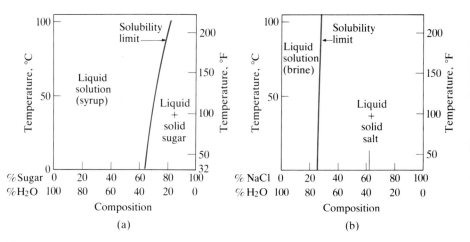

Figure 10–1.1 (a) Solubility of sugar in water. The limit of sugar solubility in water is shown by the solubility curve. Note that the sum of the sugar and water content at any point on the abscissa is 100%. (b) Solubility of NaCl in H_2O.

Solutions versus Mixtures

Different *components* can be combined into a single material by means of *solutions* or of *mixtures*.* Solid solutions have already been discussed in Sections 4–2, 5–1, and 7–4, and we are all familiar with liquid solutions. The composition of solutions can vary, because (1) one atom may be substituted for another in the phase structure, or (2) atoms may be placed in the interstices of the structure. The solute does not change the structural pattern of the solvent. A mixture, on the other hand, contains more than one phase (structural pattern). Sand and water, rubber with a carbon filler, and tungsten carbide with a cobalt binder are three examples of mixtures. In each of these examples there are two different phases, each with its own atomic arrangement.

It is, of course, possible to have a mixture of two different solutions. For example, in a lead–tin solder, one phase is a solid solution in which tin has replaced some of the lead in the fcc structure, and the other phase has the structure of tin (body-centered tetragonal). At elevated temperatures, lead atoms may replace a limited number of tin atoms in the bct structure. Thus, just after an ordinary 60–40 solder (60% Sn–40% Pb) solidifies, it contains two structures, each a solid solution.

Liquid Solubility Limits

Figure 10–1.1(a) shows the *solubility limit* of ordinary sugar in water; the curve is a *solubility curve*. All compositions shown to the left of the curve will form only one phase, because all the sugar can be dissolved in the liquid phase—*syrup*. With the higher percentages of sugar shown to the right of the curve, however, it is impossible to dissolve the sugar completely, with the result that

---

* A *solution* is a phase with more than one component; a *mixture* is a material with more than one phase.

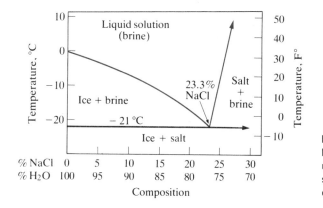

Figure 10–1.2 Solubility of NaCl salt in brine (right upward-sloping line) and solubility of ice in brine (left curve).

we have a mixture of two phases, solid sugar and liquid syrup. This example shows the change of solubility with temperature and also demonstrates a simple method for plotting temperature (or any other variable) as a function of composition. From left to right, the abscissa of Fig. 10–1.1(a) indicates the percentage of sugar. The percentage of water may be read directly from right to left, since the total of the components must, of course, equal 100%.

Figure 10–1.1(b) shows the solubility limit of NaCl in water. As in Fig. 10–1.1(a), there is a region that contains a liquid solution; this particular liquid is called a *brine*. There is a solubility limit that increases with temperature. Beyond the solubility limit there is a region of liquid plus solid salt. Figure 10–1.2 reveals additional features of the H_2O–NaCl *system*. Here the extremes of the abscissa are 100% H_2O (0% NaCl) and 30% NaCl (70% H_2O). Note from the figure that (1) the solubility limit of NaCl in a brine solution decreases with decreasing temperature, (2) the solubility limit of H_2O in a brine solution also decreases with decreasing temperature, and (3) intermediate compositions have melting temperatures lower than that of either pure ice (0°C or 32°F) or pure salt (800.4°C). Facts (1) and (3) are well known and fact (2), the less familiar limited solubility of ice in the aqueous liquid, can be verified by a simple experiment. A salt and water solution, e.g., sea water with 1.7% NaCl, can be cooled to less than 0°C (<32°F) and, according to Fig. 10–1.2, it will still be entirely liquid until minus 1°C (30.2°F) is reached. This is in agreement with observations of any arctic saline sea.* When such a salty liquid is cooled below -1°C, ice crystals will form and, because the solution cannot contain more than 98.3% H_2O at that temperature, these ice crystals must separate from the liquid. At minus 20°C (-4°F), the maximum H_2O content possible in a brine solution is 77% (23% NaCl) as can be verified by making a slush at that temperature and separating the ice from this liquid; the ice will be nearly pure H_2O and the remaining liquid will be saltier (i.e., lower in H_2O) than the original brine solution.

* There can be a slight variation if the salt content is not exactly 1.7%.

Eutectic Temperatures and Compositions	The two solubility curves of Fig. 10–1.2 cross at −21°C (−6°F), and at 76.7 H_2O–23.3 NaCl. This is the lowest temperature at which the brine solution remains liquid. It is called the *eutectic temperature*. The composition of this low-melting liquid is the *eutectic composition*.

Observe in Fig. 10–1.2 that the brine coexists with a solid phase below each solubility curve. On the H_2O-rich side, the mixture is ice and brine; on the other side of the eutectic, the mixture is solid NaCl and brine. A 2-phase mixture of solid H_2O (ice) and solid NaCl (salt) exists below the eutectic temperature.

Eutectic Reactions

A brine of eutectic composition (~77 H_2O–23 NaCl) changes from a single liquid solution into two solid phases as it is cooled through the eutectic temperature:

$$\text{Liquid (23\% NaCl)} \xrightarrow[\text{Cooling}]{-21°C} \text{Ice (0\% NaCl)} + \text{Salt (100\% NaCl).} \qquad (10\text{–}1.1)$$

Heating reverses the reaction. This eutectic reaction may be generalized to

$$L_2 \underset{\text{Heating}}{\overset{\text{Cooling}}{\rightleftharpoons}} S_1 + S_3, \qquad (10\text{–}1.2)$$

where the subscripts imply progressively higher amounts of one of the components.

Eutectic Alloys

It was discovered early in historic times that low-melting alloys could be made by selecting appropriate mixtures of two or more metals. One such alloy that is most familiar to the reader is an ordinary 60–40 *solder* (60 Sn–40 Pb). As seen in Fig. 10–1.3, this alloy closely approximates the eutectic composition that melts at 183°C. This solder has wide use in electrical circuitry because metallic connections may be made with a minimum of heating. If the solder contained more lead (say, 80 Pb–20 Sn), the liquid would become saturated with lead at 280°C during cooling. There would then be a 100°C range (280°C to 183°C) in which there would be a mixture of liquid and solid.

Solid Solubility Limits

Figure 10–1.3 differs from Fig. 10–1.2 in that the lead-rich solid can dissolve tin atoms into its structure. (Ice cannot dissolve measurable amounts of NaCl into its crystalline structure.) The amount varies from 29 a/o tin (19 w/o) in the lead-rich solid at 183°C to 8 a/o (5 w/o) tin at 300°C. Also, the tin-rich solid may contain 2.5 w/o (1.5 a/o) lead atoms at 183°C. By convention, these two solid phases have been labeled α and β, respectively.

The solid solubility limit has a maximum at the eutectic temperature. Both above and below 183°C, the amount of tin that is soluble in the lead-rich, fcc α decreases. Likewise, a maximum of 2.5 w/o lead is soluble in the tin-rich, bct β at 183°C.

Atomic percent tin, a/o

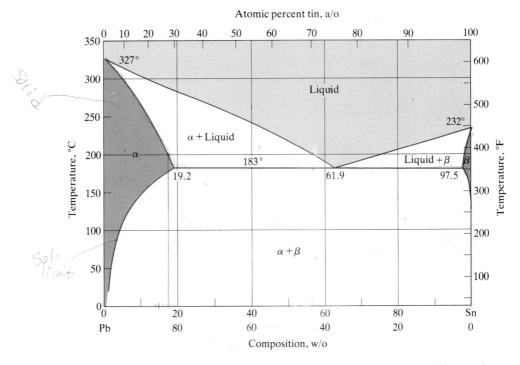

Figure 10–1.3   Pb–Sn diagram. Such a diagram indicates the phase compositions and permits the calculation of phase quantities for any lead–tin mixture at any temperature. (Adapted from *Metals Handbook*, American Society for Metals.)

Phase Names and/or Labels

In Chapters 4 and 5, we gave names to certain solid-solution phases. *Brass* is an fcc phase of zinc in copper, and *sterling silver* is a solid-solution phase of copper in fcc silver. Later we will give the name *ferrite* to the phase of bcc iron when it contains solutes; likewise, *austenite* is the common name for fcc iron when it contains added elements.

Greek-letter labels are used more commonly than names, for simplicity's sake. Thus, α and β are *phases* in the Pb–Sn *system*, while lead and tin are *components* of the Pb–Sn system.* Although the α phase has the structure of pure lead, α is not necessarily pure lead.

Example 10–1.1

According to Fig. 10–1.1(a), a syrup may contain only 67% sugar (33% water) at 20°C (68°F), but 83% sugar at 100°C (212°F). One hundred grams of sugar and 25 grams of water are mixed and boiled until all the sugar is dissolved. During cooling, the solubility limit is exceeded, so that (with time) excess sugar separates from the syrup. If equilibrium is attained, what is the weight ratio of syrup to excess sugar at 20°C?

---

* Admittedly, the label α appears in almost all systems as the first-named phase. However, this does not prove to be a problem. (The reader is aware that x is repeatedly used in algebraic calculations without special labeling confusion.)

Procedure. Basis: 100 g sugar + 25 g H_2O = 80% sugar + 20% water. All of the sugar is dissolved at first, to give a single phase of syrup (Fig. 10–1.1a). When the syrup cools, the *solubility limit* for the 80–20 sugar–water composition is reached at 87°C (189°F). Below this temperature, some of the sugar is rejected (precipitated) from solution.

Calculation. At 20°C, there are x g of syrup and $(125 - x)$ g of excess sugar.

Sugar balance: $0.67x + (1.0)(125 - x) = 100$ g

$$x = 75.75 \text{ g syrup.}$$

Or H_2O balance: $0.33x + (0)(125 - x) = 25$ g

$$x = 75.75 \text{ g syrup.}$$

Excess sugar: $125 - x = 49.25$ g.

Ratio: Syrup/excess = 75.75 g/49.25 g = 1.54.

Comments. Typically, some *supercooling* is encountered before the excess phase (in this case, sugar) starts to separate. In fact, in this case, supercooling can proceed to room temperature, so that the start of separation may be delayed considerably. Similar supersaturation commonly occurs in metals and other materials. ◀

Example 10–1.2

A brine contains 9% NaCl (91% H_2O by weight). How many grams of water (per 100 g brine) must be evaporated before the solution becomes saturated at 50°C (122°F)?

Solution. Basis: 100 g brine = 9 g NaCl + 91 g H_2O. From Fig. 10–1.1(b), solubility of NaCl in brine = 27% at 50°C.

Grams saturated brine = 9 g/0.27 = x/0.73;

$$x = 24.3 \text{ g } H_2O \text{ for saturation;}$$

$$91 \text{ g} - 24.3 \text{ g} = 66.7 \text{ g to be evaporated.}$$

Alternatively, 9 g of the salt is equal to 27% of the brine remaining after evaporation:

$$(0.09)(100 \text{ g}) = 0.27(100 - y);$$

$$y = 66.7 \text{ g to be evaporated.}$$

Comment. By convention, compositions of liquids and solids are reported in weight percent, unless stated otherwise. ◀

Example 10–1.3

A brine with 40 g H_2O and 10 g NaCl is cooled to −10°C (14°F). (a) This brine is placed in a beaker. How much salt can be dissolved in it? (b) An identical 40–10 brine is placed in a second beaker. How much ice can be added to it without exceeding the solubility limit?

Procedure. From Fig. 10–1.2, the brine can range from 13% to 25% NaCl (or from 75% to 87% H_2O) at −10°C. In (a), calculate the total salt (= 10 g + x g); in (b), calculate the total H_2O (= 40 g + y g).

Calculations

a)   g NaCl = 0.25(50 + x) = 10 + x;

$$x = 3.33 \text{ g NaCl added.}$$

b) $\qquad$ g $H_2O = 0.87(50 + y) = 40 + y$;

$$y = 26.9 \text{ g } H_2O \text{ added.} \blacktriangleleft$$

Example 10-1.4

Salt is spread on the street when the temperature is $-10°C$ (14°F). Part of the ice melts, and the brine that forms becomes saturated with H_2O. Ninety grams of this brine is splattered over your car. How many grams of NaCl come with it?

Solution. From Fig. 10-1.2, the solubility limit at $-10°C$ is 87% H_2O in brine. The balance is NaCl (13%).

$$0.13(90 \text{ g}) = 11.7 \text{ g NaCl.}$$

Comments. It is recommended that the reader obtain a short, transparent millimeter rule. This will permit a more accurate reading of the solubility limits. Many phase diagrams are drawn to an accuracy of $\pm 1\%$ on a larger scale and then photo-reduced. Thus, your accuracy of calculation is generally limited by your reading of the graphs.

$\blacktriangleleft$

10-2

PHASE DIAGRAMS (EQUILIBRIUM DIAGRAMS)

Figure 10-1.3 is the *phase diagram* for the lead–tin system. This diagram can be used as a "map" from which the phases present at any particular temperature or composition can be read if the alloy is at *equilibrium,* i.e., if all possible reaction has been completed.

For example, at 50% tin and 100°C, the phase diagram indicates two solid phases: α is a lead-rich solid solution with some dissolved tin; β is almost pure tin with very little dissolved lead. At 200°C, an alloy of 10% tin and 90% lead lies in an area that is entirely α phase. It is a solid solution of lead with some tin dissolved in it. At the same temperature, but for 30% tin and 70% lead, the phase diagram indicates a mixture of two phases—liquid and the α solid solution; if this latter alloy were heated to a temperature of 300°C, it would become all liquid.

The phase fields in equilibrium diagrams, of course, depend on the particular alloy systems being depicted. When copper and nickel are mixed, the phase diagram is as shown in Fig. 10-2.1. This phase diagram is comparatively simple, since only two phases are present. In the lower part of the diagram, all alloys form only one solid solution and therefore only one crystal structure. Both the nickel and the copper have face-centered cubic structures. Since the atoms of each are nearly the same size, it is possible for nickel and copper atoms to replace each other in the crystal structure in any proportion at 1000°C. When an alloy containing 60% copper and 40% nickel is heated, the solid phase exists until the temperature of about 1235°C (2255°F) is reached. Above this temperature and up to 1280°C (2336°F) the solid and liquid solutions coexist. Above 1280°C only a liquid phase remains.

Freezing Ranges

As shown in the foregoing phase diagrams, the range of temperatures over which freezing occurs varies with the composition of the alloy. This situation influences the plumber, for example, to select a high-lead alloy as a "wiping"

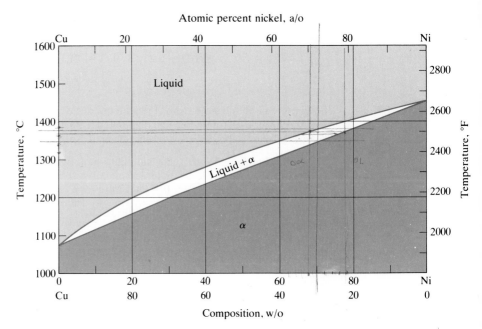

Figure 10–2.1   Cu–Ni diagram. All solid alloys contain only one phase. This phase is fcc. (Adapted from *Metals Handbook*, American Society for Metals.)

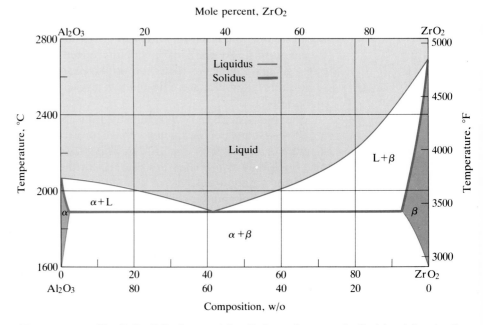

Figure 10–2.2   The Al_2O_3–ZrO_2 diagram. Like all phase diagrams, the liquidus delineates the lowest temperatures for solely liquid. The solidus is the upper limit for completely solid. The freezing range of solid + liquid lies between the two. The two meet at the eutectic and where single phases melt. (Adapted from Alper et al., Amer. Ceram. Soc.)

solder when a solder is needed that will not freeze completely at one temperature. If one chooses an 80 Pb–20 Sn solder, the freezing range is from 280°C to 183°C as compared with only 190°C to 183°C for a 60 Sn–40 Pb solder.

The terms *liquidus*, the locus of temperatures above which all compositions are liquid, and *solidus*, the locus of temperatures below which all compositions are solid, are applied in this connection. Every phase diagram for two or more components must show a liquidus and a solidus, and an intervening freezing range. Whether the components are metals or nonmetals (Fig. 10–2.2), there are certain locations on the phase diagram where the liquidus and solidus meet. For a pure component, this point lies at the edge of the diagram. When it is heated, a pure material will remain solid until its melting point is reached and will then change entirely to liquid before it can be raised to a higher temperature.

The solidus and liquidus must also meet at the eutectic. In Fig. 10–1.3, the liquid solder composed of 61.9% tin and 38.1% lead is entirely solid below the eutectic temperature and entirely liquid above it. At the eutectic temperature three phases can coexist [$(\alpha + \text{Liq} + \beta)$ in the Pb–Sn system].

Isothermal Cuts A traverse across the phase diagram at a constant temperature (*isotherm*) provides a simple sequence of alternating 1- and 2-phase fields. Consider the SiO_2–Al_2O_3 diagram of Fig. 10–2.3 at 1650°C (3000°F), and you will see that the

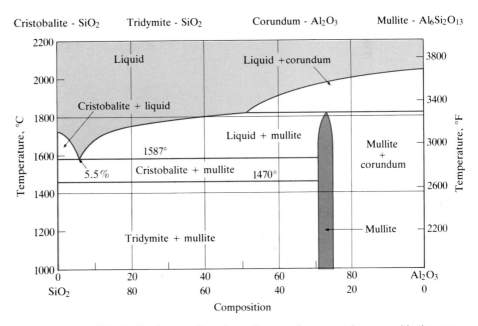

Figure 10–2.3 SiO_2–Al_2O_3 diagram. The phase diagrams for nonmetals are used in the same manner as those for metals. The only difference is the longer time required to establish equilibrium. (Adapted from Aksay and Pask, *Science*.)

sequence is 1–2–1–2–1–2–1. With pure SiO_2, only *one* phase exists (named cristobalite). It holds negligible amounts of Al_2O_3 in solid solution.* Therefore, a second phase (liquid) appears with the addition of Al_2O_3. The *two*-phase region contains cristobalite and liquid. Between 4% Al_2O_3 (96% SiO_2) and 8% Al_2O_3 (92% SiO_2), the liquid can dissolve all of the SiO_2 and Al_2O_3 that is present, so just *one* phase exists. Beyond 8% Al_2O_3 (<92% SiO_2), the solubility limit of the liquid for Al_2O_3 is exceeded, and solid mullite precipitates. The *two* phases liquid and mullite coexist.† The solid-solution range of mullite is from 71% Al_2O_3 (29% SiO_2) to 75% Al_2O_3 (25% SiO_2). Only *one* phase is stable in this range, because it can accommodate both the SiO_2 and Al_2O_3 that are present. A *two*-phase field, mullite and corundum (Al_2O_3) follows and extends to within a line's width of the right side of the phase diagram. With only Al_2O_3, this *one* phase is called corundum.

Example 10–2.1

Sterling silver, an alloy containing approximately 92.5% silver and 7.5% copper (Fig. 10–2.4), is heated slowly from room temperature to 1000°C (1830°F). What phase(s) will be present as heating progresses?

Answer

Room temperature at 740°C	$\alpha + \beta$
740°C to 810°C	Only α
810°C to 900°C	α + liquid
900°C to 1000°C	Only liquid ◀

Example 10–2.2

A combination of 90% SiO_2 and 10% Al_2O_3 is melted at 1800°C and then cooled extremely slowly to 1400°C. What phase(s) will be present in the cooling process?

Answer. (See Fig. 10–2.3.)

1800°C to 1700°C	Only liquid
1700°C to 1587°C	Liquid + mullite ($Al_6Si_2O_{13}$)
1587°C to 1470°C	Mullite + cristobalite (SiO_2)
<1470°C	Mullite + tridymite (SiO_2)

Comments. There must, of course, be three phases (liquid + mullite + cristobalite) at 1587°C as the material passes from the (Liq + Mul) field to the (Cri + Mul) field. Likewise, there are three phases at 1470°C.

The cooling will have to be extremely slow, because the process of changing the strong Si–O bonds from one structure to another is very slow.

Pure silica has three common polymorphs: cristobalite and tridymite at high temperatures, and quartz at low temperatures. ◀

Example 10–2.3

Refer to Fig. 10–2.4 for the Ag–Cu system. (a) Locate the liquidus and the solidus. (b) How many phases are present where the two meet?

---

* There is a slight solubility, but it is so low that we are not able to show it on the phase diagram.

† When 8% Al_2O_3 is just slightly exceeded there will be very little mullite. When the right side of this 2-phase field is approached, very little liquid remains.

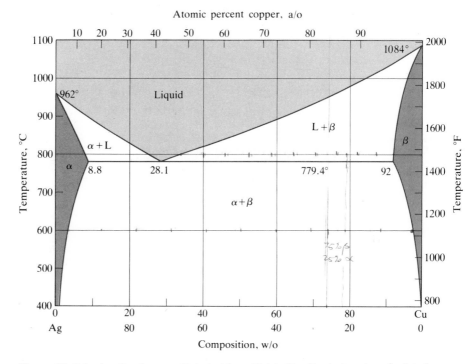

Atomic percent copper, a/o

Figure 10–2.4 Ag–Cu diagram. (Adapted from *Metals Handbook,* American Society for Metals.)

Answers

a) Liquidus: 962° at 100% Ag, to 779°C at 71.9% Ag (28.1% Cu), to 1084° at 100% Cu. Solidus: 962° at 100% Ag, to 779°C at 91.2% Ag (8.8% Cu), remaining at 779°C to 92% Cu (8% Ag), to 1084 at 100% Cu.

b) With a single component (only Ag *or* only Cu), two phases are present (solid + liquid) where the liquidus and solidus meet.

   With two components (Ag *and* Cu), three phases (α + liquid + β) are present where the liquidus and solidus meet (at the eutectic).

Comment. There are always two phases present in the temperature range between the liquidus and solidus of a 2-component, or binary, phase diagram.

Example 10–2.4   Refer to the Al–Mg diagram (Fig. 10–5.4) later in the chapter. Apply the 1–2–1–2– · · · rule (a) at 500°C; (b) at 200°C.

Answers

a)

	α	$(\alpha + L)$	Liq	$(L + \varepsilon)$	ε
% Mg	0	11	30	76	92.5	100
% Al	100	89	70	24	7.5	0

b) | α |(α + β) | β | (β + β') | β' | (β' + γ) | γ | (γ + ε) | ε |
  % Mg 0   3   35 37   ~41   ~42   49   ~56   96 100
         | | | | | | | | | |

Comment. The dashed lines on the phase diagram are the best estimates based on present information.   ◄

10-3

CHEMICAL COMPOSITIONS OF PHASES

In addition to serving as a "map," a phase diagram shows the chemical *compositions* of the phases that are present under conditions of equilibrium after all reaction has been completed. This information, along with information on the amount of each phase (Section 10–4), constitutes very useful data for the scientists and engineers who are involved with materials development, selection, and application in product design.

One-phase Areas

The determination of the chemical composition of a single phase is automatic. *It has the same composition as the alloy.* This is to be expected; since only liquid is present in a 60 Sn–40 Pb alloy at 225°C (Fig. 10–3.1), the liquid has to

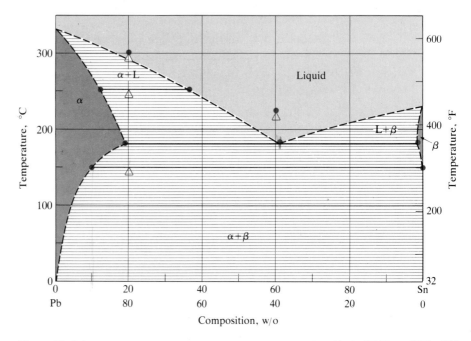

Figure 10–3.1 Compositions of phases (Pb–Sn alloys, Fig. 10–1.3). At 150°C, an 80Pb–20Sn alloy contains α and β. The chemical composition of α is dictated by the solubility curve. At this temperature, the solubility limit is 10% Sn (and 90% Pb) in the fcc α phase. See the text.

have the same 60–40 composition. This also holds when the location in the phase diagram involves a single-phase solid solution.

You will observe that we reported the chemical composition of the individual phase (and of the total alloy) in terms of the *components*—in this case, lead and tin.

Two-phase Areas

The determination of the chemical compositions of two phases can be handled on a rote basis. We will do this first; however, the rationale will follow in the next paragraph.

The chemical compositions of the two phases are located at the *two ends* of the *isotherm*, or *tie-line*, across the two-phase area. To illustrate, take an 80 Pb–20 Sn solder at 150°C. As indicated in Fig. 10–3.1, α has a chemical composition of 10 w/o tin (and therefore 90 w/o lead). The composition of the β is nearly 100 w/o tin. Other isotherms on Fig. 10-3.1 permit us to read the chemical compositions of the two phases of any Pb–Sn alloy at any temperature.

The basis for the above procedure is simply that the *solubility limit* for tin in α at 150°C is 10 w/o. Our alloy exceeds this with its composition of 20 w/o tin. Therefore, α is saturated with tin and the excess tin is present in β. Conversely, the solubility limit for lead in β is <1 w/o; therefore, almost all of the lead must be in a phase other than β—specifically, in α.

Three-phase Temperatures and Eutectic Reactions

A liquid that has the analysis of the eutectic composition (38.1 Pb–61.9 Sn, when we consider the Pb–Sn system) separates into two solid phases (α and β) at the eutectic temperature (183°C). Thus, at this temperature *only,* three phases can be in equilibrium. If this alloy is heated, the two solid phases of this solder melt into a one-phase liquid. We can write this reversible reaction as

$$\text{L(61.9\% Sn)} \underset{}{\overset{183°C}{\rightleftharpoons}} \alpha(19.2\% \text{ Sn}) + \beta(97.5\% \text{ Sn}). \qquad (10\text{–}3.1)$$

The tin analyses are shown for the three phases that are in equilibrium at 183°C.*

The reaction of Eq. (10–3.1) is called an *eutectic reaction* and involves a liquid and two solids. The more general form (presented earlier) is

$$\text{L}_2 \underset{\text{Heating}}{\overset{\text{Cooling}}{\rightleftharpoons}} \text{S}_1 + \text{S}_3, \qquad (10\text{–}1.2)$$

where the 1,2,3 subscripts refer to progressively increasing contents of one of the two components.

Study Aid (phase diagrams)

Some people find it easier to introduce phase diagrams with a eutectic-free system (Cu–Ni). This pattern is followed in Topic X of *Study Aids for Introductory Materials Courses*. This is then followed by a topic (XI) that presents

---

* We could have used 38.1% Pb, 80.8% Pb, and 2.5% Pb for L, α, and β, respectively, rather than the tin values.

eutectics. This study alternative is recommended for those who initially find phase diagrams confusing.

Example 10–3.1

Consider the Cu–Ni system (Fig. 10–2.1) and at 1300°C. (a) What is the solubility limit of copper in solid α? (b) Of nickel in the liquid? (c) What are the chemical compositions of the phases in a 45 Cu–55 Ni alloy at 1300°C?

Answer. (a) 42 w/o Cu   (b) 45 w/o Ni   (c) α: (42 Cu–58 Ni); L: (55 Cu–45 Ni).

Comment. It is the solubility limit that determines the chemical compositions of the phases in a two-phase area. ◄

Example 10–3.2

An alloy of 40 Ag–60 Cu (Fig. 10–2.4) is cooled slowly from 1000°C to room temperature. (a) What phase(s) will be present as cooling progresses? (b) Indicate their compositions. (c) Write the eutectic reaction.

Answer

a) and (b)

Temperature	α	Liquid	β
1000°C	—	40 Ag–60 Cu	—
800	—	66 Ag–34 Cu	8 Ag–92 Cu
780	91.2 Ag–8.8 Cu	71.9 Ag–28.1 Cu	8 Ag–92 Cu
600	96.5 Ag–3.5 Cu	—	2 Ag–98 Cu
400	99 Ag–1 Cu	—	Near 100 Cu
20 (extrapolated)	Near 100 Ag	—	Near 100 Cu

c)   Liquid (71.9% Ag) $\overset{780°C}{\rightleftarrows}$ α(91.2% Ag) + β(8% Ag).

Comments. The liquid becomes saturated with copper at about 890°C. The first β to separate at this temperature has the composition of 7 Ag–93 Cu. ◄

10–4

QUANTITIES OF PHASES

In addition to (1) identifying the stable, or equilibrium phases (Section 10–2) and (2) obtaining their chemical compositions from the phase diagrams (Section 10–3), we can (3) determine the *quantity* of each phase that is present under equilibrium conditions. This will be useful to us later when we consider the properties of multiphase materials in Chapter 11.

One-phase Areas

Again, as in Section 10–3, we have an automatic situation. With 225 g of an 80 Pb–20 Sn alloy at 300°C, only liquid is present, and its quantity is 225 g. It is equally valid to state that all (or 100%) of the alloy is liquid. In that manner we do not have to specify the exact weight of the alloy that is present.

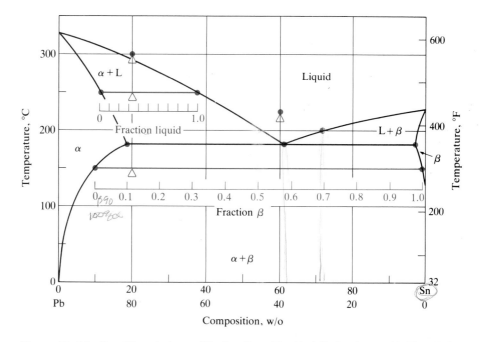

Figure 10–4.1   Quantities of phases (Pb–Sn alloys, Fig. 10–1.3). As observed in Fig. 10–3.1, an 80Pb–20Sn alloy contains α and β at 150°C. By interpolation, the fraction of β is 0.11. This same alloy contains 0.33 liquid at 250°C (and 0.67 α). See text.

Two-phase Areas

We can use a rote basis to indicate the quantities (or the *quantity fractions*) of the phases that are present in the two-phase areas of phase diagrams. We will do this first; the rationale will follow. The quantities of the two phases are determined by *interpolating the composition of the alloy along the tie-line between the compositions of the two phases.* To illustrate, we will again take an 80 Pb–20 Sn solder at 150°C. As indicated in Fig. 10–4.1, the chemical composition of the alloy (80 Pb–20 Sn) is at a position which is 0.11 of the distance between the chemical composition of α (90 Pb–10 Sn) at this temperature, and the chemical composition of β (<1 Pb and ~100 Sn). Therefore, of the total amount of solder, the quantity fraction of β is 0.11 (and 0.89 of α) at 150°C.* It is equally appropriate to report this as 89% α and 11% β.

As another example, consider the same alloy at 250°C. On the basis of Fig. 10–4.1, we have α (88.5 Pb–11.55 Sn) and L (63 Pb–37 Sn). The chemical composition of this alloy as a whole (80 Pb–20 Sn) is $\frac{1}{3}$ of the distance between the chemical composition of α and the chemical composition of the liquid. Therefore, of the total amount of solder at 250°C, the quantity fraction of liquid is $\frac{1}{3}$ and of α is $\frac{2}{3}$.

---

* Thus if we have 225 grams of solder, there would be ~200 grams of α and ~25 grams of β.

You will observe that we have reported these quantity figures in terms of *phases*—in these cases, α and β, or α and L. This is in contrast to the chemical compositions that were reported in terms of the *components*—Pb and Sn.

To follow the above procedure, we would need to have a stretchable scale, because the length of the isotherm between the two solubility limit curves continuously varies. Interpolation by using a millimeter scale offers a simple alternative.* The alert student will also be quick to suggest interpolating on the basis of the numbers on the abscissa of the graph.

Materials Balance

The rationale for the above interpolation procedure originates as a materials balance. By this we mean that the whole is equal to the sum of the parts. At 250°C, and with the above example, the lead in the total alloy is equal to the lead in the α plus the lead in liquid.

To illustrate, let us use 600 g of our same 80 Pb–20 Sn solder at 250°C. There are 480 g of Pb (and 120 g of Sn). We can consider that there are A grams of α (88.5 Pb–11.55 Sn) and L grams of liquid (63 Pb–37 Sn). Those compositions are located in Figs. 10–3.1 and 10–4.1. Of course

$$A + L = 600 \text{ g.}$$

Thus, on the basis of the chemical compositions of the two phases and the total alloy, the lead balance is

$$0.885A + 0.63L = 0.80(A + L), \tag{10–4.1}$$

or, in this case, $A = 2L$. This means that we have 200 g of liquid and 400 g of α—the same $\frac{1}{3}$ and $\frac{2}{3}$ values we calculated previously.

A generalization of Eq. (10–4.1) is

$$C_x(X) + C_y(Y) = C(X + Y), \tag{10–4.2}$$

where C_x and C_y are the chemical *compositions* of one of the components of phases x and y respectively; C is the composition of the total alloy. Likewise, X and Y are the *quantities* of the phases x and y. Algebraic rearrangement of Eq. (10–4.2) gives either

$$\frac{X}{X + Y} = \frac{C_y - C}{C_y - C_x}, \tag{10–4.3a}$$

or

$$\frac{Y}{X + Y} = \frac{C_x - C}{C_x - C_y}. \tag{10–4.3b}$$

The value of $X/(X + Y)$ is the quantity fraction of x; and, by like token, $Y/(X + Y)$ is the quantity fraction of y. Equation (10–4.3) is the *inverse lever*

---

* A small, transparent mm scale is recommended.

rule, which is commonly presented in other materials texts as a means for calculating the *quantity fractions of* the phases.*

● **The Three-phase Special Case**

Lead–tin solders have three phases when they are equilibrated at 183°C, which is the eutectic temperature. In this special case, we *cannot* calculate exactly the quantity fractions of α, β, and the liquid that are present. We can calculate that a 70 Pb–30 Sn alloy contains 0.86 α at 182°C (and 14% β). We also can determine that the same solder contains $\frac{1}{4}$ liquid (and 75% α) at 184°C. At the eutectic temperature, between 182 and 184°C, the β and some of the α react on heating to give a eutectic liquid. In the process, 1000 grams of solder would have 110 g of α (that is, 860 g − 750 g) and 140 g of β consumed to form 250 g of eutectic liquid (61.9 Pb–38.1 Sn). Therefore, with three phases at 183°C, we can only indicate that the quantity of α is between 860 and 750 grams; the quantity of β, between 140 and 0 grams; and of liquid, between 0 and 250 grams.†

Example 10–4.1

Consider a 90 Cu–10 Sn bronze. (a) What phases are present at 300°C? What are their chemical compositions? What is the fraction of each? (b) Repeat for 600°C. (c) At what temperature is there $\frac{1}{3}$ liquid? $\frac{2}{3}$ liquid?

Procedure.   Refer to Fig. 10–5.2, the Cu–Sn diagram. Read the phases and their compositions directly. Interpolate for the fractions, either with an mm scale or by using the abscissa values.

Answers

a) α: 93 Cu–7 Sn $(37–10)/(37–7) = 0.9$,
    ϵ: 63 Cu–37 Sn $(10–\ 7)/(37–7) = 0.1$.

b) α: 90 Cu–10 Sn   1.0.

c) By using an mm scale: $\frac{1}{3}$ L at 900°C, $\frac{2}{3}$ L at 965°C. ◀

Example 10–4.2

Using lead, make a materials balance that shows A, which is the quantity of α; and B, which is the quantity of β, in 75 grams of a 70 Pb–30 Sn solder at 182°C.

Solution.   From Eq. (10–4.2), and using the lead analyses of Fig. 10–1.3,

$$0.808A + 0.025B = 0.70(A + B).$$

$$A + B = 75 \text{ grams}.$$

---

* You may use this lever rule for calculating the quantity fraction of the phases, or, if simpler, interpolate according to Fig. 10–4.1. If you use the abscissa as a scale for interpolation, Eq. (10–4.3) is followed indirectly. In any event, observe that the chemical composition of the alloy is at the "center of gravity" between the compositions of the two contributing phases. This is the fulcrum of the lever; the majority phase is on the *shorter* arm of the lever.

† This special case of three phases for 2-component systems has an analog in 1-component systems. For example, with H_2O at 0°C (32°F), it is impossible to calculate the fraction that is ice and the fraction that is liquid water.

Solving simultaneously,

$$A = 64.7 \text{ g}, \quad B = 10.3 \text{ g}.$$

Comments. A materials balance using tin, $0.192A + 0.975B = 0.30(A + B)$, gives the identical answer.

The fractions of the two phases are:

$$A/(A + B) = 0.86; \quad B/(A + B) = 0.14. \quad ◀$$

Example 10–4.3

As a function of temperature, make a histogram of the quantity of the phases in 10.0 grams of a 40 Ag–60 Cu alloy (Fig. 10–2.4).

Procedure. Plot grams of β as dark bars, grams of α as medium bars, and grams of Liq as light bars. Use 50°C intervals.

Sample Calculations. (Shown on Fig. 10–4.2 as large dots.) The chemical compositions of the phases were previously determined for Example 10–3.2. At 890°C and above, all liquid; therefore, 0 grams of β. By interpolating,

At 800°C	$0.45(10 \text{ g}) = 4.5 \text{ g}$
780°+	$0.50(10) = 5.0$
780°−	$0.615(10) = 6.15$
600°	$0.60(10) = 6.0$
400°	$0.595(10) = 5.95$
20°	$0.60(10) = 6.0.$

Comment. Note the discontinuity in quantities at the eutectic temperature (780°C), which is the special case where three phases are present simultaneously. ◀

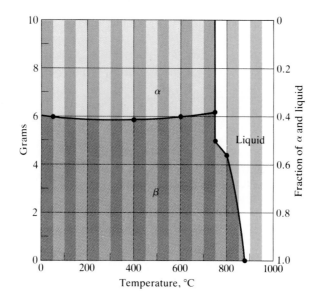

Figure 10–4.2 Quantity of phases (10 g of 40Ag–60Cu alloy). The quantity of β was determined by interpolation (Example 10–4.3).

β–dark bars
α–medium bars
L–light bars

The sums of the phases must equal 10 g.

10–5

COMMERCIAL ALLOYS AND CERAMICS

The reader may be relieved to learn that most commercial alloys have compositions that lie in the simpler parts of the phase diagrams. For example, the compositions of most brasses lie in the single-phase α area of Fig. 10–5.1. Likewise, the common bronzes contain less than 10% tin. There is little commercial interest in the more complex appearing central areas* of the Cu–Sn system (Fig. 10–5.2). In Chapter 11, we will consider alloys such as 95 Al–5 Cu, 90 Al–10 Mg, 90 Mg–10 Al (Figs. 10–5.3 and 10–5.4) because each forms one phase at elevated temperatures but crosses a solubility limit curve during cooling. By controlling the rate of separation of the second phase, the engineer

Figure 10–5.1 Cu–Zn diagram. (Adapted from *Metals Handbook*, American Society for Metals.)

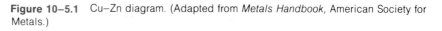

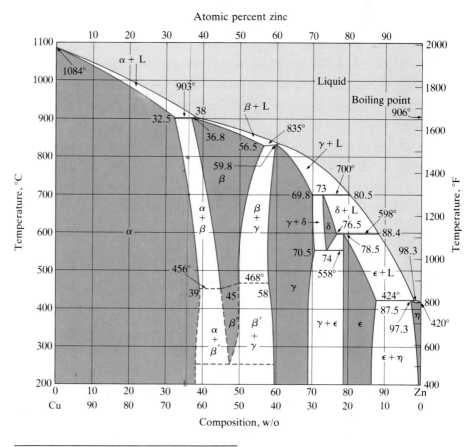

---

* Although more complex in appearance, all areas are either 1-phase or 2-phase. We have already considered these. Thus, for any alloy we can determine (1) *what* phase(s), (2) their chemical *compositions,* and (3) the *quantity* of each phase.

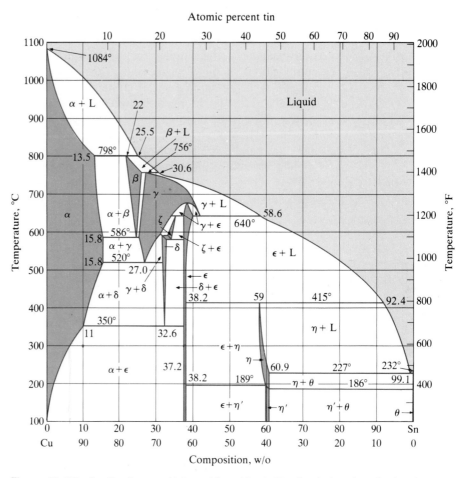

Figure 10–5.2 Cu–Sn diagram. (Adapted from *Metals Handbook*, American Society for Metals.)

can greatly increase the strength of the alloy. This is of major interest (Chapter 11). The Al–Si system (Fig. 10–5.8) reveals the commercial basis for purifying semiconducting and related materials (Example 10–5.1).

In the next two sections we shall examine the Fe–Fe$_3$C phase diagram for two reasons. First, steels are iron–carbon alloys that are very prominent in any technical civilization. Secondly, steels can serve as a prototype for heat-treating procedures. Suitable heat treatments and desired microstructures (and therefore desired properties) can be designed efficiently only with a knowledge of the phase diagram.

Ceramists utilize phase diagrams as do metallurgists.* However, within this text, we include only five: Al$_2$O$_3$–ZrO$_2$ (Fig. 10–2.2), Al$_2$O$_3$–SiO$_2$ (Fig. 10–

---

* See, for example, the 5-volume compendium, *Phase Diagrams for Ceramists,* American Ceramic Society.

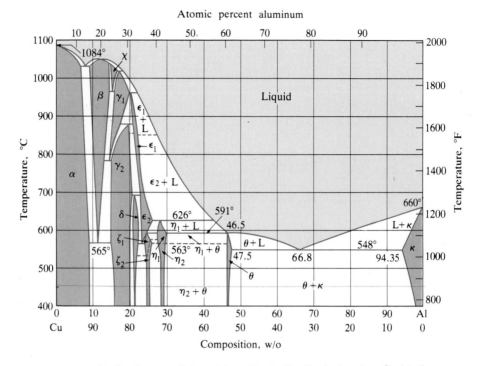

Figure 10–5.3 Al–Cu diagram. (Adapted from *Metals Handbook*, American Society for Metals.)

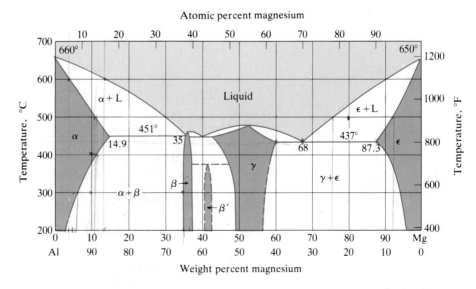

Figure 10–5.4 Al–Mg diagram. (Adapted from *Metals Handbook*, American Society for Metals.)

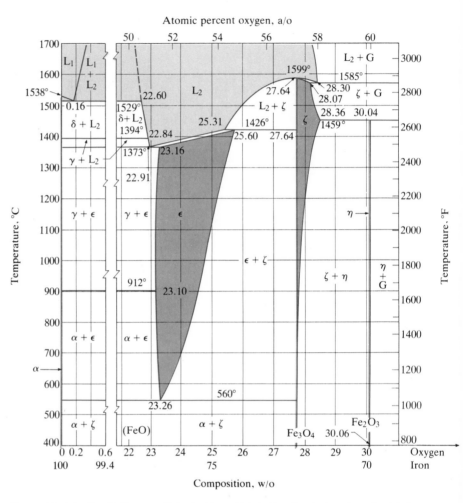

Figure 10–5.5 Fe–O diagram. (Adapted from *Metals Handbook,* American Society for Metals.)

2.3), Fe–O (Fig. 10–5.5), FeO–MgO (Fig. 10–5.6), and BaTiO$_3$–CaTiO$_3$ (Fig. 10–5.7).* The first involves very refractory, i.e., high-melting, oxides that have a eutectic temperature just short of 2000°C. The second, Al$_2$O$_3$–SiO$_2$, is pertinent to clay-base ceramics since the better grade clays are approximately 40 Al$_2$O$_3$–60 SiO$_2$ after processing. The Fe–O diagram shows a nonstoichiometric range of Fe$_{1-x}$O (ϵ of Fig. 10–5.5) that was discussed in Chapter 4. The FeO–

---

* Commonly, ceramic products involve three or more components, for example, MgO–Al$_2$O$_3$–SiO$_2$ and K$_2$O–Al$_2$O$_3$–SiO$_2$ in electrical porcelains. Although important, this text will not consider the more complex ternary phase diagrams.

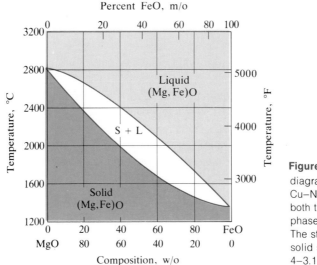

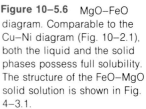

Figure 10–5.6 MgO–FeO diagram. Comparable to the Cu–Ni diagram (Fig. 10–2.1), both the liquid and the solid phases possess full solubility. The structure of the FeO–MgO solid solution is shown in Fig. 4–3.1.

MgO diagram (Fig. 10–5.6) reveals a complete solid solution series below the solidus temperatures. This compares directly with the Cu–Ni system in Fig. 10–2.1. The BaTiO$_3$–CaTiO$_3$ phase diagram (Fig. 10–5.7) shows how the addition of CaTiO$_3$ affects the temperature for the change of tetragonal BaTiO$_3$ (α) to cubic BaTiO$_3$ (β) (cf. Figs. 7–4.1 and 9–3.1).

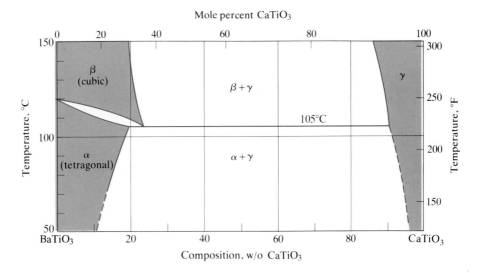

Figure 10–5.7 BaTiO$_3$–CaTiO$_3$ diagram. (Adapted from DeVries and Roy, American Ceramic Society.)

• **Example 10–5.1**

The Al–Si diagram (Fig. 10–5.8) has the solubility limit exaggerated for Al in β because, at the eutectic temperature (577°C), β will hold a maximum of only 0.17 w/o Al (and 99.83% Si).

a) Assume a linear solid solubility curve; what is the aluminum analysis of β at 1300°C?

b) An alloy containing 98 Si–2 Al is equilibrated at 1300°C, and the liquid removed. What fraction of the aluminum is removed?

c) The remaining solid is remelted at 1450°C. At what temperature will the new solid start to form during cooling?

d) What is its chemical analysis?

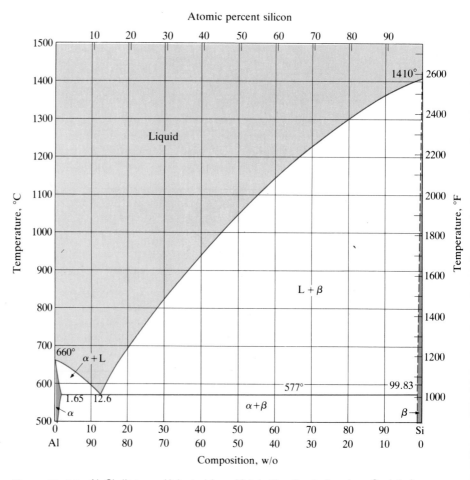

Figure 10–5.8 Al–Si diagram. (Adapted from *Metals Handbook*, American Society for Metals.)

Procedure and calculations

a) Being linear, and using ratios,

$$\frac{0.17\%}{(1410 - 577°C)} = \frac{C_\beta}{(1410 - 1300°C)},$$

$$C_\beta = 0.023 \text{ w/o Al.}$$

b) From Fig. 10–5.8, $C_L = 21$ w/o Al. Using 100 g,

$$L = [100 \text{ g}][(2 - 0.023)/(21 - 0.023)] = 9.4 \text{ g Liq};$$

$$\beta = 90.6 \text{ g.}$$

Fraction aluminum removed:

$$\frac{(9.4 \text{ g})(0.21)}{(100 \text{ g})(0.02)} = 0.99.$$

c) The upper end of the liquidus curve of Fig. 10–5.8 is approximately linear. Therefore, by using ratios again,

$$\frac{21}{(1410 - 1300°C)} = \frac{0.023}{(1410 - T°C)},$$

$$T = 1410 - 0.124°C.$$

d) Repeating (a),

$$\frac{0.17\%}{(1410 - 577°C)} = \frac{C'_\beta}{0.124°C},$$

$$C'_\beta = 0.00003 \text{ w/o Al.}$$

Comment. By modifying this type of procedure into a process called *zone refining*, the metallurgist is able to produce silicon with $<1/10^9$ aluminum for controlled-purity semiconductor materials. ◀

● **Example 10–5.2**

An alloy (5.8 kg) contains 6.7 Ag–93.3 Cu. It is melted and then cooled to 950°C. The first solid (β) appeared at ~1050°C and contained ~2% Ag. At 950°C, the separating solid contains ~6% Ag; however, the cooling is sufficiently rapid so that the initial solid did not change composition. As a result the average β composition is ~4% Ag. (Diffusion is faster in the liquid, so its composition at 950°C is ~26% Ag.)

a) How much liquid is present?

b) Another step of cooling to 800°C permits solidification to continue in a comparable manner (and without changing the composition of the previous solid). What products are present at 800°C that were not present at 950°C?

c) Give the approximate amounts.

Procedure. (a) The alloy (6.7 Ag–93.3 Cu) contains liquid (26–74) and solid β that averages ~(4–96). Therefore, we use the average composition and interpolate. (b) From (a) we know that the average β separating between 950°C and 800°C will have >6% Ag and <8% Ag, the equilibrium composition. Assume 7 Ag–93 Cu.

Estimates

a) Liquid at 950°C: $\dfrac{(6.7 - \sim4)}{(26 - \sim4)}$ (5.8 kg) $\approx$ 0.7 kg   (and 5.1 kg β).

b) Liquid ($\sim$65% Ag); solid (6%–8% Ag).

c) The average composition of the second generation of β is $\sim$7% Ag. It solidified from the 0.7 kg liquid that contained $\sim$26% Ag.

    Liquid at 800°C: $\left(\dfrac{26 - \sim7}{65 - \sim7}\right)$ (0.7 kg) $\approx$ 0.2 kg liquid.

    Therefore,   added $\beta \approx$ 0.5 kg   (or 5.6 kg total β).

Comment. This is an example of segregation (see Fig. 11–4.1). It is normal in any solidification process unless opportunity is given for diffusion.* Of course, the cooling is continuous rather than the 2-step sequence of our illustration. ◀

10–6

PHASES OF THE IRON–CARBON SYSTEM

Steels, which are primarily alloys of iron and carbon, offer illustrations of the majority of reactions and microstructures available to the engineer for adjusting material properties. Also, the iron–carbon alloys are among the prominent structural engineering materials.

The versatility of the steels as engineering materials is evidenced by the many kinds of steel that are manufactured. At one extreme are the very soft steels used for deep-drawing applications such as automobile fenders and stove panels. At the other extreme are the hard and tough steels used for gears and bulldozer blades. Some steels must have abnormally high resistance to corrosion. Steels for such electrical purposes as transformer sheets must have special magnetic characteristics so that they may be magnetized and demagnetized many times each second with low power losses. Other steels must be completely nonmagnetic, for such applications as wrist watches and minesweepers. Phase diagrams can be used to help explain each characteristic described above.

Pure iron changes its crystal structure twice before it melts. As discussed in Section 3–4, iron changes from bcc to fcc at 912°C (1673°F). This change is reversed at 1394°C (2540°F) to form the bcc structure again. The bcc polymorph then remains stable until iron melts at 1538°C (2800°F).

Ferrite (or α-iron)

The structural modification of pure iron at room temperature is called either α-iron or *ferrite*. Ferrite is quite soft and ductile; in the purity that is encountered commercially, its tensile strength is less than 310 MPa (45,000 psi). It is a ferromagnetic material at temperatures under 770°C (1418°F). The density of ferrite is 7.88 Mg/m³ (=7.88 g/cm³).

---

* An introduction to "delayed equilibrium" is presented in *Study Aids for Introductory Materials Courses*. This is recommended for the reader who wants to look more closely at segregation, etc.

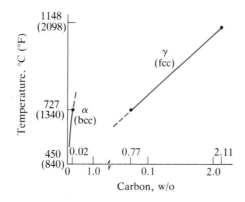

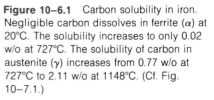

Figure 10–6.1   Carbon solubility in iron. Negligible carbon dissolves in ferrite (α) at 20°C. The solubility increases to only 0.02 w/o at 727°C. The solubility of carbon in austenite (γ) increases from 0.77 w/o at 727°C to 2.11 w/o at 1148°C. (Cf. Fig. 10–7.1.)

Since ferrite has a body-centered cubic structure, the spaces among atoms are small and pronouncedly oblate and cannot readily accommodate even a small spherical carbon atom. Therefore, solubility of carbon in ferrite is very low (<1 carbon per 1000 iron atoms). The carbon atom is too small for substitutional solid solution in ferrite, and too large for extensive interstitial solid solution (Section 4–2).

Austenite (or γ-iron)

The face-centered modification of iron is called *austenite,* or *γ-iron*. It is the stable form of pure iron at temperatures between 912°C and 1394°C. Making a direct comparison between the mechanical properties of austenite and ferrite is difficult because they must be compared at different temperatures. However, at its stable temperatures, austenite is soft and ductile and consequently is well suited to fabrication processes. Most steel forging and rolling operations are performed at 1100°C (2000°F) or above, when the iron is face-centered cubic. Austenite is not ferromagnetic at any temperature.

The face-centered cubic structure of iron has larger interatomic spacings than does ferrite. Even so, in the fcc structure the holes are barely large enough to crowd the carbon atoms into the interstices, and this crowding introduces strains into the structure. As a result, not all the holes can be filled at any one time (~6% at 912°C). The maximum solubility is only 2.11% (9 a/o) carbon (Fig. 10–6.1). By definition, steels contain less than 1.2% carbon; thus steels may have their carbon completely dissolved in austenite at high temperatures.

δ-iron

Above 1394°C (2540°F), austenite is no longer the most stable form of iron, since the crystal structure changes back to a body-centered-cubic phase called *δ-iron*. δ-iron is the same as α-iron except for its temperature range, and so it is commonly called δ-ferrite. The solubility of carbon in δ-ferrite is small, but it is measurably larger than in α-ferrite, because of the higher temperature.

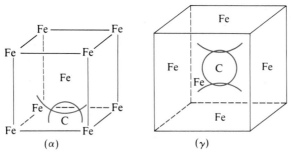

Figure 10–6.2 See Example 10–6.1. Although there is a more efficient iron packing factor in the fcc γ than in the bcc α (0.74 vs. 0.68), the interstices within γ are larger (and fewer). As a result, the carbon atoms are less crowded in the γ than in the α. This permits a higher solubility in the fcc structure (Fig. 10–6.1.)

Iron Carbide

In iron–carbon alloys, carbon in excess of the solubility limit must form a second phase, which is most commonly iron carbide (*cementite*).* Iron carbide has the chemical composition of Fe_3C. This does not mean that iron carbide forms discrete molecules of Fe_3C; it simply means that the crystal lattice contains iron and carbon atoms in a three-to-one ratio. Fe_3C has an orthorhombic unit cell with 12 iron atoms and 4 carbon atoms and thus has a carbon content of 6.7 weight percent. Its density is 7.6 Mg/m^3 (=7.6 g/cm^3).

As compared with austenite and ferrite, cementite is very hard. The presence of iron carbide with ferrite in steel greatly increases the strength of the steel (Section 11–1). However, because pure iron carbide is nonductile, it cannot adjust to stress concentrations; therefore it is relatively weak by itself. (Compare with nonductile ceramic materials in Section 7–6.)

Example 10–6.1

Carbon atoms sit in the $\frac{1}{2},\frac{1}{2},0$ positions in bcc iron.

a) Using $r_C = 0.077$ nm and $R_\alpha = 0.124$ nm, how must the nearest iron atoms be displaced to accommodate the carbon?

b) A carbon atom sits in the $\frac{1}{2},\frac{1}{2},\frac{1}{2}$ position in fcc iron. Using $R_\gamma = 0.127$ nm (Table 2–4.1) for the radius of γ-iron, how much crowding is present?

Procedure. Make a sketch for easier visualization (Fig. 10–6.2). In (α), the nearest iron atoms are at $\frac{1}{2},\frac{1}{2},\frac{1}{2}$ and $\frac{1}{2},\frac{1}{2},-\frac{1}{2}$. In (γ) the nearest iron atoms are at fcc positions.

Calculation

a)
$$a = 4(0.124 \text{ nm})/\sqrt{3} = 0.286 \text{ nm};$$

$$\Delta = (0.124 \text{ nm} + 0.077 \text{ nm}) - (0.286 \text{ nm})/2 = 0.06 \text{ nm}.$$

This increases the center-to-center distance of iron by $2(0.06)/0.286 \approx 40\%$.

b)
$$a = 4(0.127 \text{ nm})/\sqrt{2} = 0.359 \text{ nm};$$

$$\Delta = (0.127 \text{ nm} + 0.077 \text{ nm}) - (0.359 \text{ nm})/2 = 0.025 \text{ nm}.$$

This increases the center-to-center distance of iron by $2(0.025)/(0.359) \approx 14\%$.

---

* Graphite is the excess carbon phase in most cast irons where both silicon and carbon contents are high (Section 14–2).

Comments. Although bcc iron has a lower packing factor than fcc, the interstices do not accommodate a carbon atom without crowding. [The low packing factor (0.68) for bcc iron arises from six 6-f sites (related to $\frac{1}{2},\frac{1}{2},0$ and $0,0,\frac{1}{2}$ locations), plus twelve 4-f sites (related to $\frac{1}{2},\frac{1}{4},0$). The fcc iron (PF = 0.74) has only four 6-f sites and eight 4-f sites per unit cell (Fig. 7–2.2).] ◀

10–7

THE Fe–Fe₃C PHASE DIAGRAM

Figure 10–7.1 presents the phase diagram between iron and iron carbide, Fe₃C. This phase diagram is the basis for the heat-treating of the majority of our steels.

If you cover the 0%–1% carbon region of Fig. 10–7.1 by your hand, you will see a close resemblance to previous phase diagrams. The eutectic composition is at 4.3 w/o (17 a/o) carbon; the eutectic temperature is 1148°C (2100°F). *Cast irons* are based on this eutectic region, since they typically contain 2%–3.5% carbon (Section 14–1). They have relatively low melting points with certain desirable processing and mold-filling characteristics.

The iron-rich γ can dissolve up to 2.1 w/o (9 a/o) carbon. As discussed in the last section, the carbon atoms are dissolved interstitially in fcc iron. *Steels*

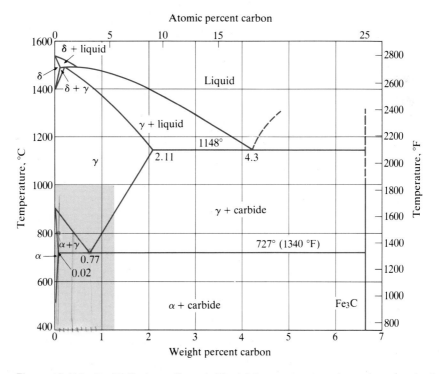

Figure 10–7.1 Fe–Fe₃C phase diagram. The left lower corner receives prime attention in heat treating steels (Fig. 10–7.3). (In calculations, 0.77% is commonly rounded to 0.8%.)

are based on this solid-solution phase. Since steels contain less than 1.2% carbon, they can be single-phase at forging and other hot-working temperatures, which are commonly in the region of 1100 to 1250°C (2000 to 2300°F), a production advantage.

In the iron-rich region (>99% Fe and <1% C), the Fe–Fe₃C diagram differs from any of the diagrams we have examined previously. This difference arises from the fact that iron is polymorphic with bcc and fcc phases. Since we are not studying steel melting and solidification in this text, we will skip the features of this low-carbon region that are found above 1400°C. We will pay considerable attention to the features of the diagram in the 700°C–900°C (1300°F–1650°F) temperature range and the 0%–1% carbon range, because it is here that the engineer can develop within the steel those microstructures that are required for the desired properties.

The Eutectoid Reaction

In Fig. 10–7.2, a comparison is made between the addition of carbon to austenite and the addition of common salt to water (Fig. 10–1.2). In each case, the addition of the solute lowers the stable temperature range of the solution. These two examples differ in only one respect: in the ice–salt system, a *liquid solution* exists above the eutectic temperature; in the iron–carbon system, a *solid solution* exists, so that a true eutectic reaction does not occur upon cooling. However, because of the analogy of this reaction to the eutectic reaction, it is called *eutectoid* (literally, eutectic-like).

$$\text{Eutectic:} \quad L_2 \underset{\text{Heating}}{\overset{\text{Cooling}}{\rightleftarrows}} S_1 + S_3; \qquad (10\text{–}1.2)$$

$$\text{Eutectoid:} \quad S_2 \underset{\text{Heating}}{\overset{\text{Cooling}}{\rightleftarrows}} S_1 + S_3. \qquad (10\text{–}7.1)$$

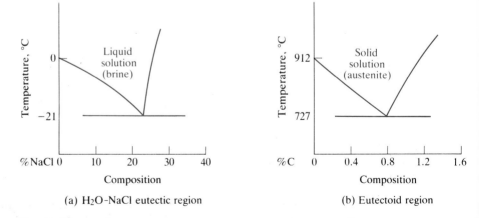

(a) H₂O-NaCl eutectic region   (b) Eutectoid region

Fig. 10–7.2 Eutectic and eutectic-like (eutectoid) regions of phase diagrams.

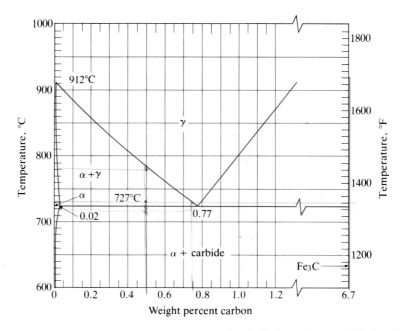

Figure 10–7.3   The eutectoid region of the Fe–Fe₃C phase diagram. (Cf. Fig. 10–7.1.) Steels with ~0.8% C are commonly called *eutectoid* steels. *Hypereutectoid* steels are above that value; *hypoeutectoid* steels, below.

The *eutectoid temperature* for iron–carbon alloys is 727°C (1340°F). The corresponding *eutectoid composition* is approximately 0.8% carbon. The specific eutectoid reaction for Fe–C alloys is

$$\gamma(0.77\%\ \text{C}) \overset{727°C}{\rightleftharpoons} \alpha(0.02\%\ \text{C}) + \text{Fe}_3\text{C}(6.7\%\ \text{C}). \qquad (10\text{–}7.2)$$

Figure 10–7.3 shows the eutectoid region in greater detail than does Fig. 10–7.1.

Study Aid (Fe–Fe₃C phase diagrams)   This short presentation in *Study Aids for Introductory Materials Courses* concisely outlines those phase relationships that are required to study steels.

Example 10–7.1   A eutectoid steel (~0.8% carbon) is heated to 800°C (1472°F) and cooled slowly through the eutectoid temperature (727°C). Calculate the number of grams of carbide that form per 100 g of steel.

Procedure. Interpolate between α (0.02% C) and Fe₃C (6.7% C) at 726°C (1339°F).

Calculation

$$\text{Carbide} = [(0.8 - 0.02)/(6.7 - 0.02)](100\ \text{g}) = 12\ \text{g} \qquad (\text{and}\ 88\ \text{g}\ \alpha).$$

Comment. Careful measurements indicate that the eutectoid composition is 0.77% carbon. However, the figure, 0.8% C, is commonly used.   ◀

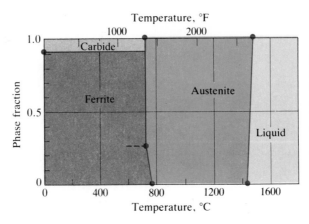

Figure 10–7.4 Phase fractions (99.4Fe–0.6C). (See Example 10–7.2.) Approximately three fourths of this steel is austenite at 728°C (1342°F). (Since the γ has a eutectoid composition at that temperature, we will observe in the next section that the steel will contain ~75% pearlite (and about one fourth proeutectoid ferrite).)

Example 10–7.2

Plot the fraction of ferrite, austenite, and carbide in an alloy of 0.60% carbon, 99.40% iron as a function of temperature.

Procedure. We could plot three separate curves for the three phases. However, since the total of the fractions at any given temperature is 1.00, modify the presentation of Fig. 10–4.2 into graph form.

Sample calculations (by interpolation)

At 728°C (1342°F),

$$\text{Ferrite} = \frac{0.77 - 0.60}{0.77 - 0.02} = 0.23.$$

At 726°C (1339°F),

$$\text{Ferrite} = \frac{6.7 - 0.60}{6.7 - 0.02} = 0.91.$$

Results are shown in Fig. 10–7.4 for other temperatures and for other phases. This form of presentation is called a *fraction chart*. ◀

10–8

AUSTENITE DECOMPOSITION

During cooling, the Fe–Fe₃C eutectoid reaction involves the simultaneous formation of ferrite α and carbide, $\overline{C}$, from the decomposition of austenite γ of eutectoid composition:

$$\gamma(\sim 0.8\% \text{ C}) \rightarrow \alpha + \overline{C}. \qquad (10\text{–}8.1)*$$

Nearly 12% carbide and more than 88% ferrite are in the resulting mixture. Since the carbide and ferrite form simultaneously, they are intimately mixed.

---

* We will use $\overline{C}$ to denote *carbide*, to avoid confusion with C for elemental carbon. For our calculations, it will be permissible to use 0.8% C as the approximate eutectoid composition.

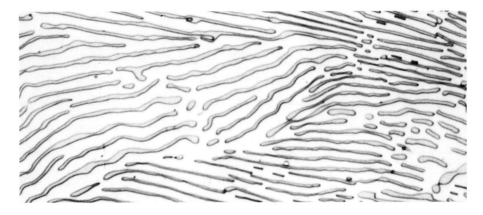

Figure 10–8.1 Pearlite (×2500). This microstructure is a lamellar mixture of ferrite (lighter matrix) and carbide (darker). Pearlite forms from austenite of eutectoid composition. Therefore the amount and composition of pearlite is the same as the amount and composition of eutectoid austenite. (Courtesy of U.S. Steel Corp.)

Characteristically, the mixture is *lamellar*; i.e., it is composed of alternate layers of ferrite and carbide (Fig. 10–8.1). The resulting microstructure, called *pearlite*, is very important in iron and steel technology, because it may be formed in almost all steels by means of suitable heat treatments.

Pearlite is a lamellar mixture of ferrite and carbide, formed by decomposing austenite of eutectoid composition. This distinction is important, since mixtures of ferrite and carbide may be formed by other reactions as well. However, the microstructure resulting from other reactions will not be lamellar (compare Figs. 10–8.1 and 11–1.4b) and consequently the properties of such mixtures will be different. (See Section 11–1.)

Since pearlite comes from austenite of eutectoid composition, the amount of pearlite present is equal to the amount of eutectoid austenite transformed (Fig. 10–8.2). We can determine this amount by calculating the fraction of γ just above the eutectoid temperature. Examples 10–8.1 and 10–8.2 demonstrate this.

Study Aid (austenite decomposition)

Since the reaction $\gamma \rightarrow \alpha + \overline{C}$ is central to the heat-treating of steel, it should be well understood by the reader as she or he proceeds to later sections of this text. The reader who needs further clarification on this solid reaction should consult Topic XV of *Study Aids for Introductory Materials Courses.*

Example 10–8.1

Determine the amount of pearlite in a 99.5% Fe–0.5% C alloy that is cooled slowly from 870°C (1600°F). Basis: 100 g of alloy.

Procedure. Since pearlite originates from austenite of eutectoid composition, determine the amount of γ just prior to the eutectoid reaction, i.e., at 727°C(+).

From 870 to 780°C: 100 g austenite with 0.5% C.

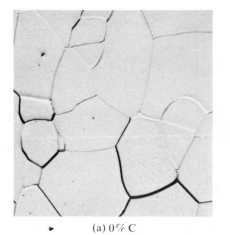

(a) 0% C

(b) 0.20% C

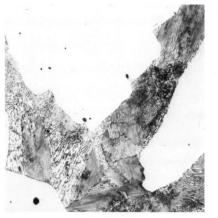

(c) 0.35% C

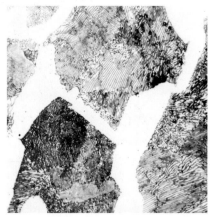

(d) 0.5% C

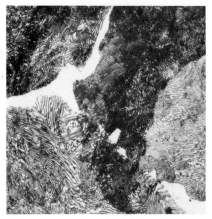

(e) 0.7% C

(f) 1.2% C

From 780 to 727°C(+):   ferrite separates from austenite and the carbon content of the austenite increases to ~0.8% C.

At 727°C(+):   composition of ferrite = 0.02% C } proeutectoid ferrite;
amount of ferrite = 38 g
composition of austenite ≈ 0.8% C } γ that transforms to pearlite.
amount of austenite = 62 g

Answer

At 727°C(−):   amount of pearlite = 62 g. (It came from, and replaced, the austenite with a eutectoid composition.)

Comments. Each of the above steps assumes sufficient time for equilibrium to be attained. Ferrite that formed above 727°C (i.e., before the eutectoid reaction) is called *proeutectoid ferrite*. That which is part of the pearlite, having been formed from austenite with the eutectoid composition, is called *eutectoid ferrite*. (See Fig. 10–8.2b–d.) ◀

Example 10–8.2

From the results of the previous example, determine the amount of ferrite and carbide present in the specified alloy (a) at 727°C(−), and (b) at room temperature. Basis: 100 g of alloy. (Some data come from Example 10–8.1.)

Procedure. There are two choices. (1) All of the carbide is in the 62 g of pearlite. Therefore, determine the amount of carbide that it contains. The balance is eutectoid ferrite (plus 38 g of proeutectoid ferrite). (2) By interpolation, determine the grams of carbide (or ferrite) in the 100 g of steel.

Calculation

a) At 727°C(−), and with the eutectoid at ~0.8% C:

Amount of carbide: $62 \dfrac{0.8 - 0.02}{6.7 - 0.02} = 7.2 \dfrac{\text{g carbide}}{100 \text{ g steel}}$;

Amount of ferrite: 62 − 7.2 = 54.8 g formed with the pearlite (eutectoid),
$\underline{38}$ g formed before the pearlite (proeutectoid),
92.8 g total/100 g steel.

Alternative calculation

Amount of carbide: $\dfrac{0.5 - 0.02}{6.7 - 0.02} = 7.2 \dfrac{\text{g carbide}}{100 \text{ g steel}}$;

Amount of ferrite: $\dfrac{6.7 - 0.5}{6.7 - 0.02} = 92.8 \dfrac{\text{g ferrite}}{100 \text{ g steel}}$.

◀ **Figure 10–8.2**   Annealed iron–carbon alloys (×500). The sequence (a) through (f) increases in carbon content; therefore there is a corresponding decrease in the amount of proeutectoid ferrite (white areas) that separated from the austenite prior to the eutectoid reaction. The gray areas are pearlite, a lamellar mixture of eutectoid ferrite and carbide. The network between the former grains of austenite in part (f) is proeutectoid carbide since the carbon content of that alloy is greater than 0.8 w/o. (Courtesy of U.S. Steel Corp.)

b) At room temperature (the solubility of carbon in ferrite at room temperature may be considered zero for these calculations),

$$\frac{0.5 - 0}{6.7 - 0} = 7.5 \ \frac{\text{g carbide}}{100 \text{ g steel}};$$

$$\frac{6.7 - 0.5}{6.7 - 0} = 92.5 \ \frac{\text{g ferrite}}{100 \text{ g steel}}.$$

Comments. The second procedure for part (a) gives the totals more directly but does not reveal the split between the two generations of ferrite, which give different properties to the steel.

As shown, additional carbide is precipitated from the ferrite below the eutectoid point because the solubility of carbon in ferrite decreases to nearly zero. This additional carbide is not part of the pearlite.

Each of these calculations assumes that equilibrium prevails. ◄

10–9

PLAIN-CARBON AND LOW-ALLOY STEELS

An important category of steels is designed for heat-treating into the γ range (*austenitizing*), followed by cooling and decomposition of the austenite either directly (or indirectly) to ferrite plus carbide ($\alpha + \overline{\text{C}}$). If a steel contains primarily iron and carbon,* the alloy is called a *plain-carbon steel*. We use the term *low-alloy steel* if we add up to five percent of alloying elements such as nickel, chromium, molybdenum, manganese, or silicon. Alloying elements are added primarily to reduce the decomposition rate of austenite to ($\alpha + \overline{\text{C}}$) during heat treatment. The result is a much harder steel, as we will see in Chapter 11.

The proeutectoid ferrite (Figs. 10–8.2b through e) deforms more readily than the ferrite within the pearlite since the former is more massive and is not reinforced by the harder, more rigid carbide. A metallurgist refers to *hypoeutectoid* steels (literally, lower carbon than the eutectoid composition) when speaking of steels with microstructures containing the separate ferrite areas. In a like manner, a *eutectoid steel* is predominantly pearlite if it is cooled slowly; and a *hypereutectoid steel* contains carbon *above* the eutectoid composition, thus producing proeutectoid carbide (Fig. 10–8.2f). Hypoeutectoid steels are more common than hypereutectoid steels.

Nomenclature for Steels

The importance of carbon in steel has made it desirable to indicate the carbon content in the identification scheme of steel types. The American Iron and Steel Institute (AISI) and the Society of Automotive Engineers (SAE) cooperated to establish a four-digit numbering code in which the last two digits designate the number of hundredths of percent of carbon content (Table 10–9.1). For example, a 1040 steel has 0.40% carbon (plus or minus a small workable range). The first two digits are simply code numbers that indicate the type of alloying

---

* Plus approximately 0.5% Mn for processing purposes.

TABLE 10–9.1

TABLE 10–9.1

Nomenclature for AISI and SAE Steels

AISI or SAE number	Composition	UNS
10xx	Plain-carbon steels*	G10xx0†
11xx	Plain-carbon (resulfurized for machinability)	G11xx0
15xx	Manganese (1.0–2.0%)	G15xx0
40xx	Molybdenum (0.20–0.30%)	G40xx0
41xx	Chromium (0.40–1.20%), molybdenum (0.08–0.25%)	G41xx0
43xx	Nickel (1.65–2.00%), chromium (0.40–0.90%) molybdenum (0.20–0.30%)	G43xx0
44xx	Molybdenum (0.5%)	G44xx0
46xx	Nickel (1.40–2.00%), molybdenum (0.15–0.30%)	G46xx0
48xx	Nickel (3.25–3.75%), molybdenum (0.20–0.30%)	G48xx0
51xx	Chromium (0.70–1.20%)	G51xx0
61xx	Chromium (0.70–1.10%), vanadium (0.10%)	G61xx0
81xx	Nickel (0.20–0.40%), chromium (0.30–0.55%), molybdenum (0.08–0.15%)	G81xx0
86xx	Nickel (0.30–0.70%), chromium (0.40–0.85%), molybdenum (0.08–0.25%)	G86xx0
87xx	Nickel (0.40–0.70%), chromium (0.40–0.60%), molybdenum (0.20–0.30%)	G87xx0
92xx	Silicon (1.80–2.20%)	G92xx0

xx: carbon content, 0.xx%.

* All plain-carbon steels contain 0.50% ± manganese, and residual amounts (<0.05 w/o) of other elements.

† The Unified Numbering System (UNS) is more extensive than the AISI–SAE numbers, since it encompasses all present *commercial* alloys (currently approaching 10,000). However, the two are compatible. Among the alloys of the UNS, plain-carbon and low-alloy steels carry the prefix G, plus a fifth digit for future variants. Thus, AISI–SAE 4017 becomes G40170. The UNS prefixes for other alloys are listed in Appendix D, together with the identification of source standards.

element that has been added to the iron and carbon. The classification (10xx) is reserved for plain-carbon steels with only a minimum amount of other alloying elements.

• **Eutectoid Shifts**

Figure 10–7.3 shows the eutectoid region for the Fe–Fe$_3$C phase diagram, i.e., with only iron and iron carbide present. The eutectoid temperature is 727°C (1340°F); the eutectoid composition is just under 0.8% carbon.

In alloy steels, the carbon atoms and the iron atoms coordinate with atoms of other types, as well as with each other. Therefore, we should not be surprised that the eutectoid carbon content and temperature are altered from the above values when other elements are present. Figure 10–9.1(a) shows a shift

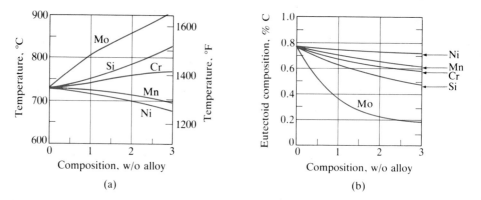

Figure 10–9.1 Eutectoids in Fe–X–C alloys. (X is the third component—Mo, Si, Cr, Mn, or Ni.) Effect of alloy additions on (a) the temperature of the eutectoid reaction, and (b) the carbon content of the eutectoid. (Adapted from ASM data.)

in the temperature from 727° as nickel, manganese, chromium, silicon, and molybdenum are added. The first two elements lower the eutectoid temperature because they preferentially form fcc solid solutions with iron (*austenite-formers*); Cr and Mo raise the eutectoid temperature because they are bcc themselves and are thus *ferrite-formers* in steels. Figure 10–9.1(b) shows that each of these alloying elements lowers the carbon content of the eutectoid composition from 0.77%. These curves are valuable because they let us predict the temperature of austenization for heat treatment of low-alloy steels.*

• **Example 10–9.1**   An SAE 1540 steel (C = 0.39%; Mn = 1.97%) is to be austenitized.

a) What temperature is required if it must be heated 30°C into the austenite range to facilitate the reaction?

b) How much pearlite will it form during cooling?

Procedure. From Fig. 10–9.1, the eutectoid is shifted to C ≈ 0.65%, and T_e = 710°C. At 0.39% C, the lower side of the γ-range is ~770°C (Fig. 10–9.2).

Answers

a) ~770°C + 30°C = 800°C (1470°F).

b) Fraction γ at 710°C = 0.6 (by interpolation between 0% and 0.65% on Fig. 10–9.2). Fraction pearlite below 710°C = 0.6 (since it formed from γ of eutectoid composition).

---

* Phase diagrams for multicomponent materials are more complex of necessity. These are introduced briefly in Topic XIV of *Study Aids for Introductory Materials Courses* and are recommended for the reader who will work with commercial engineering materials.

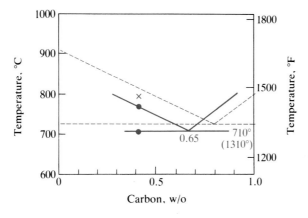

Figure 10–9.2 Eutectoid shift (Example 10–9.1). With nearly 2% Mn, the eutectoid composition drops to 0.65% carbon, and the eutectoid temperature drops to 710°C (1310°F). Therefore, there is more γ from which pearlite forms.

Comments. For our purposes we can assume that solubility lines that outline the γ field do not change slope as small amounts of alloying elements are added. (This assumption will not hold for high alloy additions.)

In the absence of 1.97% Mn, 810°C + 30° would have been required to austenitize; and only half of the steel would be pearlite. ◀

Review and Study

SUMMARY

The purposes of this chapter were two-fold: (1) To understand how to extract necessary data from phase diagrams, so we can select, control, and modify multiphase materials, and (2) to know the Fe–Fe$_3$C system so we can use it as a prototype for the solid-state microstructures of Chapter 11, and for their control.

1. Phase diagrams contain curves of *solubility limits*. The intervening areas of the two-component diagram are either single-phase *solutions* or *mixtures* of two phases. Two solubility curves cross at a *eutectic* to provide a low-melting liquid. The eutectic reaction is

$$L_2 \underset{\text{Heating}}{\overset{\text{Cooling}}{\rightleftarrows}} S_1 + S_3. \qquad (10\text{–}1.2)$$

2. Phase diagrams show *"what phases"* at various composition-temperature combinations.

3. In a single-phase area of a phase diagram, the phase has the same composition as the total material. In a two-phase area, the two compositions are located at the two ends of the isothermal *tie-line*. Chemical compositions are reported in terms of the *component* percentages.

4. With more than one phase, the fraction of each phase is determined by *interpolation along the tie-line* between the two-phase boundaries. (Some people call this the lever rule.) The amounts are reported in terms of *phase* percentages (or fractions).

5. *Commercial alloys* are commonly located in the simpler parts of the phase diagrams. However, any part of a complicated diagram may be "read" by checking (i) *what phases* exist, (ii) *phase compositions* (when more than one is present) at the two ends of the isothermal tie-line, and (iii) *phase amounts* by interpolation along the isothermal tie-line.

6. Phases in the iron-carbon system include (i) *ferrite,* or α (bcc), (ii) *austenite,* or γ (fcc), and (iii) *carbide* (Fe_3C). The solubility of carbon is greater in fcc iron than in bcc iron.

7. The *Fe–Fe₃C diagram* is basic to all property control in steels. The reader should become very familiar with the part below 1000°C and with <1% carbon, known as the *eutectoid* region (Fig. 10–7.3).

   The eutectoid reaction in this system is

$$\gamma \ (0.77\% \ C) \underset{\text{Heating}}{\overset{\text{Cooling}}{\rightleftharpoons}} \alpha \ (0.02\% \ C) + Fe_3C(6.7\% \ C). \qquad (10\text{–}7.2)$$

The generalized *eutectoid reaction* is

$$S_2 \underset{\text{Heating}}{\overset{\text{Cooling}}{\rightleftharpoons}} S_1 + S_2. \qquad (10\text{–}7.1)$$

8. *Pearlite* is a *lamellar* mixture of *ferrite* and *carbide* formed by transforming *austenite* of *eutectoid composition*.

9. *Plain-carbon* steels contain iron with up to one percent carbon (plus minor amounts of manganese for process control). *Low-alloy* steels contain a total of up to five percent of other metallic elements. These shift the eutectoid temperature, and lower the eutectoid carbon content to less than 0.77% C. The AISI-SAE composition code identifies the carbon content by the final two numbers.

PHASE-DIAGRAM INDEX

TERMS AND CONCEPTS

Austenite (γ) Face-centered cubic iron, or iron alloy based on this structure.

Austenite decomposition Eutectoid reaction that changes austenite to (α + carbide).

Austenization Heat treatment to dissolve carbon into fcc iron, thereby forming austenite.

Carbide (C) Compound of metal and carbon. Unless specifically stated, it refers to iron-base carbide (Fe_3C).

Component (phases) The basic chemical substances required to create a chemical mixture or solution.

Equilibrium The state at which net reaction ceases (because the minimum free energy has been reached).

Eutectic composition Analysis of liquid solution phase with the minimum melting temperature (at the intersection of two solubility curves).

Eutectic reaction $L_2 \underset{\text{Heating}}{\overset{\text{Cooling}}{\rightleftarrows}} S_1 + S_3$.

Eutectic temperature Temperature of the eutectic reaction at the intersection of two solubility curves.

Eutectoid composition Analysis of the solid solution phase with the minimum decomposition temperature (at the intersection of two solid solubility curves).

Eutectoid reaction $S_2 \underset{\text{Heating}}{\overset{\text{Cooling}}{\rightleftarrows}} S_1 + S_3$.

• **Eutectoid shift** The change in temperature and the carbon analysis of the eutectoid reaction arising from alloying element additions.

Eutectoid temperature Temperature of the eutectoid reaction at the intersection of two solid solubility curves.

Ferrite (α) Body-centered cubic iron, or an iron alloy based on this structure.

Ferrite, eutectoid Ferrite that forms (along with carbide) during austenite decomposition.

Ferrite, proeutectoid Ferrite that separates from austenite above the eutectoid temperature.

Isotherm Line of constant temperature.

Lever rule (inverse) Equation for interpolation to determine the quantity of phases in an equilibrated mixture.

Liquidus The locus of temperatures above which only liquid is stable.

Materials balance Mathematical calculation of "the whole equals the sum of the parts."

Mixture Combination of two phases.

Nomenclature, (AISI–SAE) See Table 10–9.1.

Pearlite (α + C) A lamellar mixture of ferrite and carbide formed by decomposing austenite of eutectoid composition.

Phase diagram Graph of phase stability areas with composition and environment (usually temperature) as coordinates.

Phase diagram, isothermal cut A constant temperature section of a phase diagram.

Phase diagram, one-phase area Part of a phase diagram containing a single unsaturated solution.

Phase diagram, two-phase area Part of a phase diagram beyond the solubility limit curves so a second phase is necessary. Temperature and the phase analyses *cannot* be varied independently.

Phase diagram, three-phase temperature Invariant temperature in a binary system at which three phases can coexist.

Phases, chemical compositions of Phase compositions expressed in terms of chemical components.

Phases, quantities of Material compositions expressed in terms of phase fractions.

Solder (Pb–Sn) Metals that melt below 425°C (~800°F) that are used for joining. Commonly, Pb–Sn alloys, but may also be other materials, even glass.

Solidus The locus of temperatures below which only solids are stable.

Solubility limit Maximum solute addition without supersaturation.

Solution A single phase containing more than one component.

Steel, eutectoid Steel with a carbon content to give 100% pearlite on annealing.

• **Steel, hypereutectoid** Steel with *more* carbon than a eutectoid steel; hence it can contain proeutectoid *carbide*.

• **Steel, hypoeutectoid** Steel with *less* carbon than a eutectoid steel; hence it can contain proeutectoid *ferrite*.

Steel, low-alloy Steel containing up to 5% alloying elements other than carbon. Phase equilibria are related to the Fe–Fe$_3$C diagram.

Steel, plain-carbon Basically Fe–C alloys with minimal alloy content.

Tie-line Isotherm across a two-phase field connecting the solubility limits of the two equilibrated phases.

FOR CLASS DISCUSSION

A_{10} From Fig. 10–1.2, suggest a method for purifying sea water.

B_{10} Arctic oceans (−1.5°C) cease melting snow when the NaCl content drops to 2.5%. Why?

C_{10} Explain why honey will "sugar" if it is stored in an open jar.

D_{10} The solidus temperature is an important temperature limit for hot-working processes of metals. Why?

E_{10} Distinguish between the chemical composition of a phase, and the quantity of a phase.

F_{10} Why does the composition of an eutectic liquid, L_2, always lie between the compositions of the two solids, S_1, and S_3?

G_{10} Why are the equilibrium compositions of two phases always at the limits of where the isotherm passes through the two-phase area?

H_{10} We can use interpolation in the horizontal isothermal direction to get the quantity fractions of the two phases. We *cannot* obtain a valid answer by interpolating along a vertical composition line. By using a 70 Pb–30 Sn solder, show why the latter doesn't work.

I_{10} Perform the algebraic rearrangement that we mentioned just before Eq. (10–4.3). (*Hint:* Solve Eq. (10–4.2) for X, to use in Eq. (10–4.3).)

J_{10} Devise a "center-of-gravity" calculation to determine the quantity fraction of a phase. Show that this is identical to the "lever rule" and the "interpolation procedure."

• **K₁₀**   Explain to a classmate why the quantity of α in an 80 Ag–20 Cu alloy is indeterminate at the eutectic temperature. However, if you have 50 g of α in 100 g of alloy, the temperature would have to be 780°C, if equilibrium exists.

L₁₀   Make a sketch of the 475°C–575°C and 10%–33% Sn region of the Cu–Sn phase diagram (Fig. 10–5.2) on an enlarged scale. Prepare a problem involving phase compositions and quantities to be solved by a classmate.

M₁₀   A long rod of silicon is "zone-refined" by passing it slowly through a short heater so that a molten zone starts at one end and moves toward its other end. It solidifies as it leaves the heated zone. Explain how this could be repeated to remove impurities (cf. Example 10–5.1).

N₁₀   Based on the comments of Example 10–6.1, explain why $D_{C \text{ in } \alpha} > D_{C \text{ in } \gamma}$, even though austenite can dissolve many times more carbon atoms than can ferrite.

O₁₀   The data of Fig. 10–7.3 will be used on numerous occasions during the balance of the text. They should be known so you don't have to look them up each time.

P₁₀   Pearlite is a 2-phase microstructure containing α and $\overline{C}$. Distinguish between phases and microstructures.

Q₁₀   The solution to Example 10–8.2(a) mentions eutectoid ferrite and proeutectoid ferrite. Based on Fig. 10–8.2(d) and Fig. 10–7.3, explain the difference.

• **R₁₀**   Predict whether tungsten (bcc) will raise or lower the temperature of austenite decomposition.

S₁₀   Refer to Fig. 4–3.2 that shows an iron oxide with an NaCl structure. Under what conditions is this defect structure stable? This iron oxide was formed by oxidizing a piece of iron at 1000°C in air. How does the composition of the scale vary through its thickness? (Turn to the Fe–O phase diagram.)

T₁₀   Devise an experiment, which can be used as a class demonstration, that will show the change of iron from ferrite to austenite. (The experiment should be reversible for repetition, and visible to those in the back of the room. Cf. Study Problem 8–2.7.)

• **U₁₀**   A phase diagram presents equilibrium conditions. Assume, however, that a 90 Al–10 Mg alloy is cooled rapidly from 600°C. How will this affect the compositions of the phases? From 400°C?

QUIZ SAMPLES   **10A**   A __**__ is a single phase with two or more components; a mixture is a material with two or more __**__ .

10B   Two __**__ curves cross at the eutectic temperature. This provides a __**__-melting liquid.

10C   The __+−__ is(are) the line(s), on a phase diagram, above which there is only __+−__. The __+−__ is(are) the line(s) at which solidification starts during cooling.

(a) solid   (b) liquid   (c) eutectic   (d) solidus   (e) liquidus

10D   Brass is a __**__-centered cubic phase of copper plus __**__ . A solid solution of copper plus tin is called a __**__ .

10E   Sterling silver _+−_ .

(a) is valuable because it is commercially pure silver   (b) is face-centered cubic   (c) has its maximum solubility of copper at the eutectic temperature   (d) None of the above

10F   A binary alloy may have three phases in equilibrium at the _+−_ temperature and at the _**_ temperature.

(a) solidus   (b) glass-transition   (c) eutectoid

10G   Interpolation may be used in a _**_ -phase field to determine the _+−_ of phases. The phases must have identical _+−_ .

(a) compositions   (b) temperatures   (c) amounts   (d) None of the above, rather identical _____

10H   In steels, _+−_ is(are) fcc; _+−_ is(are) bcc.

(a) α (b) γ (c) $\overline{C}$ (d) δ (e) austenite   (f) ferrite   (g) carbide   (h) pearlite

10I   The Fe–Fe$_3$C eutectoid temperature is _**_ °C. The Fe–Fe$_3$C eutectoid composition is _**_ %C. The Fe–Fe$_3$C eutectoid reaction *during heating* is _**_ .

10J   _**_ is a lamellar mixture of _**_ .

10K   A 99 Fe–1 C alloy may contain proeutectoid _**_ . A 99.5 Fe–0.5 C alloy may contain proeutectoid _**_ .

• **10L**   The alloying element(s), _+−_ , is(are) ferrite former(s), as is low-temperature iron; therefore, _+−_ the eutectoid temperature. The element(s), _+−_ , lower the carbon content of the eutectoid.

(a) Ni   (b) Mn   (c) Cr   (d) Mo   (e) increasing   (f) not affecting   (g) decreasing

STUDY PROBLEMS

10–1.1   A syrup contains equal quantities of water and sugar. How much more sugar can be dissolved into 100 g of the syrup at 80°C?

Answer: 138 g added

10–1.2   A molten lead–tin solder has a eutectic composition. Fifty grams are heated to 200°C. How many grams of tin may be dissolved into this solder?

10–1.3   A 90 Pb–10 Sn* solder is to be melted at 200°C by adding more tin. What is the minimum amount of tin that must be added per 100 grams of this high-lead solder?

Answer: 104.5 g additional tin

10–1.4   One ton of salt (NaCl) is spread on the streets after a winter storm. The temperature is −15°C (5°F). How many tons of ice will be melted by the salt?

10–1.5   (a) Eight grams of salt (NaCl) are added to 32 g of ice. Above what minimum temperature will the ice completely melt? (b) Twenty *more* grams of ice are added

_____

* The phase-diagram index follows the Summary, p. 416.

to the above brine. Above what minimum temperature will this ice completely melt?

Answer: (a) $-17°C$ (1.5°F)   (b) $-10°C$ (14°F)

10-1.6 The solubility limit of zinc in α-brass is 35% at 780°C. It is desired to make 100 kg of 65-35 brass by melting some 85-15 brass and some zinc. How much of each should be melted in the crucible?

10-2.1 A 90 Cu-10 Sn* bronze is cooled slowly from 1100°C to 20°C. What phase(s) will be present as the cooling progresses?

Answer: $1100 \rightarrow 1010°C$, L; $1010 \rightarrow 830°C$, (L + α); $830°C \rightarrow 340°C$, α; $340 \rightarrow 20°C$, ($\alpha + \epsilon$)

10-2.2 A 65 Cu-35 Zn* brass is heated from 300°C to 1000°C. What phase(s) are present at each 100°C interval?

10-2.3 An alloy contains 90 Pb-10 Sn.* (a) What phases are present at 100°C? 200°C? 300°C? (b) Over what temperature range(s) will there be only one phase?

Answer: (a) 100°C, $\alpha + \beta$; 200°C, α; 300°C, α (admittedly slight amount) + L (b) Only α, 150-270°C; only L above 305°C.

10-2.4 A eutectic mixture of Al_2O_3 and ZrO_2* is molten at 2400°C. (a) At what temperature(s) will there be only one phase? (b) Two phases? (c) Three phases?

10-2.5 Locate the solidus on a Cu-Zn diagram.*

Answer: It lies immediately below all the x + L fields with horizontal lines at 903°, 835°, 699°, 598°, and 424°C.

10-2.6 Locate the liquidus and the solidus as they extend from 1538°C at 100% Fe (0% O) and 1599°C for Fe_3O_4. (*Note:* There is a 2-liquid field where liquid metal separates from liquid oxide.) (See Fig. 10-5.5.)

10-2.7 Show the sequential changes in phases when the composition of an alloy is changed from 100% Cu to 100% Al (a) at 700°C, (b) at 450°C, (c) at 900°C. (See Fig. 10-5.3.)

Answer: (c) At 900°C: α, $\alpha + \beta$, β, $\beta + \gamma_1$, γ_1, $\gamma_1 + \epsilon_1$, ϵ_1, $\epsilon_1 + L$, L

10-2.8 (a) Show the sequential changes in phases when the composition of an alloy is changed from 100% Mg to 100% Al at 300°C. (b) At 400°C. (See Fig. 10-5.4.)

10-2.9 Some of the fields in the Cu-Sn diagram* are not labeled. At 625°C, what field lies (a) between β and γ? (b) Between γ and ζ?

Answer: Use the 1-2-1-2- $\cdots$ sequence to obtain the answer.

10-2.10 What fields lie to the right of γ_2 in the Al-Cu diagram?*

---

* The phase-diagram index follows the Summary, p. 416.

10–3.1 Refer to Fig. 10–5.4. At 500°C, what is the solubility limit of magnesium (a) in solid α? (b) In liquid? (c) What are the chemical compositions of the phase(s) in a 40 Al–60 Mg alloy at 500°C? (d) a 20 Al–80 Mg alloy?

Answer: (a) 11% Mg (b) 76% Mg (c) Liq: 40 Al–60 Mg
(d) Liq: 24 Al–76 Mg; ϵ: 7.5 Al–92.5 Mg

10–3.2 Refer to Fig. 10–2.2. At 2000°C, what is the solubility limit of Al_2O_3 (a) in the liquid? (b) In β? (c) What are the chemical compositions of the phase(s) in a 20 Al_2O_3–80 ZrO_2 ceramic at 1800°C? at 2000°C?

10–3.3 What are the chemical compositions of the phase(s) in Study Problem 10–2.2?

Answer: 300°, 400°, 500°, 600°, 700°: α(65 Cu–35 Zn); 800°C: α(66 Cu–34 Zn) and β(61 Cu–39 Zn); 900°C: α(67.5 Cu–32.5 Zn) and β(~63 Cu–37 Zn); 1000°C: Liq (65 Cu–35 Zn)

10–3.4 What are the chemical compositions of the phase(s) in Study Problem 10–2.1?

10–3.5 (a) Write the eutectic reaction for one of the two eutectics in the Al–Cu system (Fig. 10–5.3). (b) Repeat for the second eutectic.

Answer: (a) Liq(91.5% Cu) $\overset{1035°}{\rightleftarrows}$ α(92.5% Cu) + β(90% Cu)

10–3.6 Write the eutectic reaction for the higher-magnesium eutectic of the Al–Mg system.

10–4.1 A 90 Al–10 Mg alloy is melted and cooled slowly. (a) At what temperature does the first solid appear? (b) At what temperature does it have $\frac{2}{3}$ Liq and $\frac{1}{3}$ α? (c) $\frac{1}{2}$ Liq and $\frac{1}{2}$ α? (d) 99+%α with a trace of liquid?

Answer: (a) 620°C (b) 610°C (c) 595°C (d) 520°C

10–4.2 Repeat Study Problem 10–4.1, but for a 10 Al–90 Mg alloy.

10–4.3 What composition of Ag and Cu will possess (a) $\frac{1}{4}$ α and $\frac{3}{4}$ β at 600°C? (b) $\frac{1}{4}$ Liq and $\frac{3}{4}$ β at 800°C?

Answer: (a) 26 Ag–74 Cu (b) 23 Ag–77 Cu

10–4.4 What composition of Al_2O_3 and ZrO_2 will possess (a) $\frac{3}{4}$ α and $\frac{1}{4}$ β at 1800°C? (b) $\frac{3}{4}$ Liq and $\frac{1}{4}$ β at 1950°C?

10–4.5 The solubility of tin in solid lead at 200°C is 18% Sn. The solubility of lead in the molten metal at the same temperature is 44% Pb. What is the composition of an alloy containing 40% liquid and 60% solid α at 200°C?

Answer: 66.8% Pb, 33.2% Sn

10–4.6 (a) At what temperature will a monel alloy (70% nickel, 30% copper) contain $\frac{2}{3}$ liquid and $\frac{1}{3}$ solid, and (b) what will be the composition of the liquid and of the solid?

10–4.7 Assuming 1500 g of bronze in Study Problem 10–2.1, what are the masses of solid phase(s) at each 100°C interval?

Answer: 1100°C, 0 g; 1000°C, slight α; 800°C, 400°C, 1500 g α; 300°C, ~1350 g α + 150 g ϵ.

10–4.8 At 175°C, how many grams of α are there in 7.1 g of eutectic Pb–Sn solder?

10–4.9 With 200 g of 65 Cu–35 Zn, how many grams of α are present at each temperature of Study Problem 10–2.2?

Answer: 300 → 700°C, 200 g α; 800°C, 180 g α; 900°C, ~100 g α; 1000°C, 0 g α

10–4.10 Two kilograms of sterling silver (92.5 Ag–7.5 Cu) are cooled slowly from 1000°C to 400°C. (a) On a graph of %Ag (ordinate) versus °C (abscissa), plot the composition of α. (b) On a graph of g of α (ordinate) versus °C (abscissa), plot the amount of α. (Data on Fig. 10–2.4 are accurate to about 1 part/100. You should be correspondingly accurate.) (c) Extrapolate your curves to 20°C.

10–4.11 Plot the amount of α in 15.8 kg of a 95 Al–5 Si alloy as a function of temperature.

Answer: 0°C, 15 kg; 576°C, 15.3 kg; 578°C, 11 kg; 600°C, 8.4 kg; >630°C, 0 kg

• **10–4.12** At what temperature can an alloy of 40 Ag–60 Cu contain 55% β?

10–5.1 An alloy of 50 g Cu and 30 g Zn is melted and cooled slowly. (a) At what temperature will there be 40 g α and 40 g β? (b) 50 g α and 30 g β? (c) 30 g α and 50 g β?

Answer: (a) ~780°C   (b) 750°C   (c) 800°C

10–5.2 (a) What are the chemical compositions of the phases in a 10% magnesium–90% aluminum alloy at 600°C, 400°C, 200°C? (b) What are the quantities of these phases at each of the temperatures in part (a)? (c) Make a materials balance for the distribution of the magnesium and aluminum in the above alloy at 600°C.

10–5.3 Make a materials balance for P grams of a 92 Ag–8 Cu alloy at 500°C (equilibrium conditions).

Answer: $0.938P$ g α, 98% Ag–2% Cu; $0.062P$ g β, 1% Ag–99% Cu

10–5.4 Make a materials balance for 100 g of a 90 Mg–10 Al alloy at 200°C (assume equilibrium).

10–5.5 A die-casting alloy of 95 Al–5 Si is cooled so that the metal contains primary (proeutectic) α and an eutectic mixture of $(\alpha + \beta)$. What fraction of the casting is primary α?

Answer: 0.7 primary α (i.e., proeutectic α)

10–5.6 How much mullite will be present in a 60% SiO_2–40% Al_2O_3 brick (10 kg) at the following temperatures under equilibrium conditions: (a) 1400°C? (b) 1580°C? (c) 1600°C?

10–5.7 (a) Determine the compositions of Al–Si alloys that will contain $\frac{1}{3}$ liquid and $\frac{2}{3}$ solid when brought to equilibrium at 600°C. (b) Give the chemical compositions of the liquids.

Answer: (a) 96 Al–4 Si and 29 Al–71 Si   (b) 91 Al–9 Si, 86 Al–14 Si

10-5.8 Based on Figs. 10–1.3 and 10–4.2, make a graph for an alloy containing 80% Pb and 20% Sn showing (a) the fraction of liquid versus temperatures, (b) the fraction of α versus temperature, and (c) the fraction of β versus temperature.

10-5.9 (a) Plot the percent Cu in the α phase of sterling silver (92.5 Ag–7.5 Cu) versus temperature. (b) Plot the fraction of α in sterling silver versus temperature.

10-5.10 Sixty kg of a 90 Cu–10 Sn bronze (to be used to make a bell) are melted and then cooled very slowly to maintain equilibrium. (a) At what temperature does the first solid appear? (b) What is the composition of that solid? (c) At what temperature is there twice as much solid as liquid? (d) At what temperature does the last liquid disappear? (e) What is the composition of that liquid? (f) What phase(s) at 600°C? Give the composition(s) and amount(s) of each phase. (g) What phase(s) at 300°C? Give the composition(s) and amount(s) of each phase.

• **10-5.11** A kilogram of sterling silver contains 75 g copper (925 g Ag). Suggest a procedure for obtaining at least 100 grams of silver from the above that will have an analysis of <2% copper. Use melting, partial solidification, and separation steps. (You may assume the liquidus and solidus are straight lines.)

10-6.1 The maximum solubility of carbon in ferrite (α) is 0.02 w/o. How many unit cells per carbon atom?

Answer: 540

10-6.2 Iron carbide has a density of 7.6 Mg/m³ (=7.6 g/cm³). What is the volume of the orthorhombic unit cell described in Section 10–6?

10-6.3 At 920°C, the maximum solubility of carbon in austenite (γ) is 1.33 w/o. How many unit cells per C atom?

Answer: 4.0

10-6.4 Based on the data of Fig. 4–6.3 and Example 3–4.1, plot an approximate volume–temperature curve for iron between 0°C and T_m. If you have to make estimates, state your assumptions.

10-7.1 Making use of the Fe–Fe₃C diagram, calculate the fraction of α and the fraction of carbide present at 700°C in a metal containing 2% carbon and 98% iron.

Answer: 0.30 carbide, 0.70 ferrite

10-7.2 (a) What phases are present in a 99.8 Fe–0.2 C steel at 800°C? (b) Give their compositions. (c) What is the fraction of each?

10-7.3 Describe the phase changes that occur on heating a 0.20% carbon steel from room temperature to 1200°C. (Ferrite, α; austenite, γ; carbide, $\overline{C}$.)

Answer: RT → 727°C ($\alpha + \overline{C}$); 727°C (α + carbide + γ); 727 → 855°C ($\alpha + \gamma$) with the amount of α decreasing to zero at 860°C; 860 → 1200°C, only γ

10-7.4 Without referring to the Fe–Fe₃C phase diagram, indicate the compositions of the phases in the previous problem.

10-7.5 Write the eutectoid reaction found in the BaTiO₃–CaTiO₃ system (Fig. 10–5.7).

Answer: β(24% CaTiO₃) $\overset{105°}{\rightleftarrows}$ α(20% CaTiO₃) + γ(90% CaTiO₃)

10–7.6 (a) Locate four eutectoids in the Cu–Sn system (Fig. 10–5.2). (b) Write the reactions for two of these.

10–7.7 Refer to Fig. 10–7.4. Make a similar presentation of phase fractions for a 99 Fe–1 C steel.

10–7.8 Repeat the previous problem for a 99.8 Fe–0.2 C steel.

10–8.1 Calculate the percent ferrite, carbide, and pearlite, at room temperature, in iron–carbon alloys containing (a) 0.5% carbon, (b) 0.8% carbon, (c) 1.5% carbon.

Answer: (a) 7.5% carbide, 92.5% α, 62% pearlite   (b) 12% carbide, 88% α, 100% pearlite   (c) 22.5% carbide, 77.5% α, 88% pearlite

10–8.2 (a) Determine the phases present, the composition of each of these phases, and the relative amounts of each phase for 1.2% carbon steel at 870°C, 760°C, 700°C. (Assume equilibrium.) (b) How much pearlite is present at each of the above temperatures?

10–8.3 Assume equilibrium. For a steel of 99.5 Fe–0.5 C, determine (a) lowest temperature of 100% γ. (b) Fraction that is γ at 730°C, and its composition. (c) Fraction that is pearlite at 720°C, and its total composition. (d) Fraction that is proeutectoid ferrite at 730°C. (e) Fraction that is proeutectoid ferrite at 720°C after cooling from 730°C.

Answer: (a) 780°C   (b,c) 0.6 at ~0.8% C (99.2% Fe)   (d,e) 0.4 at 0.02% C

10–8.4 Substitute a 99 Fe–1 C steel for the previous problem and answer.

10–8.5 (a) Determine the amount of pearlite in a 99.1 Fe–0.9 C alloy that is cooled slowly from 870°C. Basis: 11 kg alloy. (b) From the results of part (a), determine the amount of ferrite and carbide at 725°C. (c) At 20°C.

Answer: (a) 10.8 kg   (b) 9.55 kg α, 1.45 kg $\overline{C}$ (c) 9.5 kg α, 1.5 kg $\overline{C}$

10–8.6 On the basis of this chapter, would you choose a high- or low-carbon steel for an automobile fender? Give reasons.

10–9.1 A steel has the following composition. Give it an AISI–SAE number: C 0.21, Mn 0.69, Cr 0.62, Mo 0.13, Ni 0.61.

Answer: AISI 8620

10–9.2 A steel has the following composition. Give it an AISI–SAE number: C 0.38, Mn 0.75, Cr 0.87, Mo 0.18, Ni 0.03.

10–9.3 Calculate the fractional quantities of the phase(s) at 700°C, 750°C, 800°C, and 900°C for the following steels: (a) 0.8 C–99.2 Fe; (b) 1.2 C–98.8 Fe; •(c) 0.6 C–0.6 Mo–98.8 Fe.

Answer: (a) 700°C, 0.88α–0.12 $\overline{C}$; 750°C, 800°C, 900°C, 1.0γ
(b) 700°C, 0.82α–0.18 $\overline{C}$; 750°C, 0.94γ–0.06$\overline{C}$; 800°C, 0.965γ–0.035 $\overline{C}$; 900°C, 1.0γ
(c) 700°C, 750°C, 0.91α–0.09 $\overline{C}$; 800°C, ~0.99γ–0.01 $\overline{C}$; 900°C, 1.0γ ($\overline{C}$ = carbide.) From Fig. 10–9.1 (for 0.6 Mo): T_e = 775°C, and C_e = 0.47%C.

• **10–9.4**   Modify Fig. 10–7.3 for a steel containing (a) 1% Mn, (b) 1% Cr, (c) 1% Ni. (The new solubility curves remain essentially parallel to the previous ones.)

• **10–9.5**   How much nickel should be added to a steel to drop the lower side of the austenite stability range in a 0.40% carbon steel as much as it is dropped by 2% Mn?

Answer: ~2.2% Ni

QUIZ CHECKS

10A	solution phases		**10H**	(b), (e)
				(a), (d), (f)
10B	solubility limit low		**10I**	727°C, 0.8% C
				$\alpha + \overline{C} \xrightarrow{\text{Heating}} \gamma$
10C	(e)		**10J**	pearlite, $\alpha + \overline{C}$
	(b)			
	(e)		**10K**	carbide
10D	face, zinc bronze			ferrite
			10L	(c), (d)
10E	(b), (c)			(e)
10F	(a), (c)			(a), (b), (c), (d)
10G	2-phase, (c)			
	(b)			

Chapter Eleven

Processes to Develop and Control Microstructures

PREVIEW

The geometric arrangement of grains and phases in a material is called its *microstructure*. Variables include size, shape, amounts and distribution of the structural features. Typically the dimensions are such that an optical microscope (up to ×2000), or even an electron microscope (up to ×50,000+), is necessary for observation.

As reactions occur within solids, atoms must break bonds, relocate, and nucleate new structures. The time requirements are very temperature-sensitive. The Arrhenius relationship (Eq. 4–7.4) is pertinent if atomic diffusion is the rate-limiting factor. Time for the nucleation of new phases can also be a factor.

Annealing (a), which is a process of extended heating followed by slow cooling, affects various materials in different manners. Therefore, we won't be able to make broad generalization about its effects. Other heat treating processing includes (b) *homogenization treatments,* (c) *precipitation processing,* and (d) *quench and tempering* sequences. The latter two (c and d) can lead to microstructures that are simultaneously strong *and* ductile, a desirable combination that commonly is mutually exclusive.

Steels are used to provide examples of commercial heat-treating processes.

CONTENTS

STUDY OBJECTIVES

1. To review the microstructural characteristics of single-phase and multiphase materials.

2. To understand the principal reactions that change the internal structure of solid materials.

3. To apply the Arrhenius relationship to reactions that are limited by diffusion, and to have a qualitative understanding of why reactions are delayed when new phases must be nucleated.

4. To distinguish between conditions that lead (a) to reactions at grain boundaries, and (b) to reactions within the grains.

5. To interpret the structural changes that occur during the annealing of various materials, and to anticipate the consequent property changes.

6. To relate the heat-treating steps of age-hardening (precipitation-hardening) and of over-aging (a) to the solubility limits of phase diagrams, and (b) to the microstructural changes of particle development and growth.

7. To understand isothermal transformation diagrams for steels so that you may (a) use them to predict reaction rates, (b) anticipate their alteration with changes in carbon content, alloy additions, and grain size, and (c) rationalize critical cooling rates that will produce all martensite or no martensite.

• 8. To relate commercial steel-treating processes to various time-temperature-transformation sequences.

• 9. To utilize hardenability curves for predicting hardness values of selected steel products, and to understand the rationale for differences in hardenability curves.

11–1

MICROSTRUCTURES *Microstructures* refer to the arrangements of grains and phases within a material. The *grains* are individual crystals. Each *phase* of a microstructure has a distinct arrangement of atoms—typically crystalline, but not exclusively so. The microstructures of interest to us will be *polycrystalline*. They may be single-phase or multiphase. For the most part, magnification is required to observe a microstructure.

Summary of Single-phase Microstructures

We limited our attention to single-phase materials in the chapters of Part I. *Single-phase microstructures* were introduced first in Section 4–4, where metallic and oxide grains were shown in photomicrographs (Fig. 4–4.8). Variables in single-phase microstructures include grain *size,* grain *shape,* and grain *orientations* (Fig. 5–1.4). Each of these affects properties; e.g., at normal temperatures fine-grained metals are stronger and less ductile than coarse-grained metals. (The opposite is true at elevated temperatures—Section 5–7.) As a second example, laminates for power transformers are processed to provide a microstructure with a $\langle 100 \rangle$ preferred orientation in order to obtain higher magnetic permeability and increased efficiency (Fig. 9–1.9).

Grain size may be indexed with a grain size number (G.S.#) that is based on 2^{n-1} as presented in Eq. (4–4.2). A more quantitative measure of grain size may be obtained from the mean-chord length, $\bar{\delta}$, which is the reciprocal of the grain-boundary intercepts per unit length P_L (Example 4–4.1). This measure is useful because the boundary surface per unit volume S_V equals $2P_L$ (Eq. 4–4.1). It is this internal surface area, rather than grain size per se, that contributes to microstructurally controlled properties.

Grain-growth in single-phase materials occurs by atom movements across the grain boundary. The net effect is that the boundaries move toward their centers of curvature, eliminating small grains (Fig. 5–7.3), and increasing the average grain size (Fig. 5–7.4). Inasmuch as the atom movements are thermally activated, grain growth occurs more rapidly at higher temperatures (Fig. 11–1.1).* Grain growth does not reverse at lower temperatures but may proceed so slowly that we do not detect any change.

Recrystallization can occur in single-phase materials after there has been cold work (Section 5–6). The microstructure develops a new "crop" of grains. The time–temperature requirements follow an Arrhenius relationship (Eq. 5–6.2). The recrystallization temperature decreases with additional cold work and greater purity, and it is lower for lower melting materials.

---

* The log–log relationship of Fig. 11–1.1 arises because the rate of growth $d\delta/dt$ is proportional to the curvatures of the grain boundaries that, in turn, are inversely related to grain size, $1/\delta^n$:

$$d\delta/dt = k'/\delta^n. \qquad (11–1.1)$$

If the initial size δ_0 is very small,

$$\delta^{n+1} = kt \qquad (11–1.2a)$$

and

$$\log \delta = \frac{1}{n+1} \log kt. \qquad (11–1.2b)$$

The slopes of the curves in Fig. 11–1.1 are $1/(n+1)$; the value of n commonly lies between two and five in commercial materials. The larger values of n accompany slower growth and commonly result from minor impurity phases. Furthermore, external surfaces eventually arrest growth as δ approaches the cross-sectional dimensions of the material.

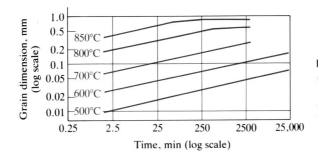

Figure 11–1.1   Isothermal grain growth (brass sheet, 1-mm thick). The logarithmic growth rate continues until the grains approach the dimensions of the product. (After J. E. Burke, AIME.)

Summary of Multiphase Microstructures

Each of the microstructural characteristics of single-phase materials has a counterpart in multiphase materials. Grain size, shape, and orientation affect properties; and grain growth occurs at elevated temperatures (Fig. 11–1.2). In addition, the microstructures of multiphase materials can vary in phase amounts, and in phase shape and distribution.

In equilibrated materials, the *phase amounts* can be predicted from appropriate phase diagrams. This was the topic of Section 10–4 and will not be repeated here. Chapter 10 did not relate properties to phase fractions. An-

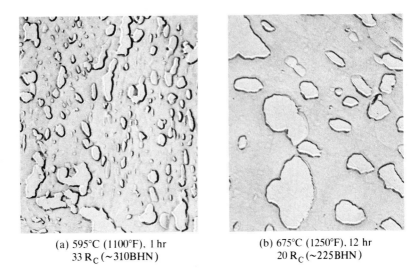

(a) 595°C (1100°F), 1 hr
33 R$_C$ (~310BHN)

(b) 675°C (1250°F), 12 hr
20 R$_C$ (~225BHN)

Figure 11–1.2   Grain growth in a two-phase microstructure (carbide particles in a tempered steel, ×11,000). Each of these samples came from the same 1080 steel. The longer time and higher temperature permitted the development of larger (and fewer) carbides in (b). (This steel was previously quenched from austenite, as discussed in Section 11–2.) (*Electron Microstructure of Steel,* American Society for Testing and Materials, and General Motors Research Laboratories.)

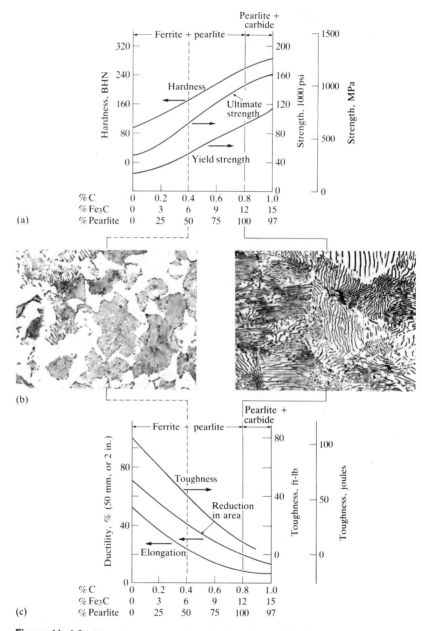

Figure 11–1.3 Properties versus microstructure (annealed, plain-carbon steels). (a) Hardness and strength versus the amount of carbon, Fe_3C, and pearlite. (b) Microstructure of 0.40% C (left) and 0.80% C (right) annealed steels (×500). (Courtesy U.S. Steel Corp.) (c) Ductility and toughness versus the amount of carbon, Fe_3C, and pearlite.

nealed steels provide an example for illustrative purposes. Figure 11–1.3 shows the increase in hardness and strengths with an increase in the carbon content of the steels. (Since the carbon is present as Fe_3C, and the carbide is present within pearlite in annealed steels, alternative scales are presented on the abscissa.) While the hard, brittle carbide increases the hardness and strength (Fig. 11–1.3a), it decreases the ductility and toughness (Fig. 11–1.3c).

Phase shape and distribution take many forms. We will use only one example, and again it will be from steels. The two microstructures of Fig. 11–1.4(a) and (b) have identical compositions; in fact, they were prepared from the same original sample and have identical fractions of ferrite (lighter) and carbide (darker). Figure 11–1.4(a) is *pearlite*, obtained in the manner that all

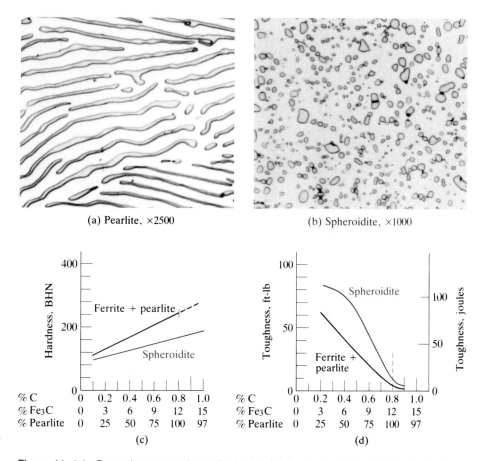

(a) Pearlite, ×2500   (b) Spheroidite, ×1000

Figure 11–1.4   Properties versus phase distribution (plain-carbon steels). (a) Pearlite (0.8% C). (b) Spheroidite (0.8% C). (c) Hardness. (d) Toughness. (Photomicrographs courtesy of U.S. Steel Corp.)

pearlite is formed from the decomposition of austenite of eutectoid composition (Section 10–8). Its microstructure contains lamellar ferrite and carbide. Figure 11–1.4(b) is *spheroidite* with spherelike carbides in a ferrite matrix. It was obtained by heating the pearlite a long time (~24 hrs) just above 700°C (~1300°F) so that the ferrite-carbide boundary could be minimized.*

The consequences of the difference in the microstructures of Figs. 11–1.4(a) and (b) are shown in parts (c) and (d). The rigid lamellar carbides of pearlite provide greater reinforcement of the soft, ductile ferrite than do the dispersed carbides of spheroidite. This accounts for the greater hardness of the former. Conversely, a crack can propagate readily through the brittle carbide layers of the pearlite by making short jumps through any intervening tough ferrite. In the spheroidite, any fracture must progress through the tough ferrite matrix for a significant percentage of its path. More energy is required for fracture and greater toughness is realized.

Other examples of multiphase microstructures and their effects upon properties will be presented later as engineering processes and applications are discussed.

Example 11–1.1

Determine the grain size of the MgO in Fig. 4–4.8(b) in terms of the mean-chord length, δ.

Procedure. The mean-chord length (average distance per grain) is the reciprocal of P_L, the grains (or boundary intercepts) per unit length. Follow the procedure used in Example 4–4.1 to estimate P_L.

Answer. From Example 4–4.1, $P_L = 68/mm$;

$$\delta = 1/(68/mm) = 0.015 \text{ mm} \qquad \text{(or 15 } \mu\text{m).} \blacktriangleleft$$

Example 11–1.2

Two identical samples were obtained from the same piece of bronze. The grain size expressed as δ was 50 μm. Sample A was heated to 650°C and required 4.5 hr to double its value of δ to 100 μm. At 700°C, sample B required only 30 minutes for the same change in grain size—50 μm to 100 μm. Estimate the time requirement at 750°C.

Procedure. Each heat treatment involves atom movements to the same extent, but at different temperatures. Therefore, we expect an Arrhenius relationship for the rates of reaction R versus T. Patterned after diffusion (Eq. 4–7.4a)

$$\ln R = \ln R_0 - E/kT. \qquad (11–1.3)$$

Since rates and time are reciprocals,

$$\ln t = C + B/T, \qquad (11–1.4)$$

where C is $-\ln R_0$, and B is E/k.

---

* The governing principle for the spheroidization of carbides in steel is the same as that which makes a drop of dew spherical. In each case, the atoms at the phase boundary possess extra energy. A sphere has the least surface per unit volume; therefore, it has less total energy.

Calculation

$$\ln 4.5 \text{ hr} = C + B/923 \text{ K} = 1.50;$$

$$\ln 0.5 \text{ hr} = C + B/973 \text{ K} = 0.70.$$

Solving simultaneously,

$$B = 39{,}500 \text{ K}; \quad C = -41.3.$$

$$\ln t = -41.3 + (39{,}500 \text{ K}/1023 \text{ K}) = -2.7;$$

$$t = 0.07 \text{ hr} \qquad \text{(or 4 min)}.$$

Comment. The grain size will *not* double by doubling the time because the growth rate decreases as the grain size increases. ◀

11–2

REACTIONS IN SOLIDS

With a few exceptions, the development of microstructures occurs within materials that are completely solid. This restricts the rates of reaction. In every case, atoms must (1) break existing bonds, and (2) move to new sites. Each of these steps requires time that is an inverse function of temperature with less time required at higher temperatures. In some reactions, (3) a new phase must be nucleated. The required time for nucleation varies with the extent of super-cooling. This means a shorter time requirement at lower temperatures.

These opposing requirements for time were first encountered in Section 6–6, where we saw that polymer *crystallization* was slow just below the melting temperature T_m, where the crystallinity was destroyed nearly as fast as crystals formed (Fig. 6–6.5). In addition, the rate of crystallization was slow at lower temperatures (near T_g), because molecular movements were restricted. The most rapid crystallization (shortest times) was at intermediate temperatures. On an empirical basis, the rate of reaction R is the product of the nucleation rate $\dot{N}$ and the growth rate $\dot{G}$

$$R = f(\dot{N}\dot{G}). \qquad (11\text{–}2.1)$$

This is shown schematically in Fig. 11–2.1

Several common solid-phase reactions are listed in Table 11–2.1. *Grain growth* and *recrystallization* follow the kinetics of Fig. 11–2.1(a) where atom movements are rate-controlling. Diffusion distances are short and a new phase does not have to be nucleated.

Solution and Precipitation

Solution and precipitation are opposite to each other. Figure 11–2.2 lets us examine them more closely as they involve solids. It is part of the Pb–Sn phase diagram originally presented in Fig. 10–1.3. At temperatures below 150°C, an alloy with 90% Pb and 10% Sn contains two phases, α and β. Above 150°C, all the tin may be contained in fcc α; consequently, as the metal is heated, solid β

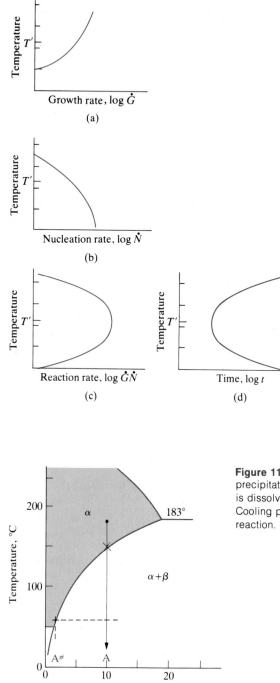

(a)

Growth rate, log $\dot{G}$

(b)

Nucleation rate, log $\dot{N}$

(c)

Reaction rate, log $\dot{G}\dot{N}$

(d)

Time, log t

Figure 11–2.1 Reactions versus temperature (schematic). (a) Growth rate, $\dot{G}$. Growth proceeds more rapidly at higher temperatures. (b) Nucleation rate, $\dot{N}$. Nucleation proceeds more rapidly with greater supercooling. (c) Reaction rate, $\dot{R} = f(\dot{G}, \dot{N})$. The most rapid reaction is at an intermediate temperature, T'. (d) Reaction time, t. Since $R = 1/t$, reaction times are shortest at T'.

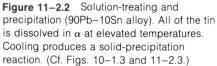

Figure 11–2.2 Solution-treating and precipitation (90Pb–10Sn alloy). All of the tin is dissolved in α at elevated temperatures. Cooling produces a solid-precipitation reaction. (Cf. Figs. 10–1.3 and 11–2.3.)

TABLE 11–2.1			

Characteristics of Selected Solid-State Reactions

		Diffusion distance (typical)		New phases
		<1 nm	>10³ nm	
1.	Grain growth	X		—
2.	Recrystallization	X		—
3.(a)	Solution		X	—
(b)	Precipitation		X	1
4.	Eutectoid decomposition		X	2
5.	Martensite formation	X		1

is *dissolved* in the solid α. *Solid precipitation* occurs when this alloy is cooled into the two-phase temperature range after being *solution-treated* above 150°C. Such precipitation is useful for various age-hardening alloys (Section 11–5). The following equations are appropriate for these two reactions in binary alloys. Note that only two phases are involved in each case; however, the composition of the solution changes from A to another value, $A^{\#}$, as the solubility limit is exceeded (Fig. 11–2.2).

$$\text{Solution-treating} \qquad \text{Solid } A^{\#} + \text{Solid } B \xrightarrow{\text{Heating}} \text{Solid } A. \qquad (11\text{–}2.2)$$

$$\text{Precipitation} \qquad \text{Solid } A \xrightarrow{\text{Cooling}} \text{Solid } A^{\#} + \text{Solid } B. \qquad (11\text{–}2.3)$$

The above reactions require diffusion. Consider another example, the composition of 95.5 w/o Al and 4.5 w/o Cu, which is a widely used aluminum alloy. As shown in Fig. 10–5.3, there are two phases, θ and κ, in equilibrium below 500°C (930°F). $CuAl_2$ comprises the θ phase, so that it contains an appreciably larger fraction of copper atoms than the κ phase. During solution treatment these copper atoms must move through the κ structure to take on random substitutional positions; during precipitation, the copper atoms must be collected onto the growing $CuAl_2$ (that is, θ) particles.

Return now to the 90 Pb–10 Sn alloy of Fig. 11–2.2 so that we might consider the time requirement for the precipitation reaction to occur. Above 150°C, only α (Pb-rich) is stable; below 150°C, some β (Sn-rich) will separate from supersaturated α if time is available. Figure 11–2.3 shows the time required for *isothermal precipitation* (50% completion), where the alloy is rapidly cooled to the selected temperature and then held at that temperature for precipitation to occur. We will observe that the required time decreases as the temperature is raised from −50°C to room temperature. In this temperature range, the precipitation rate is limited by diffusion. At about 50°C, the pattern is

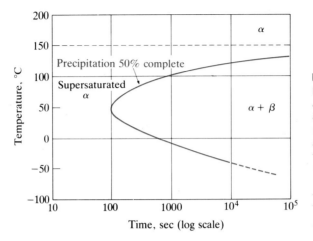

Figure 11–2.3 Isothermal precipitation (90Pb–10Sn). Above 150°C, α can contain all of the tin. The rate of β precipitation varies with the amount of supercooling below the saturation temperature. Finally, the equilibrium of Fig. 10–1.3 is attained. (Adapted from data by H. K. Hardy and T. J. Heal, *Progress in Metal Physics* 5, Pergamon Press.)

reversed and eventually runs off the long-time end of the graph just below the solubility limit of 150°C.

This C-type curve may be compared with Fig. 6–6.5. Above the knee of the isothermal precipitation curve, the rate is not limited by diffusion, because the atoms can diffuse relatively fast (50°C $\approx 0.6T_m$). Rather, the limitation arises from the time required to nucleate the new phase β. Nucleation occurs most readily at the boundaries of the α grains (Fig. 11–2.4b). Tin atoms move over many atomic distances to these sites.

Below the knee of the isothermal precipitation curve, the rate of precipitation is limited by the diffusion of the tin atoms. They cannot diffuse readily; furthermore, the supersaturation is great. Therefore, new β particles will nucleate wherever possible—at point imperfections such as vacancies and intersti-

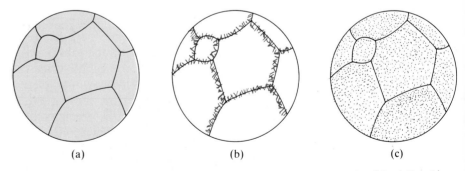

(a)   (b)   (c)

Figure 11–2.4 Solid-phase precipitation (schematic). (a) Supersaturated solid solution, for example, 90Pb–10Sn rapidly cooled from 180°C to 20°C. (b) Grain boundary precipitation. This requires long-range diffusion to the grain boundaries, where nucleation occurs more readily. (c) Intragranular precipitation. Many minor imperfections throughout the grains nucleate the new phase; therefore diffusion distances are shorter than in structure (b). Higher precipitation temperatures favor structure (b); lower, structure (c).

tials, along dislocation lines, and adjacent to impurities. These occur throughout the grains. Therefore, we see a different microstructure (Fig. 11–2.4c versus 11–2.4b).

Eutectoid Reactions   These reactions are rather common. In fact, there are four eutectoids in the Cu–Sn system (Fig. 10–5.2), as well as being present in some of the other phase diagrams of Chapter 10. Each has the general reaction

$$\text{Solid 2} \underset{\text{Heating}}{\overset{\text{Cooling}}{\rightleftharpoons}} \text{Solid 1 + Solid 3.} \qquad (11\text{--}2.4)$$

When Solid 2 is of eutectoid composition, the eutectoid reaction occurs at a single temperature—the eutectoid temperature. On cooling, the reaction goes from a phase of intermediate composition to two phases, whose compositions are on opposite sides of the eutectoid (Eq. 10–7.2).

If Solid 2 is offset from the eutectoid composition, one of the two solubility-limit curves is encountered first, so that one phase starts to separate ahead of the other. (For example, in a 0.4 C–99.6 Fe alloy, ferrite separates ahead of carbide; in a 1 C–99 Fe alloy, carbide separates first as indicated in Figs. 10–7.3 and 10–8.2f.)

Our prototype for a eutectoid reaction is the *isothermal decomposition of austenite* (Section 10–8). The time requirement for this isothermal decomposition will be considered in Section 11–6 where we will observe a C-type curve that is somewhat similar to the isothermal precipitation curve of Fig. 11–2.3 for a Pb–Sn alloy, and to the crystallization curve for a polymer (Fig. 6–6.5). Currently, we will focus our attention on the mechanism of pearlite growth.

Pearlite growth involves the simultaneous formation of ferrite, α, and carbide, $\overline{C}$. It starts at the austenite grain boundaries. The lamellae of the two phases grow inward. As this occurs, the carbon must be segregated as shown in Fig. 11–2.5. If the cooling rate is slow, the carbon can diffuse greater distances

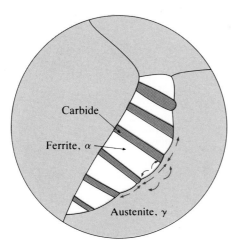

Figure 11–2.5 Pearlite formation. Carbon must diffuse from the eutectoid austenite (~0.8%) to form carbide (6.7%). The ferrite that is formed has negligible carbon.

Carbide

Ferrite, α

Austenite, γ

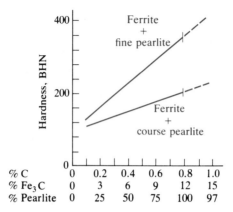
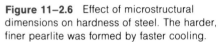

% C	0	0.2	0.4	0.6	0.8	1.0
% Fe₃C	0	3	6	9	12	15
% Pearlite	0	25	50	75	100	97

Figure 11–2.6 Effect of microstructural dimensions on hardness of steel. The harder, finer pearlite was formed by faster cooling.

and a *coarse pearlite* (thicker layers) is formed. If the cooling rate is accelerated, diffusion is limited to shorter distances. As a result, more (and thinner) lamellae are formed to give a *fine pearlite*.

The two pearlites just discussed (coarse and fine) have different properties as shown in Fig. 11–2.6. With faster cooling and finer lamellae, the pearlite is harder and stronger. Conversely, a slowly cooled steel with coarse pearlite is not as hard. The basis for this difference lies in the fact that the normally ductile ferrite cannot deform when it is adjacent to the rigid carbide. It undergoes *plastic constraint* because the two phases have a strong bond across their interface. The effect is lost only a few micrometers from the phase boundary. The finer pearlite has more phase boundary area per unit volume because there are more lamellae of the two phases. Thus, there is more plastically constrained ferrite. Figure 11–2.7 relates hardness to the boundary area S_V.

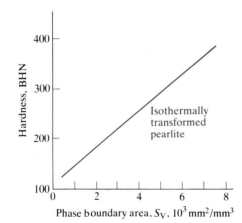

Figure 11–2.7 Strength versus microstructure (1080 pearlitic steel). The steel is harder with more ferrite-carbide boundary area (thinner lamellae). (F. Rhines, *Metal Progress.*)

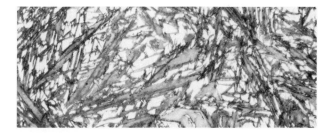

Figure 11–2.8 Martensite (×1000). The individual grains are plate-like crystals with the same composition as that of the grains in the original austenite. (Courtesy of U.S. Steel Corp.)

Martensite Formation

As indicated by Eq. (10–8.1), austenite decomposes to ferrite plus carbide ($\alpha + \overline{C}$). This assumes that there is time for the carbon to diffuse and become concentrated in the carbide phase and depleted from the ferrite. If we quench austenite very rapidly, Eq. (10–8.1) can be detoured. We can indicate this as follows:

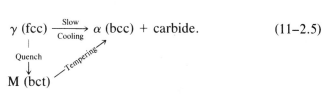

$$\gamma \text{ (fcc)} \xrightarrow[\text{Cooling}]{\text{Slow}} \alpha \text{ (bcc)} + \text{carbide.} \qquad (11\text{--}2.5)$$

This alternate route to form (α + carbide) involves a *transition phase* of *martensite*, M (Fig. 11–2.8). This polymorphic phase of iron is not stable because, given an opportunity, martensite will proceed to form ($\alpha + \overline{C}$). As a result, we do not see martensite on the Fe–Fe$_3$C diagram (Fig. 10–7.1). However, martensite is a very important phase, as we shall soon see.

Martensite forms at temperatures well below the eutectoid (but still above room temperature), because the fcc structure of austenite becomes sufficiently unstable. It changes spontaneously to a body-centered structure in a special way that does not involve diffusion, but results from a shearing action. All the atoms shift in concert, and no individual atom departs more than a fractional nanometer from its previous neighbors. Being diffusionless, the change is very rapid. All of the carbon that was present remains in solid solution. With more than 0.15 w/o carbon, the resulting body-centered structure is tetragonal (bct) rather than cubic.

Since martensite of steels commonly has a noncubic structure, and since carbon is trapped in the lattice, slip does not occur readily. Therefore, this martensite is hard, strong, and brittle. Figure 11–2.9 shows a comparison of the hardness of martensite with that of pearlite-containing steels as a function of carbon content. This enhanced hardness is of major engineering importance, since it provides a steel that is extremely resistant to abrasion and deformation.

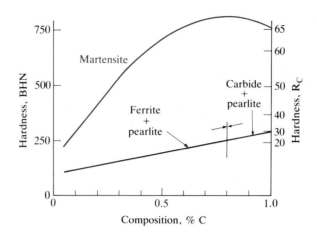

Figure 11–2.9 Hardness of annealed iron–carbon alloys (α + carbide) and martensite versus carbon content. This difference in hardness is the reason for the quenching of steel.

Tempered Martensite The existence of martensite as a *metastable* phase that contains carbon in solid solution in a bct structure does not alter the iron–carbide phase diagram (Fig. 10–7.1). With sufficient time at temperatures below the eutectoid temperature, the supersaturated solution of carbon in iron continues its progress to the more stable ferrite and carbide (Eq. 11–2.5). This process is known commercially as *tempering*:

$$\underset{\text{(Martensite)}}{M} \longrightarrow \underset{\text{(Tempered martensite)}}{\alpha + \text{carbide}}. \tag{11–2.6}$$

The resulting ($\alpha + \overline{C}$) microstructure is not lamellar like that of the pearlite, which we previously observed, but contains many dispersed carbide particles (Fig. 11–2.10) because there are numerous nucleation sites within the martensitic steel. This *tempered martensite** is much tougher than the metastable martensite, making it a more desirable product although it may be slightly softer.

The microstructure of tempered martensite becomes coarser with more time and/or higher tempering temperatures. This can be observed in Fig. 11–1.2 where we see the growth (*coalescence*) of the carbides into larger (and fewer) particles. The accompanying change in hardness is shown in Fig. 11–2.11. The initial martensite hardness (65 R_C) is not retained by tempered martensite. The coarsening of the carbides reduces the hardness to less than 50 R_C in a matter of seconds if the martensite is tempered at 600°C. At 400°C, the same softening takes approximately an hour. The hardness will drop to only 60 R_C within the first hour at 200°C.† The relationships of Fig. 11–2.11 are logarithmic, because

---

* Note that tempered martensite does not have the crystal structure of martensite. Rather, it is a two-phase microstructure containing *ferrite* and *carbide*. This microstructure originates by the decomposition of martensite.

† At 20°C, tempering occurs almost infinitely slowly. In fact, archeologists have found artifacts of early tools that are still martensitic after centuries have lapsed.

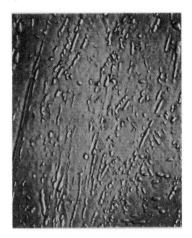

Figure 11–2.10 Tempered martensite (eutectoid steel, ×15,000). The steel was previously quenched to form martensite, which is a body-centered tetragonal (bct) phase and was >60 R_C. It was them tempered for 1 hour at 425°C (800°F). The tempered martensite is a two-phase microstructure containing carbide particles (light) in a matrix of ferrite (dark). Initially the martensite was very hard and brittle; however, with the above heating, the *tempered martensite* is now 44 R_C, but much tougher. (A. M. Turkalo, General Electric Co.)

tempering involves atom movements, and the required activation energy is thermally controlled.

The engineer adjusts the heat treatment process to meet the required specifications. Greater strengths accompany the higher hardnesses; however, there is a concurrent loss of ductility and toughness.

Example 11–2.1   Calculate the density of a solder (eutectic) that has been equilibrated at 20°C.

Procedure. From Fig. 10–1.3, the solid solder contains β and α that are essentially pure tin and lead, respectively. Therefore, from Appendix B, $\rho_\beta = 7.3$ Mg/m^3 and $\rho_\alpha \approx 11.3$ Mg/m^3 (or 11.3 mg/mm^3).

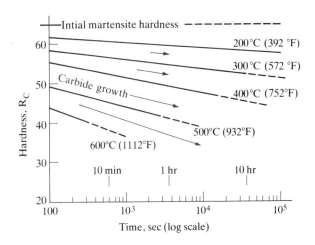

Figure 11–2.11 Hardness of tempered martensite (1080 steel quenched to 65 R_C). Softening occurs as the carbide particles grow (and decrease in numbers), giving greater intervening ferrite distances (Fig. 11–1.2). An Arrhenius pattern (Eq. 4–6.6) is followed in the early stages of softening.

Calculation. Basis: 100 mg = 61.9 mg β + 28.1 mg α.

$$\left.\begin{array}{ll}\beta: & 61.9 \text{ mg}/(7.3 \text{ g/mm}^3) = 8.48 \text{ mm}^3 \\ \alpha: & 28.1 \text{ mg}/(11.3 \text{ g/mm}^3) = 3.37 \text{ mm}^3\end{array}\right\} = 11.85 \text{ mm}^3 \text{ total}$$

$$\rho = 100 \text{ mg}/(11.85 \text{ mm}^3) = 8.44 \text{ mg/mm}^3 \qquad\qquad (\text{or } 8.44 \text{ Mg/m}^3).$$

Alternative. From the above calculation, the volume fractions are

$$f_\beta = 8.48 \text{ mm}^3/(11.85 \text{ mm}^3) = 0.715,$$

$$f_\alpha = 3.37 \text{ mm}^3/(11.85 \text{ mm}^3) = 0.285.$$

We can set up a *mixture rule* on a volume basis

$$\rho = \rho_\alpha f_\alpha + \rho_\beta f_\beta \qquad\qquad (11\text{–}2.7)$$
$$= (0.285)(11.3 \text{ g/cm}^3) + (0.715)(7.3 \text{ g/cm}^3)$$
$$= 8.44 \text{ g/cm}^3 \qquad\qquad (\text{or } 8.44 \text{ Mg/mm}^3). \quad \blacktriangleleft$$

Example 11–2.2   An Al–Cu alloy has 2 atom percent copper in solid solution κ at 550°C. It is quenched, then reheated to 100°C, where θ precipitates (Fig. 10–5.3). The θ (CuAl$_2$) develops many *very small* particles throughout the alloy so that the average interparticle distance is only 5.0 nm. (a) Approximately how many particles form per mm^3? (b) If, by extrapolating Fig. 10–5.3, we assume that negligible copper remains in κ at 100°C, how many copper atoms are there per θ particle?

Procedure. (a) Since particles are 5 nm apart, there is ~1 particle/(5 nm)^3. (b) We must determine the total number of Cu atoms per unit volume. This number equals 2 a/o of the atoms. At 100°C, essentially all of the Cu atoms are in the θ particles.

Calculation

a) ~1 θ/(5 × 10^{-9} m)^3 = 8 × 10^{24} θ/m^3 (or ~8 × 10^{15} θ/mm^3).

b)
$$\text{Atoms/m}^3 = 4/a^3 = 4/[4(\sim\!0.143 \times 10^{-9} \text{ m})/\sqrt{2}]^3$$
$$= 6 \times 10^{28}/\text{m}^3.$$

$$\text{Cu atoms/m}^3 = (0.02)(6 \times 10^{28}/\text{m}^3)$$
$$= 1.2 \times 10^{27}/\text{m}^3.$$

$$\text{Cu/particle} = \frac{1.2 \times 10^{27} \text{ Cu/m}^3}{\sim\!8 \times 10^{24} \theta/\text{m}^3} \approx 150 \text{ Cu}/\theta \text{ particle.}$$

Comment. This microstructure is approximately what we will encounter in precipitation hardening (Section 11–5). $\blacktriangleleft$

Example 11–2.3   The bct unit cell dimensions are a = 0.2845 nm and c = 0.2945 nm for martensite that contains 0.8 w/o carbon (3.6 a/o C). The lattice constant of austenite of the same composition and temperature is 0.3605 nm. How much volume change occurs when $\gamma \rightarrow$ M?

Calculation. Basis: 4 Fe atoms, or 2 u.c. M, and 1 u.c. γ.

$$\frac{\Delta V}{V_\gamma} = \frac{2(0.2845 \text{ nm})^2(0.2945 \text{ nm}) - (0.3605 \text{ nm})^3}{(0.3065 \text{ nm})^3}$$

$$= +0.018 \qquad\qquad (\text{or } 1.8 \text{ v/o}).$$

Comments. This volume change leads to residual stresses within quenched steel. For example, the surface of a steel gear transforms to martensite first (while the center is still a hot, deformable austenite). Shortly later, the center transforms to martensite with an expansion in volume. This stresses the surface, placing the martensite in tension (and compresses the transforming austenite).

Compare and contrast this with the tempering of glass (Figs. 7–6.4 and 7–6.5). ◄

Example 11–2.4

Compare the interphase boundary areas (mm^2/mm^3) in the two tempered steels of Fig. 11–1.2 ($\times 11{,}000$).

Solution. Use Eq. (4–4.1) and the perimeter of the photomicrograph. In part (b)

$$P_L = (\sim 20)/(210 \text{ mm}/11{,}000) \approx 10^3/\text{mm};$$

$$S_V \approx 2{,}000 \text{ mm}^2/\text{mm}^3.$$

In part (a)

$$S_V = 2(\sim 50)/(210 \text{ mm}/11{,}000)$$
$$\approx 5{,}000 \text{ mm}^2/\text{mm}^3.$$

Comments. The tempered martensite with the greater interphase boundary area is harder. In order to make direct comparisons, such as hardness, other microstructural features, such as grain shape and phase quantities, must be comparable. ◄

11–3

ANNEALING PROCESSES

The term *annealing* originated from craftsmen. They found benefits in heating some materials to elevated temperatures, after which the materials were generally cooled slowly (as opposed to quenching).

The purposes and benefits of annealing vary from material to material. Glass is annealed and slowly cooled to reduce the possibility of delayed cracking due to residual stresses. Annealing does not change the hardness of the glass product that we use. Cold-worked brass is annealed to soften it. We now know that the brass is recrystallized during annealing (Section 5–6). Of course, this cannot happen in glass because glass is noncrystalline, both before and after annealing (Section 7–5). Annealing leads to greater ductility in cast iron, and we often assume that softening is synonymous with ductility. Such is the case with the brass just cited; however, annealed steels and other multiphase alloys are less ductile than is possible by other heat treatments.

Annealing cannot be defined in terms of the resulting properties. Rather, it is a process by which *the material undergoes extended heating and is slowly cooled*. We must look at the individual materials to predict the properties that result from this heat treatment.

The following outline will point out the purposes and results of several common annealing treatments, together with their technical requirements. The "shop name" of the heat treatment is used.

Anneal

Material: Glass.

Purpose: To remove residual stresses, and to avoid thermal cracking. (Initial and final product have the same hardness.)

Procedure: Varies with glass composition since the glass transition temperature T_g must be approached to allow stress relief. Glass technologists select a temperature at which the viscosity η drops to 10^{12} Pa·s (10^{13} poises). This *annealing point* permits stress relaxation within a few minutes without deformation (Section 7–7). The glass must be cooled slowly past the *strain point* where the viscosity is $10^{13.5}$ Pa·s ($10^{14.5}$ poises). The slow cooling avoids the reintroduction of new thermal stresses. Below the temperature of the strain point where there has been a $\times 30$ increase in viscosity, cooling can be rapid since new residual stresses are not introduced.

Microstructural changes: None.

Stress Relief

Materials: Any metal, but steels in particular since they have a discontinuous volume change during transformation (Examples 3–4.1 and 11–2.3).

Purpose: To remove residual stresses (as in annealing glass); however, most metals are not subject to thermal cracking. Rather, they are subject to distortion and warping, if machining removes residual stresses nonsymmetrically.

Procedure: A few minutes just below the recrystallization temperature, which is ~600°C (~1100°F) for steels. Allowance must, of course, be made for the centers of large masses to attain this temperature.

Microstructural changes: None.

Recrystallization (or anneal*)

Materials: Cold-worked metals.

Purpose: To soften by removing strain hardening.

Procedure: $0.3 T_m$ to $0.6 T_m$, chosen to give a time compatible with the production sequences. Recrystallization is more rapid in pure metals than in alloys. It is more rapid with more highly strained metals (Section 5–6). (Steel sheet and wire should not be recrystallized above the eutectoid temperature, unless special care is taken for slow cooling. Otherwise, brittle martensite could form.)

Microstructural changes: New grains (Fig. 5–6.1).

Full Anneal

Materials: Steels.

Purpose: To soften prior to machining.

Procedure: Austenization 25°C–30°C (50°F) above the stability of the last ferrite (Fig. 11–3.1). This is followed by furnace cooling, so that the austenite decomposes into a coarse pearlite. The product is sufficiently soft so

---

* Depending on the equipment used, this may be called *box anneal, continuous anneal, process anneal,* etc.

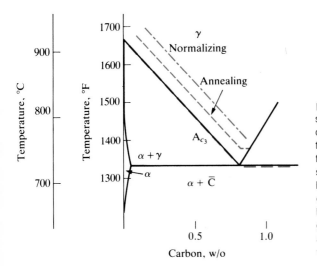

Figure 11–3.1 Annealing and normalizing (plain-carbon steels). The heat-treating temperature varies with the carbon content. For annealing, the temperature is sufficiently high to ensure complete disappearance of ferrite. The steel is then cooled slowly to form a coarse pearlite and a relatively soft product. For normalizing (Section 11–4), the steel is heated somewhat higher to promote more rapid atom diffusion and microstructural uniformity (only austenite). Excessive heating, however, would permit undesirable grain growth. Following austenization, the normalized steel is air-cooled to produce uniformly fine pearlite. Low hardness values are not a primary objective. (See Fig. 11–4.2a.)

that the steel may be machined. Annealed steel has less ductility than is possible with other heat treatments, but this is desirable since it leads to better chip formation during machining. (This is a relatively expensive operation because it ties up the furnace for a considerable period of time.)

Microstructure formed: Coarse pearlite (Fig. 11–1.4a).

Spheroidization

Materials: High-carbon steels, e.g., ball bearings.

Purpose: To toughen an otherwise brittle steel. In certain steels, it is necessary to have a high carbide content for wear resistance; however, as pearlite, it would have very little toughness (Fig. 11–1.4d).

Procedure: If initially pearlite, 16–24 hours just below the eutectoid temperature ($\sim$700°C). If initially martensite, 1–2 hours at that temperature (Fig. 11–1.2).

Microstructure formed: Spheroidite (Fig. 11–1.4b).

• **Malleabilization**

Material: Cast iron. (See Chapter 14.)

Example 11–3.1

Select the heat treatment to improve the machinability of a 1030 steel.

Solution. Since a combination of low ductility and low hardness are required, use a full anneal.

Temperature:	855°C (25°C–30°C into range of complete γ).
Time:	depends on size of product.
Cooling:	furnace cooled to produce coarse pearlite.

Comment. Heating to a higher temperature will require extra fuel and may cause undesirable grain growth. ◄

11–4

HOMOGENIZATION PROCESSES	A uniform product is desirable in most cases.* Greater uniformity can sometimes be achieved through appropriate heat treatments. "Shop names" for these processes include *homogenization, soaking,* and *normalizing.*

Soaking

Materials: Cast metals

Purpose: To make the composition more uniform. During solidification, the first solid to form is not the same as the overall composition.† This is illustrated in Fig. 11–4.1(a) for a 96 Al–4 Cu alloy. The appropriate phase diagram (Fig. 10–5.3) reveals that κ, the very first solid to form at ~650°C has only 1% copper; the copper content of the solid gradually increases as the temperature falls. At the eutectic temperature, the liquid contains 33% copper (and κ contains ~5.5% copper).

Procedure: Temperature as high as possible short of forming liquid or producing excessive grain growth. Some atoms must diffuse long distances compared to the atomic scale (~0.1 mm, or ~10^6 unit-cell distances). It is therefore necessary to heat the metal considerably above the recrystallization temperatures so that homogenization may occur in a reasonable length of time.

Microstructural changes: Greater homogeneity, and a closer match to the phase diagram.

Normalizing

Material: Steel

Purpose: To produce a uniform, fine-grained microstructure.

Procedure: Austenization, 50°C–60°C (100°F) into the range of complete austenite, followed by air-cooling (Fig. 11–3.1). The air-cooling avoids excessive proeutectoid segregation. Since low hardnesses are not required, the product is removed from the furnace and air-cooled, permitting the furnace to receive new products.

Microstructure: Fine pearlite, and absence of massive proeutectoid ferrite.

These processes as they pertain to steels are summarized in Fig. 11–4.2. The reference heat-treating temperatures vary with the steel composition (Figs. 10–7.3 and 10–9.1).

---

* Of course there are exceptions. A carburized gear is intentionally nonuniform in order to provide a hard, wear-resistant, high-carbon surface over a tougher steel core. Likewise, an asphalt road is nonuniform on a microscale since it is a mixture of particles of rock within a soft matrix that has a markedly different composition. Even so, the engineer will specify good mixing to provide "uniform heterogeneity."

† Cf. Example 10–5.2.

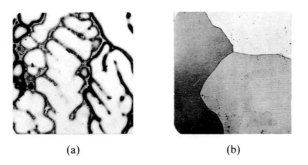

(a)   (b)

Figure 11-4.1 Solidification segregation and homogenization (96Al-4Cu). Segregation may arise from rapid solidification, during which there has been insufficient time for diffusion to provide equilibrium (Fig. 10-5.3). (a) The cast alloy has a copper-rich residual (dark) between the growing crystals. (b) The same alloy after "soaking." The copper has diffused until it is uniform throughout each grain. (These grains are oriented differently; therefore, they reflect light differently to give the several shades of gray. Within each grain there is general homogeneity.) (Courtesy of Aluminum Company of America.)

Example 11-4.1

Let 100 g of a 96 Al-4 Cu alloy equilibrate at 620°C (1150°F), forming the κ and liquid, as shown by the phase diagram (Fig. 10-5.3). The alloy is then cooled rapidly to 550°C (1020°F) with no chance for the initial solid to react. A liquid phase will still be present.

a) What will its composition be?

b) How many grams will there be of this final liquid?

Procedure. Determine the amount and the composition of the 620°C liquid. The solid that is present at 620°C has "no chance to react." Therefore, it can be ignored in this problem. Therefore, as the liquid is cooled to 550°C, it behaves as a separate alloy and forms a new solid-liquid pair. What is the composition of the liquid in this second generation pair? How much liquid?

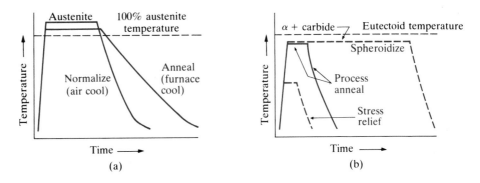

Figure 11-4.2 Steel heat-treatment processes (schematic). (a) Austenization processes. (b) Subeutectoid processes.

Calculations. At 620°C, and from Fig. 10–5.3,

$$\text{Liq:} \qquad \text{88 Al–12 Cu,} \qquad \qquad \kappa: \qquad \text{98 Al–2 Cu.}$$

$$\text{Grams liquid} = 100 \text{ g} \left(\frac{98 - 96}{98 - 88} \right) = 20 \text{ g liquid.}$$

At 550°C, and with only the 20 g of alloy reacting,

$$\kappa: \qquad \text{94.4 Al–5.6 Cu.}$$

a) $$\qquad\qquad\qquad \text{Liq:} \qquad \text{67 Al–33 Cu.}$$

b) $$\qquad\qquad\qquad \text{Grams liquid} = 20 \text{ g} \left(\frac{94.4 - 88}{94.4 - 67} \right) = 4.6 \text{ g liquid.}$$

Comments. The final 4.6 g of liquid produces local areas with high copper segregation (33% Cu). There will also be nearly 80 g of metal with only 2% Cu.

Cooling may be considered to take place in a series of small steps, in which the solute in the residual liquid is progressively concentrated with solute. Conversely, the initial solid is purer than average. ◀

11–5

PRECIPITATION PROCESSES

Precipitation-hardening (age-hardening)

A very noticeable increase in hardness may develop during the *initial stages of precipitation* from a supersaturated solid solution (Section 11–2 and Eq. 11–2.3). In fact, the *start* of the precipitation in Fig. 11–2.3 can be detected by this increased hardness. (This *precipitation-hardening* is commonly called *age-hardening* because it develops with time.) The prime requirement for an alloy that is to be age-hardened is that solubility decreases with decreasing temperature, so that a supersaturated solid solution may be obtained as shown in Figs. 11–2.2 and 11–5.1(a). Numerous metal alloys have this characteristic.

The process of age-hardening involves a *solution treatment* (Eq. 11–2.2) followed by a *quench* to supersaturate the solid solution. Usually the quenching is carried to a temperature where the precipitation rate is exceedingly slow. After the quench, the alloy is reheated to an intermediate temperature* at which precipitation is initiated in a reasonable length of time. These are the two steps *XA* and *AB* in Fig. 11–5.1, and Table 11–5.1.

Observe the enhanced properties for the age-hardened alloy (*XAB*) in Table 11–5.1, as compared to annealing the same alloy (*XD*). The former has several times the yield strength of the annealed material and at the same time possesses greater ductility. As a result, the toughness is increased markedly by age-hardening. Greatest ductility is obtained when only one phase is present, i.e., after solution-treatment (plus quenching to preserve the single phase).

An interesting example of the utility of the age-hardening process is the way it is used in airplane construction. Aluminum rivets are easier to drive and fit more tightly if they are soft and ductile, but in this condition they lack the

---

* But below the "knee" of the C-curve (Fig. 11–2.3), to produce *intra*grain precipitation (Fig. 11–2.4c).

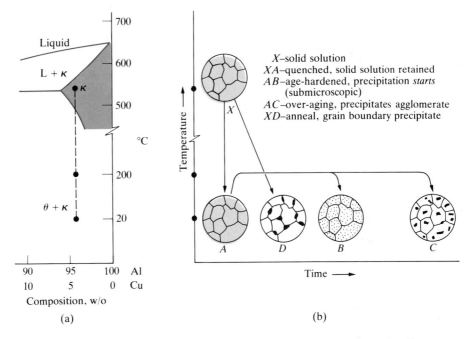

Figure 11–5.1   Age-hardening process (96% Al–4% Cu alloy). See Table 11–5.1. The precipitates are still submicroscopic at the time of maximum hardness.

desired strength. Therefore the manufacturer selects an aluminum alloy that can be quenched as a supersaturated solution, but that will age-harden at room temperature. The rivets are inserted while they are still relatively soft and ductile, and they harden further after they have been riveted in place. Since hardening sets in fairly rapidly at room temperature, there arises the practical problem of delaying the hardening process if the rivets are not to be used almost immediately after the solution treatment. Here advantage is taken of the known effects of temperature on the reaction rate. After the solution treatment the

	TABLE 11–5.1					
	Properties of an Age-hardenable Alloy (96% Al–4% Cu)					
	Treatment (See Fig. 11–5.1)	**Tensile strength**		**Yield strength**		**Ductility, % in 5 cm (2 in.)**
		MPa	**(psi)**	**MPa**	**(psi)**	
XA	Solution-treated (540°C) and quenched (20°C)	240	(35,000)	105	(15,000)	40
XAB	Age-hardened (200°C, 1 hr)	415	(60,000)	310	(45,000)	20
XAC	Over-aged	~170	(25,000)	~70	(10,000)	~20
XD	Annealed (540°C)	170	(25,000)	70	(10,000)	15

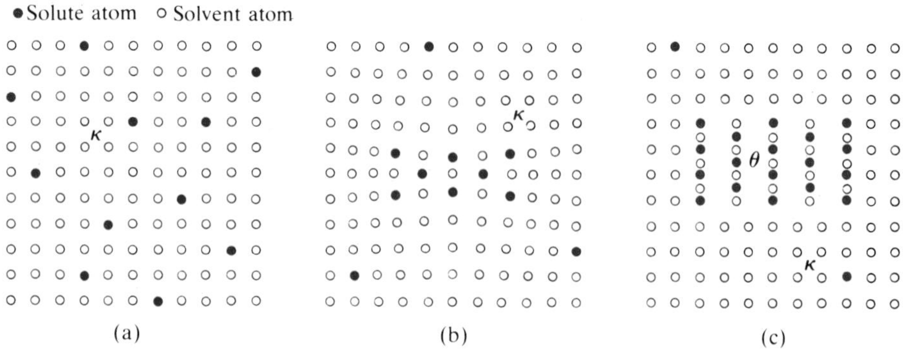

●Solute atom   ○Solvent atom

(a)   (b)   (c)

Figure 11–5.2 Age-hardening mechanism. (a) κ solid solution. (b) Age-hardened; the θ precipitation has been initiated. Since the two structures are coherent at this stage, there is a stress field around the precipitate. (c) Over-aged. There are two distinct and noncoherent phases, κ and θ. With limited numbers of solute atoms, maximum interference to dislocation movements occurs in part (b). (A. G. Guy and J. J. Hren, *Elements of Physical Metallurgy*, Addison-Wesley.)

rivets are stored in a refrigerator, where the lower temperature will delay hardening for reasonable lengths of time.

Detailed studies have produced the following interpretation of the age-hardening phenomenon. The supersaturated atoms (Cu atoms in Example 11–2.2 and Fig. 11–5.1"*B*") tend to accumulate along specific crystal planes in the manner indicated in Fig. 11–5.2(b). The concentration of the copper (solute) atoms in these positions lowers the concentrations in other locations, producing less supersaturation and therefore a more stable crystal structure. At this stage, the copper atoms have not formed a phase that is wholly distinct; a *coherency* of atom spacing exists across the boundary of the two structures. Dislocation movements proceed with difficulty across these distorted regions, and consequently the metal becomes harder and more resistant to deformation under high stresses.

Over-aging

A continuation of the local segregation process over long periods of time leads to true precipitation and *over-aging,* or softening. For example, the development of a truly stable structure in an alloy of 96% aluminum and 4% copper involves an almost complete separation of the copper from the fcc aluminum at room temperature. According to Fig. 10–5.3, nearly all the copper forms $CuAl_2$ (θ in Fig. 11–5.2c). Because the growth of the second phase provides larger areas that have practically no means of slip resistance, a marked softening occurs.

Figure 11–5.3 shows data for the aging and over-aging of a commercial aluminum alloy (2014). The initial hardening is followed by softening as the resulting precipitate is agglomerated. Two effects of the aging temperature may be observed: (1) precipitation, and therefore hardening, starts very quickly at

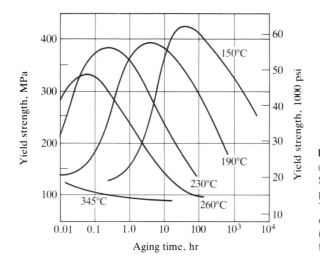

Figure 11–5.3 Aging and over-aging (2014 aluminum). Softening occurs as the precipitated particles grow. This proceeds more rapidly at elevated temperatures. (*Aluminum,* American Society for Metals.)

higher temperatures; (2) over-aging, and therefore softening, occurs more rapidly at higher temperatures. These two phenomena overlap to affect the maximum hardness that is attained. Lower temperatures permit greater increases in hardness, but longer times are required.

Combined Hardening   Occasionally it is desirable to combine two methods of hardening. The cold-working of an alloy that has previously been age-hardened increases the hardness still further. However, there are some practical difficulties encountered in this process. Age-hardening increases resistance to slip and therefore increases the energy required for cold-working, and it also decreases ductility so that rupture occurs more readily during cold-working. A possible alternative is to cold-work the metal prior to the precipitation-hardening treatment. The metal is cold-worked more readily, and the age-hardening reaction occurs at a lower temperature because the dislocations serve as nuclei for the precipitation. However, the temperature of the aging process that follows cold-working may relieve some of the strain hardening and cause a slight loss in hardness. Although it does not produce hardnesses as great as those obtained from the reverse order, the final hardness is greater than that developed by using either method alone (Table 11–5.2).

• High-strength,
Low-alloy (HSLA)
Steels
Low-alloy steels have undergone some recent developments that have led to significantly higher strength products. Their development has provided the engineer with *structural* steels that have yield strengths of greater than 500 MPa (>70,000 psi) in contrast to earlier structural steels of only half that figure.

The HSLA steels gain their strength from a very small grain size and finely distributed precipitates within the ferrite. In the latter respect, they may be compared to precipitation-hardened aluminum alloys, which have just been

TABLE 11–5.2

Tensile Strengths of a Strain- and Age-hardened Alloy (98% Cu–2% Be)

Annealed, 870°C	240 MPa	35,000 psi
Solution-treated, 870°C and cooled rapidly	500	72,000
Age-hardened only	1200	175,000
Cold-worked only (37%)	740	107,000
Age-hardened, then cold-worked*	1380	200,000
Cold-worked, then age-hardened	1340	195,000

* Cracked

discussed. The origin of the precipitate is significantly different, however. If "carbide-formers" such as vanadium or niobium are present, along with 0.05%–0.2% carbon, we can still obtain a single phase (austenite) at hot-rolling temperatures. Thus beams, pipe, etc., can be readily processed. However, as the proeutectoid ferrite forms during austenite decomposition, there is a concurrent precipitation of alloy carbides. The precipitate forms not because of the decreasing solubility curve for *one* phase but rather because these alloy carbides are less soluble in α than in γ.

Unlike quenched and tempered steels (Section 11–7), these HSLA steels have the advantage of being weldable and of maintaining their mechanical properties without subsequent heat treatment. As a result, they are attractive to those engineers who design cars, major structures, pressurized pipelines, etc.

Example 11–5.1

Use information in Fig. 11–5.3 to estimate the temperature required to reach the maximum hardness for that aluminum alloy in 10,000 hours (~14 months).

Procedure. Assume that the Arrhenius relationship applies (Eq. 11–1.4). Use the peaks of the 150°C curve (30 hr) and the 260°C curve (3 min or 0.05 hr).

Calculation. At 150°C,

$$\ln t = \quad \ln 30 = C + B/423 = 3.4.$$

At 260°C,

$$\ln t = \ln (0.05) = C + B/533 = -3.0.$$

Solving simultaneously,

$$C = -27.6, \quad B = 13,100 \text{ K}.$$

$$\ln 10^4 \text{ hr} = 9.21 = -27.6 + 13,100/T,$$

$$T = 356 \text{ K} = 83°C.$$

Comments. We may check our Arrhenius assumption at 230°C (503 K) and 190°C (463 K):

$$\ln t_{230} = -27.6 + 13{,}100/503 = -1.56, \ t = 0.2 \ \text{hr};$$

$$\ln t_{190} = -27.6 + 13{,}100/463 = 0.69, \ t = 2 \ \text{hr}.$$

The agreement with the experimental data in Fig. 11–5.3 is not exact, but reasonable; however, an extrapolation to still longer times or lower temperatures is an approximation. ◄

11–6

STEELS: $\gamma \rightarrow (\alpha + \bar{C})$ REACTION RATES

The next three sections use steel as a prototype for the use of heat-treating processes to control microstructures and hence properties. Steel has been chosen as the example for two reasons. (1) Steel is a very versatile material. As stated earlier in this text, it may be easily formed into products; it may also be made hard and strong enough to cut other steels. (2) Because steels are widely used, their nature and behavior are better known than are most other materials.

Isothermal Transformation of $\gamma \rightarrow (\alpha + \bar{C})$

The time requirement for austenite decomposition ($\gamma \rightarrow \alpha + \bar{C}$) has been studied in considerable detail because of its practical importance. Figure 11–6.1 shows time–temperature data for this reaction in a eutectoid steel (AISI-SAE 1080). The left solid curve t_s is the time required for the decomposition to *start*. The right solid curve t_f is the time to *finish* the $\gamma \rightarrow (\alpha + \bar{C})$ reaction. These

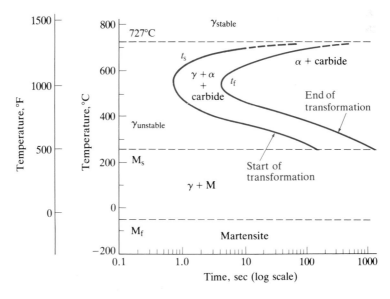

Figure 11–6.1 Isothermal-transformation curves for austenite decomposition (SAE 1080). (Adapted from U.S. Steel Corp. data.)

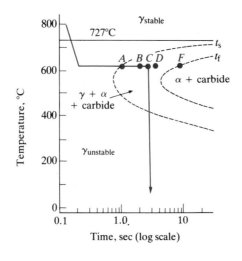
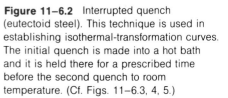

Figure 11–6.2 Interrupted quench (eutectoid steel). This technique is used in establishing isothermal-transformation curves. The initial quench is made into a hot bath and it is held there for a prescribed time before the second quench to room temperature. (Cf. Figs. 11–6.3, 4, 5.)

curves are called *isothermal-transformation diagrams,* or I-T *diagrams.** The data for Fig. 11–6.1 were obtained as follows. Small samples of eutectoid steel were heated into the austenite temperature range sufficiently long to assure complete transformation to austenite. These wire-size samples were then *quenched* to a lower temperature (e.g., 620°C) and held isothermally (at a constant temperature) for varying lengths of time before being quenched further to room temperatures (Fig. 11–6.2). The change $\gamma \rightarrow (\alpha + \overline{C})$ was not observed in samples held at 620°C for less than one second, and complete transformation to α + carbide was not observed until after more than 10 seconds had elapsed (Figs. 11–6.3 through 11–6.5). Similar data were obtained for other temperatures until the completed diagram shown in Fig. 11–6.1 was established.

The I-T diagram shows that the austenite transformation occurs slowly, both at high temperatures (close to the eutectoid) and at low temperatures. It is slow at higher temperatures because there is not enough supercooling to readily nucleate the new ferrite and carbide from the previous austenite. The austenite decomposition is slow at lower temperatures because diffusion rates are slow and therefore carbon separation from the ferrite into the carbide is slow. At intermediate temperatures, the nucleation is sufficiently rapid and the atomic migration is fast enough to quickly start and complete the reaction.

These curves may be compared with the isothermal precipitation curve of Fig. 11–2.3. That was a single curve simply because only the midpoint of the

---

* They may also be called C-curves (because of their shape) or TTT curves (**Temperature–Time–Transformation**).

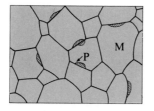

Figure 11–6.3 The beginning of austenite transformation at 620°C (1150°F) ($\gamma \rightarrow \alpha$ + carbide). After 1 second (point A of Fig. 11–6.2) the change to pearlite P is underway. M is metal that has not transformed to α + carbide.

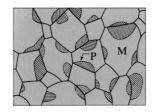

Figure 11–6.4 Transformation 25% complete at 620°C (1150°F). (Point B in Fig. 11–6.2.)

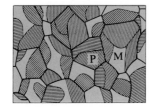

Figure 11–6.5 Transformation 75% complete at 620°C (1150°F). (Point D in Fig. 11–6.2.)

reaction was shown. A similar 50% transformation curve could be drawn on Fig. 11–6.1; it would be about halfway between the t_s and t_f curves.*

With extremely fast cooling (severe quenching), it is often possible to miss the "knee" of the curve for the beginning of the transformation, and to cool the steel to room temperature without the formation of ferrite and carbide, because the decomposition is detoured through the martensite (Eq. 11–2.5). In fact, this is the purpose of quenching steel in regular heat-treating operations.

The example just considered is the transformation of a eutectoid steel (1080) when there has been no advance separation of ferrite (or cementite) from the austenite prior to the formation of the pearlite ($\alpha + \bar{C}$). Figure 11–6.6(a)

---

* The interpolation is not linear, however. The reaction quartiles (25% and 75%) lie close to the 50% point because of the sigmoidal nature of the reaction curves. (See Fig. 5–6.4.) Thus, points B and D of Fig. 11–6.2, which are represented by Figs. 11–6.4 and 11–6.5, lie closer to C than to t_s and t_f.

shows a similar transformation diagram for an SAE 1045 steel. Two features differ from those in Fig. 11–6.1:

1. Some ferrite may separate from the austenite above the eutectoid temperature. This could be predicted from the phase diagram shown at the right side of Fig. 11–6.6.

2. The isothermal transformation of 0.45 carbon steel occurs somewhat faster than the transformation of eutectoid steel. Comparison of the "knees" of the two curves shows this difference: the higher carbon steel starts to transform in about one second; in the 0.45% carbon steel, the reaction starts sooner. In fact, in the latter case the reaction occurs sufficiently fast so that we are not able to measure the time at the "knee" of the curve with the interrupted-quench technique described above. A lower carbon content permits faster reaction, since part of the transformation delay is associated with the movement of the carbon atoms.

Isothermal-transformation curves show the start and finish of a reaction with *time* as a variable. Therefore, they are *not* equilibrium diagrams such as we encountered in Chapter 10. However, they show one thing in common with the equilibrium diagrams. The far right side of an I-T diagram represents extended periods of time. Therefore, the phases that are present should match the phases shown in a phase, or an equilibrium diagram. As an example, the phases

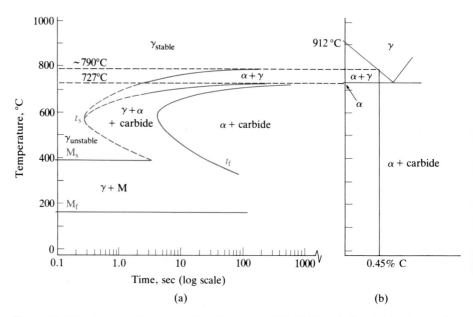

Figure 11–6.6 Isothermal-transformation diagram for SAE 1045 steel. The stable phases of the phase diagram (right) are not achieved immediately. However, in this steel, the austenite decomposition reaction is faster than in a eutectoid steel (Fig. 11–6.1).

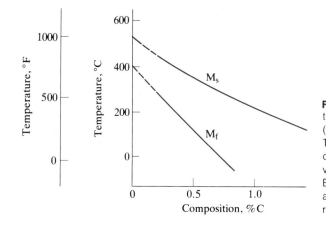

Figure 11–6.7 Martensite transformation temperatures (plain-carbon steels). Transformation is first detected at M_s and is virtually complete at M_f. Between M_s and M_f, austenite is retained as a result of induced stresses.

eventually produced according to the I-T diagram for 1045 steel (Fig. 11–6.6a) match those found at 0.45% carbon on the Fe–Fe$_3$C diagram (Fig. 11–6.6b).

Also observe that an I-T diagram is for a particular material and a specific reaction. Thus, once $\alpha + \overline{C}$ is formed from austenite, that I-T diagram has no further use. This may be illustrated by the $\gamma \rightarrow \alpha + \overline{C}$ reaction in 1080 steel (Fig. 11–6.1). If the austenite is quenched to 550°C and held there 10 seconds, *the reaction is complete.* Now, if we were to lower the temperature to 300°C, we must *not* expect to make continued use of the curves shown in Fig. 11–6.1. Specifically, the $\alpha + \overline{C}$ *cannot go back to γ at 300°C.* That would be contrary to the phase diagram. The only way to get austenite again is to raise the temperature into the austenite field of Fig. 10–7.3.

Martensite Temperatures

At the base of Figs. 11–6.1 and 11–6.6 are shown the temperatures at which the martensite reaction (Eq. 11–2.5) starts, M_s, and finished, M_f, when the steel is quenched rapidly enough to miss the "knee" of the C-curve. Since diffusion is not involved,* the process of the martensite formation is not time-dependent. However, M_s and M_f depend on the amount of carbon that is present (Fig. 11–6.7). As a result, a water-quench does not complete the $\gamma \rightarrow M$ reaction in high-carbon alloys such as those used in tool steels because M_f drops below 0°C. Thus, they possess *retained austenite* after quenching.

Retarding Austenite Transformation

From the foregoing discussion, we see that temperature obviously affects the isothermal decomposition of austenite γ to ferrite plus carbide ($\alpha + \overline{C}$). Other factors modify the time requirements, too. We will consider two: (1) *austenite grain size,* and (2) *alloy retardation.*

As sketched schematically in Figs. 11–6.3 through 11–6.5, the ferrite and carbide formation from the austenite starts at the grain boundary. A steel with a

---

* Austenite changes to martensite by shear (Section 11–2).

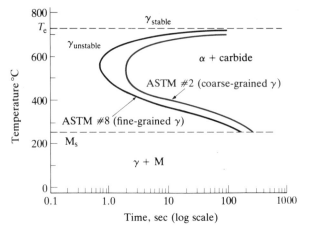

Figure 11–6.8
Grain-boundary nucleation (eutectoid steel). The effect of austenite grain size on the *start* of austenite decomposition is shown for eutectoid steels. The steel with the fine-grained austenite has more grain-boundary area from which the ($\gamma \rightarrow \alpha + \overline{C}$) reaction can start.

small austenite grain size offers more grain-boundary area per unit volume on which decomposition can be nucleated than does a coarse-grained steel. This is demonstrated in Fig. 11–6.8 where the two curves are for fine-grained (G.S. #8) steel and coarse-grained (G.S. #2) steel. The time requirements for the *start* of transformation differs by a factor of three at almost all temperatures.

The alloy retardation effect is even more pronounced. As an example, only 0.25% molybdenum (Fig. 11–6.9) delays the start of isothermal transformation by a factor of four at temperatures below 650°C (1200°F). This delay occurs because not only the carbon but also the molybdenum must be relocated when the austenite decomposes. Most of the Mo goes into the carbide. When silicon is the alloying element, most of it goes into the ferrite. Since all alloying ele-

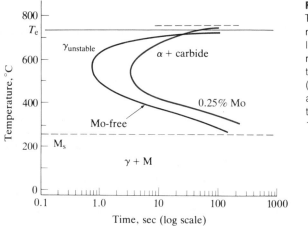

Figure 11–6.9
Transformation retardation. Molybdenum, like other alloying elements, retards the *start* of transformation of austenite. (Note: The addition of Mo also raises the eutectoid temperature, T_e (Fig. 10–9.1).)

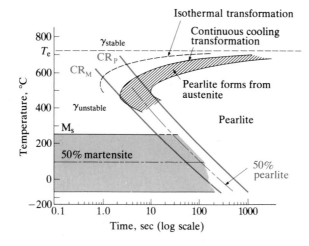

Figure 11–6.10
Continuous-cooling transformation (eutectoid steel). Transformation temperatures and times are displaced from the isothermal-transformation curve for the same steel. (Cf. Fig. 11–6.1.) CR_M = minimum cooling rate for 100% martensite. CR_P = maximum cooling rate for 100% pearlite.

ments—Cr, Ni, Si, Mn, Mo, Ti, W, etc.—diffuse more slowly in iron than carbon does (Fig. 4–7.3), we find that all low-alloy steels (Table 10–9.1) have their isothermal transformation curves shifted to the right. In fact, it is easy to cool many low-alloy steels to low temperatures and miss the "knee" of the transformation curve entirely, thus forming 100 percent martensite.

Continuous-cooling Transformation (CCT)

Isothermal-transformation diagrams are convenient for interpretation, because we can hold the effect of temperature constant during the austenite decomposition. In practice, however, we more commonly encounter continuous cooling: (1) a hot piece of steel is removed from the furnace and air cooled, or (2) the steel is quenched in water. In neither is there an isothermal holding period while the $(\alpha + \bar{C})$ forms.

Let us consider the progress of austenite decomposition with different reaction rates of continuous cooling. A severe quench will miss the "knee" of the transformation curve, with the result that the austenite is changed to martensite rather than to pearlite (i.e., $\alpha + \bar{C}$). Slow cooling permits pearlite to form; however, the start of decomposition occurs after a longer time (and therefore at a lower temperature) than for isothermal transformation, simply because part of the time was spent at higher temperatures where reactions were initiated more slowly. Thus the isothermal-transformation curves are displaced downward and to the right for *continuous-cooling* transformation (Fig. 11–6.10).

There are two important *critical cooling rates* for the continuous-cooling decomposition of austenite; these are included in Fig. 11–6.10. The first is the cooling rate, CR_M, which just misses the "knee" of the transformation curve; *more* rapid cooling rates to the left of this curve produce *only martensite*. The second is the cooling rate, CR_P, which produces no martensite; *less* rapid cooling rates to the right of the CR_P curve produce *only pearlite*. In a eutectoid steel (0.8 C–99.2 Fe), these two critical cooling rates are approximately 140°C/s and 35°C/s, respectively, through 700°C (1300°F), which is in the eutectoid temperature range. Both of these rates are significantly slower in any steel that contains alloying elements because the alloy content slows down the $\gamma \rightarrow (\alpha + \overline{C})$ reaction (Fig. 11–6.9).

Study Aid (transformation rates in steels)

The isothermal transformation (or time–temperature–transformation) diagram is presented somewhat differently in *Study Aids for Introductory Materials Courses* than in this text. Some students may find that this second presentation will be helpful.

Example 11–6.1

Three AISI–SAE 1045 steel wires underwent the following thermal steps in the indicated sequences. Give the phases after each step and their approximate chemical analyses. (The symbol # means that the wire was held at that temperature until equilibrium was reached.)

		Time held
Wire (a)	1) Heated to 820°C (1510°F)	#
	2) Quenched to 560°C (1040°F)	0
	3) Held at 560°C	1 min
	4) Reheated to 820°C	#
Wire (b)	1) and 2) Same as wire (a), 1) and 2)	#
	3) held at 560°C	1 s
	4) Quenched to 430°C (805°F)	0
Wire (c)	1) Heated to 730°C (1345°F)	#
	2) Quenched to 430°C (805°F)	0
	3) Quenched to 330°C (625°F)	10 s
	4) Held longer at 330°C	#

Procedure. For equilibrium, use the phase diagrams. For isothermal transformation, use the I-T diagram. A quench does not allow for diffusion; however, γ can change to martensite below the M_s.

Answers

Wire (a) 1) γ 0.45 C–99.55 Fe.
 2) Same, but austenite is metastable (Fig. 11–6.6).
 3) α Negligible carbon; $\overline{C}$ 6.7% carbon.
 4) γ Same as (1).

Wire (b) 3) γ(0.45% C) + [α(negligible carbon) + $\overline{C}$(6.7% C)].
 4) Same as (3). ($\alpha + \overline{C}$) will *not* revert to austenite below the eutectoid temperature.

Wire (c) 1) α(0.02% C) + γ(0.8% C). See the phase diagram.
 2) Note that nothing is expected to happen to the ferrite as it cools. The austenite is of eutectoid composition; therefore, we should turn to Fig. 11–6.1 for its transformation. At zero time, unstable γ is present.
 3) Still, α(0.02% C) + unstable γ(0.8% C).
 4) $\alpha + \overline{C}$(6.7% C). ◀

Example 11–6.2   A thin (<0.5 mm), one-gram sample of 1045 steel undergoes the following steps during a heat treating process. (The symbol # means equilibrium was reached.)

	Time held
1) Heated to 730°C	#
2) Quenched to 550°C	10 s
3) Quenched to 100°C	0 s

What phases are present after each step? What is their carbon content? Approximately how much of each phase?

Procedure. Follow the same procedure as in Example 11–6.1.

Solution and comments

    1) 0.44 g α(0.02% C), and 0.56 g γ(0.8% C)

We now have a mixture of α and γ. Upon quenching, nothing happens to the ferrite (except getting colder). The austenite is 1080 steel on a microscopic scale. Therefore, we must use the I-T diagram for 1080 (Fig. 11–6.1), and not the 1045 diagram (Fig. 11–6.6).

    2) 0.44 g α(0.02% C);
        0.56 g $\gamma \to$ 0.5 g α(0.02%), and 0.06 g $\bar{C}$(6.7% C).
        Total α = 0.94 g α.

We are now through with Fig. 11–6.1, since the γ-decomposition is complete.

    3) 0.94 g α(0.02% C), and 0.06 g $\bar{C}$(6.7% C).

The quench to 100°C does not permit the last, small amount (0.02%) of carbon to separate from the ferrite. ◄

• Example 11–6.3   By what ratio was the grain size changed to decrease the start of transformation in Fig. 11–6.8 from the left curve (G.S. = #8) to the right curve (G.S. = #2)? Repeat for the boundary area.

Solution. Refer to Eq. (4–4.2). With ×100 and G.S. #8,

$$N = 2^{8-1} = 128 \text{ grains per } (0.0645 \text{ mm}^2).$$

With ×100 and G.S. #2,

$$N = 2^{2-1} = 2 \text{ grains per } (0.0645 \text{ mm}^2);$$

$$N_8/N_2 = 128/2 = 64.$$

Let δ be a representative linear dimension of a grain, which varies inversely with the square root of the number of grains:

$$\delta_2/\delta_8 = \sqrt{N_8/N_2} = \sqrt{64} = 8.$$

For the boundary area, assume either a cubelike or a spherelike grain, where

$$\text{Boundary area} = 3/\delta. \qquad \text{(See comments.)}$$

Therefore,

$$B_2/B_8 = (3/\delta_2)/(3/\delta_8) = 1/8.$$

Comments. The surface area/volume ratio of a cube is $6a^2/a^3 = 6/a$; the same ratio of a sphere is $4\pi r^2/(4/3)\pi r^3 = 6/d$. The boundary area/volume is $3/\delta$ since each boundary is shared by two adjacent grains. Of course, the grains are neither spheres nor cubes, but something in between; thus the B/V ratio remains $3/\delta$. ◀

• 11–7

**STEELS:
COMMERCIAL
HEAT-TREATING
PROCESSES**

Since austenite can decompose in several different ways, the engineer has a choice of different microstructures. These are summarized in Table 11–7.1 and Fig. 11–7.1. The resulting properties vary significantly.

Annealing

This has already been considered in Section 11–3 (also see Fig. 11–4.2a). The resulting properties are shown in Fig. 11–1.3 for plain-carbon steels.

Quenching

Cooling rates faster than CR_M of Fig. 11–6.10 give hard (and relatively brittle) martensite, by avoiding the transformation to $(\alpha + \overline{C})$. This is the purpose of quenching steels.

Austenite is denser than martensite (and also denser than ferrite plus carbide). This presents a problem with a direct austenite-to-martensite quench, because the slower-cooling central region of a gear or other steel part transforms and expands after the quickly cooled surface has formed brittle martensite. Hence, cracking can occur if the steel is larger than sheet or wire dimensions, particularly if the carbon content is greater than 0.5%. This is an added reason for alloy additions to the steel. With more time available in which to

---

TABLE 11–7.1

Transformation Processes for Steels*

Process	**Purpose**	**Procedure**	**Phase(s)**
Annealing	To soften	Slow cool from γ-stable range	α + carbide
Quenching	To harden	Quench more rapidly than CR_M	Martensite†
Interrupted quench	To harden without cracking	Quench, followed by slow cool from M_s to M_f	Martensite†
Austempering	To harden without forming brittle martensite	Quench, followed by isothermal transformation above the M_s	α + carbide
Tempering	To toughen (usually with minimal softening)	Reheating of martensite	α + carbide

* Cf. Fig. 11–7.1.

† Steels containing martensite must be toughened by the tempering process.

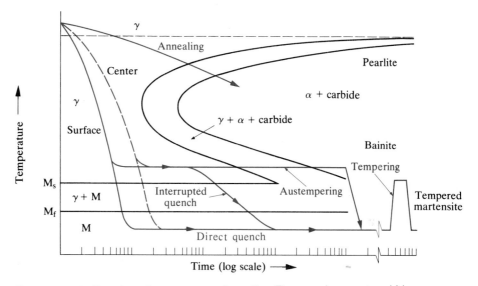

Figure 11–7.1 Transformation processes. *Annealing:* The normal $\gamma \rightarrow \alpha$ + carbide transformation occurs. *Direct quench:* Martensite forms, first in the surface, then in the center. Severe stresses result. *Interrupted quench:* Time is available for the surface and center to transform nearly simultaneously, thus avoiding the quench-cracking found in direct quenching. *Tempering:* Both the direct and the interrupted quench must be followed by a tempering process to complete the transformation. *Austempering:* Quenching avoids pearlite formation, but the $\gamma \rightarrow \alpha$ + carbide transformation may still occur above the M_s. The resulting microstructure is bainite (Fig. 11–7.2).

form martensite (Fig. 11–6.9), an alloy steel can be cooled more slowly so that the surface and center transform more or less concurrently; thus it is possible to avoid the sharp volume change differentials and the resulting stresses that promote cracking.

Interrupted Quench   By examining the isothermal-transformation diagram, we can see an alternative to the direct quench just described. That is an interrupted quench (also called *martempering* or *marquenching*).

In this process, the steel is quenched rapidly past the "knee" of the transformation curve to avoid $(\alpha + \overline{C})$ formation, but the cooling is interrupted just above the M_s temperature. Cooling is then continued at a slow rate through the martensite range to ambient temperatures, so that the surface and the center of the steel may transform more or less simultaneously, thus avoiding quenching cracks. Slower cooling is possible at these lower temperatures because the $(\alpha + \overline{C})$ transformation is delayed, while the martensite forms directly with the drop in temperature.

This process is more complicated from the production viewpoint because the cooling rate must be shifted from a quench to a "hold," and then to a slow cooling rate. As with the earlier direct quench, martensite is the product, and this must be tempered to secure toughness.

TABLE 11–7.2

Effects of Heat Treatments (SAE 4140)

Microstructure	Tensile strength		Toughness	
	MPa	(psi)	J	(ft-lb)
Annealed, (α + $\overline{C}$)				
Lamellar carbides	655	(95,000)	55	(40)
Spheroidite, (α + $\overline{C}$)				
Large "spherical" carbides in ferrite matrix	480	(70,000)	110	(80)
Martensite, M	~1400	~(200,000)	<3	<(2)
Tempered martensite, (α + $\overline{C}$) Dispersed carbides in ferrite matrix:				
500°C (930°F), 1 hr	1275	(185,000)	55	(40)
600°C (1110°F), 1 hr	1035	(150,000)	110	(80)

Tempering

Martensite has the attribute of being exceptionally hard. It is also very brittle, because it is nonductile, when it contains carbon. This is not surprising since hardness and ductility are generally inversely related. (Compare Figs. 11–1.3a and 11–1.3c, where we see that hardness and strength increase with carbon content while toughness and ductility decrease.) Fortunately, by tempering martensite, we can develop toughness faster than the hardness and strength decrease.*

Table 11–7.2 shows that a 4140 steel is relatively tough in both the annealed and the spheroidized conditions, but only moderately strong. Martensite, although very hard and basically strong, lacks toughness; therefore it may crack readily. However, tempered martensite can have a relatively high strength *and* good toughness if it is heat-treated appropriately. This is achieved because the brittle martensite is replaced by a fine dispersion of rigid carbide particles within a tough ferrite matrix (Fig. 11–2.10). The carbide particles arrest dislocation movements and thereby prevent slip, and they strengthen the alloy in a way similar to the precipitation hardening of Section 11–5. At the same

---

* This fact was discovered by trial and error several centuries ago. Its governing principles have only become understood during the lifetime of the present-day metallurgist.

    Tempering is certainly among the more important discoveries of nature that have affected the technological evolution of man. We can anneal steels so they can be machined. In turn, *these steels* can be hardened and tempered so they can machine *other steels*. A technological "bootstrap operation" such as this is not possible to the same degree of efficiency with other materials, because they lack the concurrent hardness and toughness. They either soften and become dull, or they break from brittleness.

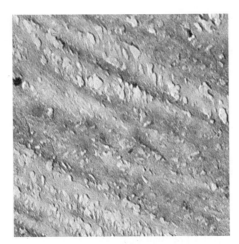

Figure 11–7.2 Bainite (×11,000). This SAE 1080 steel was austenitized, then transformed isothermally at 260°C (500°F) to form a fine dispersion of carbides in a ferrite matrix. The hardness is 57 R_C. (Courtesy General Motors and ASTM, *Electron Microstructure of Steel*.)

time, the ductile ferrite can deform locally at points of stress concentration (Fig. 7–6.2) to blunt the tip of the cracks that start to form.*

• **Austempering**

The final transformation process that we will consider for austenite decomposition is that of *austempering*. In this treatment, the austenite is allowed to transform *isothermally* to ferrite and carbide just above the M_s temperature. This requires a quench to avoid transformation to pearlite at higher temperatures. The advantage of austempering is that transformation occurs by a combination of shear and diffusion, to give a fine dispersion of carbides in ferrite and thus a strong, tough product. Also, quench-cracking is avoided because the reaction (and volume change) takes place at constant temperature. In many respects the product, *bainite* (Fig. 11–7.2), is similar to tempered martensite, and the physical properties of the two microstructures are closely related.

Study Aid (microstructures of steel)

The goal of heat treatments is to gain a desired microstructure. This topic in *Study Aids for Introductory Materials Courses* focuses on the ways different microstructures are developed in steels.

Example 11–7.1

(a) A small piece of 1045 steel is heated to equilibrium at 850°C, quenched to 650°C, held for 5 s, and then quenched to 20°C. What phases are present after each step? (b) A small piece of 1080 steel is heated to 800°C, quenched to 100°C, reheated to 290°C, and held 1 min. What phases are present after each step?

---

* Metallurgists are aware that extended heating around 500°C (~900°F) will embrittle a steel because there are secondary reactions that can occur among the alloying elements. Therefore, they require that the steel be cooled rapidly through this temperature region if tempering is performed at higher temperatures. They also know that the toughness has a transition temperature (Section 13–2) that varies with the austenite grain size and other factors.

Procedure. For equilibrium, check the phase diagram (Fig. 11–6.6b). For isothermal transformation, use the I-T diagram. A quench does not allow time for diffusion; however, γ can change to martensite below the M_s.

Answer

a) At 850°C γ_{stable}
   After 5 sec at 650°C $\gamma_{unstable} + \alpha +$ carbide
   After quenching to 20°C $M + \alpha +$ carbide*

b) At 800°C γ_{stable}
   After quenching to 100°C $\gamma_{unstable} + M$
   After 1 min at 290°C $\gamma_{unstable} + M^*$

Comment. With additional time at 290°C, both γ and M will transform to $\alpha +$ carbide.
◀

Example 11–7.2

Two drill rods (4-mm dia.) of 1080 steel receive the following sequences of treatments:

	(1)	(2)
Austenitize	775°C	775°C
Quench to	275°C	275°C
Held at 275°C for	45 sec	45 min
Cooled to 30°C in	30 sec	30 sec

a) Indicate the phase(s) that will be present after each step.

b) What microstructure does each final product have?

c) What process name does each have?

Procedure. Same as for Example 11–7.1.

Answer

Rod (1)   a) γ, $(\gamma)_{unstable}$, $(\gamma)_{unstable}$, martensite $+ (\gamma)$;
            b) martensite, plus retained austenite;
            c) interrupted quenching (or marquenching).

Rod (2)   a) γ, $(\gamma)_{unstable}$, $\alpha + \overline{C}$, $\alpha + \overline{C}$;
     • b) bainite;
     • c) austempering.

Comments. The hardness of the martensite for Rod (1) is $65R_C$; however, it must be tempered to become toughened. The final product loses some of that hardness (Fig. 11–2.11). When bainite is formed at 275°C (525°F), its hardness is $\sim 50R_C$. It does not need to be tempered because it already contains a fine dispersion of carbides within the ferrite matrix. ◀

• **Example 11–7.3**

There is still 5 v/o austenite present after quenching a 1080 steel. This retained austenite is present as small residual grains ($\delta \approx 1\ \mu m$) within a martensite matrix. What pressure must be overcome within the metal for one of these small grains of this retained austenite to complete its transformation to martensite?

---

* The reactions of Eq. (11–2.5) are *not* reversible below the eutectoid temperature.

Procedure. This last austenite must expand as it changes to martensite. In Example 11–2.3, we calculated a 1.8 v/o expansion. For calculation purposes, assume that these final γ grains transform to martensite and then are compressed (-1.8 v/o) to the confines of the surrounding martensite. The hydrostatic pressure is proportional to compression—$P_h = K(\Delta V/V)$ as presented in Eq. (5–3.5). In Example 5–3.2, the value of K for steel was calculated to be 162,700 MPa (or 23,600,000 psi).

Calculation

$$
\begin{aligned}
P_h &= K(\Delta V/V) \\
&= (162{,}700 \text{ MPa})(-0.018) \\
&= 2930 \text{ MPa compression} \qquad\qquad (\text{or } 425{,}000 \text{ psi}).
\end{aligned}
$$

Comments. This assumes that the adjacent metal is absolutely rigid. Actually, it relaxes some and reduces the pressure slightly. Even so, the compressive stresses are sufficient to stop the ($\gamma \to$ M) reaction short of completion. ◄

11–8

STEELS: HARDENABILITY

It is important to distinguish between *hardness* and *hardenability: Hardness* is a measure of resistance to plastic deformation. *Hardenability* is the "ease" with which hardness may be attained.

Figure 11–8.1 shows the maximum possible *hardnesses* for increasing amounts of carbon in steels; these maximum hardnesses are obtained only when 100% martensite is formed. A steel that transforms rapidly from austenite to ferrite plus carbide has low *hardenability* because ($\alpha + \overline{\text{C}}$) is formed at the expense of the martensite. Conversely, a steel that transforms very slowly from austenite to ferrite plus carbide has greater hardenability. Hardnesses nearer the maximum can be developed with less severe quenching in a steel of high hardenability, and greater hardnesses can be developed at the center of a piece of steel even though the cooling rate is slower there.

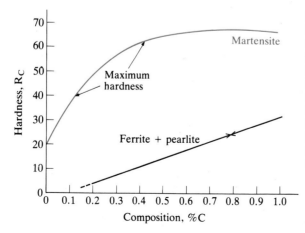

Figure 11–8.1 Maximum hardness versus carbon content of plain-carbon steels, showing maximum hardnesses arising from martensite compared with hardness developed by pearlitic microstructures. To produce maximum hardness, the reaction $\gamma \to \alpha$ + carbide must be avoided during quenching.

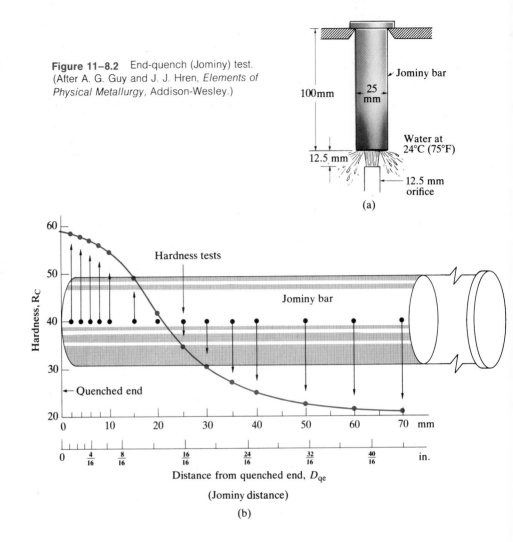

Figure 11–8.2 End-quench (Jominy) test. (After A. G. Guy and J. J. Hren, *Elements of Physical Metallurgy*, Addison-Wesley.)

Jominy bar

100mm

25 mm

Water at 24°C (75°F)

12.5 mm

12.5 mm orifice

(a)

Hardness tests

Jominy bar

Quenched end

Distance from quenched end, D_{qe}

(Jominy distance)

(b)

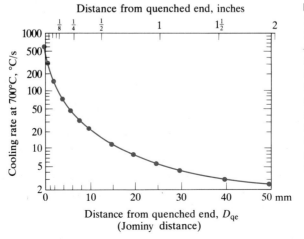

Distance from quenched end, inches

Distance from quenched end, D_{qe}
(Jominy distance)

Figure 11–8.3 Cooling rates (at 700°C) versus the distance, D_{qe}, from the quenched end of a Jominy bar. (Since the cooling rate decreases continuously as the temperature drops, 700°C is selected as a reference for comparisons. It is approximately the eutectoid temperature for most low-alloy steels and therefore critical for the $\gamma \rightarrow \alpha + \overline{C}$ reaction.)

Hardenability Curves

For any given steel, there is a direct and consistent relationship between hardness and cooling rate. However, the relationship is highly nonlinear. Furthermore, the theoretical bases for quantitative analyses are complex.* Fortunately, it is possible to use a standardized test that lets the engineer make necessary predictions of hardnesses for many applications in a minute or two, and hardness comparisons between steels at a glance. This is the *Jominy end-quench test*. In this test, a round bar with a standard size is heated to form austenite and is then end-quenched with a water stream of specified flow rate and pressure, as indicated in Fig. 11–8.2(a). Hardness values along the cooling-rate gradient are determined on a Rockwell hardness tester, and a *hardenability curve* is plotted (Fig. 11–8.2b).

The quenched end is cooled very fast and therefore has the maximum possible hardness for the particular carbon content of the steel that is being tested. The cooling rates at points behind the quenched end are slower (Fig. 11–8.3), and consequently the hardness values are lower (Fig. 11–8.2b). The cooling-rate data of Fig. 11–8.3 are generally valid for all types of plain-carbon and low-alloy steels since they have comparable values for density, heat capacity, and thermal conductivity—the three properties that affect thermal diffusivity.†

Figure 11–8.4 shows hardenability curves for several common grades of steels. They are plots of hardness versus cooling rates. The rates are shown in °C/s on the upper abscissa. In general, however, it is more convenient to use the "distance from the quenched-end," or D_{qe} (called the Jominy distance), because it can be plotted directly from the laboratory data. We will use this simplified procedure.‡

Observe several things from Fig. 11–8.4. The low-alloy steels (4140 and 4340) have greater hardenability than the plain-carbon steels, i.e., for a *given cooling rate* their hardnesses are nearer the maximum possible. Specifically, for a 0.40% C steel, the maximum hardness is ~60R_C as indicated in Fig. 11–8.1. At $D_{qe} = 10$ mm (where CR = 25°C/s), the hardnesses of 4340 and 4140 are 55R_C and 53R_C, respectively; the hardness of 1040 steel is only 26R_C. Expectedly, higher-carbon steels are harder (1060 vs. 1040 vs. 1020); this is true with rapid cooling rates ($D_{qe} = 0$ mm) as well as with slow cooling rate ($D_{qe} = 30$ mm). Finally, observe the second curve for the 1060 steels. The coarser-grained steel (#2) has higher hardenability as a direct result of the slower decomposition of the austenite. (Cf. Fig. 11–6.8.) This means that given a

---

* Variables include each and every alloying element and/or impurity, grain size, and austenitizing temperature. Also recall from the discussion of Fig. 11–6.10 that the cooling rates are measured at 700°C. This rate decreases at lower temperatures and approaches zero before cooling is complete.

† Stainless-type steels do not follow the pattern shown in Fig. 11–8.3, since their high-alloy contents reduce their thermal conductivities significantly without a comparable effect on density and/or heat capacity. However, these steels are seldom quenched for hardness requirements.

‡ This means that identical distances from the quenched-end on two different hardenability curves have the same specific cooling rates.

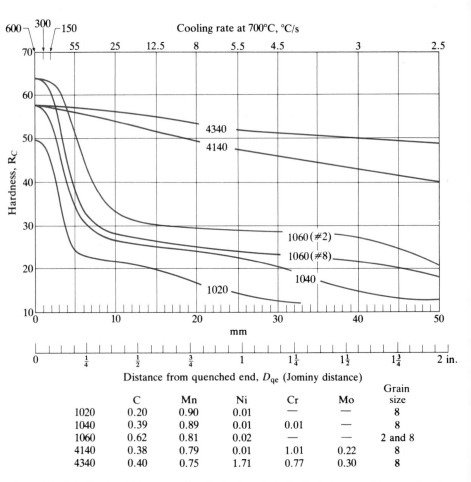

	C	Mn	Ni	Cr	Mo	Grain size
1020	0.20	0.90	0.01	—	—	8
1040	0.39	0.89	0.01	0.01	—	8
1060	0.62	0.81	0.02	—	—	2 and 8
4140	0.38	0.79	0.01	1.01	0.22	8
4340	0.40	0.75	1.71	0.77	0.30	8

Figure 11–8.4 Hardenability curves for six steels with the indicated compositions and grain sizes. The steels were end-quenched as shown in Fig. 11–8.2(a). In commercial practice, the hardenability curve of each type of steel varies because of small variations in composition. As a result, hardenability tests are commonly made for each heat of steel that is produced for quench-and-temper applications. (Adapted from U.S. Steel data.)

cooling rate, coarse austenite produces more martensite than fine-grained austenite.*

Use of Hardenability Curves

End-quench hardenability curves are of great practical value because (1) if the cooling rate of a steel in any quench is known, the hardness may be read directly from the hardenability curve for that steel, and (2) if the hardness at any point can be measured, the cooling rate at that point may be obtained from the hardenability curve for that steel.

---

* The other steels of Fig. 11–8.4 also have different hardenability curves for other grain sizes. A major advantage of the Jominy end-quench test is that a test specimen can be made from the same steel as that being processed into gears, tools, etc. Therefore, the exact hardenability curve is known. This means that variables such as grain-size, and minor composition differences, are automatically considered.

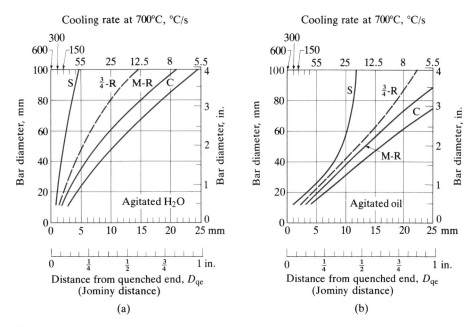

Figure 11–8.5   Cooling rates in round steel bars quenched in (a) agitated water and (b) agitated oil. Top abscissa, cooling rates at 700°C; bottom abscissa, equivalent positions on an end-quench test bar. (C, center; M-R, mid-radius; S, surface; Dashed line, approximate curve for $\frac{3}{4}$-radius.) The high heat of vaporization of water produces a severe quench in that quenching medium.

Figure 11–8.4 presents the end-quench hardenability curve for an AISI–SAE 1040 steel with the grain size and composition indicated.* The quenched end has nearly maximum hardness for 0.40% carbon steel because the cooling was very rapid and only martensite was formed. However, close behind the quenched end, the cooling rate was not rapid enough to avoid some ferrite and carbide formation, and so maximum hardness was not attained at that point.

In the laboratory, it is also possible to determine the cooling rates within bars of steel. Table 11–8.1, for example, shows the cooling rates at eutectoid temperatures for the surfaces, mid-radii, and centers of 75-mm (~3-in.) rounds quenched in mildly agitated water and oil. These cooling rates were determined by thermocouples embedded in the bars during the quenching operation. Similar data may be obtained for bars of other diameters. These data are summarized in Fig. 11–8.5.

By the use of the data of Fig. 11–8.5 and a hardenability curve, the *hardness traverse* that will exist in a steel after quenching may be predicted. For

---

* These data apply to this 1040 composition (and grain size). A slight variation is possible in the chemical specifications of any steel (e.g., in a 1040 steel, C = 0.37-0.44, Mn = 0.60-0.90, S = 0.05, P = 0.04, and Si = 0.15-0.25). As a result, two different 1040 steels may have slightly different hardenability curves.

	TABLE 11–8.1			

Cooling Rates in a 75-mm (~3-in.) Diameter Steel Bar (at 700°C)

	Agitated water quench		**Agitated oil quench**	
Position	**°C/s**	D_{qe}*	**°C/s**	D_{qe}*
Surface	~100	3	~20	11
¾-radius	27	9	9.5	18
Mid-radius	14	14	7.5	21
Center	11	17†	5.5	25

* Distance from the quenched end of a Jominy bar that has the same cooling rate at 700°C (Jominy distance).

† Observe that during a water-quench the center of a 75-mm diameter bar, which is 37 mm from the surface, cools at the same rate as when D_{qe} = 17 mm, and therefore much faster than the steel that is 37 mm from the quenched end of a Jominy bar (11°C/s vs. 3.5°C/s). Of course, heat is removed radially from the 75-mm bar, but primarily from one end of the Jominy bar.

example, the center of the 75-mm (~3-in.) round bar quenched in oil has a cooling rate of 5.5°C per second. Since the center of this large round bar has the same cooling rate as a Jominy test bar at a point 25 mm (~1-in.) from the quenched end, the hardnesses at the two positions will be the same. Thus if the bar is 1040 steel (Fig. 11–8.4), the center hardness will be $23R_C$. Figure 11–8.4 also shows that the following center hardnesses may be expected for 75-mm bars of other oil-quenched steels (cooled at 5.5°C per second):

AISI–SAE:	1040	4140	1020	4340	1060 (G.S. #8)	1060 (G.S. #2)
R_C:	23	47	14	52	24	28

Several determinations of quenched hardnesses are given in examples that follow shortly.

Tempered Hardness   The results of Examples 11–8.3 and 11–8.4 are hardnesses of quenched steel. As indicated in Fig. 11–2.11, the hardness decreases with continued tempering because the carbide particles coalesce. The data of that figure are for a plain-carbon eutectoid steel (1080). Alloy steels temper more slowly.* Data for tem-

* For example, a "high-speed" tool steel contains elements such as V, Cr, W, and Mo that make very stable carbides. These carbide particles grow and coalesce appreciably more slowly than Fe_3C does in a plain-carbon steel. Thus, tool steels can be used at higher temperatures (higher-speed operation) before overtempering, softening, and consequent destruction of the cutting edge becomes critical.

pering rates are available in metallurgical books for use by the engineer who must specify heat-treating processes for specific steels.

Study Aid (hardenability)

This topic is presented somewhat differently in *Study Aids for Introductory Materials Courses* (Topic XVIII) than in this section. Therefore, the reader is referred to this paperback if more clarification is needed.

Example 11–8.1

Determine the cooling rate for the center and mid-radius of a 20-mm (0.79-in.) diameter round steel bar when quenched in agitated water. (The bar is long enough so there is no end effect.)

Procedure. Use Fig. 11–8.5, which has plots of cooling rates versus bar diameters and radial positions (a) for water quenching and (b) for oil quenching. Since the cooling rate (CR) decreases with temperature (and therefore time), we commonly specify the rate as the temperature passes 700°C (upper abscissa). Alternatively, and more conveniently, since most CR plots are highly nonlinear, we index the cooling rates to positions on the standardized Jominy end-quench test. (Bottom abscissa)

Answer. From Fig. 11–8.5(a) for water quenching, and using the upper abscissa,

$$CR_{center} = \sim 60°C/s \qquad \text{(at 700°C).}$$

Alternatively, using the lower abscissa,

$$CR_{center} = D_{qe} \text{ of 4.5 mm.}$$

From Fig. 11–8.3, we see that

$$D_{qe} \text{ of 4.5 mm} = 60°C/s \qquad \text{(at 700°C).}$$

Similarly,

$$CR_{mid\text{-}rad} = D_{qe} \text{ of 2.5 mm} = \sim 100°C/s \qquad \text{(at 700°C).}$$

Comments. Observe that the center of the 20-mm bar is 10 mm from the surface of the bar. However, the cooling rate is much faster during the quench in this location than it would be at 10 mm from the end of the Jominy bar, where the cooling rate is 25°C/s. It is faster because heat is extracted in all radial directions and not just in the lengthwise direction. Furthermore, there is a reservoir of heat that must pass through the 20-mm point of the end-quenched bar. ◀

Example 11–8.2

What is the quenched hardness at a point 5 mm from the surface of a 40-mm diameter bar of 4140 steel that was quenched in agitated oil?

Procedure. Since the radius is 20 mm, this is a ¾-radius point. Obtain the CR as in Example 11–8.1. With the CR known, we can check the hardness for the 4140 steel in Fig. 11–8.4.

Answer. From Fig. 11–8.5,

$$D_{qe} = 9^{+} \text{ mm} \qquad \text{(and CR} \approx 27°C/s).$$

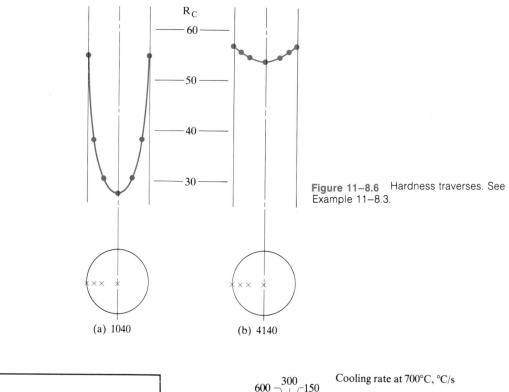

Figure 11–8.6   Hardness traverses. See Example 11–8.3.

(a) 1040

(b) 4140

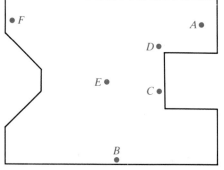

Figure 11–8.7   V-bar cross section. See Example 11–8.4.

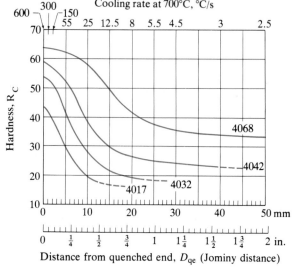

Distance from quenched end, D_{qe} (Jominy distance)

Figure 11–8.8   Hardenability curves for 40xx steels. Except for carbon content, the composition is the same for each. Additional carbon gives harder martensite and harder $\alpha + \overline{C}$, according to Fig. 11–8.1.

From Fig. 11–8.4,

$$\text{Hardness} = 54R_C.$$

Comment. Data may be interpolated from Figs. 11–8.4 and 11–8.5. In doing so, it is recommended that D_{qe} be used rather than °C/s, because the latter is highly nonlinear. ◀

Example 11–8.3

Sketch the *hardness traverses* for two steel rounds quenched in water; each is 38 mm (1.5 in.) in diameter, with AISI–SAE 1040 and 4140 compositions, respectively.

Procedure. Follow the procedure of Example 11–8.2, but for the four radial positions. With an agitated quenching liquid, the hardness traverse will be symmetrical across the center.

Answer

	From Fig. 11–8.5(a)		From Fig. 11–8.4		
	Approximate cooling rate at 700°C		Cooling rate at 700°C	AISI–SAE 1040	AISI–SAE 4140
Position	°C/s	D_{qe}	D_{qe}		
Surface	200	1.5 mm	1.5 mm	55R_C	57R_C
¾-radius	75	4 mm	4 mm	38R_C	56R_C
Mid-radius	55	5 mm	5 mm	34R_C	55R_C
Center	35	7.5 mm	7.5 mm	28R_C	54R_C

Comments. The hardness traverses for the two steels of Example 11–8.3 are shown in Fig. 11–8.6. Although the surface hardnesses of the two are very similar, the difference in their hardenability produces a higher center hardness for the AISI–SAE 4140 steel. As indicated, this steel has a higher alloy content, which slows down the transformation of austenite to ferrite and carbide. Consequently, more of the hard martensite can form. ◀

Example 11–8.4

Figure 11–8.7 shows the points in the cross section of a V-bar of AISI–SAE 1060 steel (G.S. #2) in which the following hardness readings were obtained after oil quenching: A—40R_C, B—36R_C, C—33R_C, D—32R_C, E—31R_C, F—63R_C. What hardness values would be expected for an identically shaped bar of AISI–SAE 4068 steel?

Procedure. The hardenability curve for the 1060 steel (G.S. #2) of Fig. 11–8.4 permits one to know the cooling rates at each point. The 4068 steels will have the same cooling rates for the same locations since the thermal diffusivity for the two steels are comparable, as pointed out in the text. Obtain the 4068 hardnesses for each D_{qe} from the 4068 hardenability curve in Fig. 11–8.8.

Answers

	From Fig. 11–8.4		From Fig. 11–8.8	
Point	AISI–SAE 1060 (G.S. #2)	Approximate cooling rate D_{qe}	Cooling rate at 700°C D_{qe}	AISI–SAE 4068
A	$40R_C$	7 mm	7 mm	$62R_C$
B	36	8	8	61
C	33	10	10	59
D	32	11	11	58
E	31	12	12	57
F	63	1	1	64

◀

Example 11–8.5

A 1020 steel rod (dia = 32 mm, or $1\frac{1}{4}$ in.) is *carburized* to 0.62% C at the surface, to 0.35% C at 2 mm below the surface, and unaltered beyond the depth of 4 mm.

a) Determine the hardness profile of the steel after quenching in water.

b) What would the profile have been without carburizing?

Procedure. Use the same procedure as in Example 11–8.3, except for the fact that the 1020 hardenability curve (Fig. 11–8.4) is replaced by 1062 and 1035 data in the carburized zone. Interpolate where necessary. Assume G.S. = #8.

Answers

	D_{qe}	(a)		(b)	
Surface	1.5 mm	0.62% C	$62R_C$	0.20% C	$47R_C$
2 mm	~2.5	0.35	~50	0.20	~40
4 mm	3	0.20	34	0.20	34
MR	4	0.20	26	0.20	26
Center	6.5	0.20	22	0.20	22

Comment. The harder, carburized surface zone (case) provides better wear resistance. The bulk of the steel remains tough. ◀

Review and Study

SUMMARY

Properties of many widely used engineering materials may be developed and changed through manufacturing processes. The most widely used processes are heat treatments that modify microstructures. *Microstructure* refers to the arrangement of grains and phases within a material. Most microstructures require optical or electron magnification procedures for visual examination.

1. The microstructures of *single-phase* materials include variations in grain *size* and in grain *shape,* as well as the extent of *preferred orientation.* The rate of *grain growth* increases rapidly at elevated temperatures.

   The microstructures of *multiphase* materials include the above variables plus the *amounts* of the phases, and their *shape* and *distribution.* Understandably, properties are a function of the phase quantities, particularly when the individual phases are markedly different. A major purpose of heat treatments is to control the shape and distribution of the minor phases.

2. Since atoms move within solids, microstructures can be changed. *Solution and precipitation* changes can be anticipated on the basis of phase diagrams. These, plus the *eutectoid* reactions that are encountered in a variety of materials, occur almost infinitely slowly at low temperatures because atomic *diffusion* is slow. Likewise, the reactions are slow with slight supercooling, because the separating phase must be *nucleated.* Reactions are fastest at intermediate temperatures.

   A rapid quench is required to form *martensite* in steel in order to avoid the usual $\gamma \to (\alpha + \overline{C})$ reactions. Its attribute of high hardness is counterbalanced by its low toughness. Fortunately, by *tempering,* it is possible to transform the hard and brittle martensite to a hard *and* tough mixture called *tempered martensite.*

3. A process of extended heating, followed by slow cooling, is called *annealing.* The microstructural changes that result depend upon the material. However, for many materials (not all), annealing produces a product that is less susceptible to brittle fracture.

4. Segregation within materials is reduced by *normalization* processes that provide time at elevated temperatures for diffusion and homogenization.

5. *Precipitation hardening,* also called age hardening, is achieved by *solution treating,* followed by rapid cooling to produce a *supersaturated* solid solution. As the second phase *starts* to separate (precipitate), the material gains significant hardness. The required time is a function of temperature.

   Over-aging and softening occur as the precipitating particles coalesce into a coarser microstructure. Thus, there are limitations to service temperatures.

6. *Isothermal-transformation* (I-T) diagrams provide a method of anticipating steel-treating processes. For example, the addition of alloying elements slow the $\gamma \to (\alpha + \overline{C})$ reaction and permit $\gamma \to M$. Thus, products that are larger than needles and razor blades may be fully hardened.

● 7. *Annealing, quenching, interrupted quenches,* and *austempering* (Fig. 11–7.1) are commercial heat treatments for steels. They can be best understood and controlled in terms of the combined use of transformation diagrams and phase diagrams.

● 8. *Hardness* is a measure of the resistance to plastic penetration; *hardenability* is the "ease" with which maximum hardness is attained. *Hardenability curves* are obtained by a simple, standardized, end-quench test that allows one to predict hardnesses in quenched steels. These are based on the fact that a given steel always develops the same (microstructure and therefore) hardness with the same cooling rate.

Age hardening See precipitation hardening.

Alloy retardation Decrease in the rate of austenite decomposition, because of alloying elements.

Annealing Extended heating, followed by controlled cooling.

Annealing (steels) For a *full* anneal, austenite is formed, then the steel is cooled slowly enough to form pearlite. (Annealing to remove strain hardening is called *process anneal*. The steel is heated just below the eutectoid temperature.)

• **Austempering** Transformation of $\gamma \rightarrow (\alpha + \overline{C})$ below the "knee" of the I-T curve to form a dispersion of carbides in a ferrite matrix.

Austenization Heat treatment to dissolve carbon into fcc iron, thereby forming austenite.

Carburize Introduction of carbon through the surface of the steel by diffusion, to change the surface properties.

Continuous cooling transformation (CCT) A thermal reaction during cooling, particularly austenite decomposition.

Critical cooling rates (CR_P and CR_M) CR_M is the slowest cooling rate, which produces only *martensite*. CR_P is the fastest cooling rate, which produces all *pearlite*.

Dispersed phase Microstructure of very fine particles within a matrix phase.

• **End-quench test (Jominy bar)** Standardized test, by quenching from one end only, for determining hardenability.

• **Hardenability** The ability to develop maximum hardness by avoiding the $(\gamma \rightarrow \alpha + \overline{C})$ reaction.

• **Hardenability curve** Hardness profile of end-quench test bar.

• **Hardness traverse** Profile of hardness values.

Homogenization Heat treatment to equalize composition by diffusion.

• **Interrupted quench** Two-stage quenching of steel that involves heating to form austenite and an initial quench to a temperature above the start of martensite formation, followed by a second (slower) cooling to room temperature.

Isothermal precipitation Precipitation from supersaturation at constant temperature.

• **Isothermal transformation** Transformation with time by holding at a constant temperature.

• **Jominy distance (D_{qe})** Cooling rate indexed to a distance from the quenched end of a Jominy bar.

Martensite (M) A phase arising from a diffusionless, shearlike phase transformation. In steels containing >0.15 w/o carbon, martensite is a hard, brittle, body-centered tetragonal phase that is supersaturated with carbon.

Microstructure Structure of grains and phases. Generally requires magnification for observation.

Microstructure, multiphase Microstructure containing two or more correlated phases.

• **Normalizing** Heating steel to ~50°C (100°F) into the austenite range so that it will contain a uniform, fine-grained microstructure.

Nucleation The start of the growth of a new phase.

Over-aging Continued aging until softening occurs.

Pearlite growth Simultaneous formation of ferrite and carbide layers from austenite by nucleation and carbon segregation.

Phase A physically homogeneous part of a materials system. (See Section 4–5.)

Phase, transition Metastable phase that forms as an intermediate step in a reaction.

Phase boundary   Compositional and/or structural discontinuity between two phases.

Plastic constraint   Prevention of plastic deformation in a ductile material by the presence of an adjacent rigid material.

Polycrystalline   Materials with more than one crystal; therefore with grain boundaries.

Precipitation   Separation from a supersaturated solution.

Precipitation hardening   Hardening by the formation of clusters prior to precipitation (also called age-hardening).

Quench   Cooling accelerated by immersion in agitated water or oil.

Rate $(R = t^{-1})$ Reaction per unit time.

Solid-phase reactions   Reactions involving microstructural changes within a solid; e.g., grain growth, recrystallization, polymorphic changes, solid solutions, precipitation, eutectoid decomposition, martensite formation.

Solution treatment   Heating to induce solid solutions.

Spheroidization   Process of making spheroidite, generally by extensive overtempering to develop spherelike carbides.

Stress relief   Removal of residual stresses by heating.

Tempered martensite   A two-phase microstructure of ferrite and carbide obtained by heating martensite.

Tempering   A toughening process in which martensite is heated to initiate a ferrite-plus-carbide microstructure.

FOR CLASS DISCUSSION

A_{11} An annealed brass is heated until the average grain dimension is 1 mm. Another sample of the same brass is cold-worked 5% and heated for the same time at the same temperature. The average grain dimension is 10 mm. A third sample is cold-worked 20%. Its grains are 3 mm after the same heat treatment. Provide a rationale for these results.

B_{11} Refer to Table 11–2.1. Point out how reaction rates are affected by the features that are cited in each category.

C_{11} Why do eutectoid reactions generally require longer times than polymorphic reactions? Why is time required for pearlite formation from austenite?

D_{11} The bct structure of martensite contains two iron atoms per unit cell and has $c = 0.291$ nm and $a = 0.285$ nm when the w/o carbon is 0.5. Point out the locations within the unit cell where the carbon atoms can reside with least strain.

E_{11} Equation 11–2.5 is irreversible as written. Under what conditions could austenite reform from $(\alpha + \overline{C})$ or from M? (*Note:* The direct reversal of $(\alpha + \overline{C})$ to M has never been observed, nor is it expected. Why?)

F_{11} Why does Fe_3C strengthen ferrite even though the carbide is so brittle that "it breaks like glass"?

G_{11} Why is fine pearlite stronger than coarse pearlite?

H_{11} Explain to a freshman chemistry student why the curves of Fig. 11–2.11 have a negative slope.

I_{11} Why do diffusion, recrystallization, stress relaxation (Section 6–7), and viscous flow (Section 7–7) have similar time–temperature relationships?

J_{11} Point out the differences between normalizing a steel and giving it a full anneal. What are the technical bases for these differences?

K_{11} Explain the data variations in Table 11–5.1.

L_{11} As a practical production matter, why is it simpler to first quench a 96 Al–4 Cu alloy to room temperature and then reheat to 100°C for aging, than to quench directly to 100°C?

M_{11} (a) Refer to the phase diagrams of Chapter 10. What alloys are candidates for precipitation-hardening? (b) Cu–Zn brass is never age-hardened. Why?

N_{11} Why does the "start" curve in Fig. 11–6.6 have two branches on its upper arm?

O_{11} The start (or finish) of any reaction is generally very difficult to determine (compared to the midpoint). Give two reasons. (*Note:* Because of this, M_s is commonly considered to be $\approx 1\%$ reaction; and M_f is commonly considered to be 99% completion.)

P_{11} Why does the austenite in a fine-grained steel transform to $(\alpha + \bar{C})$ faster than in a coarse-grained steel?

Q_{11} Why does the austenite of a 2% Cr steel transform to $(\alpha + \bar{C})$ less rapidly than the austenite in an 0.7% Cr steel?

R_{11} Why did we *not* use Fig. 11–6.6 for wire (c), step 3, of Example 11–6.1?

S_{11} Compare and contrast: tempered martensite, pearlite, • bainite, martensite, spheroidite, ferrite, carbide.

T_{11} Assume CR_M and CR_P for a steel are 75°C/s and 15°C/s, respectively. What phases will be present if the steel is cooled at 10°C/s? At 20°C/s? At 50°C/s? At 100°C/s?

U_{11} Assume the same CR_M and CR_P as in T_{11}. If this steel is cooled at 10°C/s, it is harder than if it is cooled at 5°C/s. Why? (They both contain only α and $\bar{C}$.)

• V_{11} Explain why low-alloy steels are commonly specified for steels that are to be heat-treated by quenching and tempering.

• W_{11} Explain the difference between hardness and hardenability.

• X_{11} Why should a steel (or the water) be agitated when a gear is water-quenched?

• Y_{11} Why does the center of a 50-mm (2-in.) diameter round steel bar quench to a greater hardness than the steel that is 25 mm behind the end of a Jominy end-quench bar? Which will quench faster?

• Z_{11} The temperature chosen for carburization is generally the same as that for normalizing. Why not use a temperature in the α-range since the carbon diffuses faster in that structure? Why not use a temperature further into the γ-range?

QUIZ SAMPLES

11A   In single-phase materials, microstructural variations include grain __**__ , __**__ , and __**__ .

11B   In multiphase materials, microstructural variations include those of 11A, plus the __**__ and __**__ of the phases.

11C   The name(s), _+−_ , is(are) used to label characteristic microstructure(s) encountered in steel, while the name(s), _+−_ label(s) phase(s).

(a) ferrite   (b) pearlite   (c) carbon   (d) spheroidite
(e) tempered martensite   (f) martensite   (g) liquid

11D   Typically, extensive supercooling _+−_ the nucleation rate and _+−_ the growth rate.

(a) increases   (b) has no effect on   (c) decreases

11E   Diffusion is limited to approximately interatomic distances for _+−_ to proceed as a solid-state reaction.

(a) solution   (b) precipitation   (c) eutectoid decomposition
(d) None of the above, rather _____

11F   Reaction products are commonly located at grain boundaries when _+−_ . These products are commonly found as dispersed particles when _+−_ .

(a) diffusion is possible   (b) there has been extreme supercooling
(c) the cooling rate is slow   (d) None of the above, rather _____

11G   When formed in a 99.5 Fe–0.5 C alloy, martensite is _+−_ .

(a) stable below the M_f temperature   (b) hard and easily fractured
(c) body-centered cubic   (d) None of the above

11H   _+−_ is(are) mixtures of ferrite and carbide that are found in steels.

(a) pearlite   (b) tempered martensite   • (c) bainite   (d) spheroidite
(e) None of the above

11I   Phase changes occur when annealing involves _+−_ .

(a) glass   (b) cold-worked brass   (c) stress relief of gears
(d) spheroidization of pearlite   (e) None of the above, rather _____

11J   _**_ and _**_ are names of processes used to homogenize metals.

11K   The pearlite layers are _**_ in a normalized steel than in one that is fully annealed because the cooling rate is more rapid.

11L   To precipitation-harden a material, the first step is to _**_ the solubility limit. The second step is to _**_ the solubility limit.

11M   Over-aging brings about a decrease in the _+−_ precipitate(s) and an increase in the _+−_ the precipitates.

(a) number of   (b) hardness of   (c) distances between
(d) volume fraction of

11N   Other things equal, the start of isothermal transformation, t_s, is sooner for steels with _+−_ .

(a) slight supercooling   (b) smaller γ grain sizes   (c) higher alloy contents
(d) lower carbon contents

11O   _**_ (labeled CCT) displaces the start and finish of austenite decomposition to _**_ times and _**_ temperatures than with isothermal transformation.

• **11P**   Martensite does not form during the steel-treating process called _+−_ .

(a) interrupted quench   (b) normalizing   (c) annealing
(d) austempering   (e) quenching

- **11Q** High strengths and toughnesses are difficult to realize simultaneously in steel when __+−__ is predominant.

  (a) spheroidite   (b) tempered martensite   (c) pearlite   (d) martensite

- **11R** A steel that resists __**__ has high hardness. One that readily attains its maximum hardness has high __**__ .

- **11S** Steels have greater hardenability if they have a __**__ grain size and if __**__ are present.

- **11T** __**__ quenching is more severe than oil quenching since the former has a very high __**__ .

STUDY PROBLEMS

11–1.1 Refer to Example 11–1.2. At what temperature will the grain size grow from 0.05 mm to 0.10 mm in 15 hours?

Answer: 625°C.

- **11–1.2** The data of Fig. 11–1.1 follow Eq. 11–1.1 independently of temperature as long as the grain dimensions δ do not approach the sample dimensions. What is the value of the grain growth exponent n for the brass in Fig. 11–1.1?

11–2.1 An alloy of 95 Al–5 Cu is solution-treated at 550°C, then cooled rapidly to 400°C where it is held for 24 hours. During that time it produces a microstructure with 10^6 particles of θ per mm^3. The θ particles ($CuAl_2$) are nearly spherical and have twice the density of κ matrix. (a) Approximately how far apart are these particles? (b) What is the average particle dimension?

Answer: (a) ~10 μm (b) 3.8 μm

11–2.2 A microstructure has spherical particles of β with an average dimension $\bar{d}$ that is 10% of the average distance $\bar{D}$ between the centers of the adjacent particles. (a) What is the volume percent of β? (b) What is the ratio $\bar{d}/\bar{D}$ with 0.5 v/o β?

11–2.3 A 90 Pb–10 Sn solder is held at 185°C until equilibrium is established. The average grain size is 10^6 μm^3. The alloy is then *rapidly* cooled to room temperature where a β precipitate forms within the initial α grains. The resulting matrix within those former α grains is 99 Pb–1 Sn ($\rho = 13.3$ Mg/m^3), and the β particles (~100% Sn) are separated by an average distance of 0.1 μm. (a) How many particles are there per original grain? (b) What is the volume fraction of β? (c) Approximately how many tin atoms are there per β particle?

Answer: (a) ~10^9 (b) 15 v/o   (c) ~5 × 10^6 Sn

11–2.4 A sterling silver (92.5 Ag–7.5 Cu) is quenched after solution treatment, then reheated to 400°C until equilibrium is attained; at equilibrium it contains a β precipitate (approximately spherical in shape) in an α matrix. (a) If the representative dimension d of the β particle is 0.1 μm, how many will there be per mm^3? If $d = 0.05$ μm? (b) What is the average distance between particles?

11–2.5 Refer to the Cu–Sn diagram in Chapter 10. Locate and state the reactions (on cooling) for four eutectoid transformations.

11–2.6 An oxide containing 75 w/o iron and 25 w/o oxygen is solution-treated at 1300°C (2370°F). It is then cooled slowly to room temperature. Compare and contrast the reactions and possible microstructures with those found after slow cooling a 1060 steel from 900°C (1650°F).

11–2.7 Refer to discussion topic D_{11}. Calculate the density of the martensite that contains 0.50 w/o C.

Answer: 7.89 Mg/m³ (=7.89 g/cm³)

11–2.8 An Al–Cu alloy contains 2.5 v/o θ (ρ = 5.5 Mg/m³) in a matrix of κ which is essentially pure Al. What is the density of the alloy?

11–2.9 The density of a Pb–Sn alloy is 10.0 Mg/m³ (=10.0 g/cm³) after equilibration at 20°C. (a) What is the volume fraction β? (b) What is the % Sn?

Answer: (a) 33 v/o β (b) 24 w/o Sn

11–4.1 (a) What temperature should be used to normalize a 1030 steel? (b) 1080 steel? (c) 1% carbon steel?

Answer: (a) 880°C

• **11–4.2** What is the normalizing temperature for a steel of Example 10–9.1?

11–4.3 Twenty kilograms of an 8 Al–92 Mg alloy are first melted and then cooled rapidly to 500°C. Since there wasn't time for diffusion in the solid, the average composition of the ε is 5% Al. (a) What is the composition of the liquid at 500°C? (Assume rapid diffusion in the liquid.) (b) How much liquid at 500°C? (c) At what temperature will γ appear? (d) What is the composition of the final liquid?

Answer: (a) 23 Al–77 Mg   (b) 3.3 kg   (c) 437°C   (d) 32 Al–68 Mg

11–5.1 Maximum hardness is obtained in a metal when the aging time is 10 seconds at 380°C, or 100 seconds at 315°C. Neither of these is satisfactory for production because we cannot be certain the parts are uniformly heated. Recommend a temperature for a maximum hardness in 1000 seconds (15 min–20 min), a time compatible with production.

Answer: ~260°C

11–5.2 How long should it take for the metal in the previous problem to reach maximum hardness at 100°C?

11–5.3 Explain why a 92% copper, 8% nickel alloy can (or cannot) be age-hardened.

11–5.4 Explain why the following alloys can (or cannot) be considered for age-hardening. (a) 97% aluminum, 3% copper. (b) 97% copper, 3% zinc. (c) 97% nickel, 3% copper. (d) 97% copper, 3% nickel. (e) 97% aluminum, 3% magnesium. (f) 97% magnesium, 3% aluminum.

11–5.5 A slight amount of age-hardening is realized when a steel (99.7 w/o Fe, 0.3 w/o C) is quenched from 700°C (1300°F) and reheated for 3 hours at 100°C. Account for the hardening.

Answer: Carbon solubility in ferrite decreases.

11–5.6 An aircraft manufacturer receives a shipment of aluminum alloy rivets that have already age-hardened. Can they be salvaged? Explain.

11–5.7 (Refer to Fig. 11–5.3.) Select a plotting procedure that shows a relationship between temperature and time for peak hardness and that lets you estimate (graphically) the time for peak hardness at 100°C. What is that time for this 2014 aluminum?

Answer: ~2000 hours

11–5.8 Estimate graphically (as in Study Problem 11–5.7) the time that would be required to attain 200 MPa (30,000 psi) at 100°C.

11–6.1 A small piece of 1080 steel is heated to 800°C, quenched to −60°C, reheated immediately to 300°C, and held 10 seconds. What phases are present at the end of this time?

Answer: Martensite (with some possible tempering to α + carbide).

11–6.2 Some 1045 steel is quickly quenched from 850°C to 400°C and held for 1 second; for 10 seconds; for 100 seconds. What phase(s) will be present at each time point?

11–6.3 Some 1045 steel is quickly quenched from 850°C to 425°C and held 5 seconds before quenching again to 20°C. (a) What phase(s) will be present just before the second quench? Give the composition of each. (b) What phase(s) will be present immediately after the second quench?

Answer: (a) γ(0.45% C), α (0.02% C), and $\overline{C}$(6.7% C)   (b) M(0.45% C), α, and $\overline{C}$

11–6.4 Repeat Study Problem 11–6.3, but change 20°C to 275°C.

11–6.5 A 1020 steel is equilibrated at 790°C (1450°F) and quenched rapidly to 400°C (750°F). How long must the steel be held at that temperature to reach the midpoint of the austenite decomposition?

Answer: ~8 seconds

11–6.6 Sketch an isothermal-transformation diagram for a steel of 1.0 w/o C and 99.0 w/o Fe.

11–6.7 Sketch an isothermal-transformation diagram for a 1020 steel.

Answer: M_s = 450°C, M_f = 290°C; α and γ are stable between 727°C and 855°C after long times; at 550°C the curve is farther to the left than in 1045 steel.

11–6.8 A small wire of AISI–SAE 1045 steel is subjected to the following treatments as *successive* steps:

1) heated to 875°C, held there for 1 hour;

2) quenched to 250°C, held there 2 seconds;

3) quenched to 20°C, held there 100 seconds;

4) reheated to 550°C, held there 1 hour;

5) quenched to 20°C and held.

Describe the phases or structures present *after each step* of this heat-treatment sequence.

11–6.9 (a) Repeat Problem 11–6.8 with steps (1), (2), (5). (b) Repeat Problem 11–6.8 with steps (1), (3), (4), (5). (c) Repeat Problem 11–6.8 with steps (1), (2), (4), (5).

• **11–6.10** Refer to Fig. 11–6.9. A steel without molybdenum is equilibrated at 740°C then quenched to 400°C and held for ~5 seconds for γ decomposition to start. What sequence would be necessary to achieve the same results if 0.25% Mo were present? Explain.

11–7.1 A small piece of 1080 steel has its quench interrupted for 20 seconds at 300°C (570°F) before final cooling to 20°C. What phase(s) will be present?

Answer: M(0.8% C), and some retained γ(0.8% C)

11–7.2 Compare and contrast: (a) martensite and tempered martensite; (b) tempered martensite and bainite; (c) bainite and spheroidite.

11–7.3 A 1080 steel is to be quenched and tempered to a hardness of $50R_C$. (a) What should the austenitizing temperature be? (b) How long should it be tempered if the tempering temperature is 400°C (750°F)?

Answer: (a) ~750°C (~1380°F)   (b) ~40 minutes

11–7.4 Draw temperature (ordinate) and time (abscissa) plots for the following heat treatments. Indicate the important temperatures, relative times, and reasons for drawing the curves as you do. (a) Normalizing a 1095 steel, contrasted with annealing the same steel. (b) Solution-treating a 95 Al–5 Cu alloy, contrasted with aging the same alloy. (c) Austempering a 1080 steel, contrasted with martempering (interrupted quench) the same steel. (d) Spheroidizing 1080 steel, contrasted with spheroidizing 10·105 steel.

• **11–7.5** A 50-mm diameter bar of 1080 steel is quenched and forms a martensite "rim" that is 2.5 mm thick. While the austenitic center is hot, it can still adapt to the $(\gamma \rightarrow M)$ volume changes (Example 11–2.3). The more slowly cooled center changes directly to $(\alpha + \overline{C})$. What hoop stress is placed on the martensite rim? (For 1080, $\rho_\gamma = 7.99$ Mg/m³ (= 7.99 g/cm³); $\rho_M = 7.85$ g/cm³; and $\rho_{(\alpha+\overline{C})} = 7.84$ g/cm³.)

Answer: 1170 MPa tension (170,000 psi)

- 11–8.1 (a) What is the cooling rate at the mid-radius of a 50-mm (1.97-in.) round steel bar quenched in agitated oil and reported in °C/s? (b) Reported as the distance from the quenched end of a Jominy bar?

  Answer: (a) ~17°C/s   (b) 13 mm

- 11–8.2 Repeat Study Problem 11–8.1 but for the ¾-radius and a water-quench.

- 11–8.3 (a) What is the quenched hardness at the mid-radius of a 50-mm (1.97-in.) round steel bar of 1040 steel quenched in agitated oil? (b) In agitated water?

  Answer: (a) $25R_C$ (b) $28R_C$

- 11–8.4 A round bar of 1040 steel has a surface hardness of $41R_C$ and a center hardness of $28R_C$. How fast were the surface and center cooled through 700°C?

- 11–8.5 What hardness would you expect the center of a 50-mm (2-in.) round bar of 1040 steel to have if it were quenched in (a) agitated oil? (b) Agitated water?

  Answer: (a) $24R_C$ (b) $26R_C$

- 11–8.6 How would the hardness traverse of the 1040 steel shown in Fig. 11–8.6 vary if it were quenched in (a) still oil? (b) Still water? (c) If it had a coarser austenite grain size? Explain.

- 11–8.7 (a) A 40xx steel is to have a hardness of $40R_C$ after it is quenched at the rate of 17°C/s. What is the required carbon content? (b) What will the hardness of this steel be if quenched twice as fast?

  Answer: (a) 0.5% C   (b) 4050 at 34°C/s ($\therefore D_{qe} = 7.5$ mm) $53R_C$

- 11–8.8 An 80-mm (3.15-in.) round 4340 steel bar is quenched in agitated oil. Plot the hardness traverse.

- 11–8.9 Plot a hardness traverse for a 1060 steel bar (diameter of 38 mm, or 1.5 in.) that is to be quenched in (a) agitated water. (b) Agitated oil. (G.S. = #2)

  Answer: (a) S: $63R_C$; MR: 48; C: 37

- 11–8.10 A 40-mm (1.6-in.) bar of 1040 steel (i.e., with diameter = 40 mm, and length $\gg$ 40 mm) is quenched in agitated water. (a) What is the cooling rate through 700°C at the surface? At the center? (b) Plot a hardness traverse.

- 11–8.11 The center hardness of six bars of the same steel are indicated below. From these data, plot the hardenability curve for the steel. (*Hint:* There should be only *one* curve.)

Diameter	Water quench	Oil quench
25-mm (1.0-in.)	$58R_C$	$57R_C$
50-mm (2.0-in.)	$55R_C$	$47R_C$
100-mm (3.9-in.)	$34R_C$	$30R_C$

  Answer: $58R_C$ at $D_{qe} = 5$ mm; $56R_C$ at 10 mm; $49R_C$ at 15 mm; $39R_C$ at 20 mm; $33R_C$ at 25 mm; $31R_C$ at 30 mm; $30R_C$ at 35 mm

- **11–8.12** A spline gear had a hardness of $45R_C$ at its center when it was made of the 4068 steel shown in Fig. 11–8.8. What hardness would you expect the same gear to have if it were made of 1040 steel?

- **11–8.13** Repeat Example 11–8.5, but with oil quenching.

    Answer: (a) S: $33R_C$; 2 mm: $\sim28R_C$; $\frac{3}{4}$R: $23R_C$; MR: $22R_C$; C: $21R_C$

- **11–8.14** A 40-mm (1.6-in.) diameter steel rod of 4017 steel has been carburized to 0.60% carbon at the surface, 0.3% C at 2 mm below the surface, and 0.17% C at 5 mm below the surface. Determine the hardness profile of the steel bar after water quenching.

QUIZ CHECKS

11A	size, shape, orientation	**11L**	heat above, quench below
11B	amounts, distributions	**11M**	(a)
11C	(b), (d), (e)		(c)
	(a), (f), (g)	**11N**	(b), (d)
11D	(a), (c)	**11O**	continuous-cooling transformation, longer, lower
11E	(d): grain growth or recrystallization or martensite formation		
11F	(a), (c)	• **11P**	(b), (c), (d)
	(b)	• **11Q**	(a), (c), (d)
11G	(b) (not a)	• **11R**	indentation (or deformation), hardenability
11H	(a), (b), (c), (d)		
11I	(e): full anneal of steels	• **11S**	larger
11J	soaking, normalizing		alloying elements
11K	thinner	• **11T**	water
			heat of vaporization

Chapter Twelve

Composites

PREVIEW

Composites contain two or more components that have been combined into a *unified material*. They include coated materials, reinforced materials, plus other combinations that take advantage of the special properties of the several contributing materials.

Commonly, the contributors to the composite can be evaluated and tested separately. This leads to the desire for *mixture rules* so that the properties of the composite may be calculated from those of the components.

We will observe two major requirements for reinforced composites in order to be an effective product. (1) The reinforcing component must have a *higher modulus of elasticity* than the matrix component. (2) There must be a strong *interfacial bond* between the reinforcement and the matrix.

CONTENTS

STUDY OBJECTIVES

1. Acquaint yourself with types of composites other than reinforced concrete and fiber-glass reinforced plastics. Examples include glass-coated steels, aluminum-coated mylar, and plastic-laminated glass.

2. Understand those mixture rules that apply to simple composites where the geometries of the reinforcement and the matrix may be defined.

3. Recognize the role of the interfacial boundary between the components in establishing a coherent composite.

12–1

MATERIAL COMBINATIONS

Composites contain two (or more) distinct materials as a unified combination. Thus, reinforced concrete is a composite of steel rods in a concrete matrix. Likewise, many sailboat hulls are made of fiber-reinforced plastics (FRP) in which the fiber is typically glass and the plastic is commonly a polyester. Other composites include products such as glass-coated steel for water heater liners, and grinding wheels with fiber-glass mesh reinforcement (Fig. 12–1.1).

Steel, which is a mixture of ferrite and carbide, is generally considered not to be a composite, because the two materials are not formulated separately but originate within a single manufacturing process. Likewise, an asphalt concrete is considered to be of a singular origin—the "hot plant." Admittedly, composites are not sharply differentiated from other mixtures.

Properties of composites are commonly studied and analyzed in terms of the properties of the contributing materials. This leads to mixture rules in which the properties are a function of the amounts and geometric distribution of each contributing material.

Mixture Rules

The simplest mixture rules are for scalar properties such as density or heat capacity. The density of a mixture, ρ_m, is

$$\rho_m = f_1\rho_1 + f_2\rho_2 + \cdots \tag{12–1.1a}$$

or, simply,

$$\rho_m = \Sigma f_i\rho_i, \tag{12–1.1b}$$

where f is the volume fraction of each contributor. Likewise, for heat capacity, c,

$$c_m = \Sigma f_i c_i. \tag{12–1.2}$$

493

Figure 12–1.1 Industrial composite product (fiber-glass-mesh reinforced grinding wheel). This triple composite includes abrasive grit, bonded with a phenolic resin, and reinforced with a 1-in. by 1-in. mesh of fiber glass roving. Each component contributes to the performance of the product. (Courtesy of the Acme Abrasive Co.)

In this calculation, f is the volume fraction if the heat capacity is in J/°C per unit volume. If the heat capacity is in J/°C per unit mass, the f must be a mass fraction.

Conductivity is directional along a gradient. Therefore, the geometry of the mixture enters the calculation. Two extreme situations are illustrated in Fig. 12–1.2. Consider the thermal conductivity coefficient k. In Fig. 12–1.2(a), the conduction is *parallel* to the structure; so the equation

$$k_\| = f_1 k_1 + f_2 k_2 + \cdots \tag{12–1.3}$$

applies. The structure is in *series* in Fig. 12–1.2(b), so heat must be conducted perpendicularly to the layers. This leads to a reciprocal mixture rule for the

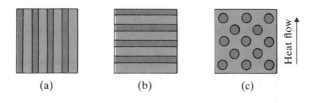

Figure 12–1.2 Conductivity versus phase distribution (idealized). (a) Parallel conductivity (Eq. 12–1.3). (b) Series conductivity (Eq. 12–1.4). (c) Conductivity through a material with a dispersed phase (Eqs. (12–1.6) and (12–1.7), when the two phases have markedly dissimilar conductivities).

conductivity of the multiphase structure:

$$1/k_\perp = f_1/k_1 + f_2/k_2 + \cdots. \tag{12–1.4}*$$

In these two equations f is the volume fraction.

The more commonly encountered microstructures involve a dispersion of two or more phases (Figs. 11–1.4b and 12–1.2(c)). Furthermore, even a lamellar structure such as pearlite has random orientation of the pearlite colonies. To a first approximation, the conductivities of these microstructures may be considered to vary linearly on a volume fraction basis. However, again modifications must be made if one of the two phases is continuous, c, and the other is dispersed, d. Where $k_c \ll k_d$,

$$k_m \approx k_c(1 + 2f_d)/(1 - f_d). \tag{12–1.6}$$

Where $k_c \gg k_d$,

$$k_m \approx k_c(1 - f_d)/(1 + f_d/2). \tag{12–1.7}$$

In these equations, f_c and f_d are the volume fractions of the continuous and dispersed phases, respectively. The dominating role of the continuous phase is evident, even though it may be minor in amount. If it has a high conductivity, it forms a route for thermal (or electrical) transport. If the continuous phase is insulative, transport is limited even though there may be major amounts of a highly conductive dispersed phase.

Not all properties and not all composites lend themselves to simplified mixture rules. For example, we find it simpler to load test phenol-formaldehyde (PF) resins that are filled with wood flour (very fine sawdust) than to formulate mixture rules for compressive strength (Fig. 12–1.3).

Example 12–1.1

Fifty w/o SiO_2 flour (very fine quartz powder) is added to a phenol-formaldehyde resin as a filler. (a) What is the density of the mixture? (b) What is the thermal conductivity?

Procedure. Determine the volume fractions of the two ingredients. Density and thermal conductivities are from Appendix C.

$$\rho_{SiO_2} = 2.65 \text{ g/cm}^3; \qquad \rho_{PF} = 1.3 \text{ g/cm}^3.$$

$$k_{SiO_2} = 0.012 \text{ W/mm·K}; \qquad k_{PF} = 0.00016 \text{ W/mm·K}.$$

Use 100 g (= 50 g SiO_2 + 50 g PF) as a basis.

---

* For electrical conductivity σ, the relationship for a series sequence is

$$1/\sigma_\perp = f_1/\sigma_1 + f_2/\sigma_2 + \cdots, \tag{12–1.5a}$$

or for resistivity ρ, the equation is

$$\rho_\perp = f_1\rho_1 + f_2\rho_2 + \cdots. \tag{12–1.5b}$$

This equation relates to the more familiar resistance R equation for series that is encountered in physics calculations.

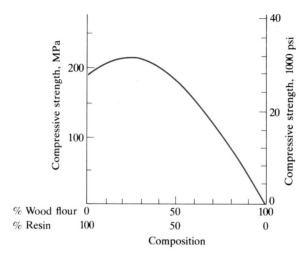

Figure 12–1.3 Strength of mixtures (wood-flour filler in phenol-formaldehyde resin). The mixture of wood flour and resin is stronger than either alone. The wood flour prevents slip in the resin; the resin bonds the particles of wood flour.

Calculation

a)

$$50 \text{ g SiO}_2 = 18.8 \text{ cm}^3 \text{ SiO}_2; \quad f_{\text{SiO}_2} = 0.33$$

$$50 \text{ g PF} = \underline{38.4 \text{ cm}^3 \text{ PF}}; \quad f_{\text{PF}} = \underline{0.67}$$

$$57.2 \text{ cm}^3 \qquad\qquad 1.0$$

$$\rho_m = 100 \text{ g}/57.2 \text{ cm}^3$$

$$= 1.75 \text{ g/cm}^3,$$

or Eq. (12–1.1),

$$\rho_m = (0.33)(2.65) + (0.67)(1.3)$$

$$= 1.75 \text{ g/cm}^3.$$

b) Since $k_{\text{PF}} \ll k_{\text{SiO}_2}$,

$$k_m \approx k_c \left(\frac{1 + 2f_d}{1 - f_d} \right),$$

$$k_m \approx 0.00016 \left[\frac{1 + 2(0.33)}{1 - 0.33} \right]$$

$$= 0.0004 \frac{\text{W/mm}}{°\text{C}} . \quad \blacktriangleleft$$

Example 12–1.2

Formulate a mixture rule for Young's modulus (longitudinal) of a "glass fishing rod." This contains longitudinally oriented glass fibers bonded with a polyester resin.

Procedure. In longitudinal loading, the strain in the two components must be equal (if they are to deform elastically as a unit). [We will ignore the lateral strains by assuming the two materials have the same Poisson's ratio.]

Derivation

$$e_{gl} = e_m = e_{pr}.$$

$$\left[\frac{F/f\,A}{E}\right]_{gl} = \frac{F/A}{E_m} = \left[\frac{F/f\,A}{E}\right]_{pr}$$

where E_m is Young's modulus of the composite mixture.

$$F_{gl} = f_{gl}E_{gl}F/E_m;$$

$$F_{pr} = f_{pr}E_{pr}F/E_m.$$

Since $F = F_{gl} + F_{pr}$,

$$F = (f_{gl}E_{gl} + f_{pr}E_{pr})\,F/E_m,$$

$$E_m = f_{gl}E_{gl} + f_{pr}E_{pr}. \qquad (12\text{–}1.8)$$

Comment. This mixture rule applies only where strains are identical. A composite with the configuration of Fig. 12–1.2(b) under vertical loading will have a different mixture rule. (See Example 12–2.2.) ◄

Example 12–1.3

A 1060 steel wire (1 mm in diameter) is coated with copper (combined diameter = 2 mm). What is the thermal expansion coefficient of the composite material?

Procedure. In the composite, $(\Delta L/L)_{st} = (\Delta L/L)_{Cu}$, and in the absence of an external load, $F_{Cu} = -F_{st}$. From Appendix C,

$$E_{st} = 205 \text{ GPa}; \qquad E_{Cu} = 110 \text{ GPa}.$$

$$\alpha_{st} = 11 \times 10^{-6}/^{\circ}\text{C}; \qquad \alpha_{Cu} = 17 \times 10^{-6}/^{\circ}\text{C}.$$

Basis for calculation, $\Delta T = +1^{\circ}\text{C}$.

Calculation

$$A_{st} = (\pi/4)(0.001 \text{ m})^2 = 0.8 \times 10^{-6} \text{ m}^2;$$

$$A_{Cu} = (\pi/4)(0.002 \text{ m})^2 - 0.8 \times 10^{-6} \text{ m}^2 = 2.4 \times 10^{-6} \text{ m}^2.$$

$$(\Delta L/L)_{st} = (\Delta L/L)_{Cu},$$

$$[\alpha\Delta T + (F/A)/E]_{st} = [\alpha\Delta T + (F/A)/E]_{Cu}, \qquad (12\text{–}1.9)$$

$$(11 \times 10^{-6})(1) + \frac{F_{st}/0.8 \times 10^{-6} \text{ m}^2}{205 \times 10^9 \text{ N/m}^2} = (17 \times 10^{-6})(1) + \frac{-F_{st}/2.4 \times 10^{-6} \text{ m}^2}{110 \times 10^9 \text{ N/m}^2};$$

$$F_{st} \approx +0.61 \text{ N} \qquad \text{(tension on heating)};$$

$$F_{Cu} \approx -0.61 \text{ N} \qquad \text{(compression on heating)}.$$

For $\Delta T = +1^{\circ}\text{C}$,

$$(\Delta L/L)_{Cu} = (17 \times 10^{-6}/^{\circ}\text{C})(1^{\circ}\text{C}) + \frac{-0.61 \text{ N}/(2.4 \times 10^{-6} \text{ m}^2)}{110 \times 10^9 \text{ N/m}^2};$$

$$\bar{\alpha} \approx 15 \times 10^{-6}/^{\circ}\text{C}.$$

Comment. The thermal expansion varies slightly with the exact type of steel. ◄

12–2

REINFORCED MATERIALS

Materials are reinforced to make them stronger. When improved strength is the major goal, the reinforcing component must have a large *aspect ratio*, that is, its length/diameter ratio must be high, so that the load is transferred across potential points of fracture. Thus, we place steel rods in a concrete structure. And we combine glass fibers and polymers for fiber-reinforced plastics (FRP).

It is obvious that the reinforcement must be the stronger component if it is to carry the load. It may be less obvious that the reinforcement must have the higher elastic modulus. Likewise, it may not be immediately apparent that the bond between the matrix and the reinforcement is critical, since it is generally necessary to transfer the load from the matrix to the fibers or rods if the reinforcement is to serve its purpose.

For the reinforcing material to carry most of the load, *the reinforcement must have a higher Young's modulus than the matrix.* Consider Fig. 12–2.1 in which a steel core reinforces an aluminum wire. When loaded in tension, the two metals must deform together. Assume a strain, e, of 0.001. With the two moduli being ~205,000 MPa (~30,000,000 psi) and ~70,000 MPa (~10,000,000 psi), respectively, and using Eq. (1–5.3), the steel develops a stress of ~205 MPa versus ~70 MPa in the aluminum:

$$e = 0.001 = (s_{st}/205,000 \text{ MPa}) = (s_{Al}/70,000 \text{ MPa}). \qquad (12\text{--}2.1)$$

Because E_{st} is ~3(E_{Al}), the stress in the steel is ~3(s_{Al}). Generally, for a two-component composite,

$$s_1/s_2 = E_1/E_2. \qquad (12\text{--}2.2)$$

The modulus of elasticity (in tension) of a fiber-reinforced plastic, E_{FRP}, may be estimated as the volume fraction average if all of the fibers are aligned parallel to the direction of loading. Consistent with the derivation in Example 12–1.2:

$$E_{FRP} = \Sigma f_i E_i, \qquad (12\text{--}2.3)$$

where f_i and E_i are the volume fractions and Young's moduli, respectively, for the components. Consider a plastic reinforced with 50 v/o of parallel glass fibers ($E = 70,000$ MPa, or 10^7 psi) and the same fraction of a plastic that has a Young's modulus of 4000 MPa (580,000 psi). The resulting composite has a Young's modulus in its longitudinal direction of ~37,000 MPa (or 5.3×10^6 psi). (Of course, the elastic moduli in the other two coordinate directions will be low and close to that of the plastic, because they are at right angles to the reinforcement.)

If the same amount of glass is incorporated as a woven fabric into the above plastic, the composite gains two-way reinforcement. However, the values drop below that of the above mixture rule (Eq. 12–2.3). Furthermore, the 45° modulus is low (Fig. 12–2.2). It is now common to use matted glass fibers to avoid this

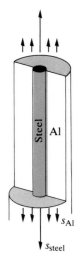

Figure 12–2.1 Stress in composites (steel-reinforced aluminum wire). The strain in the two must be equal. Therefore Eq. (12–2.2) applies.

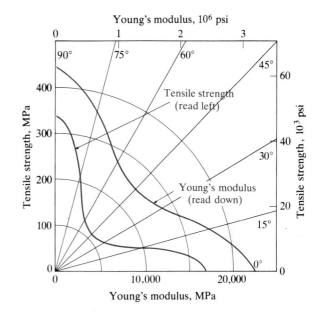

Figure 12–2.2 Directional properties (cross-laminated, glass-reinforced epoxy). The reinforcing fibers are woven in two right-angle directions. Therefore, there is high strength and good rigidity in these directions. The corresponding values are low at 45°. (Adapted from Broutman, *Modern Composite Materials*, Addison-Wesley.)

anisotropy in composite sheet products. The glass fibers of the reinforcing mats possess a sufficiently random distribution to give uniform elastic moduli and therefore uniform load distributions in two dimensions.

Example 12–2.1

Refer to the composite of Example 12–1.3. The yield strength of the steel is 280 MPa (40,000 psi); that of the copper is 140 MPa (20,000 psi). (a) If this composite is loaded in tension, which metal will yield first? (b) How much load, F, can the composite carry in tension without plastic deformation? (c) What is Young's modulus for the composite?

Procedure. From Example 12–1.3,

$$A_{st} = 0.8 \times 10^{-6} \text{ m}^2; \quad A_{Cu} = 2.4 \times 10^{-6} \text{ m}^2.$$

$$E_{st} = 205 \text{ GPa}; \quad E_{Cu} = 110 \text{ GPa}.$$

The elastic strain must be equal in the two metals.

Calculation

$$(s/E)_{st} = e_{st} = e_{Cu} = (s/E)_{Cu}$$

$$s_{st} = s_{Cu} (205,000 \text{ MPa})/(110,000 \text{ psi}) = 1.86 \ s_{Cu}.$$

a) With a 1.86 stress ratio, the steel is stressed 260 MPa when the copper is stressed 140 MPa. Therefore, the copper yields first.

b)
$$F_{total} = F_{Cu} + F_{st}$$
$$= (140 \times 10^6 \text{ N/m}^2)(2.4 \times 10^{-6} \text{ m}^2)$$
$$+ (260 \times 10^6 \text{ N/m}^2)(0.8 \times 10^{-6} \text{ m}^2)$$
$$= 540 \text{ N} \qquad\qquad \text{(or 55 kg on earth)}.$$

c) From Eq. (12–2.3),
$$\overline{E} = (fE)_{st} + (fE)_{Cu}$$
$$= 0.25 \,(205{,}000 \text{ MPa}) + 0.75 \,(110{,}000 \text{ MPa})$$
$$= 130{,}000 \text{ MPa}.$$

Alternatively,

$$e = (s/E)_{st} = [F/(A_{st} + A_{Cu})]/\overline{E}.$$
$$\overline{E} = (E/s)_{st}[540 \text{ N}/(3.2 \times 10^{-6} \text{ m}^2)]$$
$$= \left[\frac{205{,}000 \times 10^6 \text{ N/m}^2}{260 \times 10^6 \text{ N/m}^2}\right]\left[\frac{540 \text{ N}}{3.2 \times 10^{-6} \text{ m}^2}\right]$$
$$= 130 \times 10^9 \text{ N/m}^2 \qquad\qquad \text{(or 130,000 MPa)}. \quad \blacktriangleleft$$

• **Example 12–2.2**

Formulate a mixture rule for Young's modulus of a laminate, when the stress is applied perpendicular ($\perp$) to the "grain" of the laminate.

Derivation. With transverse loading, $s_1 = s = s_2$.
$$\overline{E}_\perp = s_\perp/e = s/(e_1 f_1 + e_2 f_2)$$
$$= 1/(f_1/E_1 + f_2/E_2),$$
$$1/\overline{E}_\perp = f_1/E_1 + f_2/E_2. \qquad (12\text{–}2.4)$$

Comment. This derivation and the one for Eq. (12–1.8) assume that the Poisson ratios for the two components are comparable. Thus there would be no secondary stresses because of differences in lateral strains. $\blacktriangleleft$

Example 12–2.3

A 2.5-mm (0.10-in.) iron sheet that is to be used as a lining within a household oven is coated on *both* sides with a glassy enamel. The final processing occurs above the 500°C (930°F) strain point (Section 7–7), to give a 0.5-mm (0.02-in.) coating. The glass has a Young's modulus of 70,000 MPa (10^7 psi) and a thermal expansion of 8.0×10^{-6}/°C.

a) What are the stresses in the glass at 20°C?

b) At 200°C? (Assume no plastic strain.)

Procedure. Since $\Delta L/L$ = thermal expansion + elastic strain, and in this case $(\Delta L/L)_{gl} = (\Delta L/L)_{Fe}$, we may write

$$\alpha_{gl} \Delta T + s_{gl}/E_{gl} = \alpha_{Fe} \Delta T + s_{Fe}/E_{Fe}.$$

Calculation

a) By using data from above and from Appendix C,

$$s_{Fe}/205{,}000 \text{ MPa} - s_{gl}/70{,}000 \text{ MPa} = (8.00 - 11.75)(10^{-6}/°C)(-480°C).$$

But $A_{Fe} = 2.5A_{gl}$ and $F_{Fe} = -F_{gl}$, so $s_{gl} = -2.5s_{Fe}$. Thus

$$s_{Fe} \left[\frac{1}{205,000} + \frac{2.5}{70,000} \right] = 0.0018.$$

Solving, we obtain

$$s_{Fe} = +44 \text{ MPa} \qquad \text{(or 6400 psi)} \qquad (+ : \text{tension});$$

$$s_{gl} = -110 \text{ MPa} \qquad \text{(or } -16,000 \text{ psi)} \qquad (- : \text{compression}).$$

b) By similar calculations for $\Delta T = (500 - 200)°C$,

$$s_{gl} = -69 \text{ MPa (or } -10,000 \text{ psi)}, \quad \text{and} \quad s_{Fe} = +27.5 \text{ MPa (or 4000 psi)}.$$

Comments. We assumed unidirectional strain. In reality, plane (i.e., two-dimensional) strain occurs. The necessary correction gives a higher stress by the factor of $(1 - \nu)^{-1}$, where ν is Poisson's ratio. These residual stresses are desirable. With the glass in compression, the composite is stronger. (See Fig. 7–6.5.) ◄

12–3
INTERFACIAL STRESSES

Civil engineers are aware that shear stresses develop at the interface between a reinforcing rod and concrete. For this reason, they specify "deformed" rods with merloned surfaces (ASTM, A305). Comparable shear stresses are encountered between fiber reinforcement and the surrounding plastic matrix in FRP. Here, however, the shear stresses are supported by a chemical rather than a mechanical bond.

Interfacial shear stresses become particularly important if the fibers are not continuous. This is illustrated in Fig. 12–3.1. In this figure, s_f represents the stress that is carried by the fiber if there are no end effects (infinite length). This corresponds to the calculation in Example 12–3.1 and depends both on the volume fraction of the reinforcement and the Young's moduli of the two materials. If the fiber is broken, however, its stress automatically drops to zero at the end of the fiber, and the load is transferred into the matrix (Fig. 12–3.1b). This transfer is by shear stresses across the interface (Fig. 12–3.1c). The shear stress is very high near the fiber ends, and the weaker matrix must carry an overload. This places a premium on long continuous fibers in load-bearing composites. It also favors greater numbers of small diameter fibers rather than fewer larger fibers, since there is more interfacial area to support the shear loads, and less chance that one broken fiber will introduce damaging flaws in the matrix. Finally, a ductile matrix will adapt more readily to the stress concentrations at fiber ends than will a brittle matrix.*

Example 12–3.1

A glass-reinforced polyvinylidene chloride rod contains 25 w/o borosilicate glass fibers. All the fibers are aligned longitudinally. What fraction of the load is carried by the glass?

---

* If the fracture path of a brittle composite is diverted along the interface, the effective crack tip is blunted, increasing the toughness. This can be advantageous in certain applications.

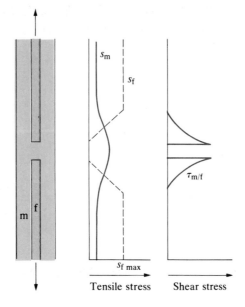

Figure 12–3.1 Stress distribution (at a break in a reinforcing fiber). The fiber stress, s_f, drops from its maximum value to zero. The load must be transferred across the interface from the fiber to the matrix by shear stresses, $\tau_{m/f}$. The matrix has to carry a higher stress, s_m, in the vicinity of the break.

Solution. Basis: 1 g. (Data from Appendix C. Use g/cm^3.)

$$\text{v/o glass} = \frac{(0.25 \text{ g})/(2.4 \text{ g/cm}^3)}{(0.25/2.4)_{gl} + (0.75)/(1.7)_{PVDC}}$$

$$= 19 \text{ v/o} (= 19 \text{ area percent});$$

$$\text{load}_{gl}/A_{gl}E_{gl} = e_{gl} = e_{PVDC} = \text{load}_{PVDC}/A_{PVDC}E_{PVDC},$$

$$\frac{\text{load}_{gl}}{\text{load}_{PVDC}} = \frac{(0.19)(70{,}000 \text{ MPa})}{(0.81)(350 \text{ MPa})} \approx \frac{98\%}{2\%}. \blacktriangleleft$$

• 12–4

HIGH-STIFFNESS COMPOSITES

Because of their lighter weight, many polymeric composites are attractive to design engineers. Although the composites may have lower ultimate tensile strength, S_u, than many metals, their strength-to-density ratios, S_u/ρ, are high and very desirable. However, reinforcements function differently in tension than in compression. In compression, the elastic modulus-to-density ratio, E/ρ, provides a better design criterion where weight is important. Table 12–4.1 shows the E/ρ ratios for a number of common structural materials, including several of those in Appendix C. It is noteworthy that the majority have comparable values. The 50-50 glass-plastic composite immediately below Eq. (12–2.3) has an E/ρ ratio of only 20,000 N·m/g. These lower values are typical for most artificial composites. A glass-reinforced plastic would have to contain more

TABLE 12–4.1			
Modulus/Density (E/ρ) Ratios of Common Materials			
	Density, Mg/m³	**Young's modulus, MPa**	**E/ρ, N·m/g**
Aluminum	2.7	70,000	26,000
Iron and steel	7.8	205,000	26,000
Magnesium	1.7	45,000	26,000
Glass (soda-lime)	2.5	70,000	28,000
Wood (spruce)	0.43	11,000	26,000
Wood (birch)	0.61	16,500	27,000
Polystyrene	1.05	2,800	2,700
Polyvinyl chloride	1.3	<4,500	<3,500
50 v/o glass-plastic*	~1.9	~37,000	~20,000
70 v/o glass-plastic*	~2.1	~50,000	~24,000

* Values for these composites will vary, depending on the plastic; however, these values are typical of most glass reinforced plastics.

than 70 v/o glass to approach the E/ρ of metals. That percentage is difficult to produce cheaply because special efforts are required to align the fibers perfectly. Thus, attention has been paid to materials with extremely high moduli for reinforcement. Some possibilities are listed in Table 12–4.2. In general, however, these materials have not been amenable to fiber production. Boron and carbon fibers show the most potential; but processing costs are high and engineering applications will be limited until these materials are more generally available.

TABLE 12–4.2			
Modulus/Density (E/ρ) Ratios of Materials Potentially Available* to Reinforce Composites			
	Density, Mg/m³	**Young's modulus, MPa**	**E/ρ, N·m/g**
Alumina	3.9	400,000	100,000
Boron	2.3	400,000	170,000
Beryllium	1.9	300,000	160,000
BeO	3.0	400,000	130,000
Carbon	2.3	700,000	300,000
Silicon carbide	3.2	500,000	160,000
Silicon nitride	3.2	400,000	120,000

* Each of these will have to be made available in fiber form at competitive costs, before widespread use may be anticipated.

Review and Study

Composites are unified combinations of two or more materials.

1. *Mixture rules* based on volume fractions may be established to calculate selected properties of composites, if the geometric distributions can be simply defined.

2. To be effective in *reinforced materials,* the reinforcement must possess a higher elastic modulus than the matrix. Nonscalar properties are anisotropic when reinforcement is not random.

3. The *interface* between reinforcement and matrix must transfer shear loads, thus calling for mechanical or chemical bonding between the two.

4. Many composites with high *strength-to-density ratios, S/ρ,* are useful for tensile loads. The *elastic modulus-to-density ratio, E/ρ,* is a better criterion where compression loads or stiffness are design considerations. For these rigidity applications, only composites with the newer ultrahigh modulus fibers hold future promise in comparison with most metals.

TERMS AND CONCEPTS

Composite   Unified combinations of two (or more) distinct materials.

High-stiffness composite   Composite with a high elastic modulus-to-density ratio, E/ρ.

Interfacial stress   Shear stresses that transfer the load between the matrix and the reinforcement.

Matrix   Principal material of a composite that envelops the reinforcement.

Mixture rules   Simplified equations to calculate properties of mixtures.

Reinforcement   Component of composites with a high elastic modulus and high strength.

FOR CLASS DISCUSSION

A_{12} Sand is ground to be used as a filler for a plastic. What properties of the composite will be affected by the amount of grinding (average size of the particles)? Unaffected?

B_{12} The shape of the aggregate affects the behavior of an asphalt concrete. Discuss: How does this involve the choice of raw materials?

C_{12} What design considerations must be given if a fiber-reinforced plastic (FRP) is to be used instead of steel for car fenders? For a drive shaft?

D_{12} A pressure tank is to be made out of plastic reinforced with continuous glass fibers. What geometric factors enter? (Size, shape, openings, etc.)

E_{12} A ''glass'' fishing rod has been broken. It ''looks like splintered wood.'' Suggest failure mechanisms.

F_{12} Glass-coated steel (porcelain enameled steel) is used as an oven liner. Why this product? How should the properties of the glass and the steel match? Be unlike?

QUIZ SAMPLES

12A   For a composite to show improved strength, the reinforcement must be stronger than the matrix. In addition, it must have a higher __**__ .

12B   To a first approximation, the elastic modulus of an oriented composite is proportional to the __**__ of each component.

12C   The properties of __**__ composites are very directional, or __**__ .

12D   The interface between the reinforcement and the matrix must be well bonded in order to withstand __**__ stresses.

• **12E**   In __**__ loading, S_u/ρ serves as basis for comparison; in __**__ loading, a better basis is E/ρ.

STUDY PROBLEMS

12–1.1   Silica flour (finely ground quartz, $\rho = 2.65$ Mg/m³, or 2.65 g/cm³) is used as a filler for polyvinyl chloride ($\rho = 1.3$ Mg/m³). (a) What volume fraction is required to give a product density of 1.70 Mg/m³? (b) What is the w/o SiO_2?

Answer: (a) 30 v/o SiO_2 (b) 47 w/o SiO_2

12–1.2   Calculate the density of a glass-reinforced phenol-formaldehyde rod, in which the glass content is 15 w/o. (A borosilicate glass is used for the longitudinal glass fibers.)

12–1.3   Estimate the thermal conductivity (longitudinal) of the reinforced plastic of Study Problem 12–1.2.

Answer: 0.0002 (W/mm²)/(°C/mm)

12–1.4   A radiation shield is made by using lead powder (90 w/o) as a filler in polystyrene, then pressing it above the T_g of the polystyrene. (a) What is the volume fraction lead? (b) What is the shield's density? (c) Estimate its thermal conductivity.

12–1.5   A cube (25 mm along each side) is made by laminating alternate sheets of aluminum and vulcanized rubber (0.5 mm and 0.75 mm thick, respectively). What is the thermal conductivity of the laminate (a) parallel to and (b) perpendicular to the sheets? (Use the data of Appendix C.)

Answer: (a) 0.09 (W/mm²)/(°C/mm)   (b) 0.0002 (W/mm²)/(°C/mm)

12–1.6   The presence of 2 v/o Al_2O_3 within aluminum as numerous ($>10^9$/mm³) very fine particles (<0.1 μm) adds significantly to the metal's strength. (a) How much (%) does the Al_2O_3 change the density? (b) How much (%) does the Al_2O_3 change the thermal conductivity? The electrical conductivity?

12–2.1   What is the longitudinal elastic modulus for a rod of borosilicate glass-reinforced polystyrene in which all of the glass fibers are oriented lengthwise? There is 80 w/o glass.

Answer: 46,000 MPa (6,500,000 psi)

12–2.2   What are the elastic moduli in the two dissimilar directions of the laminate of Study Problem 12–1.5?

12–2.3 A glass-reinforced plastic rod (fishing pole) is made of 67 v/o borosilicate glass fibers in a polystyrene matrix. What is the thermal expansion coefficient?

Answer: $\sim 4 \times 10^{-6}/°C$

12–2.4 An AISI–SAE 1040 steel wire (cross section 1 mm^2) has an aluminum coating, so that the total cross-sectional area is 1.2 mm^2. (a) What fraction of a 450 N load (100-lb$_f$) will be carried by the steel? (b) What is the electrical resistance of this wire per unit length?

12–2.5 A 2.5-mm (0.10-in.) steel wire coated with 0.5 mm of copper [total dia. = 3.5 mm (or 0.14 in.)] is loaded with 4450 N ($\sim$1000 lb$_f$). (a) What is the elastic strain? (b) How much strain occurs if the total diameter is composed of 1040 steel? (c) If the total wire is composed of copper?

Answer: (a) 0.003

12–2.6 The copper-coated steel (1020) wire of the previous problem is stress-relieved at 400°C (750°F) and is cooled rapidly to 10°C (50°F). (a) Which metal is in tension? (b) What is the stress?

12–2.7 A 2.5-mm (0.1-in.) steel wire (ρ = 200 ohm·nm) is to be copper-coated so its resistance is 3.3×10^{-3} ohm/m ($\sim$0.001 ohm/ft). How thick should the coating be?

Answer: 0.5 mm (0.02 in.)

QUIZ CHECKS

12A elastic modulus

12B volume fraction

12C oriented, anisotropic

12D shear

12E tension, compression

Chapter Thirteen

Performance
of Materials
in Service

PREVIEW

Every material that fails in service has had its structure altered. This will include plastics that have been fractured by overload, aluminum alloys weakened by excessive heating, rubbers that have degraded by ultraviolet light, steels corroded by sea water, etc. The scientist and engineer must identify and anticipate the changes that can occur if service failures are to be avoided.

This chapter addresses four types of failure: (1) Failure from *fracture* due to the imposition of a load on a notch-sensitive material. (2) Failure by *over-heating* that leads to creep and stress-ruptures. (3) Failure due to neutron *radiation* that damages materials atom by atom and electron by electron. Finally, (4) failure by *corrosion*, which is commonly a surface alteration through an electrochemical attack.

Of course, there are other failure modes, but these will suffice to point out the correlation of performance with structural changes and resulting property modification.

CONTENTS

STUDY OBJECTIVES

1. To realize that failure is not always catastrophic but simply the point where the product is no longer capable to fulfill its intended purpose. Failure prevention commonly involves multiple inputs and alternatives for materials selection.

2. To develop the concept of the stress intensity factor, or fracture toughness, as a design requirement vis-à-vis strength, hardness, ductility, etc.

• 3. To appreciate the role of elevated temperatures upon the rates of microstructural change and thus property changes of materials in service.

• 4. To be able to anticipate the nature of property changes in a material that has been exposed to neutron radiation.

5. To know the roles of the anode, cathode, and electrolyte in metallic corrosion.

6. To be acquainted with the principal methods of minimizing corrosion.

13–1

**SERVICE
REQUIREMENTS**

If engineers are to ''adapt materials and/or energy for society's needs'' as stated in Chapter 1, they must consider the performance of their product in service, be it an automobile, bridge, computer, device, etc. This involves not simply the ''handbook properties'' of the purchased materials and components but also considerations for how a material will change while performing its task. For example, if the bond for an 0.1-mm wire at the edge of a computer chip (Fig. 13–1.1) cracks and fails during service, the 256 K integrated chip is affected. Likewise, a cutting tool may be specified and manufactured with a desired microstructure of tempered martensite, and with the required hardness (Fig. 13–1.2a). However, a temporary absence of a cutting fluid, or the use of excessive cutting speeds, can produce heating, over-tempering, and softening at the cutting tip. In turn, the tool dulls and fails (Fig. 13–1.2b)—a sequence that can occur within seconds.

Failure

Product failure need not necessarily be catastrophic, nor involve fracture, leakage, or excessive wear. Failure is simply the point at which *the product is no longer capable to fulfill its intended purpose*. Thus, failures range from the fracture of a welded bridge beam to the dulling of a rear-view mirror in an automobile; from potholes in streets to the oxidized contact points of a micro-switch in a home thermostat; and from a punctured tire to a faded color print. In some products, failures are nearly instantaneous; the intended functions of other products are lost gradually, extending over a period of months or years.

Since failure of a product or device occurs within the material, it is natural to assume that there was a deficiency of that material. However, the failure of some products arises from improper design judgments or unanticipated service conditions; in others from misuse or from deficient maintenance. Insufficient quality control during the manufacturing process also may be a factor leading to failure. In any event, the design and manufacturing engineers must anticipate

Figure 13–1.1 Service performance (input-output ports to a very large scale integrated (VLSI) computer chip). Although the failure rate of VLSI chips is only 0.00001 as great as for equivalent vacuum tube circuits, quality control is still paramount because of the multiplicity of operations. Poor bonds between the wires and the chip ports account for the majority of service failures. (Reprinted with permission from the American Society for Testing and Materials, 1916 Race Street, Philadelphia, Pa. 19103.)

the effects of service conditions upon their products, and upon the materials that comprise those products. The more the engineer knows about the nature of failure, the easier it is to control it.

We shall examine four types of failure that arise from the numerous situations that are encountered in service. They are failure by fracture, failure at

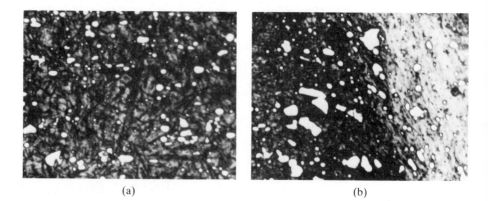

(a)   (b)

Figure 13–1.2 Structural change during service (high-speed tool steel, ×500). (a) Microstructure before use. (b) Microstructure after cutting at excessive speeds. The temperature of the tip increased sufficiently to permit over-tempering and softening, which lead to failure. (Courtesy of Crucible Research Center.)

elevated temperatures, radiation damage, and corrosion. These originate from mechanical loading, thermally induced reactions, alteration by radiation, and chemical interactions of a material with its environment.

Design Considerations

Many factors affect the design engineer's choice of material for a product. Several examples will illustrate.

1. The strongest material may not withstand impact loads because the high-strength materials generally are not tough materials. A trade-off is necessary.

2. A material with an ultimate corrosion resistance (e.g., gold) is too expensive to solve the automotive corrosion problem. Stainless steel could solve the problem (Fig. 13–1.3) but it's not practical because the supply of the necessary chromium is too limited and must be imported. As a result, economic and political factors must be introduced.

3. Aluminum is cheaper than copper as an electrical conductor, but it requires special techniques for splicing and terminal connectors to avoid high resistances and possible melting. Should aluminum conductors be used in residential products? In commercial installations?

4. Age-hardened aluminum is widely used in airplane construction because it has one of the highest strength/weight ratios. The engineer must know its over-aging characteristics and the service conditions before choosing aluminum for a structural component of a space shuttle.

These examples, as well as all other technical designs, require multiple considerations before choosing the specified material. As noted, not all of the decisions are technical.

Figure 13–1.3 Stainless steel alternative (1936 car, with original body). Materials availability, international political factors, processing complications, and the customer's willingness to pay the cost differential all affect the materials selection decisions. (Courtesy Allegheny Ludlum Steel Corp.)

13–2

FRACTURE

The ultimate mechanical failure is fracture. We commonly categorize fracture as being either ductile or brittle (i.e., nonductile). Little energy is required to break brittle materials, such as glass, polystyrene, and some of the cast irons. Conversely, tough materials, such as rubber and many steels, absorb considerable energy in the fracture process. The contrast is important since the service limit in many engineering products is not the yield or ultimate strength; rather it may be the energy associated with fracture propagation.

Brittle fracture requires energy to separate atoms and expose new surface along the fracture path. *Ductile failure* requires not only the energy just cited but much additional energy to deform plastically the material ahead of the fracture.

One measure of toughness is the area under the stress–strain curve (Fig. 13–2.1). With no plastic deformation, that area is $se/2$, or $Ee^2/2$, and the energy is solely elastic. The calculation is not as simple for the ductile material in Fig. 13–2.1(b), because the plastic strain of the deformable material far exceeds the limited elastic strain. The result is a much higher energy consumption before fracture. The units of energy as just described are the product of stress and strain:

$$(\text{N/m}^2)(\text{m/m}) = \text{joule/m}^3 \qquad (\text{or ft·lbs/in}^3)$$

i.e., joules per unit volume. In reality, the energy consumption is very nonuni-

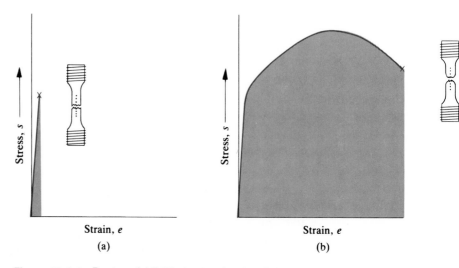

Figure 13–2.1 Fracture. (a) Brittle fracture involves little or no plastic deformation. (b) Ductile fracture requires energy for plastic deformation. Toughness, the energy requirement, is equal to the area under the s–e curve.

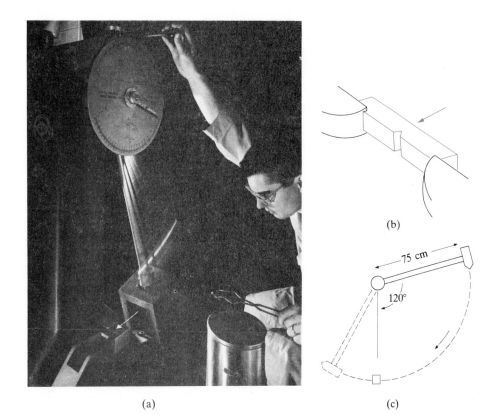

(a)   (b)

(c)

Figure 13–2.2 Toughness test. The notched test specimens—arrow in (a) and sketched in (b)—is broken by the impact of the swinging pendulum (c). The amount of energy absorbed is calculated from the arc of the follow-through swing (Example 13–2.1). (Courtesy of U.S. Steel Corp.)

form within the fractured test piece; negligible energy is absorbed in undeformed regions, while a major quantity is absorbed in the vicinity of the fracture. Furthermore, this distribution is markedly affected by the size and shape of the test specimen. Notches are especially critical in determining the energy requirements for fracture.

Toughness Tests   Figure 13–2.2 shows an impact tester that has been widely used to obtain toughness data. The energy absorbed in the fracture is calculated from the height of the follow-through swing of the heavy pendulum (Example 13–2.1). A Charpy V-notch test specimen (Fig. 13–2.2b) is commonly used for comparative purposes. It should be noted, however, that these values are a function of size and shape as well as of the materials being compared.

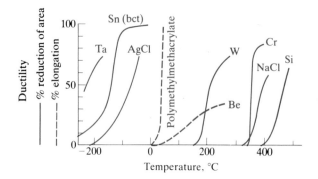

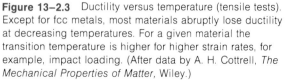

Figure 13–2.3 Ductility versus temperature (tensile tests). Except for fcc metals, most materials abruptly lose ductility at decreasing temperatures. For a given material the transition temperature is higher for higher strain rates, for example, impact loading. (After data by A. H. Cottrell, *The Mechanical Properties of Matter*, Wiley.)

Ductility-transition Temperature

Many materials exhibit an abrupt drop in ductility and toughness as the temperature is lowered. In glass and other amorphous materials, this change corresponds to the glass-transition temperature. Of course, metals are crystalline and do not have a glass-transition temperature. However, they may have a *ductility-transition temperature*, T_{dt} (Fig. 13–2.3), that divides a lower temperature regime where the fracture is said to be nonductile from a higher temperature range where considerable plastic deformation accompanies failure.

Figure 13–2.4 shows the energy absorbed during impact fracture by two different steels as a function of temperature. As with the glass-transition temperature (Section 6–5), the ductility-transition temperature is not exact, but it depends on size and shape, the rate of loading, presence of impurities, etc. However, the engineer will be quick to choose Steel C over Steel B of Fig. 13–2.4 if it is to be used in a welded ship in the North Atlantic winter waters. A crack, once started in the steel with a high ductility transition temperature,

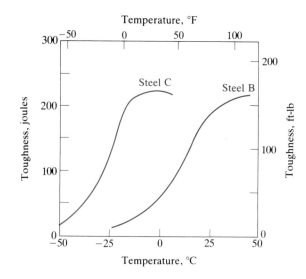

Figure 13–2.4 Toughness transitions. For each steel, there is a marked decrease in toughness at lower temperatures. The transition temperature is significantly lower for Steel C (fine-grained) than for Steel B (rimmed). (Adapted from Leslie, Rickett, and Lafferty, *Trans. AIME*.)

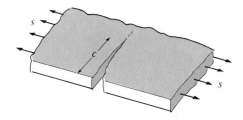

Figure 13–2.5 Fracture stresses (Mode I).*
Stresses that open an edge crack in tension
introduce an intensity factor, K_I, that
increases with $\sqrt{\pi c}$ as indicated in Eq.
(13–2.1). If K_I exceeds a critical value, K_{Ic},
the crack will advance catastrophically.

could continue to propagate with a low energy fracture until the ship is broken apart. There were several unfortunate naval catastrophies that occurred before the design engineer learned how to minimize stress concentrations, and metallurgists found that fine-grained steels had lower ductility transition temperatures than coarse-grained steels.

Fortunately, fcc metals do not possess an abrupt change in ductility as the temperature is lowered. Thus, aluminum, copper, and some stainless steels can be used in cryogenic applications. Unfortunately these metals are too costly to replace steels in large structures, pipelines, etc.

Fracture toughness, K_{Ic}

Fracture always starts at some geometric irregularity that introduces a stress concentration. It may be a keyway in a shaft, a corner of a hatch on a ship, a scored line on a piece of window glass, or a crack in the edge of a structural plate (Fig. 7–6.2).

A common mode of stressing that produces fracture is sketched in Fig. 13–2.5, which shows how tension forces open an existing edge crack.* An index of the stresses at the crack front—called the *stress intensity factor, K_I,*— may be calculated as follows:

$$K_I = s\alpha\sqrt{\pi c}. \qquad (13–2.1)$$

In this equation, s is the stress in the absence of a crack (the stress ordinarily calculated for design), c is the depth of an edge crack (or half the length of an interior crack), and α is a dimensionless geometric factor. The latter is near 1.0 for plate products as sketched in Fig. 13–2.5, in which the crack is a small fraction of the total plate width. The units of K_I are MPA·$\sqrt{m}$ (or psi·$\sqrt{in.}$).

There are several design implications from the above relationships. Cracks can be tolerable, providing $s\alpha\sqrt{\pi c}$, that is, K_I, does not exceed the critical stress intensity factor K_{Ic} for the material.† Conversely, the material will frac-

---

* The loading in Fig. 13–2.5 is called Mode I. Other modes include: II, shear forces parallel to the crack surface and perpendicular to the crack front; and III, shear forces parallel to the crack surface as well as to the crack front.

† This is comparable to saying that a material can tolerate a load without plastic deformation, providing F/A, that is, s, does not exceed the critical yield stress S_y for the material.

Figure 13–2.6 Fatigue fracture (14-cm (5½-in.) steel shaft). Fracture progressed slowly from the set-screw hole at the top through nearly 90% of the cross section before the final rapid fracture (bottom). (Courtesy of H. Mindlin, Battelle Memorial Institute.)

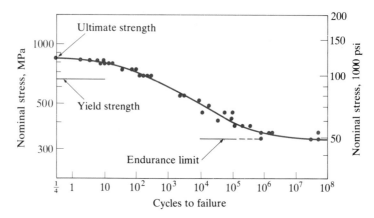

Figure 13–2.7 *S–N* curve (SAE 4140 normalized steel). *S–N* = cyclic *stress* versus *number* of cycles to failure. At the endurance limit, the number of cycles becomes indeterminately large. (Adapted from R. E. Peterson, *ASTM Materials Research and Standards*.)

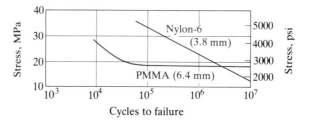

Figure 13–2.8 *S–N* curves (Nylon-6 and polymethylmethacrylate). Design specifications must call for lower stresses (*S*) when a greater number (*N*) of stress reversals are encountered. (Adapted from M. N. Riddell et al., *Polymer Science and Engineering*.)

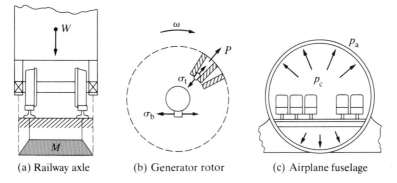

(a) Railway axle   (b) Generator rotor   (c) Airplane fuselage

Figure 13–2.9 Examples of cyclic loading. (a) Axle of rail car. (b) Rotor of generator during starting and stopping. (c) Pressurization and depressurization of plane. The latter may encounter only a few thousand cycles; however, as indicated by Fig. 13–2.7, the yield strength cannot be used by designers. (Courtesy of *ASTM*, R. E. Peterson, "Fatigue in Metals," *Materials Research and Standards.*)

ture if $s\alpha\sqrt{\pi c}$ exceeds K_{Ic} (called *fracture toughness*), even though S_y or S_u are not exceeded. The engineers must look at both the strength and the fracture toughness. In general, materials with higher strengths have lower critical flaw sizes. This means that the condition for fracture may be encountered before the yield stress is reached, and then full advantage could not be taken of the high strength. (See Example 13–2.2.)

Fatigue

There are many documented examples of eventual failure of rotating shafts on power turbines and on other mechanical equipment, which had initially performed satisfactorily for long periods of time. Figure 13–2.6 shows one such fracture.*

The stresses that a material can tolerate under cyclic loadings are much less than under static loading. The tensile strength can be used as a guide in design only for structures that are in service under static loading. The number of cycles N that a metal will endure decreases with increased stresses S. Figure 13–2.7 is a typical *S–N* curve for fatigue fracture of steels. In designing for unlimited cyclic loading, it is necessary to restrict the stresses to values below the *endurance limit* of this curve. Similar data for two plastics are shown in Fig. 13–2.8.

Figure 13–2.9 shows three examples of cyclic loading. The axle of a train has many sinusoidal stress cycles. Examples of low-cycle stresses are found in the rotor of a generator that is used as a "topping" unit to meet peak demand for electricity. It starts and stops up to 1000 times per year, each time introducing a stress cycle at the base of the winding slot and at the arbor. Likewise an

---

* The early explanation was that the metal became "tired" and failed from fatigue. We now know that fatigue fracturing is a result of very localized microstructural movements that lead to crack propagation.

TABLE 13–2.1

Surface Finish versus Endurance Limit
(SAE 4063 steel, quenched and tempered to 44R$_C$)*

Type of finish	Surface roughness		Endurance limit	
	μm	μin.	MPa	psi
Circumferential grind	0.4–0.6	16–25	630	91,300
Machine lapped	0.3–0.5	12–20	720	104,700
Longitudinal grind	0.2–0.3	8–12	770	112,000
Superfinished (polished)	0.08–0.15	3–6	785	114,000
Superfinished (polished)	0.01–0.05	0.5–2	805	116,750

* Adapted from M. F. Garwood, H. H. Zurburg, and M. A. Erickson, "Correlation of Laboratory Tests and Service Performance," *Interpretation of Tests and Correlation with Service*, Amer. Soc. Metals.

airplane fuselage has stresses imposed during pressurization following each take-off; these are removed as it returns to the ground.

Fatigue cracks usually start at the surface where bending or torsion cause the highest stresses to occur and where surface irregularities introduce stress concentrations. As a result, the endurance limit is very sensitive to surface finish (Table 13–2.1).

A close examination of the early stages of crack development shows that microscopic and irreversible slip occurs within individual grains.* There is a gradual reduction in ductility in the regions of these slip planes, causing microscopic cracks to form. This may be after only 10% or 20% of the eventual fatigue life. The cracks progress slowly through the remaining cycles. Eventually the crack exceeds the critical size to raise the stress intensity factor to K_{Ic} so that a final catastrophic fracture occurs (Fig. 13–2.6).

Any design factor that concentrates stresses can lead to premature failure. We have already seen, in Table 13–2.1, that surface finish is important. Keyways and other notches (Fig. 13–2.6) are also critical. Finally the generous use of fillets is recommended in engineering design as shown in Fig. 13–2.10.

Example 13–2.1

An impact pendulum on a testing machine weighs 10 kg and has a center of mass 75 cm from the fulcrum. It is raised 120° and released. After the test specimen is broken, the follow-through swing is 90° on the opposite side. How much energy did the test material absorb?

* This will cause extrusions and intrusions on the external surface of the grains even when highly polished. Then these further concentrate stresses. See Fig. 21–20, Van Vlack, *Materials Science for Engineers* (Reading, Mass.: Addison-Wesley).

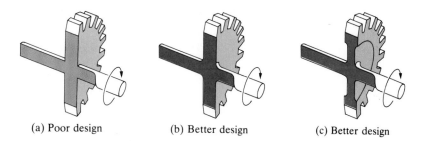

(a) Poor design   (b) Better design   (c) Better design

Figure 13–2.10   Design of fillet. The use of generous fillets is recommended in mechanical engineering design. It should be observed that (c) is a better design than (a), even with some additional material removed. Of course, if too much metal is removed, failure may occur by mechanisms other than fatigue.

Calculation. Refer to Fig. 13–2.2.

$$\Delta E = 10 \text{ kg}(9.8 \text{ m/s}^2)(0.75 \text{ m})[\cos(-120°) - \cos 90°]$$
$$= -36.8 \text{ J} \qquad\qquad (lost \text{ by the pendulum})$$
$$= +36.8 \text{ J} \qquad\qquad (absorbed \text{ by the sample}).$$

Comments. In general, materials have more toughness at high temperatures than at low. In fact, there is an abrupt decrease in the toughness of many steels as they are cooled below ambient temperatures. The temperature at which this discontinuous decrease occurs is called the *ductility-transition temperature* (Fig. 13–2.4). ◄

Example 13–2.2

A steel has a yield strength, S_y, of 1100 MPa (160,000 psi), an ultimate strength, S_u, of 1200 MPa (174,000 psi), and a fracture toughness, K_{Ic}, of 90 MPa·$\sqrt{m}$ (82,000 psi·$\sqrt{in}$.). (a) A plate has a 2-mm (0.08-in.) edge crack. Will it fail by fracturing before it fails by yielding? (b) What are the deepest tolerable cracks that may be present without fracturing before yielding? [Assume that the geometric factor, α, equals 1.1 for steel plates loaded in tension as in Fig. 13–2.5.]

Procedure. Determine the stress s required to raise the stress intensity factor K_I to the critical value of 90 MPa·$\sqrt{m}$. In (b) use 90 MPa·$\sqrt{m}$ and 1100 MPa (K_{Ic} and S_y, respectively) to calculate the critical limiting crack length.

Calculation. Use Eq. (13–2.1).

a)
$$s = (90 \text{ MPa·}\sqrt{m})/(1.1\sqrt{0.002\pi \text{ m}})$$
$$= 1030 \text{ MPa}.$$

This stress of 1030 MPa (150,000 psi) to cause fracture is less than S_y; so this steel will fracture before it yields.

b)
$$c\pi = [(90 \text{ MPa·}\sqrt{m})/(1100 \text{ MPa})(1.1)]^2$$
$$c = 0.0017 \text{ m} \qquad\qquad (\text{or } 1.7 \text{ mm}). \quad ◄$$

• 13–3

In general, materials deform more readily at elevated temperatures. This occurs because plastic deformation very commonly arises from dislocation movements that involve a continual displacement of atoms to new neighbors (Fig. 5–4.5). Of course, these rearrangements proceed more readily and with less stress at higher temperatures.

Several microstructural changes that accompany increases in temperature also lead to softening. These include recrystallization, grain growth, overaging, and over-tempering. The service temperature of a strain-hardened metal is limited by T_R, the recrystallization temperature; and, of course, T_R is a function of time at temperature (Fig. 5–6.4). Likewise, a precipitation-hardened alloy of Fig. 11–5.3 will start losing its hardness and strength in only a few hours if used at 200°C. This temperature sensitivity, as well as the over-tempering described in Fig. 13–1.2, occurs because the finely dispersed particles that restrict dislocation movements coalesce. Not only are there fewer particles for blocking, but there are larger regions of the soft matrix phase (nearly pure fcc Al in Fig. 11–5.3, and bcc ferrite in the tool steel of Fig. 13–1.2).

Extended High-temperature Service (crystalline materials)

We saw in Section 5–7 that *creep* can occur at elevated temperatures independently of dislocation movements. The mechanism involves atom diffusion to and from grain boundaries (Fig. 5–7.7). The grain size and grain-boundary orientation are important factors.

Implied in the name, creep is a slow mechanism of strain. The rates range from a few percent per hour at high stress levels, or at high temperatures, down to less than 10^{-4}%/hr* (Fig. 13–3.1). These are small; however, consider their importance when designing a steam power plant or nuclear reactor, which must be in high-temperature service for many years. Likewise, creep becomes important in gas turbines and in equipment that must be operated without change in dimensions at high stresses and elevated temperature to maximize energy-conversion efficiencies.

We may plot strain as a function of time for various stresses and temperatures. Figure 13–3.2 is schematic. When a metal is stressed, it undergoes immediate elastic deformation, which is greater when either the stress or the temperature is high. In the first short period of time (Stage 1), it makes additional, relatively rapid, plastic adjustments at points of stress concentration along grain boundaries and at internal flaws. These initial plastic adjustments give way to a slow, nearly steady rate of strain that we define as the *creep rate ė*. This second stage of *steady creep* continues over an extended period of time, until sufficient strain has developed so that a necking-down and area-reduction starts. With this change in area at a constant load, the rate of strain accelerates until rupture occurs (Stage 3). If the load could be adjusted to match the

---

* ~1%/year.

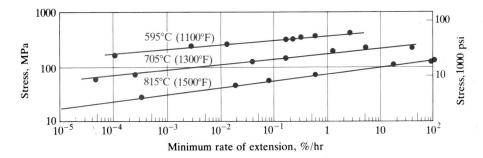

Figure 13–3.1 Creep data (type 316 stainless steel). Higher stresses and higher temperatures increase the creep rate. (From Shank, in McClintock and Argon, *Mechanical Behavior of Metals*, Addison-Wesley.)

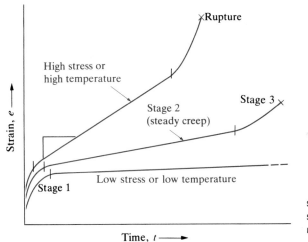

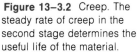

Figure 13–3.2 Creep. The steady rate of creep in the second stage determines the useful life of the material.

reduction in area and thus maintain a constant stress, the creep rate of Stage 2 would continue until rupture.

In Fig. 13–3.2 the following relationships are shown schematically: (1) The steady creep rate increases with both increased temperature and stress. (2) The total strain at rupture also increases with these variables. (3) The time before eventual failure by *stress-rupture* is decreased as the temperatures and applied stresses are increased (Fig. 13–3.3).

Thermal Stability (amorphous materials)

Polymers and glasses are rigid below their transition temperature, T_g (Section 6–5). As the temperature is raised beyond the glassy range, creep and distortion begin to be significant, particularly when a stress is applied. For this reason, *heat distortion* tests have been developed. For example, ASTM D648

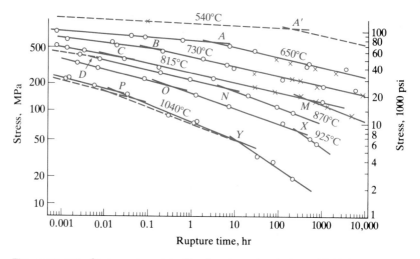

Figure 13–3.3 Stress-rupture data. The time to rupture is less with higher stresses and higher temperatures. (After N. J. Grant, American Society for Metals.)

heats a sample bar, 5 in. × 0.5 in. × 0.5 in., at the rate of 2°C per minute while loaded in flexure. The temperature at which a deflection of 0.010 in. (0.25 mm) develops is called the heat-distortion temperature. Of course, this temperature varies with the load. Therefore, a calculated "outer fiber" stress of either 66 psi (0.455 MPa) or 264 psi (1.82 MPa) is commonly chosen. This gives a reproducible distortion temperature for any given polymer that, as expected, varies significantly between products, even of the same type, because fillers, plasticizers, molecular weight, etc., are additional factors.

Although viscoelastic deformation is accelerated as the temperature exceeds T_g, this softening does not break the primary covalent bonds within the molecule. Under more severe conditions, however, these bonds may be ruptured. Of course, any resulting change in structure affects the properties. Extreme heat can degrade a polymer by breaking bonds and by oxidation.

The most obvious *degradation* of polymers is charring. If the side groups or the hydrogen atoms of a vinyl polymer (Table 6–1.1) are literally torn loose by thermal agitation, only the backbone of carbon atoms remains. We also see this occur with the carbohydrates of our morning toast or with charred wood. It is generally to be avoided in a commercial product.* Carbonization is accelerated in the presence of air, because oxygen reacts with the hydrogen atoms along the side of the polymer chain.

The degradation just described is accentuated by oxygen. In addition, oxygen can have other effects. For example, many rubbers are vulcanized with

---

* Under controlled conditions, a graphite fiber can be formed from a polymer fiber. Such fibers hold considerable promise as a high-temperature reinforcement for composites (Section 12–4).

only 5–20 percent of the possible positions anchored by sulfur cross-links. This permits the rubber to remain deformable and "elastic." Over a period of time, the rubber may undergo further cross-linking by oxygen of the air. The result is identical to Fig. 6–3.4, except that oxygen is the connecting link rather than sulfur. Naturally the rubber becomes harder and less deformable with an increased number of cross-links.

Several factors accelerate the oxidation reaction just described.

1. Oxygen in the form of ozone, O_3, is much more reactive than normal O_2.
2. Ultraviolet light can provide the energy to break existing bonds so that the oxidation reaction can proceed.
3. The existing bonds are broken more readily when the molecules are stressed.

Because of these features, tires commonly contain carbon black or similar light absorbers to decrease the oxidation rate. Applying the same principle, accelerated testing procedures for product stability commonly expose the polymer to ozone and/or ultraviolet light.

• 13–4
RADIATION DAMAGE

Material selection is the key to successful nuclear-reactor design. First, the reactors must be "fail-safe" in case there are any operational problems. Secondly, radiation will alter the internal structure of many materials. Since the property changes that accompany these alterations are generally (but not always) undesired, we speak of *radiation damage*. While we will focus much of our attention on neutron radiation, other types of radiation such as α-particles (He^{2+} ions), β-rays (energetic electrons), and γ-rays (high-frequency photons) are also present.

Unlike elevated temperatures, which energize all the atoms within a solid, radiation processes may focus relatively large amounts of energy locally. As sketched in Fig. 13–4.1, a *neutron* possessing a number of electron volts may collide with an atom, dislodging it from its crystal site. Typically, the neutron has sufficient energy remaining after the first collision to ricochet to other atoms, dislodging as many as a dozen or more before being finally captured by the nucleus of an atom. In the meantime, considerable damage has occurred within the crystal structure. Slip processes are thereby inhibited, with the result that hardness and strength increase as shown in Figs. 13–4.2 and 13–4.3. There is a concurrent decrease in ductility and toughness.

The property changes caused by neutron irradiation of crystalline materials are similar to the changes that arise during cold work (Section 5–5). Similarly, *annealing* will soften and toughen materials to achieve *damage recovery*. Despite these comparisons, however, the structural mechanisms are basically different, because point defects are mainly responsible for radiation damage,

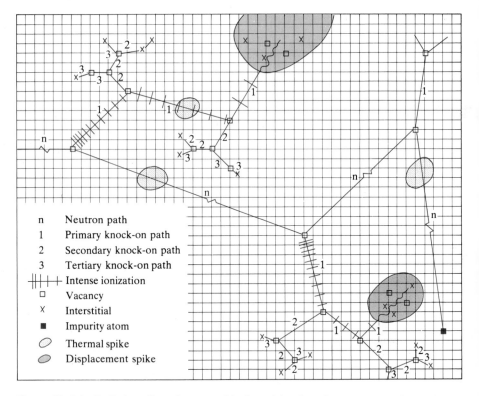

Figure 13–4.1 Radiation effects in a crystal lattice arising from the neutron entering at the left. (C. O. Smith, *Nuclear Reactor Materials*, Addison-Wesley.)

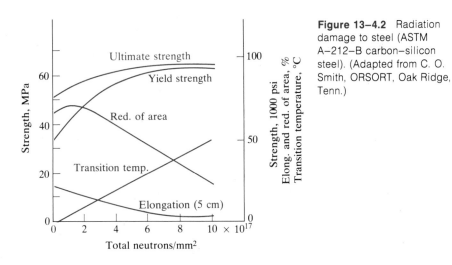

Figure 13–4.2 Radiation damage to steel (ASTM A–212–B carbon–silicon steel). (Adapted from C. O. Smith, ORSORT, Oak Ridge, Tenn.)

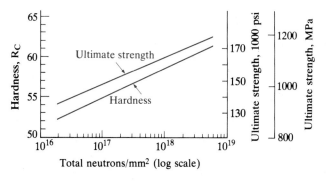

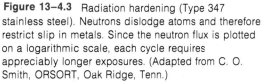

Figure 13–4.3 Radiation hardening (Type 347 stainless steel). Neutrons dislodge atoms and therefore restrict slip in metals. Since the neutron flux is plotted on a logarithmic scale, each cycle requires appreciably longer exposures. (Adapted from C. O. Smith, ORSORT, Oak Ridge, Tenn.)

but linear defects are mainly responsible for strain hardening. This difference is reflected in the fact that lower annealing temperatures are required to remove radiation damage than to remove strain hardening.

Scission (chain splitting) is common within polymeric materials when they are exposed to radiation, either of the particulate type (neutrons, α-rays, etc.) or of the electromagnetic type (γ-rays, x-rays, etc.). After scission, the broken ends of the chains and the exposed side radicals possess reactive sites (Section 6–2). In the simplest case, the reactive sites rejoin. More commonly, however, there are sufficient displacements so that other types of reactions can occur; for example, a chain half can be grafted onto another chain to produce *branching,* with consequent changes in properties (Fig. 6–4.2).

By noting several items from earlier chapters, we should expect that *electrical properties* are sensitive to radiation. (1) Recall first that disordered structures have shorter mean free paths for electron movement (Section 8–2). Thus any radiation damage that introduces imperfections into the crystal structure of a metal increases the electrical resistivity. (2) Also recall from Section 8–4 that photoconduction occurs when an electron is excited across the energy gap. Therefore, irradiation by photons in the ultraviolet-to-γ-ray range can increase the number of intrinsic carriers (both electrons and holes) by several orders of magnitude. Hence, the resistivity of semiconductors is decreased. (3) Finally, recall from Section 9–4 that vacancies in some ionic solids may produce color centers, which modify dielectric and optical behaviors.

Table 13–4.1 presents examples of property changes that are attributable to neutron radiation.

Example 13–4.1

Assume that all the energy required to produce scission in a polyethylene molecule comes from a photon (and that none of the energy is thermal).

a) What is the maximum wavelength that can be used?

b) How many eV are involved?

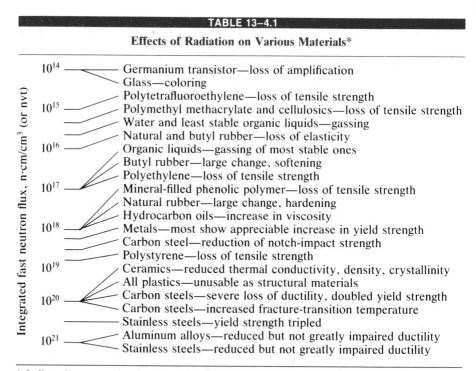

TABLE 13–4.1
Effects of Radiation on Various Materials*

Integrated fast neutron flux, n·cm/cm³ (or nvt)

10^{14} — Germanium transistor—loss of amplification
— Glass—coloring
10^{15} — Polytetrafluoroethylene—loss of tensile strength
— Polymethyl methacrylate and cellulosics—loss of tensile strength
— Water and least stable organic liquids—gassing
— Natural and butyl rubber—loss of elasticity
10^{16} — Organic liquids—gassing of most stable ones
— Butyl rubber—large change, softening
— Polyethylene—loss of tensile strength
10^{17} — Mineral-filled phenolic polymer—loss of tensile strength
— Natural rubber—large change, hardening
— Hydrocarbon oils—increase in viscosity
10^{18} — Metals—most show appreciable increase in yield strength
— Carbon steel—reduction of notch-impact strength
— Polystyrene—loss of tensile strength
10^{19} — Ceramics—reduced thermal conductivity, density, crystallinity
— All plastics—unusable as structural materials
10^{20} — Carbon steels—severe loss of ductility, doubled yield strength
— Carbon steels—increased fracture-transition temperature
— Stainless steels—yield strength tripled
10^{21} — Aluminum alloys—reduced but not greatly impaired ductility
— Stainless steels—reduced but not greatly impaired ductility

* Indicated exposure levels are approximate. Indicated changes are at least 10%. The table is reprinted from C. O. Smith, *Nuclear Reaction Materials*. Reading, Mass.: Addison-Wesley.

Procedure. The energy of the C–C bond is listed in Table 2–2.1 as 370,000 J/(0.6 × 10^{24}). This energy must be related to the photon frequency. From introductory chemistry or physics, $E = h\nu = hc/\lambda$.

Answer

a)
$$\lambda = \frac{hc}{E} = \frac{(0.66 \times 10^{-33} \text{ J·s})(3 \times 10^8 \text{ m/s})}{370{,}000 \text{ J}/0.6 \times 10^{24}}$$

$$= 0.32 \times 10^{-6} \text{ m} \qquad \text{(or 320 nm)}.$$

b)
$$(370{,}000 \text{ J}/0.6 \times 10^{24})/(0.16 \times 10^{-18} \text{ J/eV}) = 3.8 \text{ eV}.$$

Comment. This is in the uV range; thus, photons of visible light will not cause scission alone. However, a visible photon can cause scission, if it hits a bond that momentarily possesses high thermal energy. Laboratory test cells use uV to accelerate stability evaluations. ◀

13–5

**CORROSION
REACTIONS**

Corrosion is the process of surface deterioration of metals and related materials. The alteration occurs by the metal, M, losing electrons and becoming a positive ion:

$$M^0 \rightarrow M^{m+} + m\ e^-. \qquad (13\text{–}5.1)$$

For the reaction to proceed, both the electrons and the metal ions must be removed.* The electrons are accepted by nonmetallic element, N, or by other metallic ions, M^+:

$$N^0 + n\ e^- \rightarrow N^{n-} \qquad (13\text{–}5.2a)$$

or

$$M^{m+} + m\ e^- \rightarrow M^0. \qquad (13\text{–}5.2b)$$

The metal ions of Eq. (13–5.1) either dissolve in the surrounding electrolyte, or combine with nonmetallic ions to form a surface deposit.

Figure 13–5.1 shows the corrosion reaction schematically. The location of the electron-loss (oxidation) reaction is called the *anode*. That of the electron-gain (reduction) reaction is called the *cathode*. The two reactions may proceed adjacently, or they may be separated by a considerable distance, provided they have a low-resistance electrical connection (for electron transfer). However, the two reactions must occur simultaneously if corrosion is to proceed. Neither can occur alone. Corrosion will stop if (1) the electrical connection is interrupted, (2) the cathode reactants are depleted, or (3) the anode products ($M^{m+} + m\ e^-$) are saturated. These three considerations will receive attention for our understanding of corrosion and its control.

Anode Reactions

All metals are subject to the oxidation reaction of Eq. (13–5.1). However, the tendencies to ionize differ from metal to metal. For example, Eq. (13–5.1) written for zinc and copper are

$$Zn \longrightarrow Zn^{2+} + 2\ e^- \qquad (13\text{–}5.1_{Zn})$$

and

$$Cu \rightarrow Cu^{2+} + 2\ e^-. \qquad (13\text{–}5.1_{Cu})$$

Both release electrons and lead to an electric potential; the value of which *cannot* be measured in isolation. However, the voltage difference between the two reactions can be measured when the metals are in standard 1-molar solutions of their own salts at 25°C (Fig. 13–5.2). In an open circuit (black switch), the voltage difference is 1.1 V. If electrical contact is made (color switch), the

---

* Otherwise, the reverse reaction becomes significant and equilibrium is established.

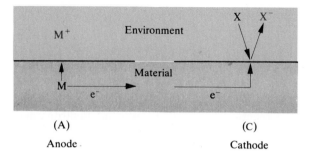

Figure 13–5.1 Corrosion (schematic). (A) Anode (Eq. 13–5.1). (C) Cathode (X + e⁻ → X⁻). The material (commonly metal) must provide a continuous electrical path between the anode and cathode. The environment (commonly a liquid electrolyte) receives the corrosion product and supplies the reactant to the cathode.

voltage difference is removed because electrons move from the zinc to the copper. (The "current" is in the opposite direction.) Thus, reaction (13–5.1$_{Zn}$) proceeds with zinc becoming the anode. Reaction (13–5.1$_{Cu}$) is reversed to reaction (13–5.2b) and copper atoms are reduced from the 1-molar solution of Cu^{2+} ions. Zinc is corroded at the anode; copper is plated onto the cathode.

We will identify the electrodes as *anode* and *cathode* to avoid the confusion that accompanies the positive and negative labels for electrodes. The *anode supplies* electrons to the *external* circuit. Conversely, the *cathode receives* electrons from the *external* circuit. This definition is appropriate for corrosion, batteries, electroplating, CRTs, and all other electronic devices. All corrosion

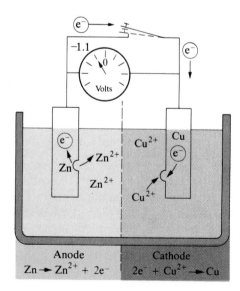

Figure 13–5.2 Galvanic cell (Zn–Cu). With the switch closed, zinc provides electrons through the external circuit to copper. A 1.1-volt potential difference develops when the circuit is opened. (Table 13–5.1 with molar solutions.)

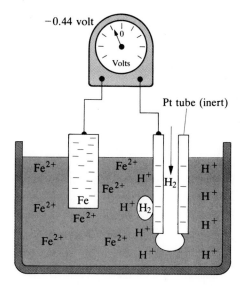

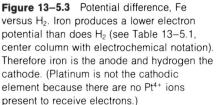

Figure 13-5.3 Potential difference, Fe versus H_2. Iron produces a lower electron potential than does H_2 (see Table 13-5.1, center column with electrochemical notation). Therefore iron is the anode and hydrogen the cathode. (Platinum is not the cathodic element because there are no Pt^{4+} ions present to receive electrons.)

occurs at the *anode*. [Electroplating onto the cathode is "corrosion in reverse" (Example 13-5.3).]

The voltage differences could be measured between all possible pairs of metals in the manner just described. However, the more common practice is to use the $H_2 \rightarrow 2\,H^+ + 2\,e^-$ reaction as a reference for other anode (electron-releasing) reactions. As shown schematically in Fig. 13-5.3, iron is more reactive (more anodic) than hydrogen to the extent that the potential difference is 0.44 volt. It is arbitrary whether we call this difference $+0.44$ V, or -0.44 V; therefore, we will use -0.44 V for the reason cited in the footnote of Table 13-5.1.

The equations of Table 13-5.1 are for anode reactions, i.e., they release electrons. A reversal of any of these reactions is electron-consuming and therefore cathodic. Since neither type of reaction can occur in isolation, but must occur as anode-cathode reaction pairs, the equations of Table 13-5.1 are called *half-cell reactions*.

As previously noted, the electrode potentials of Table 13-5.1 require an electrolyte that is a 1-molar solution (25°C). The actual potential varies with concentration and temperature according to the *Nernst* relationship,

$$\mathcal{E} = \mathcal{E}_{298} + (kT/n)(\ln C) \qquad (13\text{-}5.3a)$$

where $\mathcal{E}_{298}$ is the potential at 298 K (25°C) as listed in Table 13-5.1. Concentration C is expressed as moles per liter, i.e., molar. Since the constant k is 86.1×10^{-6} eV/K, Eq. (13-5.3a) can be condensed to

$$\mathcal{E} = \mathcal{E}_{298} + (0.0257/n)\ln C \qquad (13\text{-}5.3b)$$

for use at ambient temperatures (25°C or 298 K). The term n is the number of electrons released per atom. For example, $n = 3$ for $Al \rightarrow Al^{3+} + 3\ e^-$.

Most electrolytes that pertain to corrosion are dilute with $C \ll 1$. Therefore, the values of Table 13–5.1 are commonly shifted in the anodic direction. For example, with $C_{Fe^{2+}} = 0.01$ M, $\mathcal{E}$ is -0.47V rather than -0.44V.

Cathode Reactions

The cathode reaction of Fig. 13–5.2 was the reduction of copper from the electrolyte containing a 1-molar Cu^{2+} solution. Of course, that reaction cannot occur if the electrolyte contains no copper ions. Under such conditions, the zinc (of Fig. 13–5.2) will still be able to corrode if the electrons can be consumed in other reactions. Possibilities include (1) other metals that are cathodic to zinc, that is, above zinc in Table 13–5.1; (2) evolution of hydrogen (since it is cathodic to zinc); (3) hydroxyl formation; and (4) water formation.

Electroplating:	$M^{m+} + m\ e^- \rightarrow M^0$	(13–5.2b)
Hydrogen evolution:	$2\ H^+ + 2\ e^- \rightarrow H_2 \uparrow$	(13–5.4)
Hydroxyl formation:	$O_2 + 2\ H_2O + 4\ e^- \rightarrow 4(OH)^-$	(13–5.5)
Water formation:	$O_2 + 4\ H^+ + 4\ e^- \rightarrow 2\ H_2O$	(13–5.6)

Each of these cathode half-cell reactions consumes electrons; each is the reverse of one of the anode half-cell reactions in Table 13–5.1. Based on Eq. (13–5.3), greater concentrations shift the electrode potential to become more cathodic. Thus, a more acid solution (higher hydrogen ion concentrations) increases the potential difference between Eqs. (13–5.4) and (13–5.1$_{Zn}$) for the corrosion of zinc.

Probably the most ubiquitous cathode reaction is Eq. (13–5.5) that forms hydroxyl ions, because the reactants include oxygen and water. Under the majority of situations, special efforts are required to avoid the presence of these components. We readily see the corrosion product when iron corrodes and OH^- ions are formed, because the iron ions combine with the OH ions to produce $Fe(OH)_3$, which is ordinary red iron rust.

Galvanic Corrosion Cells

Corrosion couples, called galvanic cells, may be categorized in three separate groups: (1) composition cells, (2) stress cells, and (3) concentration cells.

A *composition* cell may be established between any two *dissimilar* metals. In each case the metal lower in the electromotive series as listed in Table 13–5.1 acts as the anode. For example, on a sheet of *galvanized* steel (Fig. 13–5.4), the zinc coating acts as an anode and protects the underlying iron even if the surface is not completely covered, because the exposed iron is the cathode and does not corrode. Any corrosion that does occur is on the anodic zinc surface. So long as zinc remains it provides protection to adjacent exposed iron.

Conversely, a *tin* coating on sheet iron or steel provides protection only so long as the surface of that metal is completely covered. However, if the surface

TABLE 13–5.1

Electrode Potentials (25°C; 1-molar solutions)

Anode half-cell reaction (the arrows are reversed for the cathode half-cell reaction)	Electrode potential used by electrochemists and corrosion engineers,* volts			Electrode potential used by physical chemists and thermodynamists,* volts
$Au \rightarrow Au^{3+} + 3\ e^-$	+1.50			−1.50
$2\ H_2O \rightarrow O_2 + 4\ H^+ + 4\ e^-$	+1.23	Cathodic	(noble)	−1.23
$Pt \rightarrow Pt^{4+} + 4\ e^-$	+1.20			−1.20
$Ag \rightarrow Ag^+ + e^-$	+0.80			−0.80
$Fe^{2+} \rightarrow Fe^{3+} + e^-$	+0.77			−0.77
$4(OH)^- \rightarrow O_2 + 2\ H_2O + 4\ e^-$	+0.40			−0.40
$Cu \rightarrow Cu^{2+} + 2\ e^-$	+0.34			−0.34
$H_2 \rightarrow 2\ H^+ + 2\ e^-$	0.000	Reference		0.000
$Pb \rightarrow Pb^{2+} + 2\ e^-$	−0.13			+0.13
$Sn \rightarrow Sn^{2+} + 2e^-$	−0.14			+0.14
$Ni \rightarrow Ni^{2+} + 2\ e^-$	−0.25			+0.25
$Fe \rightarrow Fe^{2+} + 2\ e^-$	−0.44			+0.44
$Cr \rightarrow Cr^{2+} + 2\ e^-$	−0.74	Anodic	(active)	+0.74
$Zn \rightarrow Zn^{2+} + 2\ e^-$	−0.76			+0.76
$Al \rightarrow Al^{3+} + 3\ e^-$	−1.66			+1.66
$Mg \rightarrow Mg^{2+} + 2\ e^-$	−2.36			+2.36
$Na \rightarrow Na^+ + e^-$	−2.71			+2.71
$K \rightarrow K^+ + e^-$	−2.92			+2.92
$Li \rightarrow Li^+ + e^-$	−2.96			+2.96

* The choice of signs is arbitrary. Since we are concerned with corrosion, we will use the middle column.

coating is punctured, the tin becomes the cathode with respect to iron, which acts as the anode (Fig. 13–5.5). The galvanic couple that results produces corrosion of the iron. Since the small anodic area must supply electrons to a large cathode surface, very rapid localized corrosion can result.

Other examples of galvanic couples often encountered are (1) steel screws in brass marine hardware, (2) Pb–Sn solder around copper wire, (3) a steel

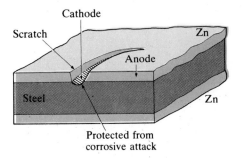

Figure 13–5.4 Galvanized steel (cross section). Zinc serves as the anode; the iron of the steel serves as the cathode. Therefore the iron is protected even though it is exposed where the zinc is scraped off.

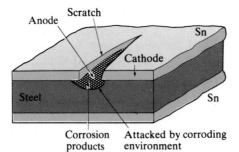

Figure 13–5.5 Tinplate (cross section). The tin protects the iron while the coating is continuous. When the coating is broken, the iron of the steel becomes the anode and is subject to accelerated corrosion.

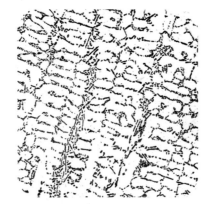

Figure 13–5.6 Galvanic microcells (Al–Si alloy). Any two-phase alloy is more subject to corrosion than is a single-phase alloy. A two-phase alloy provides anodes and cathodes. (Courtesy of Aluminum Company of America.)

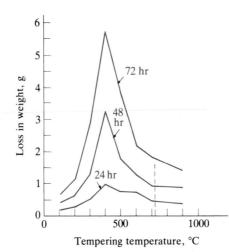

Figure 13–5.7 Microcells and corrosion. After quenching, only martensite exists. After intermediate-temperature tempering, many small galvanic cells exist as a result of the fine (α + carbide) structure in tempered martensite. After high-temperature tempering, the carbide is agglomerated and fewer galvanic cells are present. (Adapted from F. N. Speller, *Corrosion: Causes and Prevention*, McGraw-Hill.)

propeller shaft in bronze bearings, and (4) steel pipe connected to copper plumbing. Each of these is a possible galvanic cell unless protected from a corrosive environment. Too many engineers fail to realize that the contact of dissimilar metals is a potential source of galvanic corrosion. In an actual engineering application, a brass bearing was once used on a hydraulic steering mechanism made of steel. Even in an oil environment, the steel acted as an anode and corroded sufficiently to permit leakage of oil through the close-fitting connection.

Galvanic cells can be microscopic in dimension, because each phase has its individual composition and structure; therefore, each possesses its own electrode potential. As a result, galvanic cells can be set up in two-phase alloys when those metals are exposed to an electrolyte. For example, the pearlite of Fig. 10–8.1 reveals the carbide lamellae because the carbide was the anode in the electrolyte that was used as an etch.* Figure 13–5.6 shows the microstructure of an Al–Si casting alloy. Again we depend on corrosion to reveal the two phases. One is the anode, the other the cathode.

Heat treatment may affect the corrosion rate by altering the microstructure of the metal. Figure 13–5.7 shows the effect of tempering on the corrosion of a previously quenched steel. Prior to tempering reactions, the steel contains a single phase, martensite. The tempering of the martensite produces many galvanic cells and grain boundaries of ferrite and carbide, and the corrosion rate is increased. At higher temperatures the coalescence of the carbides reduces the number of galvanic cells and the number of grain boundaries, which decreases the corrosion rate markedly.

When only a single phase is present, the corrosion rate of an age-hardenable aluminum alloy is low (Fig. 13–5.8), but the corrosion rate is significantly increased with precipitation of the second phase. Still greater agglomeration of the precipitate once again decreases the rate, but never to as low a level as in the single-phase alloy. The maximum corrosion rate occurs in the over-aged alloy.

Stress cells do not involve compositional differences; rather, dislocations, grain boundaries, and highly stressed regions. As shown in Fig. 4–4.9, where the grain boundaries had been etched (i.e., corroded), the atoms at the boundaries between the grains have an electrode potential different from that of the atoms within the grains; thus an anode and a cathode were developed (Fig. 13–5.9). The grain-boundary zone may be considered to be stressed, since the atoms are not at their positions of lowest energy.

The effect of internal stress on corrosion is also evident after a metal has been *cold-worked*. A very simple example is shown in Fig. 13–5.10(a), where strain-hardening exists at the bend of an otherwise annealed wire. The highly

---

* The etch was 4% picral. The carbides are darkened because a corrosion reaction product remains on the surface. The electrode potentials of ferrite and carbide are sufficiently close so that with other electrolytes their cathodic and anodic roles may be interchanged.

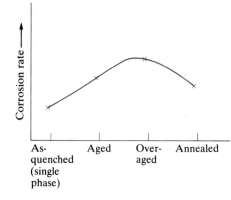

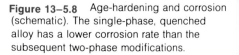

Figure 13–5.8 Age-hardening and corrosion (schematic). The single-phase, quenched alloy has a lower corrosion rate than the subsequent two-phase modifications.

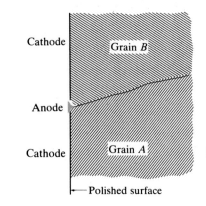

Figure 13–5.9 Grain-boundary corrosion. The grain boundaries served as the anode because the boundary atoms have a higher energy. (Cf. Fig. 4–4.9.)

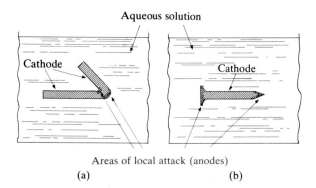

Figure 13–5.10 Stress cells. In these two examples of strain hardening, the anodes are in the more highly cold-worked areas. The electrode potential of a deformed metal is higher than that of an annealed metal.

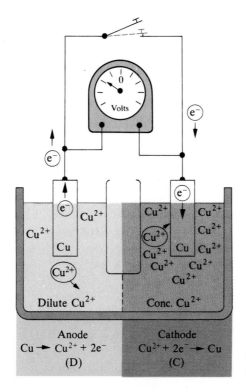

Figure 13–5.11 Concentration cell. When the electrolyte is not homogeneous, the less concentrated area becomes the anode.

Below the figure:

Anode	Cathode
Cu → Cu²⁺ + 2e⁻	Cu²⁺ + 2e⁻ → Cu
(D)	(C)

cold-worked metal serves as the anode and the unchanged metal as the cathode.*

The engineering importance of the effects of stress on corrosion is plain. When engineering components must be used in a corrosive environment, the presence of stress may significantly accelerate the corrosion rate.

Concentration cells arise from differences in electrolyte compositions. According to the Nernst equation (13–5.3), an electrode in a dilute electrolyte is anodic with respect to a similar electrode in a concentrated electrolyte. We may view this in terms of Fig. 13–5.11 and Eq. (13–5.7):

$$Cu^0 \underset{Conc.}{\overset{Dilute}{\rightleftarrows}} Cu^{2+} + 2\,e^-. \qquad (13\text{–}5.7)$$

The metal on side (D) of the figure is in the more dilute Cu^{2+} solution. Therefore, reaction (13–5.7) readily proceeds to the right. The metal on side (C) is in a solution with a higher concentration of Cu^{2+}. Therefore, reaction (13–5.7)

---

* *Corrosion in Action,* published by the International Nickel Company, demonstrates (with illustrative experiments) the effect of cold work on galvanic corrosion.

more readily plates copper on that electrode. The electrode in the concentrated electrolyte is protected and becomes the cathode; the electrode in the dilute electrolyte undergoes further corrosion and becomes the anode.

The concentration cell accentuates corrosion but it accentuates it where the concentration of the electrolyte is lower.

Concentration cells of the above type are frequently encountered in chemical plants, and also under certain flow-corrosion conditions. However, in general, they are of less widespread importance than are *oxidation-type concentration cells*. When oxygen in the air has access to a moist metal surface, corrosion is promoted. However, the most marked corrosion occurs in the part of the cell with an oxygen deficiency.

This apparent anomaly may be explained on the basis of the reactions at the cathode surface, where electrons are consumed. Equation (13–5.5) is restated below because it indicates the role of O_2 (at the cathode) in promoting corrosion in oxygen-free (anode) areas:

$$2\ H_2O + O_2 + 4\ e^- \rightleftarrows 4(OH)^-.$$

Since this cathode reaction, which requires the presence of oxygen, removes electrons from the metal, more electrons must be supplied by adjacent areas that do not have as much oxygen. The areas with less oxygen thus serve as anodes.

The oxidation cell accentuates corrosion but it accentuates it where the oxygen concentration is lower. This generalization is significant. Corrosion may be accelerated in apparently inaccessible places such as cracks or crevices, and under accumulations of dirt or other surface contaminations (Fig. 13–5.12) because the oxygen-deficient areas serve as anodes. This frequently becomes a self-aggravating situation, because accumulation of rust or scale restricts the access of oxygen and establishes an anode, to promote still greater accumulation. The result is localized *pitting* due to nonuniform corrosion (Fig. 13–5.12), and the useful life of the product is thereby reduced to a greater extent than the weight loss would indicate.

Example 13–5.1

Determine the electrode potential (with respect to hydrogen) for a chromium electrode in a solution containing 2 g of Cr^{2+} ions per liter (25°C).

Procedure. Table 13–5.1 applies to 1-molar electrolyte at 25°C. This solution has (2/52.0) moles of Cr/liter. We use the Nernst equation (13–5.3) to adjust for the concentration.

Calculation

$$\mathscr{E} = -0.74\ V + (0.0257/n)\ln (2/52.0)$$
$$= -0.78\ V. \blacktriangleleft$$

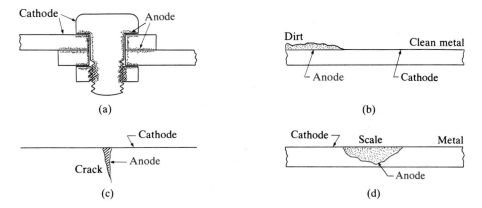

Figure 13–5.12 Oxidation cells. Inaccessible locations with low oxygen concentrations become anodic. This situation arises because the mobility of electrons and metal ions is greater than that of oxygen or oxygen ions.

Example 13–5.2

You are informed that the voltage difference is 1.2 V when silver and cadmium are tested in a standard cell similar to that in Fig. 13–5.2 (1-molar solutions of Ag^{2+} and Cd^{2+}). You do not know whether it is +1.2 or −1.2 V. Where should the $Cd \rightarrow Cd^{2+} + 2\ e^-$ half-cell reaction be placed in Table 13–5.1?

Analysis. Since $Ag \rightarrow Ag^{2+} + 2\ e^-$ is +0.80 V, the $Cd \rightarrow Cd^{2+} + 2\ e^-$ reaction must be

$$+0.80 + 1.20 = +2.00\ V, \qquad \leftarrow or \rightarrow \qquad +0.80 - 1.20 = -0.40\ V.$$

The +2.0 value would make cadmium more noble than gold (+1.5 V), an unexpected result. The −0.4 V value appears more realistic. An alternate check would be to couple Cd with another metal, say nickel, and see whether the standard potential difference is 0.15V (or 2.25 V). ◀

Example 13–5.3

Electroplating is described as "corrosion in reverse." How thick will a nickel-plated layer be if a current density of 500 A/m² is used for an hour on a 16 cm × 43 cm part?

Procedure. Determine the charge and the number of electrons involved. From this, calculate the grams per hour of nickel and the thickness. Use 1 cm² as the basis.

Calculation

$$(500\ A/m^2)(3600\ sec)/(10^4\ cm^2/m^2) = 180\ C/cm^2$$

$$(180\ A \cdot s/cm^2)(1\ Ni^{2+}/2\ el)/(0.16 \times 10^{-18}\ A \cdot s/el) = 5.6 \times 10^{20}\ Ni^{2+}/cm^2$$

$$(5.6 \times 10^{20}\ Ni^{2+}/cm^2)(58.7\ g/0.6 \times 10^{24}\ Ni) = 0.055\ g/cm^2$$

$$(0.055\ g/cm^2)/(8.9\ g/cm^3) = 0.006\ cm \qquad (or\ 60\ \mu m).$$

Comment. The 16 cm × 43 cm surface will receive 38 g of Ni. ◀

13–6

SERVICE FACTORS AFFECTING CORROSION

Corrosion can be minimized by isolating the metal surface from its environment. A layer of paint, a nickel-plated surface, and a vitreous enamel coating separate the metal from the electrolyte that supplies the cathode reactants and receives the anode products (Fig. 13–5.1). The choice of protective coatings depends on the application as summarized in Table 13–6.1.

Corrosion can also be restricted if galvanic couples are avoided. It is the responsibility of the design engineer to ascertain that unlike metals are not in contact in places where corrosion is critical. Alternatively, unlike metals, which of course have different electrode potentials, may be insulated from each other. For example, residential plumbing codes commonly require that a teflon $(C_2F_4)_n$ washer be used in the coupling where copper plumbing joins steel pipes in order to separate the two metals.

Passive Surface Films

Corrosion rates are typically more rapid for those metals at the anodic end of the electrode potential scale (Table 13–5.1), particularly if they are coupled with a cathode reaction at the other end of the scale. However, there are major exceptions. Aluminum, for example, appears to be relatively inert to many service conditions. We even use it for boat hulls. Also, stainless steels are stainless because of their chromium content. Yet, Cr, like Al, is more anodic in Table 13–5.1 than is iron.

When aluminum is the metal of Fig. 13–5.1, it readily forms Al^{3+} ions. Any oxygen in the environment first reacts with the released electrons to produce O^{2-} ions; also the Al^{3+} and O^{2-} ions react with each other at the metal surface

TABLE 13–6.1

Comparison of Inert Protective Coatings

Type	Example	Advantages	Disadvantages
Organic	Baked "enamel" paints	Flexible Easily applied Cheap	Oxidizes Soft (relatively) Temperature limitations
Metal	Noble metal electroplates	Deformable Insoluble in organic solutions Thermally conductive	Establishes galvanic cell if ruptured
Ceramic	Vitreous enamel, oxide coatings	Temperature resistant Harder Does not produce cell with base	Brittle Thermal insulators

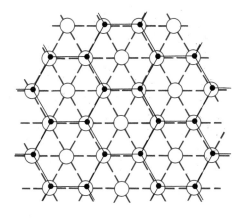

Figure 13–6.1 Al–Al$_2$O$_3$ coherency. Color—pattern of aluminum on {111} planes of the metal. Black—pattern of aluminum on the (0001) plane of the oxide. The Al-to-Al distances differ only slightly. Therefore, the boundary atoms can be part of both structures.

to form a thin, invisible surface layer of Al$_2$O$_3$. The Al$_2$O$_3$ is protective because it has tightly bonded, stable Al^{3+}-to-O^{2-} bonds, and also the Al$_2$O$_3$ is *coherent* with the underlying metal. That is, the crystal structure of the Al$_2$O$_3$ nearly matches the {111} planes in the aluminum (Fig. 13–6.1). Not only is an Al$_2$O$_3$ film protective, it also reforms almost immediately if it is ruptured. This *passive* (nonreactive) film loses its protection in alkaline solutions that dissolve Al$_2$O$_3$.

Anodizing is a commercial process that uses the electrical current to force the formation of a thicker Al$_2$O$_3$ film than occurs naturally. It can be built up to provide protection from saline solutions, such as sea water, which can destroy the thinner, natural Al$_2$O$_3$ film. Furthermore, by introducing the appropriate ions into the processing electrolyte, the Al$_2$O$_3$ (sapphire) film can be colored blue, red, etc., to meet any appearance requirements.

Stainless steels are stainless because their surfaces are passivated by a thin chromium oxide film. Reverting to Fig. 13–5.1, the chromium (12%–25% of the stainless steel) reacts to form Cr ions before the iron reacts. If oxygen is present, it immediately claims the released electrons and forms an extremely thin (~monolayer) chromium oxide film that halts all further reaction. The reaction is called *passivation* and gives a very high degree of protection.

If, however, the service conditions are reducing and no oxygen is present, the above passive layer can be destroyed. Without that protection, the steel becomes active and the corrosion of Cr-bearing steels proceeds rapidly. Table 13–6.2 presents a *galvanic series* of common alloys that indicates their relative position in the cathode-anode spectrum. Observe that the Cr-containing steels are listed twice, (A) in their activated and (P) in their passivated condition. Obviously, the designer must anticipate the possibility that these stainless steels may encounter reducing (low-oxygen) conditions in service. When this occurs, the steels are no longer stainless.

TABLE 13–6.2

Galvanic Series of Common Alloys*

Graphite	Cathodic	Nickel—A
Silver		Tin
12% Ni, 18% Cr, 3% Mo steel—P		Lead
20% Ni, 25% Cr steel—P		Lead–tin solder
23 to 30% Cr steel—P		12% Ni, 18% Cr, 3% Mo steel—A
14% Ni, 23% Cr steel—P		20% Ni, 25% Cr steel—A
8% Ni, 18% Cr steel—P		14% Ni, 23% Cr steel—A
7% Ni, 17% Cr steel—P		8% Ni, 18% Cr steel—A
16% to 18% Cr steel—P		7% Ni, 17% Cr steel—A
12 to 14% Cr steel—P		Ni-resist
80% Ni, 20% Cr—P		23 to 30% Cr steel—A
Inconel—P		16 to 18% Cr steel—A
60% Ni, 15% Cr—P		12 to 14% Cr steel—A
Nickel—P		4 to 6% Cr steel—A
Monel metal		Cast iron
Copper–nickel		Copper steel
Nickel–silver		Carbon steel
Bronzes		Aluminum alloy 2017-T
Copper		Cadmium
Brasses		Aluminum, 1100
80% Ni, 20% Cr—A		Zinc
Inconel—A		Magnesium alloys
60% Ni, 15% Cr—A	Anodic	Magnesium

* Adapted from C. A. Zapffe, *Stainless Steels,* American Society for Metals. A–active; P–passivated.

Inhibitors provide another variant of a passive surface film. Best known of these are the rust inhibitors that are used in automobile radiators (commonly as an additive with the antifreeze). These inhibitors contain chromate or similar highly oxidized ions that are adsorbed onto the metal surface. The protection received is similar to that of the passive film on stainless steel, except for the fact that it cannot be regenerated from the metal. New inhibitor must be added if the film is flushed away.

Galvanic Protection   Service protection against corrosion may be achieved in some applications by arranging for the product to be the cathode. The galvanized steel of Fig. 13–5.4 provided a preview for us. That steel was coated with zinc, which is anodic to iron. Therefore, the zinc anode corrodes and the steel, being the cathode, is protected. Several adaptations of these *sacrificial anodes* are shown in Fig. 13–6.2. They are replaceable after the anode has been spent.

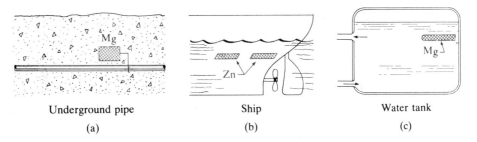

Figure 13–6.2   Sacrificial anodes. (a) Buried magnesium plates along a pipeline. (b) Zinc plates on ship hulls. (c) Magnesium bar in an industrial hot-water tank. Each of these sacrificial anodes may be easily replaced. They cause the *equipment* to become a cathode.

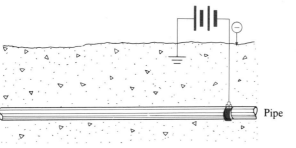

Figure 13–6.3   Impressed voltage. A small dc voltage will provide sufficient electrons to make the equipment a cathode.

Galvanic protection can also be provided with a dc current that feeds electrons into the metal that must be protected against corrosion (Fig. 13–6.3). In a pipeline, for example, the *impressed current* stops, and even reverses the corrosion reaction,

$$M^0 \rightleftarrows M^{n+} + n\,e^-. \qquad (13–5.1)$$

Local Corrosion

Corrosion seldom progresses uniformly into a metal surface. Rather, it acts locally. In part, this is due to the microscopic sized anodes and cathodes in what appears to be a homogeneous material. For example, the grain boundary is anodic to the grain proper (Fig. 13–5.9). Likewise, carbides and ferrite form microscopic anode–cathode pairs within an otherwise uniform steel product.

These initial heterogeneities can lead to accelerated local corrosion. Figure 13–6.4 is a modification of Fig. 13–5.1. It shows that the anode reaction that corrodes the metal penetrates behind the original surface. To illustrate, a microspot that temporarily serves as an anode is deepened. This local area now has less access to oxygen that could produce a protective film. As a result, it remains as the anode and is corroded further, establishing a permanent corrosion pit that ''bores'' its way into the metal. The more exposed areas become

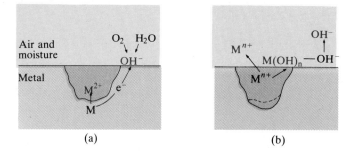

(a)   (b)

Figure 13–6.4 Pit corrosion (schematic). If initiated by any type of nonuniformity, the corroded spot becomes more anodic. (a) Electrons can move to the available oxygen and water (Eq. 13–5.5). (b) The corrosion continues in the oxygen-free area. The metal ions and the hydroxyl ions either diffuse into the electrolyte or combine to form a hydroxide, e.g., iron rust. The debris-filled pit continues to deepen by localized corrosion.

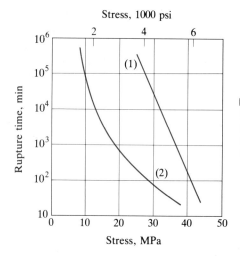

Figure 13–6.5 Accelerated failure (polymers, typical). Curve (1): rupture time in a dry environment; curve (2): rupture time in an organic solvent. (Adapted from Alfrey and Gurnee, *Organic Polymers*, Prentice-Hall.)

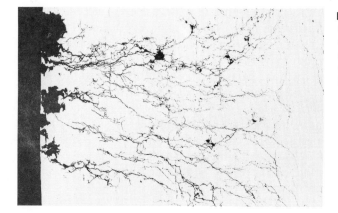

Figure 13–6.6 Stress-corrosion cracking (316 stainless steel, boiling NH_4Cl). This type of failure is common when metals are stressed in the presence of an electrolyte. (×150.) (Courtesy of A. W. Loginow, U.S. Steel Corp.)

542

the cathode. *Pit corrosion* of this origin is a common mode of failure in sheet aluminum products, because holes eventually penetrate the full thickness of the metal.

The examples of Fig. 13–5.12 also illustrate localized corrosion, because the oxidation cell develops its anode in the regions that are inaccessible to O_2. In this condition called *crevice corrosion*, the smaller metal ions can diffuse out of the crevice to the surface to combine with the oxide (or OH^-) ions. Most commonly, rust spots on car fenders that have not been damaged otherwise originate as crevice corrosion. Not uncommonly, they start on the underside of the fender adjacent to weld spots or metal bends that collect dirt and locally exclude oxygen.

Stress Corrosion

A stressed material corrodes more rapidly. Conversely, fracture or rupture occurs more readily in a corrosive environment. This is observed for all types of materials. Figure 13–6.5 shows the rupture time of a plastic as a function of stress, with and without the presence of an organic solvent. Glass is subject to *static fatigue* in which delayed fracture may occur at constant loads in the presence of water vapor. Finally, stress corrosion is well known in metals that are loaded by tension in corrosive environments (Fig. 13–6.6).

Stress corrosion proceeds through a metal as nonductile cracks (even in ductile metals). Higher stresses shorten the time for fracture (Fig. 13–6.7). The corrosive environments that lead to stress corrosion are usually highly specific; for example, electrolytes with Cl^-, NO_3^-, and OH^- ions generally produce fracture across the grains of the metal (Fig. 13–6.6).

The explanation of stress corrosion is not fully understood. One possibility is that corrosion is extremely rapid at the tip of a flaw where the stress intensity

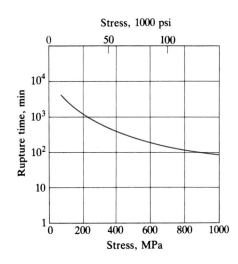

Figure 13–6.7 Stress corrosion (metal in electrolyte). Failure is accelerated at higher stresses. The testing procedure is shown in Fig. 13–6.8(b).

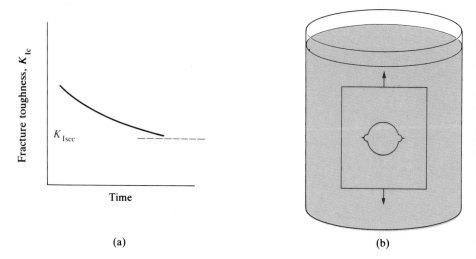

Figure 13–6.8 Fracture toughness (in an electrolyte). (a) The value of K_I drops with time in the electrolyte (schematic). The design limit is K_{Iscc}, the critical stress intensity factor for stress corrosion cracking. (b) Test method (schematic). Cracks are initiated to known depths before loading the specimen in the chosen electrolyte.

factor is high. Thus, the crack penetrates faster than the tip is blunted by plastic deformation, unlike the situation in normally ductile materials.

The fracture toughness K_{Ic} of a metal drops with time in a corrosive media as shown in Fig. 13–6.8(a). The long-term value is labeled K_{Iscc}, the stress-intensity factor for *stress corrosion cracking*. Figure 13–6.8(b) shows schematically the procedure for simultaneously testing corrosion and fracture toughness.

Design engineers must specify stress levels far below the static yield strength if their products are to be used in corrosive environments. Naturally, the sensitivity to stress corrosion varies from material to material.

Example 13–6.1

Aluminum oxide, Al_2O_3, is hexagonal with the Al^{3+} ions spaced 0.267 mm apart. (See Fig. 13–6.1.) Compare this distance with these center-to-center distances on the {111} planes of metallic aluminum. For fcc Al, $a = 0.4049$ nm.

Procedure. Neighboring atoms "touch" on the {111} planes of an fcc metal (Fig. 3–7.7). Thus,

$$d = 2R = a\sqrt{2}/2$$
$$= 0.286 \text{ nm}.$$

$$\Delta d/d = (0.286 - 0.267 \text{ nm})/0.286 \text{ nm}$$
$$= 0.066.$$

Comment. With less than 7% difference, the aluminum atoms at the Al_2O_3–Al interface can be common to both structures. This coherency provides a tight bond between the two phases. ◄

Example 13–6.2   The fracture toughness for the steel of Example 13–2.2 drops by 40% when it is tested in sea water. (See K_{Iscc} in Fig. 13–6.8a.) What crack length can be tolerated and still retain yielding as the mode of failure?

Procedure. The stress cannot exceed 1100 MPa; however, any crack must be short enough to stay under a stress intensity factor of 54 MPa·$\sqrt{m}$, that is, 60% of K_{Ic}.

Calculation

$$c\pi = [K_{Iscc}/(S_y)\alpha]^2;$$

$$c = [54 \text{ MPa}\cdot\sqrt{m}/(1100 \text{ MPa})(1.1)]^2/\pi$$

$$= 0.6 \text{ mm.} \blacktriangleleft$$

Review and Study

SUMMARY

1. A material has failed if it is no longer capable of fulfilling its intended purpose. The causes of *failure* include improper design judgments, unexpected service conditions, misuse, deficient maintenance, as well as insufficient quality control during the manufacturing process. The mechanism of failure may be by fracture, by excessive temperature, by radiation exposure, or by chemical reactions with the environment (corrosion).

2. Fracture may be ductile or brittle, i.e., with or without plastic deformation. The former is *tougher*. Toughness tests measure the energy for the fracture of standardized test specimens. Steels and a number of other materials possess a *ductility-transition temperature*, T_{dt}, below which fracture occurs with little energy absorption.

   The *stress-intensity factor* (K_I of Eq. 13–2.1) relates the nominal stress and the depth of a crack to the stress concentration at the crack tip. *Fracture toughness*, K_{Ic}, is the critical value of K_I that permits crack propagation. It is a property of a material, and commonly dictates the stress limits on high-strength materials that have low ductility.

   Fatigue is a time-delayed fracture that is commonly associated with cyclic loading. More stress cycles reduce the allowable design stress. Some materials (but not all) have an *endurance limit* below which unlimited cycling is tolerable.

• 3. It is common for materials to lose strength at elevated temperatures. Reasons include easier atom (and dislocation) movements, and microstructural coarsening that accompanies over-aging and over-tempering. Atomic diffusion permits *creep* of metals stressed in high temperatures. The *creep rate* increases with increases in temperature and/or stresses.

Stress relaxation and creep occur in polymers, particularly in thermoplastics. High temperatures promote *degradation* in which covalent bonds are broken. In the extreme, *charring* occurs and only the carbon "skeleton" remains.

- 4. Radiation exposure can produce *radiation damage* if the radiation knocks atoms or electrons out of their normal locations. This hardens and embrittles metals, generally weakens polymers, and alters the conductivity of both metals and semiconductors. *Damage recovery* occurs by heating. The accompanying atom movements partially restore the former undamaged structures.

5. Corrosion occurs at the anode; electrons are supplied to the cathode. The most common of several cathode reactions is

$$O_2 + 2\ H_2O + 4\ e^- \rightarrow 4\ (OH)^-. \qquad (13\text{--}5.5)$$

Oxygen accelerates corrosion, but corrosion occurs in the oxygen-deficient areas. The types of galvanic cells include

	Anode	Cathode
	Base metal	Noble metal
Zn versus Fe	Zn	Fe
Fe versus H_2	Fe	H_2
H_2 versus Cu	H_2	Cu
	Higher energy	Lower energy
Boundaries	Boundaries	Grain
Stresses	Cold-worked	Annealed
Stress corrosion	Stressed areas	Nonstressed areas
	Lower conc.	Higher conc.
Electrolyte	Dilute solution	Concentrated solution
Oxidation	Low O_2	High O_2
Dirt or scale	Covered areas	Clean areas

6. Corrosion is minimized by *protective coatings* and by isolating anodes from cathodes.

Corrosion is restricted by surface *passivation*. Al_2O_3 forms on aluminum; chromium forms monolayers of oxide on stainless steels; and rust inhibitors provide adsorbed chromate surfaces on metals. Thus, while oxygen accelerates corrosion by its cathode reaction (Eq. 13–5.5), it can also restrict corrosion by isolating the anode.

Corrosion is reversed by *impressed currents* and by *sacrificial anodes* that are expendable.

Corrosion is commonly localized; *pitting* and *crevice corrosion* are the result.

Corrosive environments reduce the fracture toughness of metals. *Stress corrosion* proceeds through metals as nonductile cracks.

TERMS AND CONCEPTS

Anode The electrode that supplies electrons to the external circuit. The electrode that undergoes corrosion. The negative electrode.

Anode reactions Electrochemical oxidation reactions.

Anodized Surface-coated with an oxide layer; achieved by making the component an anode in an electrolytic bath.

Cathode The electrode that receives electrons from the external circuit. The electrode on which electroplating is deposited. The positive electrode.

Cathode reactions Electrochemical reduction reactions.

Cell, composition Galvanic cell between electrodes of different compositions.

Cell, concentration Galvanic cell arising from nonequal electrolyte concentrations. (The more dilute solution produces the anode.)

Cell, galvanic An electrochemical cell containing two dissimilar metals in an electrolyte.

Cell, oxidation Galvanic cell arising from nonequal oxygen potentials. (The oxygen-deficient area becomes the anode.)

Cell, stress Galvanic cell arising from a plastically deformed anode.

Corrosion Surface deterioration by electrochemical reaction.

Crevice corrosion Localized corrosion in an oxidation cell.

• **Degradation** Reduction of polymers to smaller molecules.

Ductility-transition temperature, T_{dt} Temperature that separates the regime of brittle fracture from the higher temperature range of ductile fracture.

Electrode potential ($\mathscr{E}$) Voltage developed at an electrode (as compared with a standard reference electrode).

Electrolyte Conductive ionic solution (liquid or solid).

Endurance limit The maximum stress allowable for unlimited cycling.

Fatigue Time-delayed fracture, commonly under cyclic loading.

Fracture, brittle Fracture with negligible plastic deformation and a minimum of energy absorption.

Fracture, ductile Fracture accompanied by plastic deformation and, therefore, by energy absorption.

Fracture toughness, K_{Ic} The critical stress intensity factor for fracture propagation.

Galvanic protection Corrosion protection by establishing the product as the cathode.

Galvanic series Sequence (cathodic to anodic) of corrosion susceptibility for common metals. Varies with passivity and with electrolyte composition. (Cf. Electrode potentials, which are for standard electrolytes.)

Galvanized steel Zinc-coated steel.

Half-cell reaction See Table 12–5.1

Hydrogen electrode Standard reference electrode with the half-cell reaction:

$$H_2 \rightarrow 2\,H^+ + 2\,e^-.$$

Impressed current Direct current applied to make a metal cathodic during service.

Inhibitor An additive to the electrolyte that promotes passivation.

Nernst equation Electrode potential as a function of electrolyte concentration and temperature.

Oxidation (general) The raising of the valence level of an element.

Passivation The condition in which normal corrosion is impeded by an adsorbed surface film on the electrode.

Pit corrosion Localized corrosion that penetrates the metal nonuniformly.

• **Radiation damage** Structural defects arising from exposure to radiation.

Reduction   Removal of oxygen from an oxide; the lowering of the valence level of an element.

Sacrificial anode   Expendable metal that is anodic to the product it is to protect.

• **Scission**   Degradation of polymers by radiation.

S–N curve   Plot of fatigue stress vs. number of stress cycles.

Stainless steel   High-alloy steel (usually containing Cr, or Cr + Ni) designed for resistance to corrosion and/or oxidation.

Stress corrosion   Corrosion accentuated by stresses; or fracture accelerated by corrosion.

Stress intensity factor   The value that relates the design limit for the nominal stress to the depth of a propagating crack (Eq. 13–2.1).

• **Stress-rupture**   Time-dependent rupture resulting from constant load (usually at elevated temperatures).

Toughness   A measure of the energy required for mechanical failure.

**FOR CLASS
DISCUSSION**

A_{13} Give examples where ductile fracture provides a ''fail-safe'' design feature.

B_{13} Transition temperatures were not given the same attention in the design of riveted ships as is now being given in the design of welded ships. Why?

C_{13} What is the meaning of the $\frac{1}{4}$ cycle on Fig. 13–2.7?

D_{13} A small hole drilled through a steel plate ahead of a crack will stop its progress until repairs can be made. Explain.

E_{13} The rotating shaft of Fig. 13–2.6 fractured as a result of tension-compression stress reversals. Give examples of potential fatigue arising from cyclic tensile stresses (no compression).

• F_{13} Explain the various stages of creep shown in Fig. 13–3.2.

• G_{13} Why are coarse-grained materials stronger than fine-grained materials at high temperatures?

• H_{13} Other things being equal, which will have the lowest creep rate: (a) steel in service with a high tensile stress and low temperature, (b) steel in service with a low tensile stress and high temperature, (c) steel in service with a high tensile stress and high temperature, (d) steel in service with a low tensile stress and low temperature? Why?

I_{13} Distinguish between anode and cathode on the basis of electron movements. How does a cathode-ray tube fit into the definitions of Section 13–5?

J_{13} What is the source of electrons in an ordinary dry cell? (The electrolyte is a gelatinous paste containing NH_4Cl. The cathode reaction changes Mn^{4+} in MnO_2 to Mn^{2+}.)

K_{13} Why do we use the hydrogen electrode as the reference electrode rather than some other element, such as lead?

L_{13} A discarded tin-coated can is dropped in a lake. Eq. (13–5.1$_{Fe}$) develops as the anodic reaction. The cathode reaction is *not* $Sn^{2+} + 2\,e^- \rightarrow Sn$. Why? What is the probable cathode reaction?

M_{13} Plumbing codes call for an insulator such as Teflon to be placed between copper tubing and steel pipe when they are coupled in a plumbing system. Explain.

N_{13} Electrical codes call for a "jumper" to be placed around the insulated coupling just described, to provide a ground. Does this defeat the purpose of the plumbing code? Explain.

O_{13} Explain why the metal in a dilute electrolyte becomes the anode to metal in a more concentrated electrolyte.

P_{13} Explain why the metal in an oxygen-deficient electrolyte becomes the anode to metal in an oxygen-enriched electrolyte.

Q_{13} If oxygen accelerates corrosion (Eqs. 13–5.5 and 13–5.6), why does the corrosion occur in the oxygen-lean area?

R_{13} How do barnacles accelerate corrosion on a ship hull?

S_{13} Why does aluminum corrode less than iron when used in kitchen utensils?

T_{13} Why is corrosion accelerated when the anode is smaller than the cathode, but not when the situation is reversed?

U_{13} A flashlight with an old drycell will dim during use. If left unused for a short time, it will shine more brightly again. Explain.

V_{13} An enterprising mechanic suggests the use of a magnesium drain plug for the crankcase of a car as a means of combating engine corrosion. Discuss.

W_{13} Undercoatings of various types are used on new cars. Under what conditions are they helpful? Under what conditions are they detrimental?

X_{13} Cite three examples of corrosion from your experience. Describe the nature of the deterioration, and account for the corrosion.

QUIZ SAMPLES

13A   Failure can arise from __+−__ as well as normal service wear and tear.

(a) misuse   (b) unanticipated service conditions   (c) poor quality control   (d) poor design judgments   (e) improper maintenance

13B   Design considerations include(s) the __+−__ of the material as well as specific properties required of the material.

(a) cost   (b) availability   (c) discard, or reuse   (d) combinations of properties   (e) environmental implications

13C   Metallic fracture is commonly categorized as __**__ or __**__ depending on the energy absorbed, and/or the ductility.

13D   Toughness is the integrated product of __**__ and strain prior to fracture. This gives units of __**__ per volume.

13E   There is a major __**__ of toughness for steels below the __**__ temperature. This characteristic is encountered in __**__-centered cubic metals, but not in __**__-centered cubic metals.

13F   Other things equal, a steel with a __**__-grain size will have a lower T_{dt}.

13G   It is necessary to standardize the geometry of test samples for toughness testing, since the required energy depends on the __**__ and __**__ of the sample.

13H   For a given nominal tensile stress, the stress intensity factor increases by a factor of __+−__ if the crack length is doubled.

(a) two   (b) square root of two   (c) four

13I   A __**__ is a stress-raiser because it focuses stresses at points with a small __**__. The stress intensity factor is proportional to the nominal stress and the __**__.

13J   Cyclic stresses can lead to __**__ if the stress level is above the __**__ .

• **13K**   During __**__ , the side atoms of a polymer are lost and only the carbon remains.

• **13L**   Radiation exposure can lead to __**__ in polymers, by which large molecules are broken into two or more smaller molecules.

• **13M**   Plain-carbon steels that are exposed to neutron radiation develop __+−__ .

(a) lower ductility transition temperature   (b) reduced yield strength
(c) reduced ductility   (d) reduced resistivity

13N   For corrosion to proceed, there must be a(n) __**__ , a cathode, and a(n) __**__ .

13O   Corrosion occurs at the __**__ . The commercial process, called __**__ , is the corrosion reaction in reverse.

13P   Reference electrolytes are __**__-molar at __**__ °C. This established a standard emf of +0.34 V for copper (Table 13–5.1), but <0.34 V in __+−__ electrolytes.

(a) more dilute   (b) more concentrated   (c) warmer, 0.1 M
(d) cooler, 0.1 M

13Q   The most common cathode reaction that leads to the rusting of iron in moist air is __**__ .

13R   The anode for each of the following pairs is __+−__ . (Unless indicated, concentration, temperature, etc., are the same.)

(a) iron vs. (b) copper
(e) dilute solution vs. (f) conc. solution
(i) annealed vs. (j) cold-worked

(c) high O_2 vs. (d) low O_2
(g) grains vs. (h) grain boundary

STUDY PROBLEMS

13–2.1   What would the follow-through angle have been in Example 13–2.1 if the initial angle of the pendulum had been 105° before testing the sample?

Answer: 76°

13–2.2   The value of K_{Ic} for a steel is 160 MPa·m$^{1/2}$ (146,000 psi·in$^{1/2}$). What is the maximum tolerable crack when the steel carries a nominal stress of 760 MPa (110,000) psi?

13–2.3   The steel of Study Problem 13–2.2, which has a 6-mm (0.24-in.) crack, can support what maximum nominal stress without the crack advancing?

Answer: 1060 MPa (or 153,000 psi)

13–2.4   Convert K_{Ic} = 60 MPa·m$^{1/2}$ to psi·in$^{1/2}$.

13–2.5 A steel has a yield strength of 550 MPa (80,000 psi), and a K_{Ic} value of 40 MPa·m$^{1/2}$ (36,500 psi·in$^{1/2}$). What will be the limiting design stress if the minimum tolerable crack is 1.5 mm (0.06 in.) and no plastic deformation is permitted?

Answer: 530 MPa (or 76,500 psi). The steel will fracture before it yields at S_y.

13–2.6 Sometimes it is assumed that if failure does not occur in a fatigue test in 10^8 cycles, the stress is below the endurance limit. A test machine is connected directly to a 1740-rpm electric motor. How long will it take to log that number of cycles?

Answer: 40 days

13–2.7 Examine a crankshaft of a car. Point out design features that alter the resistance to fatigue.

• **13–3.1** The following data were obtained in a creep–rupture test of Inconel "X" at 815°C (1500°F): (a) 1% strain after 10 hr, (b) 2% strain after 200 hr, (c) 4% strain after 2000 hr, (d) 6% strain after 4000 hr, (e) "neck-down" started at 5000 hr and the rupture occurred at 5500 hr. What was the creep rate?

Answer: 0.001%/hr

• **13–3.2** Design considerations would permit a pressure tube to have 3 l/o strain during a year of service. What maximum creep rate is tolerable when reported in the normal %/hr figures?

• **13–3.3** Raw polyisoprene (i.e., nonvulcanized natural rubber) gains 2.3 w/o by being cross-linked by oxygen of the air. What fraction of the possible cross-links are established?

Answer: 0.10

• **13–3.4** A chloroprene rubber gains 3 w/o by oxidation. Assume that this oxygen produced cross-linkage. What fraction of the possible anchor points (cross-linkages) contain oxygen atoms?

• **13–4.1** The average energy of a C–Cl bond is 340 kJ/mole (81 kcal/mole) according to Table 2–2.1. Will visible light [400 nm (violet) to 700 nm (red)] have enough energy to break the average C–Cl bond?

Answer: 350 nm required

• **13–4.2** (a) What frequency and wavelength must a photon have to supply the energy necessary to break an average C–H bond in polyethylene? (b) Why can some bonds be broken with slightly longer electromagnetic waves?

13–5.1 How many coulombs are required to plate each gram of nickel from an Ni^{2+} electrolyte?

Answer: 3300 coul

13–5.2 A metallic reflector that has 1.27 m^2 of surface is being electroplated. The current is 100 amp. (a) How many grams of Cr^{2+} must be added to the electrolyte per hour of plating? (b) What thickness will form per hour?

13–5.3 The electrodes of a standard galvanic cell are nickel and magnesium. What potential difference will be established?

Answer: 2.1 volts (with nickel as the cathode)

13–5.4   What copper concentration (g/l) is required in an electrolyte for it to have an electrode potential of 0.32 V (with respect to hydrogen)?

13–5.5   In Fig. 13–5.2, copper is replaced by gold and zinc by tin. Based on the half-cell reactions of Table 13–5.1, will the weight of tin corroded per hour be greater or less than the weight of gold which is plated?

Answer: 10 w/o more gold

13–5.6   Two pieces of metal, one copper and the other zinc, are immersed in sea water and connected by a copper wire. Indicate the galvanic cell by writing the half-cell reactions (a) for the anode and (b) for the cathode; also, by indicating (c) the direction of the electron flow in the wire, and (d) the direction of the "current" flow in the electrolyte. (e) What metal might be used in place of copper so that zinc changes polarity?

13–5.7   Consider Eqs. (13–5.2b), (13–5.4), (13–5.5), (13–5.6), and the corrosion of iron. Which, if any of these, provide the prevalent cathode reaction if (a) iron is in water along a seashore? (b) Iron is in a copper sulfate solution? (c) Iron is a can containing a carbonated beverage? (d) Iron is the blade of a rusty spade? (e) Iron is a scratched car fender?

13–6.1   Pit corrosion punctured the hull of an aluminum boat after 12 months in sea water. The aluminum sheet was 1.1-mm thick. The average diameter of the hole is 0.2 mm. (a) How many atoms were removed from the pit per second? (b) What was the corrosion current density?

Answer: (a) 6.6×10^{10}/s   (b) 1 A/m^2

13–6.2   With a current density of 0.1 A/m^2, how long will it take to corrode an average of 0.1 mm from the surface of aluminum?

13–6.3   An average current density of 10^2 amp/m^2 is used to build up an anodized coating on aluminum. The Al$_2$O$_3$ coating is to be 1-μm thick. What time is required?

Answer: 215 s

13–6.4   A dry cell with a zinc anode (cell wall) provides 15 amperes of current from its 300 cm^2 of surface. How long will it take for 10% of the 1-mm thick wall to corrode (if one assumes uniform corrosion)?

QUIZ CHECKS

13A	(a), (b), (c), (d), (e)		**13F**	fine
13B	(a), (b), (c), (d), (e)		**13G**	size, shape
13C	ductile, brittle		**13H**	(b)
13D	stress		**13I**	notch (or flaw)
	energy			radius of curvature
13E	loss, ductility-transition			square root of crack depth
	body		**13J**	fatigue, endurance limit
	face		• **13K**	carbonization (charring)

- **13L** scission
- **13M** (c)
  13N anode, electrolyte
  13O anode
  electroplating

13P one, 25°C
(a), (c), (d)

13Q $2 H_2O + \frac{1}{2} O_2 + 4 e^- \rightarrow 4 (OH)^-$

13R (a), (d), (e), (h), (j)

Three Widely Used, More Complex Materials

The final three chapters of this text consider three specific types of materials, all being generally familiar to the reader, and none of which have simple structures. These are

14 Cast Irons—the least expensive metal
15 Concrete—the most widely used of all current materials (weight basis)
16 Wood—the most widely used of all current materials (volume basis)

Thus, each of these materials holds right for a spotlight in current technology. Were it not for time limitations, all of them could command detailed attention in a materials course. However, with time restrictions, it is recognized that some courses may not introduce all three.

Each of the above materials is appreciably more complex than the copper, polyethylene, and MgO that were used earlier as prototypes of metals, polymers, and ceramics, respectively. However, in each case the same structure ⇔ property ⇔ performance relationships hold; in fact, this materials science and engineering approach becomes even more useful because the materials are more complex.

Chapter Fourteen

Cast
Irons

PREVIEW

Cast irons are the *least expensive of all metals*. This is because iron is the second most plentiful metallic resource. Only aluminum exceeds it; but aluminum requires major amounts of energy for extraction. Cast iron requires less refining than steel. Finally, as the name implies, molten cast iron is easily cast into complex shapes (Fig. 14–1.1) and easily machined to required tolerances.

It will be necessary to modify some of our previous understandings of the iron-carbon phase diagram that we gained in Chapter 10, since Fe_3C *dissociates to iron and graphite*. This fact, however, opens the door to a wider variety of cast metal products.

With cast irons, the readers will have the opportunity to use previous knowledge to "design a material" through the use of compositional and processing alternatives.

CONTENTS

STUDY OBJECTIVES

1. Distinguish among various types of cast iron on the basis of microstructure.
2. Be able to detail with an Fe–C phase diagram how the ferritic and pearlitic versions of gray, nodular, and malleable irons may be produced.
3. Relate selected properties to the nature of the microstructure.
4. Become familiar with the terms and concepts used in technical communications about cast irons.

14-1

INTRODUCTION

In principle, cast irons are eutectic alloys of iron and carbon. Thus, they are relatively low melting ($\sim$1200°C, or 2200°F). This is advantageous because they are easily melted, requiring less fuel and more easily operated furnaces. Also, the molten metal will easily fill intricate molds completely. These characteristics lead to an inexpensive material and considerable versatility in product design (Fig. 14–1.1).

Cast irons are somewhat more complex than the simple eutectic alloy just described. For example, most cast irons contain 1%–3% silicon. This is partially a consequence of the fact that silicon is retained with the iron in the

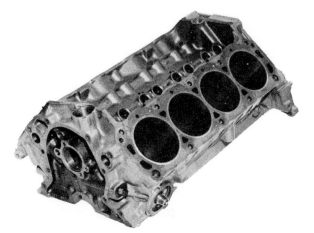

Figure 14–1.1 Cast-iron engine block (V-8). Not only is cast iron one of the most available and least expensive alloys, it has the advantage of castability into intricate designs, thus requiring a minimum of machining before assembling the final engine (Fig. 1–1.1). (Courtesy of the Ford Motor Company.)

559

production process, and additional efforts would be required to remove it. More important, however, is the role of silicon in the final product. First, silicon increases the strength of the ferrite within the cast iron. Second, with silicon, the low-melting eutectic is achieved with 2%–3.5% carbon rather than the 4.3% carbon indicated in Fig. 10–7.1. Finally, the silicon leads to a decomposition of the carbide to iron and graphite.

14–2

IRON–CARBON PHASE DIAGRAM

The phase diagram of Fig. 10–7.1 was based on iron and iron carbide (Fe_3C). Although this Fe–Fe_3C diagram serves many purposes for the study of steels (Fe + less than ~1 w/o C), it cannot be used for cast irons. The compound, Fe_3C is only metastable. That is, Fe_3C will eventually change to other phases.* In this case, the Fe_3C dissociates to iron and graphite:

$$Fe_3C \rightarrow 3\ Fe + C\ (graph). \qquad (14\text{–}2.1)$$

Since Fe_3C is not stable, we must revise the phase diagram to be an Fe–C diagram rather than an Fe–Fe_3C diagram. This has been done in Fig. 14–2.1. Fortunately, the difference is very slight—with only one significant change. Graphite replaces Fe_3C! Thus, at 400°C, the amount of ferrite in cast iron containing 3.5% carbon is

$$\alpha = (100 - 3.5)/(100 - {\sim}0)$$
$$= 0.965,$$

and *not*

$$\alpha = (6.7 - 3.5)/(6.7 - {\sim}0)$$
$$= 0.48.$$

The former uses 100% C as the composition of graphite; the latter uses 6.7% C as the composition of Fe_3C.

Observe the 6° and 11° increases from Fig. 10–7.1 for the eutectic and eutectoid temperatures, respectively. These increases produce a slight decrease in the solubility limit on the right side of the austenite field—2.11% C to 2.08% C, and 0.77% C to 0.68% C.

---

* We have encountered metastable phases previously. For example, diamond is supposedly "forever"; however, given an opportunity it will alter to form graphite. Diamond is stable only when it is subjected to very high pressures and temperatures. Likewise, martensite and glass are two metastable phases. Martensite will temper to (α + $\overline{C}$), and glass will crystallize if given an opportunity. Admittedly, the time requirements at 20°C may be almost infinitely long.

We used the Fe–Fe_3C diagram of Figs. 10–7.1 and 10–7.3, because under normal steel treating conditions (low Si, and short or moderate heating times) we do not observe graphite. Thus, Fe_3C is the carbon-containing phase that is present in the microstructure of steels and influences their properties.

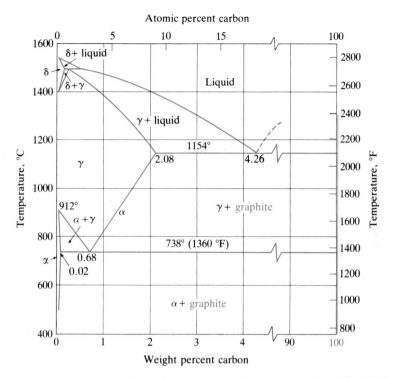

Figure 14–2.1 Fe–C phase diagram. The Fe–Fe₃C phase diagram (Fig. 10–7.1) is not truly an equilibrium diagram, because Fe₃C eventually decomposes to iron and graphite, particularly with the higher carbon contents of cast iron. The only significant difference between this diagram and Fig. 10–7.1 is the presence of graphite as the carbon-rich phase beyond the solubility-limit curves for carbon in ferrite and for carbon in austenite.

Example 14–2.1

Determine the phases, their compositions, and the number of pounds of each in a 60-lb iron casting (97Fe–3C). (a) at 1200°C, (b) at 900°C, (c) at 600°C. (Assume complete equilibrium.)

Procedure. This is a Chapter 10-type problem, but use Fig. 14–2.1, since Fe₃C is only metastable. Either interpolate, or use the lever rule (Eq. 10–4.3).

Answers

a) At 1200°C, γ: 1.9% C   (0.9/2.0)(60 lb) = 27 lb;

L: 3.9   (1.1/2.0)(60 lb) = 33 lb.

b) At 900°C,   γ: 1.25% C   (97/98.75)(60 lb) = 58.9 lb;

Gr: 100   (1.75/98.75)(60 lb) = 1.1 lb.

c) At 600°C,   α: ~0% C   (97/100)(60 lb) = 58.2 lb;

Gr: 100   (3/100)(60 lb) = 1.8 lb.   ◀

14–3

GRAPHITIZATION   Iron carbide forms readily in steels, even though it is only metastable. The dissociation to iron and graphite (Eq. 14–2.1) is infinitely slow at ambient temperatures. As expected, the reaction rate increases with temperature. The reaction rate is also increased when more carbon is present. Thus, cast irons (2%–3.5% carbon) graphitize more readily than do steels with less than one percent carbon. Finally, the 1 to 3 percent silicon that is commonly present also facilitates the dissociation of Fe_3C to iron and graphite.

Below the eutectoid temperature, 738°C (1360°F), Eq. (14–2.1) can be written more specifically as

$$Fe_3C \rightarrow 3\ Fe(\alpha) + C\ (graph). \tag{14–3.1a}$$

Above the eutectoid temperature:

$$Fe_3C \rightarrow 3\ Fe(\gamma) + C\ (graph). \tag{14–3.1b}$$

Of course the compositions of the α and γ products differ in their carbon contents. In fact, the γ of Eq. (14–3.1b) has a eutectoid composition if reacted just above 738°C. That γ is nearly identical with austenite that produces pearlite or is quenched to martensite. This opens the possibility for producing the wide variety of microstructures that are available for commercial cast irons.

Example 14–3.1   A 98 Fe–2C cast iron can be given the following processing steps.

1. Equilibrate at 1150°C (2100°F).
2. Equilibrate at 740°C (1365°F).
3. Air-cool from 740°C to 35°C (95°F).
4. Water-quench from 740°C to 35°C.
5. Cool *very* slowly from 740°C to 35°C.

Five different heat-treating sequences are chosen from the above steps.

a) 1
b) 1 + 2
c) 1 + 2 + 3
d) 1 + 2 + 4
e) 1 + 2 + 5

Describe the microstructure after each of the five sequences.

Procedure. Use the Fe–C equilibrium diagram (Fig. 14–2.1) for steps (1) and (2). However, note that the austenite at 740°C is essentially identical to the austenite in a steel. With only 0.7% carbon, it behaves like a steel, unless ample time is provided for any Fe_3C to dissociate.

Answers

a) Single-phase microstructure of austenite (2% C).

b) Two-phase microstructure of γ (0.7% C) and graphite.

c) Pearlite and graphite.

d) Two-phase microstructure of martensite (0.7% C) and graphite.

e) Two-phase microstructure of ferrite (nil C) and graphite.

Comments. The products of (c), (d), and (e) are called pearlitic cast iron, martensitic cast iron, and ferritic cast iron, respectively. (The latter will form more readily if 1%–2% silicon is present.) ◄

14–4

ENGINEERING CAST IRONS

Gray Cast Iron

A cast iron with a high silicon content (~2% Si) graphitizes so readily that Fe_3C never forms. Graphite flakes form within the metal during solidification. This is shown in Fig. 14–4.1 where we see the cross sections of graphite flakes in the polished 2-dimensional section of the metal. When this metal is fractured in tension, the fracture path progresses from flake to flake because the micalike graphite is very weak in tension. Thus, a very high percentage of the fractured surface is graphite and is gray in appearance. Hence, the name, *gray iron*.

Gray iron has almost negligible ductility because the graphite flakes are present; however, it has the merit of being an inexpensive metal. Furthermore, with these graphite flakes present, a gray iron is an excellent absorber of vibrational energy. In engineering terms, its *damping capacity* is high. Thus,

Figure 14–4.1 Gray cast iron (×100). Carbon is present as graphite *flakes*. The high carbon content gives a eutectic composition that permits easy casting. However, the strength and ductility are less than for plain-carbon steels (J. E. Rehder, dec., Canada Iron Foundries.)

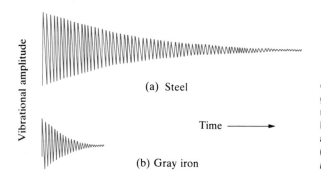

(a) Steel

Time ⟶

(b) Gray iron

Figure 14–4.2 Relative damping capacity of steel and gray iron. The vibrations decay much faster in the gray iron because the graphite flakes absorb the vibrational energy. (After Wallace, *Metals Engineering Qtrly.*)

this type of metal finds extensive use as a base for machine tools and heavy equipment (Fig. 14–4.2).

Gray iron may be heat-treated in a variety of ways to give pearlitic, ferritic, martensitic, and bainitic microstructures to the metal. For example, if a gray iron is heated to ~750°C (1380°F), the resulting phases are austenite (0.75% C) and graphite. Moderately slow cooling produces *pearlite* from the austenite as described with Fig. 11–2.5. The carbide forms more rapidly than does the graphite because less extensive diffusion is required. Specifically, $(\alpha + \overline{C})$ can form without major iron diffusion. In contrast, the graphitic areas within the solid must be completely depleted of iron.

A quench from 750°C to low temperatures produces *martensite* from the austenite. Alternatively, a quench from 750°C to intermediate temperatures, followed by austempering (Section 11–7), produces *bainite*. These alternatives lead to corresponding variations in strengths and hardnesses for pearlitic gray iron, martensitic gray iron, and bainitic gray iron (Table 14–4.1), much as expected from our previous knowledge of pearlite, martensite, and bainite.

Still another microstructure is available to the engineer if the silicon is increased slightly to facilitate complete graphitization of gray iron at ~700°C (~1300°F), because the resulting microstructure is ferrite and graphite. This *ferritic* gray iron is a soft, machinable cast iron, which meets many needs at low cost. It must be cooled slowly through 700°C to avoid all carbide formation.

Nodular (ductile) Cast Irons

Careful laboratory work revealed that the microstructure of easily graphitized cast irons (Fig. 14–4.1) could be replaced by that shown in Fig. 14–4.3 if small additions of magnesium or cerium are introduced into the molten metal. The resulting graphite spheroids (or nodules) drastically affect the ductility of the cast iron (Table 14–4.1). Whereas a normal gray iron has essentially no ductility, this *nodular iron* may have 10%–20% elongation. Although an explanation that accounts for this microstructural change is very complex, the practical

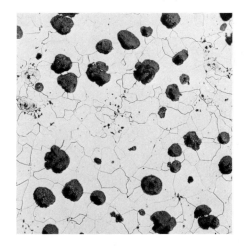

Figure 14–4.3 Nodular cast irons (×100). Carbon is present as graphite *nodules*. Unlike gray cast iron, nodular cast iron is ductile (Table 14–4.2). The graphite forms *during* solidification in both gray cast iron and nodular cast iron.

effects are dramatic, because it permits a cast iron to be used in applications such as crankshafts where a brittle failure would be catastrophic.

Nodular iron, like gray iron, may be treated to be ferritic, pearlitic, or to contain tempered martensite. The former is the more ductile, the latter produces an exceptionally strong cast product (Table 14–4.1).

Malleable Cast Iron   We have just seen that cast irons with 3%–3.5% C and 2%–2.5% Si graphitize readily. Typical steels (<1% C and <0.25% Si) do not graphitize under normal circumstances. Expectedly, intermediate compositions behave in an intermediate fashion; however, the usefulness of the product is highly dependent upon how the engineer manipulates the composition, the cooling rates, and subsequent heat treatments. For example, a cast iron with 2.25% carbon and 1% silicon will produce a *white cast iron* if it is solidified rapidly, i.e., if it forms the metastable Fe_3C during freezing and is not given an opportunity for graphitization as a solid.

The above white cast iron* finds applications in certain wear-resistant parts, because the large amount of carbide (~30 v/o of the product) withstands abrasion well. However, this metal finds wider use if it is graphitized to form *malleable iron*. By annealing a white cast iron, the Fe_3C will decompose to iron and graphite. Since the reaction occurs in the absence of a liquid, the resulting graphite is neither flakelike (Fig. 14–4.1) nor spheroidal (Fig. 14–4.3). Rather, it grows as "clusters" within the solid metal (Fig. 14–4.4). This graphite is not "cracklike" on a submicroscopic scale as are the edges of graphite flakes. Therefore, the metal retains some ductility (Table 14–4.2) and is more malleable than gray iron. Hence, the name.

---

* Its fracture is metallic white, rather than gray.

<table>
<tr><td colspan="6" align="center">**TABLE 14–4.1**</td></tr>
</table>

Properties of Unalloyed Cast Irons*

Cast iron	Type	Minimum mechanical properties			Typical uses
		S_u	S_y	Elong. (5 cm)	
Gray (3.2 C–2 Si)	Pearlitic	275 MPa (40,000 psi)	240 MPa (35,000 psi)	<1%	Engine blocks
	Martensitic	550 (80,000)	550 (80,000)	nil	Wearing surfaces
	Bainitic	550 (80,000)	550 (80,000)	nil	Cam shafts
	Ferritic†	172 (25,000)	138 (20,000)	<1%	Pipe, machine bases
Nodular‡ (3.5 C–2.5 Si)	Ferritic	413 MPa (60,000 psi)	275 MPa (40,000 psi)	18	Pipe
	Pearlitic	550 (80,000)	380 (55,000)	6	Crankshafts
	Tempered martensite	825 (120,000)	620 (90,000)	2	Special machine parts
Malleable (2.2 C–1 Si)	Ferritic	365 MPa (53,000 psi)	240 MPa (35,000 psi)	18	Hardware
	Pearlitic	450 (65,000)	310 (45,000)	10	Railroad equipment
	Tempered martensite	700 (100,000)	550 (80,000)	2	Railroad equipment
White (3.5 C–0.5 Si)	As cast (pearlitic)	275 MPa (40,000 psi)	275 MPa (40,000 psi)	nil	Wear-resistant products

* Adapted from Flinn and Trojan, *Engineering Materials and Their Applications,* and from Amer. Soc. Metals.

† Ferritic gray cast iron generally has ~3.5% C and ~2.5% Si.

‡ Also called *ductile* cast iron.

<table>
<tr><td colspan="3" align="center">**TABLE 14–4.2**</td></tr>
</table>

Ductility of Ferritic Alloys

Type	Ductility % elong. (5 cm)	See figure
Ferritic iron	40–50	10–8.2(a)
Ferritic nodular C-I	15–20	14–4.3
Ferritic malleable C-I	15–20	14–4.4
Ferritic gray C-I	<1	14–4.1

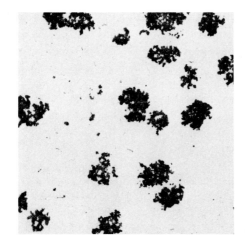

Figure 14–4.4 Malleable cast iron (×100). Carbon is present as graphite clusters (*temper carbon*). As a result, there is significant ductility (Table 14–4.2). The graphite clusters are formed by the solid reactions of Eqs. (14–3.1a) and (14–3.1b). (From ASM Committee on Nodular Iron, "Nodular Cast Iron," *Metal Handbook* Supplement, 1A, American Society for Metals.)

As indicated with Eqs. (14–3.1a) and (14–3.1b), it is possible to produce either ferritic malleable iron or pearlite malleable iron. Furthermore, appropriate quenching and tempering can lead to other relatively strong and inexpensive castings usable in railway and similar equipment.

The reader is asked to observe one limitation on designs that specify malleable iron castings. Since the solidification must be sufficiently fast to form a white iron initially, these castings cannot have thick sections. Otherwise, graphite flakes would form during the time required for solidification. Depending on the silicon content and the thermal conductivity of the mold wall, the maximum thickness of malleable iron castings is normally limited to 20 mm–30 mm (~1 in.).

Example 14–4.1 Plot the ferrite and austenite contents as a function of temperature for a cast iron containing 2.5% carbon, 1% silicon, and the balance iron:

   a) as a white cast iron,

   b) as a carbide-free, malleable cast iron.

Procedure. The silicon dissolves in the ferrite and austenite. Therefore, we will approximate the composition as being 97.5 Fe–2.5 C.

Calculation

   a) Use Fig. 10–7.1 for the Fe–Fe$_3$C diagram (since there is no graphite).

At 700°C (and below), $\alpha = (6.7 - 2.5)/(6.7 - 0) = 0.63$

At 800°C, $\gamma = (6.7 - 2.5)/(6.7 - 1.0) = 0.74$

Etc., See Fig. 14–4.5.

Temperature, °F

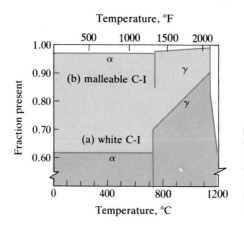

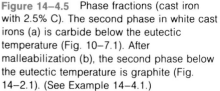

Figure 14–4.5   Phase fractions (cast iron with 2.5% C). The second phase in white cast irons (a) is carbide below the eutectic temperature (Fig. 10–7.1). After malleabilization (b), the second phase below the eutectic temperature is graphite (Fig. 14–2.1). (See Example 14–4.1.)

b) Use Fig. 14–2.1 for the Fe–C diagram (with graphite present).

At 700°C (and below), $\alpha = (100 - 2.5)/(100 - 0) = 0.975$

At 800°C, $\gamma = (100 - 2.5)/(100 - 0.95) = 0.985$

Etc.,   See Fig. 14–4.5.

Comments. Silicon, like all elements, will shift the eutectoid to lower carbon contents (Section 10–9). This decreases the carbon solubility in austenite somewhat. Therefore, the curves in Fig. 14–4.5 for γ are slightly high. ◀

Example 14–4.2

Indicate the heat-treating sequence to produce a nodular iron (3 C–2 Si–95 Fe) that will contain tempered martensite. Rationalize any factors that receive different considerations than if the metal were a 1080 steel.

Procedure and Comments

a) Melting, ~1300°C (~2400°F). The metal must be inoculated with magnesium (or cerium) to produce graphite nodules during solidification. The melting temperature is lower than steel since more carbon is present.

b) Austenization, 800°C (1475°F). Since there is 2% silicon present, the eutectoid temperature and composition are ~775°C and ~0.55%, respectively (Fig. 10–9.1). Austenite (~0.55% C) + graphite (100% C).

c) Oil-quench to ambient. The 2% Si increases the hardenability appreciably. An oil-quench is thus possible and will reduce the probability of cracking. Martensite (~0.55% C) + graphite (100% C).

d) Temper as if it were a 1055 steel. Ferrite (nil C) + carbide (6.7% C) + graphite (100% C). ◀

Example 14–4.3

Compare the densities of (a) the white cast iron and (b) ferritic malleable cast iron of Example 14–4.1. With 1% Si, the densities are ~7.8 g/cm³, 7.6 g/cm³, and ~2.2 g/cm³, respectively, for α, Fe₃C, and graphite.

Procedure. The compositions at 20°C are essentially the same as they are at 700°C in Example 14–4.1. Use 100 g as a basis.

Calculations

a) White cast iron: 0.63α and 0.37 Fe_3C (Example 14–4.1).

$$(63 \text{ g } \alpha)/(7.8 \text{ g/cm}^3) = 8.08 \text{ cm}^3 \text{ } \alpha;$$

$$(37 \text{ g } Fe_3C)/(7.6 \text{ g/cm}^3) = 4.87 \text{ cm}^3 \text{ } Fe_3C;$$

$$\rho_{wci} = 100 \text{ g}/(8.08 + 4.87 \text{ cm}^3)$$
$$= 7.72 \text{ g/cm}^3 \qquad \text{(or 7.72 Mg/m}^3\text{).}$$

b) Ferritic malleable: 0.975α and 0.025 graph (Example 14–4.1).

$$(97.5 \text{ g } \alpha)/(7.8 \text{ g/cm}^3) = 12.50 \text{ cm}^3 \text{ } \alpha;$$

$$(2.5 \text{ g graph})/(2.2 \text{ g/cm}^3) = 1.14 \text{ cm}^3 \text{ graph};$$

$$\rho_{mci} = 100 \text{ g}/(12.50 + 1.14 \text{ cm}^3)$$
$$= 7.33 \text{ g/cm}^3 \qquad \text{(or 7.33 Mg/m}^3\text{).}$$

Comment. There is a volume expansion during malleabilization that must be considered in processing control. ◄

Review and Study

1. *Cast irons* have higher carbon contents and lower melting temperatures than steels. As such, they are the most available and inexpensive of all alloys.

2. Iron carbide, Fe_3C, is only *metastable* because it eventually dissociates to iron and graphite. Thus, the true *equilibrium diagram* between iron and carbon is that of Fig. 14–2.1.

3. Iron carbide dissociates more rapidly (1) at elevated temperatures, (2) at higher carbon contents (>2 w/o), and (3) in the presence of silicon (1–2 w/o). Above the eutectoid temperature, the reaction

$$Fe_3C \rightarrow \gamma + \text{graphite.}$$

Below that temperature,

$$Fe_3 \rightarrow \alpha + \text{graphite.}$$

4. *Gray* cast iron contains flake graphite that forms during solidification. In contrast, the graphite of *nodular* (ductile) cast iron is present as spherulites. As a result, the latter product is both stronger and more ductile.

White cast iron, which is processed to contain Fe_3C rather than graphite, finds some application as a wear-resistant alloy. More commonly, it is annealed to produce *malleable* cast iron with graphite clusters. Thus, the ductility is increased for much wider usage.

Each of the above cast irons can be heat treated to produce ferritic, pearlitic, or martensitic variants so that a wide range of properties is available (Table 14–4.1).

TERMS AND CONCEPTS

Annealing (cast irons)   Heating to graphitize Fe_3C.

Cast iron   Fe–C alloy, sufficiently rich in carbon to produce a eutectic liquid during solidification. In practice, this generally means more carbon than can be dissolved in austenite (>2%).

Damping capacity   Attenuation of mechanical vibrations.

Graphitization   Decomposition of iron carbide into iron and carbon (Eq. 14–2.1).

Gray cast iron   Cast iron with graphite flakes and, therefore, a gray fracture surface.

Iron-carbon diagram   Equilibrium phase diagram of iron and carbon; therefore without Fe_3C (Fig. 14–2.1).

Malleable cast iron   Cast iron that undergoes graphitization after solidification. Graphite is present as "clusters," also called temper carbon.

Nodular cast iron   Cast iron in which graphite spherulites form during solidification. Sometimes called ductile iron because of its properties.

Temper carbon   Graphite present as "clusters" in malleable cast iron. (Formed after solidification.)

White cast iron   Cast iron with Fe_3C rather than graphite.

FOR CLASS DISCUSSION

A_{14} Indicate several reasons why cast iron is one of the least expensive metallic materials.

B_{14} Outline the processing differences that lead to gray cast iron, to nodular cast iron, and to malleable cast iron.

C_{14} Outline the processing differences that lead to ferritic, to pearlitic, and to bainitic malleable iron.

D_{14} Can a pearlitic cast iron be changed to a ferritic cast iron without remelting the metal? If so, how? Can a ferritic cast iron be changed to a pearlitic cast iron without remelting the metal? If so, how?

E_{14} Can a gray cast iron be changed to a malleable cast iron without remelting the metal? If so, how? Can a malleable cast iron be changed to a white cast iron without remelting the metal? Why, or why not?

F_{14} Based on microstructures, compare the expected ductilities of ferritic gray cast iron, ferritic malleable cast iron, ferritic steel, ferritic nodular cast iron. Explain your answer.

G_{14} Explain why it is impossible to make large castings out of malleable cast iron.

H_{14} Steels that contain silicon are poor choices for applications that must withstand high temperatures for extended periods of time. Suggest why.

I₁₄ Cast irons are usually brazed rather than welded. A brazing alloy is commonly a copper-base filler metal that bonds to the cast iron but doesn't melt it. Why not weld it as we do steels?

QUIZ SAMPLES

14A Cast irons differ from steels by having a __+−__ .

(a) higher percentage of iron   (b) lower melting temperature
(c) higher elastic modulus   (d) better vibration absorption

14B A pearlitic gray cast iron has less graphite than a __+−__ with the same carbon content.

(a) pearlitic nodular cast iron   (b) ferritic gray cast iron
(c) Neither of the above two

14C As a class, the __**__ cast irons have less ductility than the others that contain graphite.

14D Graphitization proceeds more rapidly when the temperature is __**__ , and when the __**__ and __**__ contents are higher.

14E Heavy castings (>50 mm thick) are impractical for __+−__ cast irons.

(a) gray   (b) nodular   (c) malleable   (d) white

STUDY PROBLEMS

14–4.1 A micrographic analysis of a gray cast iron shows 12 v/o graphite and 88 v/o ferrite. Estimate the carbon content.

Answer: 3.7% C and 96.3% (Fe + Si)

14–4.2 Estimate the density of a ferritic gray cast iron (3.5% C).

14–4.3 A white iron contains 0.5% silicon and 3% carbon. (a) What is the approximate carbon content of γ if the metal has been malleabilized at 800°C? (b) How much graphite will form at that temperature?

Answer: (a) ~0.9% C   (b) ~2% graphite

14–4.4 Estimate the density of the pearlite malleable iron from Study Problem 14–4.3. (Use necessary data from Example 14–4.3.)

• **14–4.5** What heat-treating sequence will produce a bainitic malleable cast iron? Give a basis for your choices.

14–4.6 What heat-treating sequence will produce a pearlitic malleable cast iron? Give a basis for your choices.

QUIZ CHECKS

14A (b), (d)	**14D**	higher
14B (b)		carbon, silicon
14C gray	**14E** (c), (d)	

Chapter Fifteen

Concretes

PREVIEW

More tons of concrete are produced and used than of any other technical material. This is due in part to its use of low cost and widely available raw materials (available in almost every country of the world). Its wide use is also a consequence of the fact that the final stages of production are located on the job site, thus simplifying logistics and facilitating the construction of large structures.

We shall learn that *concrete* is the final monolithic product, while *cement* is the bonding material. Further, portland cement hardens and bonds the aggregate by a *hydration reaction* and not by drying.

The properties of portland cement concrete originate from its structure, which includes the aggregate mixture, and the very complex, semiamorphous, hydrated chemical bond. Furthermore, the voids from entrained air and from hairline shrinkages have pronounced effects upon service performance.

Asphalt concrete uses a petroleum bitumen as the cement for the aggregate.

CONTENTS

STUDY OBJECTIVES

1. Recognize the features that characterize the aggregate and particle packing.
2. Know the general nature of the hardening process for portland cement.
3. Be able to calculate unit mixes of concrete.
4. Identify factors that lead to concrete failures.

15-1

AGGREGATES

The major components of concrete are the *aggregates* of the mix. Of course, the sand and gravel used for aggregate must be durable materials, such as quartzite, crushed limestone, or basalt, having greater strength than what is expected of the final concrete. If not, and slate (or a similar friable rock) is present, failure will occur through the aggregate, even with the best formulated cement additions. Concrete codes specify aggregate quality, for example, ASTM–C 88. In addition, the engineer requires a measure of aggregate size. This is obtained by *sieve analyses*. Two series of screens are in common use for sieve analyses (Table 15–1.1) and are generally comparable. The U.S. Series of sieve sizes is most widely used for concrete codes. The Tyler Series, which is based on $\sqrt{2}$, has wider use in the ceramic industry.

An average particle size has only limited significance. For example, the distribution could be very narrow, as is typical of beach sands, or very broad, such as river sands, and still have the same *mean* size (Fig. 15–1.1). Rather, the engineer is commonly interested in what fractional quantity is retained on successive screens. Several such analyses are shown in Fig. 15–1.2.

Packing density and porosity are also important. Figure 15–1.3 shows several subdivisions in volume that affect the amount of water and/or a cement paste that can be associated with the aggregate. *Closed* (inside) pores are impermeable. These may be encountered in artificial light-weight aggregates and in some crushed basalt aggregates. *Open* pores are permeable voids. They include capillaries, fine striations, and reentrant shape irregularities of the aggregate particle. In addition, there can be major amounts of *interparticle* porosity among sand, gravel, etc., which is open pore space.

These pores lead to several identifiable volumes. The *true volume* V_{tr} is that volume that is inherent to the material. The *apparent volume* V_a is the sum of the true volume and the closed-pore volume V_{cp}. The *bulk volume* V_b includes not only the closed-pore volume (and the true volume) but also the open-pore

575

TABLE 15–1.1

Sieve Series Used in Particle Analyses (openings)

	U.S. Series				Tyler Series		
	1.5	38	mm	1.50	in.	1.5	38	mm	1.50	in.

		U.S. Series					Tyler Series		
	1.5	38	mm	1.50 in.	1.5	38	mm	1.50 in.
	1	25.4		1.00	1	26.7		1.050
	3/4	19.1		0.75	3/4	18.9		0.742
	1/2	12.7		0.50	1/2	13.3		0.525
	3/8	9.5		0.375	3/8	9.4		0.371
					No. 3	6.7		0.263
No.	4	4.75		0.187	4	4.7		0.185
					6	3.3		0.131
	8	2.38		0.0937	8	2.36		0.093
					10	1.66		0.065
	16	1.19		0.0469	14	1.18		0.046
					20	0.83		0.0328
	30	0.59		0.0232	28	0.59		0.0232
					35	0.42		0.0164
	50	0.297		0.0117	48	0.295		0.0116
					65	0.208		0.0082
	100	0.150		0.0059	100	0.147		0.0058
						*		*

* Series continues to No. 400 with an opening of 0.037 mm (0.0015 in.); steps are on the basis of $\sqrt{2}$.

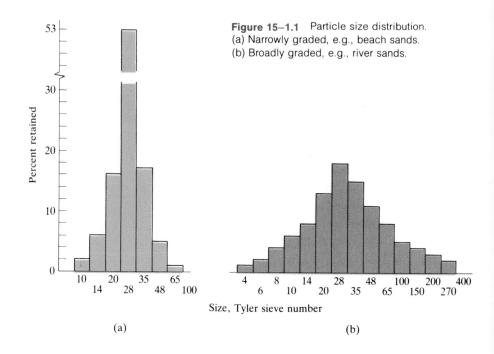

Figure 15–1.1 Particle size distribution.
(a) Narrowly graded, e.g., beach sands.
(b) Broadly graded, e.g., river sands.

Size, Tyler sieve number

(a)   (b)

Sieve No. (U.S. series)

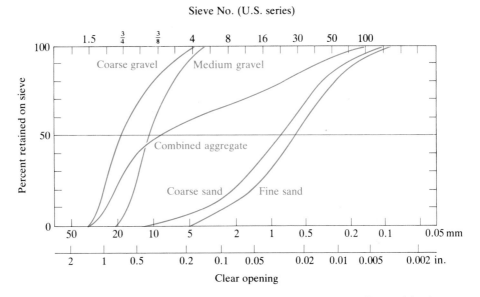

Figure 15–1.2 Analyses of typical aggregates used for concrete mixes. The combined aggregate is a mixture of coarse gravel (60%) and sand (40%). The combination fills space more completely than gravel alone or sand alone. (Data adapted from Troxell, Davis, and Kelly, *Composition and Properties of Concrete.*)

volume V_{op}. With this *total volume*, we must define our basis. If only the aggregate particles are considered, the open-pore volume is limited to the small cracks and capillaries. If our basis involves bulk handling, the total volume must include the relatively large amount of interparticle porosity.

These several volumes lead to two principal porosities, true (or total) and apparent (or open):

$$Total\ porosity = \frac{open\text{-}pore\ volume + closed\text{-}pore\ volume}{total\ volume}. \qquad (15\text{--}1.1)$$

$$Apparent\ porosity = \frac{open\text{-}pore\ volume}{total\ volume}. \qquad (15\text{--}1.2)$$

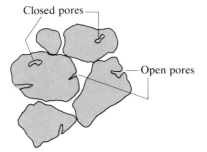

Closed pores

Open pores

Figure 15–1.3 Pore space (schematic). Closed pores are inside pores and, therefore, impermeable. Open pores are permeable voids. Open pores include space between particles (interparticle porosity), plus small cracks and capillaries that extend into the aggregate particles.

In either case, we should identify the basis of the total volume—either the particles of aggregate, or the volume that includes the unfilled voids between aggregate particles.

There are also three principal densities:

$$\textit{True density} = \text{mass/true volume.} \tag{15–1.3}$$

$$\textit{Apparent density} = \text{mass/apparent volume} \tag{15–1.4}$$
$$= \text{mass/(true volume + closed-pore volume)}$$
$$= \text{mass/(total volume} - \text{open-pore volume).}$$

$$\textit{Bulk density} = \text{mass/total volume.} \tag{15–1.5}$$

Here again, we should identify our basis for the total volume, as to whether we are considering only the aggregate particles or are including the interparticle space present that accompanies bulk materials.

Example 15–1.1

A sample of wet gravel is dried to remove surface water, but not the absorbed water. It weighs 8.91 kg. In this condition, its bulk volume (excluding interstitial pores) of the gravel particles was determined to be 3570 cm³ by water displacement. After thorough drying, its weight is 8.79 kg.

a) What is its apparent porosity?

b) The apparent density of the gravel particles?

Procedure. Determine the open-pore volume by the water loss during drying. The apparent volume is $V_b - V_{op}$.

Calculation

a)
$$V_{op} = (8910 - 8790 \text{ g})/(1 \text{ g/cm}^3) = 120 \text{ cm}^3,$$
$$\text{Apparent porosity} = 120 \text{ cm}^3/3570 \text{ cm}^3 = 3.4 \text{ v/o.}$$

b)
$$V_a = 3570 \text{ cm}^3 - 120 \text{ cm}^3 = 3450 \text{ cm}^3;$$
$$\rho_a = 8790 \text{ g}/3450 \text{ cm}^3 = 2.55 \text{ g/cm}^3 \qquad (\text{or } 2.55 \text{ Mg/m}^3).$$

Comment. We can also calculate the bulk density of the gravel particles, 8790 g/3570 cm³ = 2.46 g/cm³ (or 2.46 Mg/m³). ◀

15–2

PORTLAND CEMENT

All concretes use a *cement** as the material to bond the aggregate into a monolithic product. The most common cement is *portland* cement (so-named because of its eighteenth-century association with the Island of Portland). Its raw materials include calcined clay and limestone, or their derivatives.

When 0.4 to 0.7 units (weight) of water are mixed with a unit of cement, the combination initially has the consistency of a paste. The strength of this hard-

---

* The term, *cement,* should refer to the material used to bond the concrete, not to the final product.

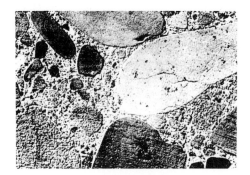

Figure 15–2.1 Cross section of concrete (×1.5). The space among the gravel is filled with sand. The space among the sand is filled with a hydrated portland cement paste. (Portland Cement Association.)

ened cement paste essentially establishes the strength of concrete, because with a strong aggregate, any failure occurs *between* the sand and/or rock particles. Therefore, in principle, a driveway that could be produced from only cement and water mixtures would be as strong as when an aggregate is present. However, it would be prohibitive on the basis of cost. Therefore, the cement–water mixture is "diluted" with the stronger, cheaper aggregate materials.

Conceptually, concrete contains gravel or crushed rock, with sand filling the interstices; then a paste of cement and water fills the space among the sand grains. In practice, some excess sand is required to ensure the complete filling of the space within the gravel (Fig. 15–2.1); likewise, approximately 10 percent more cement paste is required than the theoretical amount for the space among the sand grains. Otherwise, some local voids could remain where mixing is not perfect.

Cement Hydration

The portland cement used in concretes is a mixture of calcium silicates and calcium aluminate that hydrate in the presence of water:

$$Ca_3Al_2O_6 + 6\ H_2O \rightarrow Ca_3Al_2(OH)_{12}; \tag{15–2.1}$$

$$Ca_2SiO_4 + x\ H_2O \rightarrow Ca_2SiO_4 \cdot x\ H_2O; \tag{15–2.2}*$$

$$Ca_3SiO_5 + (x + 1)H_2O \rightarrow Ca_2SiO_4 \cdot x\ H_2O + Ca(OH)_2, \tag{15–2.3}$$

In each case, the hydrated product is less soluble in water than was the original cement. Therefore, in the presence of water, the above reactions are ones of solution and reprecipitation.

Each of these reactions emphasizes that cement does not harden by drying, but by the chemical reaction of hydration. In fact it is necessary to keep concrete moist to ensure proper *setting*. The hydration reactions described above

---

* This reaction may be more correctly stated as

$$2\ Ca_2SiO_4 + (5 - y + x)H_2O \rightarrow Ca_2[SiO_2(OH)_2]_2 \cdot (CaO)_{y-1} \cdot x\ H_2O + (3 - y)Ca(OH)_2, \tag{15–2.4}$$

where x varies with the partial pressure of water and y is approximately 2.3. Equation (15–2.3) may also be modified accordingly.

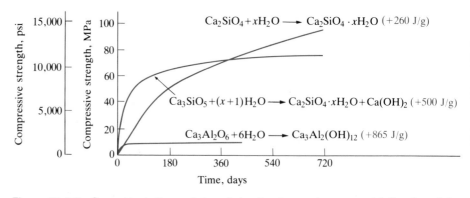

Figure 15–2.2   Cement hydration and strength (portland cement components). For strength to develop, water must be present for hydration. Concrete does *not* set by drying. Heat is given off during hydration as indicated. Its removal may require special provisions in massive structures, such as dams.

release heat, as shown in Fig. 15–2.2. Engineers take advantage of this in cold climates in which it may be necessary to pour concrete at temperatures that are slightly below the freezing temperature of water. Conversely, the *heat of hydration* presents a problem when concrete is poured into massive structures such as dams. In these extreme cases, special cooling is required, and a refrigerant is passed through embedded pipes that are left in place and later act as reinforcement.

The variables we have discussed suggest that the engineer may utilize various kinds of cement to obtain different characteristics. For example, a cement with a larger fraction of tricalcium silicate (Ca_3SiO_5) sets rapidly and thus gains strength early.* In contrast, in cases in which the heat of reaction might be a complication, and more setting time is available, a Ca_2SiO_4-rich cement is used. The American Society for Testing and Materials (ASTM) lists the following types of portland cement: (1) Type I, used in general concrete construction in which special properties are not required. (2) Type III, used when an early-strength concrete is desired. (3) Type IV, used when a low heat of hydration is required. There are also two other types (II and V) for special applications in which sulfate attack is a design consideration.

Example 15–2.1   A concrete mix contains 585 kg of sand and gravel, 100 kg of cement, and 55 kg of water. Assume the cement is (a) all tricalcium aluminate, $Ca_3Al_2O_6$, (b) all tricalcium silicate, Ca_3SiO_5, (c) all dicalcium silicate, Ca_2SiO_4. What is the temperature rise during hydration, given that the average specific heat of the concrete mix is 0.8 J/g·°C (or 0.2 cal/g·°C)? Assume that half of the heat is lost.

---

* A *high, early-strength cement* can also be achieved by grinding the cement more finely.

Solution. Basis: 1 g cement = 7.4 g mix. From the data in Fig. 15–2.2,

a) $(7.4 \text{ g/g cement})(0.8 \text{ J/g·°C})\Delta T = (865 \text{ J/g cement})(0.5)$;

$$\Delta T = 73°C.$$

b) $\Delta T = 0.5(500 \text{ J})/(5.92 \text{ J/°C}) = 42°C.$

c) $\Delta T = 0.5(260 \text{ J})/(5.92 \text{ J/°C}) = 22°C.$

Comments. Not only do $Ca_3Al_2O_6$ and Ca_3SiO_5 release more heat than does Ca_2SiO_4, they react in a relatively short period of time. Therefore, a smaller fraction of the heat escapes from a large structure during the early stages of setting. The temperature rise may become critical unless only small fractions of $Ca_3Al_2O_6$ and Ca_3SiO_5 are present in the cement. ◄

15–3
CONCRETE MIXES

Gravel (or crushed rock), sand, cement, and water are mixed in specific ratios called a *unit mix*. The unit mix is commonly based on the amount of cement used. Thus, the engineer may specify a mix of 3.1 gravel–2.6 sand–1 cement–0.55 water on a weight basis (or these may be converted into appropriate volumes).

We will observe first that the cement/water ratio is critical for the strength of the concrete (Fig. 15–3.1). Since excess water does not enter the hydration reaction, it occupies space and precludes solid-to-solid bonding. If the excess water eventually evaporates, hairline cracks remain. Therefore, the mix is designed with the water/cement ratio as low as possible. However, there is a limit. There must be enough water present to make the concrete workable, and to allow it to fill the forms completely without voids. Vibration procedures help with form filling. Also the incorporation of air bubbles (~1 mm dia.) into the

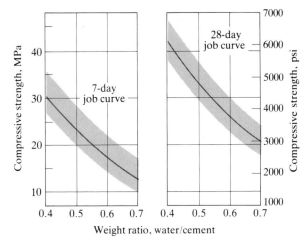

Figure 15–3.1 Strength of concrete versus water/cement ratio (Type I portland cement, moist-cured at 20°C (68°F)). There must be enough water to provide workability to the mix and to ensure that the forms are filled without voids. Additional water simply requires extra space and means less bonding. (After Cernica, *Fundamentals of Reinforced Concrete*, Addison-Wesley.)

concrete through the use of appropriate *air-entraining agents* increases the workability (and therefore lowers the required amount of water). These spherical voids do not weaken the concrete as much as the capillary voids that remain after excess water is evaporated.*

The small bubbles of entrained air also increase the resistance of the concrete to deterioration resulting from the freezing and thawing of pore water. As a result, almost all concrete road pavements of northern latitudes have air entrainment specifications.

We may calculate the volume of a concrete mix from the volumes of the aggregate, cement, and water additions. As demonstrated in Example 15–3.2, we assume that the sand will (more than) fill the space among the gravel and that the cement/water parts will more than fill the passages between the sand. Finally, an adjustment may be made for the volume of the entrained air.

The density and porosity of the concrete product may be related to the various volumes of that product—true, apparent, and total (Eqs. 15–1.1 to 15–1.5). In turn, density and porosity are significant for properties. For example, the heat losses through a wall are a function of the true volume (and of the total porosity), while a wall's permeability to water is related to the amount of apparent (open) porosity.

Example 15–3.1

The true density of a concrete to which an air-entraining agent was added is 2.80 Mg/m³ ($=174.7$ lb/ft³). However, a dry core [152 mm × 102 mm in diameter (6.0 in. × 4.01 in.)] of this concrete weighs only 2.982 kg (6.56 lb). The same core weighs 3.08 kg (6.776 lb) when saturated with water, and 4.322 kg (9.51 lb) when suspended in water.

a) What are the percents of the open and closed pores of the concrete product?

b) What are the bulk and apparent densities of the concrete product?

Procedure. In this case the bulk volume can be calculated from the geometry, since the core is a cylinder. Alternatively, the bulk volume V_b may be calculated from the buoyancy.

Calculations

$$\text{Bulk volume} = \pi r^2 h = \pi(51 \text{ mm})^2(152 \text{ mm})(1 \text{ cm}/1000 \text{ mm}^3)$$
$$= 1{,}242 \text{ cm}^3 \qquad\qquad (\text{or } 75.4 \text{ in}^2).$$

$$\text{Buoyancy} = 4322 \text{ g} - 3080 \text{ g} = 1242 \text{ g};$$

$$V_b = 1242 \text{ g}/(1 \text{ g/cm}^3) = 1242 \text{ cm}^3.$$

$$\text{True volume} = m/\rho_{tr} = [(2982 \text{ g})/2.8 \times 10^6 \text{ g/m}^3)](10^6 \text{ cm}^3/\text{m}^3)$$
$$= 1065 \text{ cm}^3 \qquad\qquad (\text{or } 65.0 \text{ in}^2).$$

$$\text{Total porosity} = (V_b - V_{tr})/V_b = (1242 \text{ cm}^3 - 1065 \text{ cm}^3)/(1242 \text{ cm}^3)$$
$$= 14 \text{ v/o}.$$

---

* Cf. the effect of graphite nodules and graphite flakes on the strength of cast iron (Section 14–4).

$$\text{Volume open pores} = m/\rho_{H_2O} = (3080 \text{ g} - 2982 \text{ g})/(1 \text{ g } H_2O/cm^3)$$
$$= 98 \text{ cm}^3 \text{ H}_2O \qquad \text{(or 5.98 in}^3\text{)}.$$

a) Apparent (open) porosity $= 98 \text{ cm}^3/1242 \text{ cm}^3 = 0.08$ (or 8 v/o).

  Closed porosity $= 14 \text{ v/o} - 8 \text{ v/o} = 6 \text{ v/o}$.

b)   Bulk density $= 2982 \text{ g}/1242 \text{ cm}^3 = 2.4 \text{ g/cm}^3$

    $= 2.4 \text{ Mg/m}^3$ (or 150 lb/ft^3).

  Apparent density $= 2982 \text{ g}/(1242 - 98 \text{ cm}^3) = 2.6 \text{ g/cm}^3$

    $= 2.6 \text{ Mg/m}^3$ (or 163 lb/ft^3). ◀

Example 15–3.2   A concrete unit mix contains 136 kg (300 lb) of gravel, 95 kg (210 lb) of sand, 42.7 kg (94 lb of portland cement, and 22.5 kg (50 lb) of water, plus 5 v/o entrained air. How much cement is required to build a low retaining wall 28.6 m × 1.07 m × 0.305 m (94 ft × 3.5 ft × 1 ft)?

	Density Mg/m^{3*}	
Material	Apparent	Bulk†
Gravel	2.60	1.76
Sand	2.65	1.68
Cement	3.25	1.50

* Multiply by 62.4 to obtain lbs/ft^3.
† As-received basis (includes interstitial volume).

Solution. Basis: 1 unit mix.

Volume of gravel $= 136{,}000 \text{ g}/(2.60 \times 10^6 \text{ g/m}^3) = 0.0523 \text{ m}^3$

Volume of sand $= 95{,}000 \text{ g}/(2.65 \times 10^6 \text{ g/m}^3) = 0.0358$
(in the space among the gravel)

Volume of cement $= 42{,}700 \text{ g}/(3.25 \times 10^6 \text{ g/m}^3) = 0.0131$
(in the space among the sand and gravel)

Volume of water $= 22{,}500 \text{ g}/(10^6 \text{ g/m}^3) \qquad = 0.0225$
(in the remaining space among all solids)
                                        ──────
                                        0.124 m^3

Total m^3/unit mix $= 0.124 \text{ m}^3/0.95 = 0.1305 \text{ m}^3$

Unit mixes $= 9.33 \text{ m}^3/0.1305 \text{ m}^3 = 71.5$

Cement required $= (71.5)(42.7 \text{ kg}) = 3050 \text{ kg}$.

Comment. Using lbs and ft^3,

$$\text{Volume of gravel} = 300 \text{ lbs}/(2.60 \times 62.4 \text{ lbs/ft}^3) = 1.85 \text{ ft}^3;$$

$$V_s = 1.27 \text{ ft}^3;$$

$$V_c = 0.46 \text{ ft}^3;$$

$$V_{H_2O} = 0.80 \text{ ft}^3;$$

$$\Sigma V = 4.38 \text{ ft}^3/0.95 = 4.6 \text{ ft}^3;$$

$$\text{Unit mixes} = 329 \text{ ft}^3/4.6 \text{ ft}^3 = 71.5.$$

$$\text{Cement required} = (71.5)(94 \text{ lb}) = 6700 \text{ lb.} \quad \blacktriangleleft$$

15–4

OTHER CEMENTS AND CONCRETES

Portland cement is made in large tonnages. Of course, other cements also exist. Gypsum plaster and lime are also hydraulic cements. For example, gypsum plaster (plaster of paris) reacts as

$$2 \text{ CaSO}_4 \cdot \text{H}_2\text{O} + 3 \text{ H}_2\text{O} \rightarrow 2(\text{CaSO}_4 \cdot 2 \text{ H}_2\text{O}). \tag{15–4.1}$$

In principle, its hardening mechanism is identical to that of portland cement since the initial compound dissolves and reprecipitates in the hydrated form.

Polymeric cements involve polymerization reactions, as is the case for silicic acid:

$$x \text{ Si(OH)}_4 \rightarrow \left[\begin{array}{c} \text{OH} \\ | \\ \text{Si} \\ | \\ \text{OH} \end{array} \right]_x + x \text{ H}_2\text{O}. \tag{15–4.2}$$

In this example the polymerization reaction is a condensation reaction (Section 6–2). The above reaction leads to an amorphous, 3-dimensional, SiO_2 structure when it is extended to involve the remaining two –OH radicals. Other, more complex polymeric reactions occur in epoxy cements, some of which produce exceptionally strong bonds.

Asphalt

Bonding may also be achieved by thermoplastic materials. Asphalt is a product of oil production. Like coal tar, asphalt is a bitumen or the residue containing the higher melting components that remain after distillation removes gaseous and fluid fractions. It does not have a specific composition; rather asphalt is a mixture of many molecular species in which hydrocarbons predominate. In effect, asphalts are naturally occurring linear polymers that become thermo-

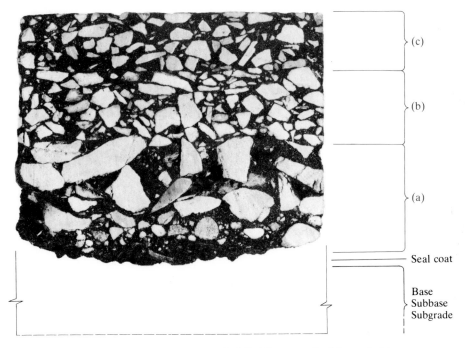

(c)

(b)

(a)

Seal coat

Base
Subbase
Subgrade

Figure 15–4.1 Core from asphalt pavement (×0.5). Like concrete, this product is agglomerated; but viscous asphalt, rather than a hydrated silicate, serves as the bond. (a) Bonding course. Sand (30 w/o) and asphalt (5 w/o) fill the interstices among the coarse stone (65 w/o) to give a rigid support above the subgrade aggregate. (b) Leveling course. The stone is smaller, otherwise the composition is similar to the preceding course. (c) Wearing course. More asphalt (6 w/o), more sand (55 w/o), and 4–5 w/o mineral filler (fly-ash) provide an impermeable, tough surface to resist traffic wear. (Core from Michigan Highway Testing Laboratory.)

plastic on heating (Section 6–3). Typically, the glass temperature is near $-20°C$ (near $-5°F$). Thus, in winter climates or under impact loading asphalt becomes brittle, and in hot summer exposures it flows.

Asphalt-bonded Concretes

Asphalt concretes are widely used to pave roads (Fig. 15–4.1), parking lots, driveways, and airports. In hot mixes, the asphalt is heated until it becomes fluid and mixed with heated aggregates at a "hot plant." It is delivered, spread, and compacted before it cools and becomes stiff. The technical variables of the product include (a) stability, (b) durability, and (c) skid resistance.

The *stability* of an asphalt concrete requires that it will not flow under load. This is achieved by quantity and shape of the crushed aggregate. The maximum amount of *crushed* rock should be present, with grading that provides smaller

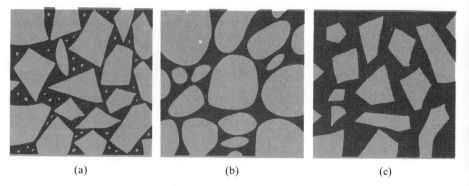

(a)   (b)   (c)

Figure 15–4.2   Asphalt concrete (schematic). (a) Ideal, with contact of crushed-rock aggregate. (b) Unsatisfactory, since rounded aggregate reduces stability. (c) Unsatisfactory, excess asphalt.

sized particles in the spaces among the coarser particles. Excess asphalt permits the aggregate to "float" and shift in the asphalt. Of course, this is undesirable. Figure 15–4.2 shows a schematic sketch of (a) an idealized asphalt concrete, (b) an unsatisfactory asphalt concrete with rounded aggregate, and (c) an unsatisfactory concrete with too much asphalt. In the first example with crushed rock, there are numerous points of contact that give "aggregate interlock" to promote mechanical stability. In both (b) and (c), the concrete can gradually flow as it is repeatedly loaded with vehicle acceleration, or with continuous traffic.

Asphalts can "age" by hardening and becoming embrittled. This can lead to cracking and to loss of cohesion in an asphalt concrete. Since this loss of *durability* is associated with evaporation and with oxidation of the asphalt, a thin-film oven test (TFOT) is commonly required (ASTM-D-1754). A thin layer (3.2 mm, $\frac{1}{8}$ in.) of asphalt is heated in air for 5 hours at 163°C (325°F) to determine changes in hardness, ductility, viscosity, weight, etc. Pavements with embrittled asphalts are more sensitive to freezing/thawing damage and to the stripping of the asphalt from the aggregate (loss of cohesion).

Skid resistance is a quality that is obviously related to asphalt concretes that are primarily used for pavements. Excess asphalt cement permits the aggregate to be forced below the surface of the pavement, thus producing an undesirable surface solely of asphalt. Excess surface asphalt must be avoided by mix control. If encountered, its effect may be partially combated by the addition of aggregate chips to the surface at a later date. The skid resistance of asphalt concrete can also decrease if the aggregate becomes polished from road traffic.

Review and Study

SUMMARY

Concretes are aggregates bonded into a monolithic product.

1. The major component of a concrete are coarse and fine aggregates consisting of gravel (or crushed rock) and sand, respectively. Size distributions are measured by sieve analyses. The *closed-pore* volume is added to the *true* volume to determine the *apparent* volume. Both the closed-pore volume and the open-pore volume (permeable pores) are added to the true volume to determine the *bulk* volume. The true, apparent, and bulk densities vary accordingly.

2. *Cement* is the bonding material for concrete. (The monolithic product is *not* cement.) The common type is *portland* cement that contains calcium silicates and calcium aluminates. They hydrate to harden, or set, the concrete. (Concrete does *not* dry to harden.) Heat is released during setting. The strength of the concrete is determined primarily by the strength of the hydrated cement.

3. The *concrete mix* may be viewed as coarse aggregates containing interstitial sand. A paste of cement and water fills the remaining volume and produces the bond. *Excess water* beyond that required for hydration and for workability weakens the concrete because it introduces hairline cracks. *Air entrainment* increases the resistance of concrete to frost cracking; therefore, additives that introduce up to 5 v/o air (as small ~1-mm "bubbles") are commonly specified for northern construction.

4. *Asphalt concretes* utilize petroleum bitumens as a bond. Angular aggregates (crushed rock) are preferred for *stability*. *Durability* tests check for changes of hardness, viscosity, weight, etc., to anticipate embrittlement.

TERMS AND CONCEPTS

Aggregate   Coarse particulate materials such as sand, gravel, and crushed rock used for construction purposes.

Air entrainment (concrete)   The incorporation of small (~1 mm) bubbles of air into concrete to give increased resistance to freezing/thawing damage (and improved workability). The bubbles are formed by the addition of small quantities of surface-active chemicals.

Asphalt   Bitumen residues from oil production.

Cement   Bonding material that adheres to the two adjacent solid surfaces.

Cement, portland   A hydraulic, calcium silicate (and aluminate) cement widely used for structural concrete.

Concrete   Agglomerate of aggregate and a hydraulic cement.

Density, apparent   Mass divided by apparent volume (material + closed pores).

Density, bulk   Mass divided by total volume, including pores.

Density, true   Mass divided by true (pore-free) volume.

Heat of hydration   Energy released as heat during the hydration reaction, for example, of portland cement.

Hydration   Chemical reaction consuming water: $\text{Solid}_1 + H_2O \rightarrow \text{Solid}_2$.

Pores, impermeable Pores without access to surrounding fluids.

Pores, permeable (open) Pores accessible to fluids.

Porosity, apparent (Permeable-pore volume)/(total volume).

Porosity, total (true) (Total pore volume)/(total volume).

Sand (for concrete) Aggregate that passes a No. 4 sieve (4.75 mm, or 0.187 in.).

Sieve series A sequence of mesh sizes for grading aggregate. The ASTM series is used predominantly for structural aggregates.

Unit mix Ratio of gravel, sand, portland cement ($=1$, or 100) and water in a concrete mix. May be based on either weight or volume.

Water absorption, % Test used to measure open porosity.

Water/cement ratio, w/c Weight ratio of the water to portland cement used in a unit mix of concrete.

Workability Flowability for a concrete mix to fill concrete forms without the presence of large voids. Measured by slump test.

FOR CLASS DISCUSSION

A_{15} Assume sand is composed of spheres. How do particles on successive Tyler screens compare in their maximum cross-sectional area? In their volumes?

B_{15} Distinguish among apparent, bulk, and true densities; between total and apparent porosities.

C_{15} If other factors such as shape are equivalent, a fine-mesh sand will have the same total porosity as a coarse-mesh sand. An exception to this generalization occurs if the particle size begins to approach the size of the container. Explain.

D_{15} Water is required to hydrate portland cement and therefore give it strength. However, excess water weakens concrete. Explain.

E_{15} Many consider that concrete dries as it hardens. If so, how does a bridge pier harden below tide level?

F_{15} Compare and contrast the composition, structure, and characteristics of a mortar for a brick wall and a concrete.

G_{15} In order to smooth out the top of a concrete patio, the amateur mason will commonly "pat" the surface of the wet concrete with a trowel. How does this change the structure and properties of the concrete? What kind of adjustment could be made to partially compensate for its effects?

H_{15} Why does a mason sometimes sprinkle the brick with water before spreading the mortar?

I_{15} Concrete blocks are sometimes made with light-weight aggregate. Suggest how a light-weight aggregate may be manufactured. What properties of these blocks warrant the premium price that they command?

QUIZ SAMPLES

15A   Apparent volume is the sum of $\underline{+-}$ volume and $\underline{+-}$ volume, or the $\underline{+-}$ volume less the $\underline{+-}$ volume.

    (a) total (or bulk)   (b) true   (c) total pore   (d) open pore
    (e) closed pore

15B   The difference between the dry weight and saturated weight of a concrete block permits us to calculate the __**__ pore volume.

15C   The apparent porosity is the amount of __**__ pore volume per __**__ volume.

15D   __**__ bonds an aggregate into a monolithic product called __**__ through a __**__ reaction.

15E   Type __**__ portland cement is used for general concrete construction.

15F   If more water is added to increase the workability of a concrete, it is also necessary to add more __**__ to maintain the required __**__ ratio.

15G   The major advantage of air-entrained concrete over regular concrete is its resistance to __**__.

15H   The strength of concrete gradually __**__-creases as setting time extends from weeks to months, and even to years.

STUDY PROBLEMS

15–1.1   A sample of crushed basalt weighed 482 g after being completely dried. The water level changed from 452 cm^3 to 630 cm^3 when it was added to a large graduated cylinder. After the water was drained away, but the absorbed water remained, the sample weighed 487 g. (a) What is the apparent volume? The total volume of the particles? (b) The open porosity of the crushed particles? (c) The apparent density?

Answer: (a) 178 cm^3; 183 cm^3 (b) 2.7 v/o   (c) 2.71 g/cm^3 (=2.71 Mg/m^3)

15–1.2   In order to determine the volume of an irregularly shaped concrete building block, three measurements were made: weight (dry) = 17.57 kg; weight (saturated with water) = 18.93 kg; weight (suspended in water) = 10.43 kg (a) What is the bulk volume of the block? (b) The apparent volume of the block?

15–1.3   The bulk density of crushed limestone is 1.827 Mg/m^3 (114 lbs/ft^3) as loaded into a truck. What is its total porosity in the truck if the true density of the limestone is 2.7 Mg/m^3?

Answer: 32 v/o

15–1.4   A small construction truck holds 3500 kg (7700 lb) of gravel when level full. (Capacity = 2 m^3 or 70.6 ft^3.) The apparent density of this gravel is measured by placing 5 kg (11 lb) of gravel into a bucket with a volume of 0.0125 m^3 (0.44 ft^3). This partially filled bucket is then filled with water. The water plus the gravel has a net weight of 15.6 kg (34.3 lb). (a) What is the apparent density of the gravel? (b) How much volume of sand should be obtained to add to this truckload of gravel to produce the greatest packing factor?

15–2.1   A cement contains 10% $Ca_3Al_2O_6$, 25% Ca_3SiO_5, and 65% Ca_2SiO_4. The temperature rises 7°C during the initial stages of setting. What percent of the total heat of hydration is retained after this initial period if the concrete mix is comparable to that in Example 15–2.1?

Answer: 11%

15–3.1 A mix of concrete comprises the following volumes (as-received basis): cement, 1; sand, 2.25; gravel, 2.85; water, 0.8. Using the data of Example 15–3.2, calculate the final volume of this mix.

Answer: 4.6 (m^3, cm^3, or ft^3 may be used)

15–3.2 A unit mix of concrete comprises 42.7 kg (94 lb) of cement, 125 kg (275 lb) of sand, 160 kg (350 lb) of gravel, and 21 kg (46 lb) of water. Using the data of Example 15–3.2, calculate the number of unit mixes required for a driveway of 20 m^3 (700 ft^3).

15–3.3 On a dry-weight basis, a unit mix of concrete was to contain 2.7 gravel, 2.2 sand, and 1 cement, plus 0.50 water. The gravel is that of Example 15–1.1, but it is not dry. Rather, the aggregate particles are saturated with water (capillaries and open pores filled). What adjustments are necessary in the unit mix?

Answer: 0.46 units of H_2O, rather than 0.50

15–3.4 The mix in Study Problem 15–3.1 is changed to cement, 1; sand, 2; gravel, 2.5; and water, 0.7. In addition, 4 v/o of entrained air is specified. What is the volume of the final mix?

QUIZ CHECKS

15A	(b), (e)
	(a), (d)
15B	open
15C	open, bulk (total)
15D	cement
	concrete, hydration

15E	I
15F	cement
	water/cement
15G	frost cracking
15H	increases

• Chapter Sixteen

Wood

PREVIEW

A *larger volume of wood* is used than of any other structural material. This distinction occurs because (1) wood is a *renewable source,* (2) it is light and *easily transported* to almost all locations of eventual use. Finally, (3) the last step of its processing can be *on the job site*.

The structure ⇔ property relationship is very obvious in wood and is well known to most people. Not only does the *grain* relate to strength, splitting, etc., there are additional microstructural features that greatly influence properties. We shall examine these.

Although the majority of wood is used directly from the sawmill, wood products are also upgraded for improved properties and performance.

CONTENTS

STUDY OBJECTIVES

1. Become familiar with the biological units that are found in the structure of wood.
2. Relate the structural anisotropy of wood to its directional properties.
3. Identify ways in which wood may be upgraded for improved properties and/or service.
4. Know the principal technical terms used to describe the structure of wood.

16-1

STRUCTURE OF WOOD

Wood has a past, a present, and a future as a material of technical importance. From pre-Stone Age and throughout history, it has been a material of structures. It is currently more widely used by society than any other material (volume basis). It is a renewable resource and, therefore, will continue to have a significant future.

Wood has three noticeable characteristics: (a) It has a high strength-to-weight ratio; (b) many of its properties are anisotropic; and (c) it is easily processed to size, even on the job.

As with all materials, the properties of wood can be related to structure. Most woods exhibit a gross annual structure, called *grain,* that arises from the repeating growth cycles during the seasons of the year (Fig. 16–1.1a). All wood contains *biological cells* with cross-sectional dimensions near the visual limit of the eye. They commonly possess an elongated spindle shape (Fig. 16–1.1b). A closer examination of these cells reveals they are surrounded by *fibers* with submicron diameters. They construct a natural composite (Fig. 16–1.1c). The principal molecular species in wood is *cellulose* (Fig. 16–1.1d). Its structure permits a significant amount of crystallinity within the fibers, thus adding to the strength.

Grain

The trunk of a tree grows by adding another ring every season. Usually this is in the annual sequence of spring and summer growing seasons; however, in tropical regions, the rings may reflect a sequence of wet and dry periods. When the growth is fast, the biological cells are thin-walled and have a large central cavity. This growth is called the *earlywood* (Fig. 16–1.2). Later in the growing season, when the growth rate slows down, the cells have thicker walls and are denser. This is *latewood*. Because of its higher density, it is stronger than the earlywood. The cyclic pattern of the earlywood and latewood produces the typical ring pattern as seen in the cross section of the tree.

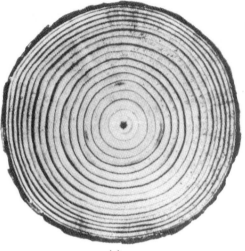

(a)

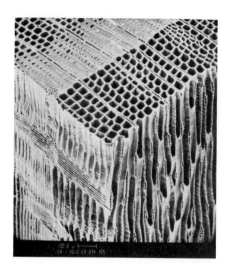

(b)

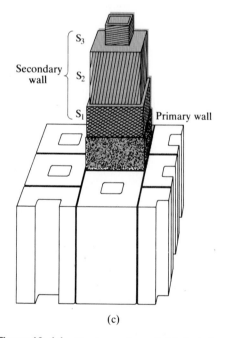

(c)

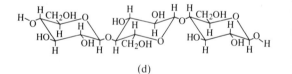

(d)

Figure 16–1.1 Structures of wood. (a) *Grain* formed by annual growth rings (American Forest Institute). (b) Biological cells (Douglas fir, ×50). The individual cells are hollow and spindle shaped. (Courtesy of W. A. Côtè, Jr., S.U.N.Y. College of Environmental Science and Forestry, Syracuse.) (c) Microfibrils in the cell wall (schematic). The three layers in the secondary wall provide reinforcement similar to that in tailored composites. (Courtesy of R. J. Thomas, North Carolina State University.) (d) Molecular (cellulose). It is a polymer, $(C_6H_{10}O_5)_n$.

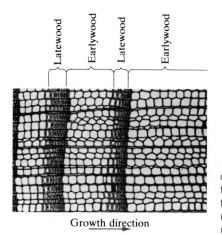

Figure 16–1.2 Microstructure of wood. The units of structure are biological cells. The cells that form with early (spring) growth are larger but have thinner walls than those that form with late (summer) growth. Cellulose (Fig. 16–1.1d) is the major molecular constituent.

These alternating layers of earlywood and latewood and the biological cells that are oriented longitudinally in the tree are primarily responsible for the *anisotropy* of wood. They account for the easy longitudinal splitting, variation of elastic moduli, thermal expansion, conductivity, etc., with direction. Because of these variations wood technologists discuss properties in terms of the *longitudinal* direction, l, the *radial* direction, r, and the *tangential* direction, t (Fig. 16–1.3a). Also, the lumbermen will distinguish between *quartersawed* wood and *plainsawed* wood (Fig. 16–1.3b and c). Of course, the cut of the lumber may have some intermediate orientation; however, it becomes sufficiently obvious to the carpenter that the properties of these two cuts differ, so that intentional selection is sometimes made between the two as they are used in construction and in cabinetmaking.

Finer Structures

The wood cells that show in Fig. 16–1.1(b) possess a structure within themselves. They are made up of several layers of *microfibrils* that spiral at various angles around the central hollow cavity. A typical microfibril has a cross sec-

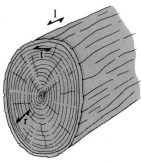
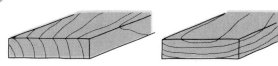

Figure 16–1.3 Directions in wood. (a) Log (l, longitudinal; r, radial; t, tangential). (b) Lumber (quartersawed). (c) Lumber (plainsawed).

(a)   (b)   (c)

tion of <1 μm. They are sketched schematically in Fig. 16–1.1(c). There are three secondary layers that account for 99% of the cell wall:

S_1	10%–22% of wall thickness	50°–70° from longitudinal;
S_2	70%–90% of wall thickness	10°–30° from longitudinal;
S_3	2%– 8% of wall thickness	60°–90° from longitudinal.

The multiple layers and the angles of pitch are nature's way of simultaneously giving strength and flexibility to the wood cell.

Cellulose is the principal molecular species in wood. It is a polymer with mers of $nC_6H_{10}O_5$ (Fig. 16–1.1d). These molecules make up the microfibrils of Fig. 16–1.1(c). Polycellulose possesses a high degree of crystallinity as a result of the presence of both –H and –OH attachments to the molecule (Fig. 16–1.1d). The degree of polymerization, n, sometimes exceeds 30,000. Cellulose accounts for ~42 percent (dry basis) of the mass of most wood.

16–2

PROPERTIES OF WOOD

Wood in a tree contains a high percentage of water, and even seasoned wood contains moisture. The amount of moisture affects all properties. In extreme cases, it can more than double the density. Up to a certain point, it increases the volume and conductivity and lowers the strength and elasticity. The moisture content varies with the surrounding environment. The interior woodwork of a house may contain as little as 6% moisture in midwinter, to 15% moisture after a humid season; a wood piling on a boat dock can possess more than 100% water (oven-dry basis) at the water line. Thus, before considering properties, it will be helpful to describe how the water is contained within the wood.

Moisture in Wood

The reference point for moisture content is oven-dried wood with 0% H_2O. Where wood is exposed to air, which always contains some water vapor, wood gains or loses moisture depending on the relative humidity of the air and the existing moisture status of the wood. It is always seeking an equilibrium. The equilibrium amount during midwinter is the 6% cited above. This *ab*sorption into the wood is really an *ad*sorption onto the surface of the microfibrils sketched in Fig. 16–1.1(c). Warm humid summer air carries more moisture; therefore, more moisture is adsorbed onto the microfibrils. The microfibrils become saturated with moisture, and expansion ceases when about 30%–35% H_2O (dry basis) is present in the wood. This is called the *fiber-saturation point*, FSP. This is an important point, because shrinkage occurs and all strength properties increase as the moisture drops below this level. Moisture in excess of the FSP enters the cavity within the cells as free water (Fig. 16–1.1c). That does not occur from atmospheric moisture during normal lumber usage. The wood would need to be submerged in water or exposed to a continuous presence of water on its surface to fill these cavities.

Each *one percent* of moisture (dry basis) below the FSP changes the dimensions of typical wood:

radial	~0.15 l/o per % moisture,
tangential	~0.25 l/o per % moisture,
longitudinal	~0.01 l/o per % moisture.

Thus, there are constant but minor volume changes in finished wood in service related to prolonged moisture fluctuations of the atmosphere.

Density

Common woods in the United States have densities in the range of 0.4 to 0.7 g/cm³. Among these woods, redwood and oak are near the limits (Table 16–2.1). However, other woods are more extreme, for example, balsa has a density of only 0.13 g/cm³, and a few tropical woods have a density greater than water and, therefore, sink in water. The data of Table 16–2.1 are on a 12% moisture basis.

Elasticity and Strength

Selected mechanical properties of several woods are given in Table 16–2.1. In general, their Young's moduli lie near the centers of the following ranges:

longitudinal	7500 MPa–16,000 MPa	(1,100,000 psi–2,400,000 psi);
radial	500 MPa– 1,000 MPa	(75,000 psi– 150,000 psi);
tangential	400 MPa– 700 MPa	(60,000 psi– 100,000 psi).

The high values for longitudinal stressing are expected since that is the alignment of the wood cells. The other two orientations are low because the cell cavity permits easy strain. The radial direction is stiffened.

The longitudinal strength in tension, $S_{t(l)}$, is commonly 20 times (and sometimes as much as 40 times) the radial strength in tension, $S_{t(r)}$. This contrast will not be surprising to those readers who have worked with wood. Test specimens for mechanical properties must be specifically cut for each orientation (Fig. 16–2.1). Unlike ductile metals, the strength in compression of wood does not

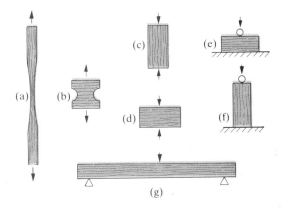

Figure 16–2.1 Test specimens of wood. (a, b) Tension. (c, d) Compression. (e, f) Hardness. (g) Modulus of rupture. The modulus of rupture is the calculated stress at lower (tension) surface.

TABLE 16–2.1				
Properties of Common Woods (12% moisture)*				
	White oak	Paper birch	Douglas fir	White pine	Redwood (old rough growth)
Density, g/cm³	0.68	0.55	0.50	0.38	0.40
Hardness (radial), N†	6,000	4,000	2,900	1,900	2,100
Young's modulus, MPa longitudinal‡ (10⁶ psi)	12,300 (1.78)	11,000 (1.60)	12,500 (1.81)	10,100 (1.46)	9,200 (1.33)
Strength MPa (psi)					
tension (radial)§	5.4 (780)	—	2.4 (350)	—	1.7 (250)
compression (long.)	51 (7400)	39 (5700)	51 (7400)	35 (5100)	42 (6100)
compression (radial)	7.4 (1100)	4.1 (590)	5.2 (750)	3.2 (460)	4.8 (700)
modulus of rupture	105 (15,200)	85 (12,300)	87 (12,600)	67 (9,700)	69 (10,000)

* Data extracted from Wood Handbook USDA, Forest Service, *Agric. Hdbk.,* No. 72.

† Force required to embed 0.444-in. (11.3-mm) ball to one half of its diameter.

‡ See text for ranges of E_r and E_t.

§ Longitudinal strength in tension $S_{t(l)}$ is commonly 20 $S_{t(r)}$.

closely match the strength in tension. Longitudinal compression strengths, $S_{c(l)}$, are lower than in tension, $S_{t(l)}$, because the denser latewood buckles and permits shear to occur. Radial strengths in compression, $S_{c(r)}$, are higher than $S_{t(r)}$ (Table 16–2.1) because the tension failure is by splitting.

Example 16–2.1

A solid wood door was made by gluing together edges of quartersawed boards of Douglas fir. That is, the width of the board is in the radial direction, and the thickness is in the tangential direction (and, of course, the length is the longitudinal direction of the wood). The door was initially trimmed to 761 mm by 2035 mm (to fit a 765 × 2050 mm opening) in late fall when the moisture content of the wood was 9 percent. Will the door "stick" when the moisture content increases to 14 percent?

Solution

$$\text{Radial change} = 0.0015/\% \times [14 - 9\%] = +0.0075.$$

$$\Delta L = (761 \text{ mm})(+0.0075) = +5.7 \text{ mm}.$$

It will not fit, since 761 + 5.7 mm is greater than 765 mm. The longitudinal change is negligible.

Comment. Suggest a design feature that may be used to minimize this problem. ◀

Example 16–2.2

A plainsawed board of Douglas fir (width is tangential) is not permitted to expand its width as its moisture content increases from 5 to 10 percent. What pressure develops?

Procedure. Assume expansion is allowed, followed by compression ($-$) to the original dimension. Consider that $E = 550$ MPa, the mid-value of those indicated in the text for the tangential modulus of elasticity.

Calculation

$$\Delta L/L = (0.0025/\%)(10 - 5\%) = +0.0125.$$

$$s = (-0.0125)(550 \text{ MPa})$$

$$= 7 \text{ MPa compression} \qquad \text{(or 1000 psi).} \quad \blacktriangleleft$$

Example 16–2.3

Water absorption into wood from moist air is initially within the wood fibers (and not into the pores). The fiber saturation point for most wood is $\sim$30 w/o (of oven-dry wood). What is the density increase of birch (oven-dry density $\sim$0.51 g/cm^3) when it is changed from 10% moisture to 20% moisture?

Procedure. The masses change from 110% to 120% of the dry weight. The volume changes must also be referenced to the dry basis.

Calculation. Basis: 1 cm^3 dry wood.

With 10% moisture	With 20% moisture
$l = 1.001$ cm.	$l = 1.002$ cm.
$r = 1.000[1 + (0.0015/\%)(10\%)]$	$r = 1.030$ cm.
$\quad = 1.015$ cm.	
$t = 1.000[1 + (0.0025/\%)(10\%)]$	$t = 1.050$ cm.
$\quad = 1.025$ cm.	
Volume $= 1.041$ cm^3;	$V = 1.084$ cm^3;
Density $= \dfrac{(0.51 \text{ g})(1.10)}{1.041 \text{ cm}^3}$	$\rho = \dfrac{(0.51 \text{ g})(1.20)}{1.084 \text{ cm}^3}$
$\quad = 0.539$ g/cm^3.	$\quad = 0.565$ g/cm^3.

$$\Delta\rho = (0.565 \text{ g/cm}^3 - 0.539 \text{ g/cm}^3)/(0.539 \text{ g/cm}^3) = 5\% \text{ increase.} \quad \blacktriangleleft$$

16–3

PROCESSED WOOD

Most wood is used as boards with rectangular cross sections. The *nominal sizes* of boards do not coincide with the actual sizes (Table 16–3.1). There are several reasons: the saw must consume 2 mm–5 mm of wood that cannot receive compensation in successive breakdowns; rough-cut lumber must be planed to the finished dimension by removing 1 mm–3 mm on each side of the board; and shrinkage occurs if the wood is cut before it is completely dried.

TABLE 16–3.1		
Construction Lumber Sizes		
Nominal size*	**Typical size (dry)**	
	inches	**mm**
1×2	$\frac{3}{4} \times 1\frac{1}{2}$	20×40
1×4	$\times 3\frac{1}{2}$	20×90
1×8	$\times 7\frac{1}{4}$	20×185
2×4	$1\frac{1}{2} \times 3\frac{1}{2}$	40×90
2×6	$\times 5\frac{1}{4}$	40×140
2×8	$\times 7\frac{1}{4}$	40×185
2×10	$\times 9\frac{1}{4}$	40×235

* Dimensional units do not need to be used, since the nominal size is only a label.

Shrinkage amounts to 0.0025 change per % moisture in the tangential direction, which is greater than the 0.0015 per % moisture in the radial direction, and leads to *warpage* in the wood product as indicated in Fig. 16–3.1. Furthermore, *twists* occur if the two ends of the board have different relative position in the original log.

Modified Wood

Plywood, laminated wood, particle board, and fiberboard are widely used wood products. Each utilizes wood that is restructured into panels. They fulfill many construction needs not obtainable from the unaltered wood.

Plywood is made by laminating separate plies alternately at right angles to one another. That is, the longitudinal directions of the grain structure are at 90°. This produces a biaxial panel—one that has comparable properties in two planar directions. (The properties are not identical in the thin, third direction; however, this does not matter since panels are generally used in planar applica-

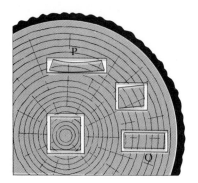

Figure 16–3.1 Shrinkage of wood. Since the structure is anisotropic, shrinkage varies with orientation. This may cause warpage if the wood is cut before it has dried completely. (Shrinkage is exaggerated in this sketch.) Q = quartersawed; P = plainsawed.

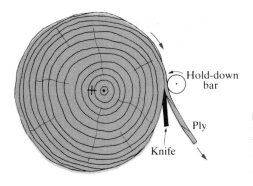

Figure 16–3.2 Cutting veneer. The wood is initially softened by steam or hot water. Since the longitudinal and tangential shrinkages differ, drying must precede the laminating step.

tions.) The plies are generally produced by rotating a log against a large knife blade. By first softening the wood with steam or hot water, and using an appropriate hold-down bar (Fig. 16–3.2), uniform veneers can be cut thinner than 1 mm when necessary.

Compared to solid wood, the plywood has added resistance to splitting and improved dimensional stability because of its cross-ply construction. This utilizes the anisotropy of wood advantageously (Section 16–2). Glues for interior plywood can sometimes be simple ones that develop strength by drying. Exterior plywood is commonly bonded with phenol-formaldehyde or similar resins that are water resistant. Since these must be thermally set at higher temperatures, exterior plywood is a more expensive product.

Laminated wood is a composite of a number of smaller boards bonded into a larger unit. Thus, it is possible to build up large beams (Fig. 16–3.3), arches, and columns, and to make special shapes, such as chair frames and aircraft propellers. The grain of the contributing boards is laid parallel. The boards can be selected to avoid knots and dried to avoid severe checking that would result in the drying of large timbers. The result is a superior product.

Fiberboard contains wood that was first pulped and then reconstituted into a panel. The product is porous and with low density when it is to be used for acoustical tile or for insulation sheathing for house construction. In contrast, hardboard panels may be made that are denser than wood and are therefore highly wear-resistant.

All panel products are made to specific sizes since additional trimming and planing are not necessary.

Example 16–3.1

A 280-mm (11-in.) tree is cut for use as a telephone pole. Initially, it contains 20 percent moisture (dry basis) and contains no cracks. One year after installing, it contains a single radial crack that is 7 mm wide at the surface, wedging to zero at the center. Estimate its moisture content. (Assume uniformity.)

Procedure. Calculate the circumference, C, based (1) on the radial and (2) on the tangential shrinkages. Moisture loss = $x\%$.

Figure 16-3.3 Laminated wood beam. Greater uniformity and larger sizes are achieved than with nonlaminated beams because the effects of knots and checks can be reduced. (Courtesy of Forest Products Laboratory, Forest Service, USDA.)

Calculations

$$\text{Original } C = \pi(280 \text{ mm}) = 880 \text{ mm};$$

$$d_{1 \text{ yr}} = 280 \, (1 - 0.0015 \, x).$$

Based on radial shrinkage:

$$C_r = 280\pi \text{ mm} \, (1 - 0.0015 \, x)$$

Based on tangential shrinkage:

$$C_t = 880 \text{ mm} \, (1 - 0.0025 \, x)$$

$$C_r - C_t = 880 \text{ mm} \, (0.001 \, x) = 7 \text{ mm}$$

$$x = 8\%$$

$$\% \text{ H}_2\text{O} = 20\% - 8\% = 12\%.$$

Comment. If the radial shrinkage had been greater than the tangential shrinkage, the cracks would have formed "parallel" to the growth rings. ◀

Review and Study

SUMMARY

Although complex, wood is sufficiently familiar so that we can readily appreciate its structure ⇔ property relationships.*

1. The structural levels are *grain* with a transverse dimension of 1 mm–10 mm, *biological cells* (100 μm to 1 mm), *fibrils* (10 nm–1 μm), and *molecules* (~1 nm). In each case longitudinal dimensions are at least 10^2–10^4 times greater than transverse dimensions.

   The grain has alternating layers of *earlywood* and *latewood,* which contain thin-walled and thick-walled cells, respectively. The cell walls contain three *layers* of microfibrils pitched at an angle to the cell axis, thus reinforcing it. *Cellulose* is the most common molecular species.

   The three principal directions of wood are labeled: l—longitudinal; r—radial, and t—tangential.

2. The properties of wood are highly anisotropic between the three principal directions. The longitudinal direction is the strongest, the most rigid (highest E), and the most stable against moisture changes. The radial properties generally (but not always) lie between the longitudinal and the tangential properties.

   Moisture absorption between the microfibrils is a function of humidity. It produces expansion until the fiber-saturation point (FSP) is reached at about 30%–35% H_2O (dry basis). Beyond that, the cell cavity is filled and the wood becomes waterlogged.

3. The nominal sizes of cut lumber are indexes of dimensions rather than engineering dimensions.

   Wood is subject to *warpage* and *cracking* as a result of moisture changes.

   Wood may be modified into plywood, laminated wood, particle board, and fiberboard to equalize the directional properties, particularly in panel products.

TERMS AND CONCEPTS

Anisotropic   Having different properties in different directions.

Cellulose   Natural polymer of $(C_6H_{10}O_5)_n$.

Earlywood   First wood growth during the growing season. Part of the wood growth ring with larger, thinner-walled cells. (Spring wood.)

Fiber-saturation point (FSP)   Upper limit of moisture adsorption onto the wood fibers. The limit of volume expansion by moisture absorption into the wood. (Excess water enters the wood pores.)

Grain (wood)   Seasonal growth pattern, alternating between earlywood and latewood.

Hardwood   Wood from deciduous (broad-leaved) trees.

Laminated wood   Wood product built up from thinner boards.

---

* A more detailed (but still elementary) discussion of the structure and properties of wood is presented in Chapter 10 of *Materials for Engineers: Concepts and Applications* (Reading, Mass.: Addison-Wesley, 1982).

Latewood   Part of the wood growth ring with smaller, thicker-walled cells. Formed late in the growing season. (Summer wood.)

Microfibrils   Discrete filamentary units of the cell wall.

Nominal size (wood)   Size label, approximating the ''as-cut'' cross-sectional dimensions in inches. Always larger than the actual finished dimensions.

Plainsawed   Wood cut with the width of the board approximately tangential to the growth rings (thickness approximately radial).

Plywood   Panels made from laminated plies; each ply cut is a longitudinal-tangential sheet; alternate plies are laid at 90°.

Quartersawed   Wood cut with the width of the board approximately radial to the growth rings (thickness approximately tangential).

Softwood   Wood from conifers. It is commonly used for construction lumber.

FOR CLASS DISCUSSION

A_{16} A denser wood absorbs more moisture in humid weather than a lighter wood. Why?

B_{16} The low latitudes do not have spring and summer. Account for the grain in their woods. (Cf. Fig. 16–1.2.)

C_{16} Account for the warpage concavity of the upper cut of wood in Fig. 16–3.1.

D_{16} Why is the longitudinal shrinkage of wood negligible compared to the shrinkage in the other directions?

E_{16} Soft woods, such as pine, are cured before milling. Why?

F_{16} Refer to an encyclopedia and report on the difference between ''soft'' woods and ''hard'' woods.

G_{16} Distinguish between particle board and fiberboard. What are the anticipated differences in properties?

H_{16} Plainsawed lumber is more subject to warping than quartersawed lumber. Why?

QUIZ SAMPLES

16A   Cellulose provides the major molecular structure in wood. Coarser structures include micro-**, __**__ cells, and __**_ .

16B   A normal pattern of growth leads to alternating layers of _+−_ and _+−_. Of the two, the _+−_ has thinner-walled cells.

(a) earlywood   (b) sap wood   (c) heart wood   (d) latewood

16C   Of the three principal directions—radial, longitudinal, and tangential—the _+−_ has the highest elastic modulus. The _+−_ has the greatest moisture expansion. The _+−_ has the greatest tensile strength. The _+−_ has the greatest density.

(a) radial   (b) longitudinal   (c) tangential   (d) None; property is not directional

16D   The cross section of a **-sawed 1×4 board will show more growth rings than a **-sawed board of the same nominal size.

16E The biological cells of most woods contain three layers of __**__ that spiral around a central cavity.

16F The principal molecular species of wood is __**__ in which the degree of polymerization may be as high as 30,000.

16G The reference for moisture content is __+−__.

(a) as-cut wood   (b) "aged" wood dried for 1 yr at 20°C
(c) moisture at the FSP   (d) None of the above, rather _____

16H Warpage occurs during the drying of a board since the shrinkage differs in the __**__ and __**__ directions. Of the two, it is greater in the __**__ direction.

16I Cracking can occur during the drying of plywood following lamination because shrinkage differs in the __**__ and __**__ directions. Of the two, it is greater in the __**__ direction.

16J Typical seasonal variations in the moisture contents of wood trim within a house are between ~15% and __**__%.

STUDY PROBLEMS

16–2.1 The bulk density of some dry wood is ~0.54 g/cm³. Estimate the percent porosity if the true density of the dry wood is 1.5 g/cm³ (=1.5 Mg/m³).

Answer: 64 v/o total porosity

16–2.2 A quartersawed board of Douglas fir was trimmed to 18 mm × 91 mm (nominal 1 × 4) when it contained 18 percent moisture. What cross-sectional dimensions will it have if it loses two thirds of its moisture?

16–2.3 The equilibrium moisture content for wood can vary from 6 percent in winter to 15 percent in the summer. Calculate the biannual dimensional change of a solid Douglas fir table top, 900 mm × 1500 mm × 20 mm, (a) when it is quartersawed (width is radial); (b) when it is plainsawed (width is tangential).

Answer: (a) $\Delta w = 12$ mm, $\Delta t = 0.4$ mm, $\Delta l =$ nil
(b) $\Delta w = 20$ mm, $\Delta t = 0.3$ mm, $\Delta l =$ nil

16–2.4 The trimmed door in Example 16–2.1 was shut, then its moisture content increased from 9 to 14 percent. What force would develop if the door is 30 mm thick? Assume that the doorway casing is rigid.

16–2.5 Assume in case (a) of Study Problem 16–2.3 that the edges of the table were clamped in the summer. Would the top split in the winter? Where ranges are given, state your assumed values.

Answer: $s = 10$ MPa > 2.4 MPa. It will split. (1500 psi > 350 psi.)

16–3.1 Birchwood veneer is impregnated with phenol-formaldehyde (Fig. 6–3.1b) to ensure resistance to water and to increase the hardness of the final product. Although dry birch weighs only 0.56 g/cm³, the true specific gravity of the cellulose–lignin combination is 1.52. (a) How many grams of phenol-formaldehyde

(PF) are required to impregnate 10,000 mm³ (0.6 in³) of dry birchwood? (b) What is the final density?

Answer: (a) 8.2 g PF   (b) 1.38 g/cm³ (=1.38 Mg/m³)

16–3.2 How much water would the table top of Study Problem 16–2.3 absorb going from winter to summer if it were made of (a) balsa (0.16 g/cm³ dry)? (b) Birch (0.56 g/cm³ dry)? (c) Explain the greater absorption of the birch.

QUIZ CHECKS

16A microfibrils, biological, grain

16B (a), (d)
          (a)

16C (b)
          (c)
          (b)
          (e)

16D quartersawed, plainsawed

16E microfibrils

16F cellulose

16G (d): oven-dry

16H radial and tangential, tangential

16I longitudinal and tangential, tangential

16J 6%

Appendixes

APPENDIX A

Constants and Conversions

Constants*

Acceleration of gravity, g	$9.80 \ldots$ m/s^2
Atomic mass unit, amu	$1.66 \ldots \times 10^{-24}$ g
Avogadro's number, N	$0.6022 \ldots \times 10^{24}$ mole^{-1}
Boltzmann's constant, k	$86.1 \ldots \times 10^{-6}$ eV/K
	$13.8 \ldots \times 10^{-24}$ J/K
	8.31 J/mole·K
Capacitivity (vacuum), ε	$8.85 \ldots \times 10^{-12}$ C/V·m
Electron charge, q	$0.1602 \ldots \times 10^{-18}$ C
Electron moment, β	$9.27 \ldots \times 10^{-24}$ A·m^2
Electron volt, eV	$0.160 \ldots \times 10^{-18}$ J
Faraday, $\mathscr{F}$	$96.5 \ldots \times 10^3$ C
Fe–Fe$_3$C eutectoid composition	0.77 w/o carbon
Fe–Fe$_3$C eutectoid temperature	727°C (1340°F)
Gas constant, R	$8.31 \ldots$ J/mole·K
	$1.987 \ldots$ cal/mole·K
Gas volume (STP)	$22.4 \ldots \times 10^{-3}$ m^3/mole
Planck's constant, h	$0.662 \ldots \times 10^{-33}$ J·s
Velocity of light, c	$0.299 \ldots \times 10^9$ m/s

Conversions*

1 ampere	$= 1$ C/s
1 angstrom	$= 10^{-10}$ m
	$= 10^{-8}$ cm
	$= 0.1$ nm
	$= 3.937 \times 10^{-9}$ in.
1 amu	$= 1.66 \ldots \times 10^{-24}$ g
1 Btu	$= 1.055 \ldots \times 10^3$ J
1 Btu/°F	$= 1.899 \ldots \times 10^3$ J/°C
1 [Btu/(ft^2·s)]/[°F/in.]	$= 0.519 \ldots \times 10^3$ [J/(m^2·s)]/[°C/m]
	$= 0.519 \ldots \times 10^3$ (W/m^2)/(°C/m)
1 Btu·ft^2	$= 11.3 \ldots \times 10^3$ J/m^2
1 calorie, gram	$= 4.18 \ldots$ J
1 centimeter	$= 10^{-2}$ m
	$= 0.3937$ in.
1 coulomb	$= 1$ A·s
1 cubic centimeter	$= 0.0610 \ldots$ in^3
1 cubic inch	$= 16.3 \ldots \times 10^{-6}$ m^3
1°C difference	$= 1.8$°F
1 electron volt	$= 0.160 \ldots \times 10^{-18}$ J
1°F difference	$= 0.555 \ldots$ °C
1 foot	$= 0.3048 \ldots$ m

* All irrational values are rounded downward.

Conversions (*continued*)

1 foot·pound$_f$	= 1.355 . . . J
1 gallon (U.S. liq.)	= 3.78 . . . $\times 10^{-3}$ m^3
1 gram	= 0.602 . . . $\times 10^{24}$ amu
	= 2.20 . . . $\times 10^{-3}$ lb$_m$
1 gram/cm^3	= 62.4 . . . lb$_m$/ft^3
	= 1000 kg/m^3
	= 1 Mg/m^3
1 inch	= 0.0254 . . . m
1 joule	= 0.947 . . . $\times 10^{-3}$ Btu
	= 0.239 . . . cal, gram
	= 6.24 . . . $\times 10^{18}$ eV
	= 0.737 . . . ft·lb$_f$
	= 1 watt·sec
1 joule/meter^2	= 8.80 . . . $\times 10^{-5}$ Btu/ft^2
1 [joule/(m^2·s)]/[°C/m]	= 1.92 . . . $\times 10^{-3}$ [Btu/(ft^2·s)]/[°F/in.]
1 kilogram	= 2.20 . . . lb$_m$
1 megagram/meter^3	= 1 g/cm^3
	= 10^6 g/m^3
	= 1000 kg/m^3
1 meter	= 10^{10} Å
	= 10^9 nm
	= 3.28 . . . ft
	= 39.37 in.
1 micrometer	= 10^{-6} m
1 nanometer	= 10^{-9} m
1 newton	= 0.224 . . . lb$_f$
1 ohm·inch	= 0.0254 . . . Ω·m
1 ohm·meter	= 39.37 Ω·in.
1 pascal	= 0.145 . . . $\times 10^{-3}$ lb$_f$/in^2
1 poise	= 0.1 Pa·s
1 pound (force)	= 4.44 . . . newtons
1 pound (mass)	= 0.453 . . . kg
1 pound/foot^3	= 16.0 . . kg/m^3
1 pound/inch^2	= 6.89 . . . $\times 10^{-3}$ MPa
1 watt	= 1 J/s
1 (watt/m^2)/(°C/m)	= 1.92 . . . $\times 10^{-3}$ [Btu/(ft^2·s)]/[°F/in.]

SI prefixes

giga	G	10^9
mega	M	10^6
kilo	k	10^3
milli	m	10^{-3}
micro	μ	10^{-6}
nano	n	10^{-9}

APPENDIX B

Table of Selected Elements (from various sources, including the ASM Handbooks.)

Element	Symbol	Atomic number	Atomic mass, amu	Orbitals 1s	2s	2p	3s	3p	Melting point, °C	Density (solid), Mg/m³ (= g/cm³)	Crystal structure, 20°C	Approx. atomic radius, nm†	Valence (most common)	Approx. ionic radius, nm‡
Hydrogen	H	1	1.0078	1					−259.14	—	—	0.046	1+	Very small
Helium	He	2	4.003	2					−272.2	—	—	0.176	Inert	—
Lithium	Li	3	6.94	He +	1				180.7	0.534	bcc	0.1519	1+	0.068
Beryllium	Be	4	9.01	He +	2				1290	1.85	hcp	0.114	2+	0.035
Boron	B	5	10.81	He +	2	1			2300	2.3	—	0.046	3+	~0.025
Carbon	C	6	12.011	He +	2	2			>3500	2.25	hex	0.077		—
Nitrogen	N	7	14.007	He +	2	3			−210	—	—	0.071	3−	—
Oxygen	O	8	15.999	He +	2	4			−218.4	—	—	0.060	2−	0.140
Fluorine	F	9	19.00	He +	2	5			−220	—	—	0.06	1−	0.133
Neon	Ne	10	20.18	He +	2	6			−248.7	—	fcc	0.160	Inert	—
Sodium	Na	11	22.99	Ne +			1		97.8	0.97	bcc	0.1857	1+	0.097
Magnesium	Mg	12	24.31	Ne +			2		650	1.74	hcp	0.161	2+	0.066
Aluminum	Al	13	26.98	Ne +			2	1	660.4	2.699	fcc	0.14315	3+	0.051
Silicon	Si	14	28.09	Ne +			2	2	1410	2.33	*	0.1176	4+	0.042
Phosphorus	P	15	30.97	Ne +			2	3	44	1.8	—	0.11	5+	~0.035
Sulfur	S	16	32.06	Ne +			2	4	112.8	2.07	—	0.106	2−	0.184
Chlorine	Cl	17	35.45	Ne +			2	5	−101	—	—	0.101	1−	0.181
Argon	Ar	18	39.95	Ne +			2	6	−189.2	—	fcc	0.192	Inert	—

Table of Selected Elements

Electron configuration subshells: 3d, 4s, 4p

Element		Z	At. wt.	Config	3d	4s	4p	MP (°C)	Density	Structure	Radius†	Valence	Ionic radius‡
Potassium	K	19	39.1	Ar +		1		63	0.86	bcc	0.231	1+	0.133
Calcium	Ca	20	40.08	Ar +		2		839	1.55	fcc	0.1976	2+	0.099
Titanium	Ti	22	47.90	Ar +	2	2		1668	4.51	hcp	0.146	4+	0.068
Chromium	Cr	24	52.00	Ar +	5	1		1875	7.19	bcc	0.1249	3+	0.063
Manganese	Mn	25	54.94	Ar +	5	2		1244	7.47	—	0.112	2+	0.080
Iron	Fe	26	55.85	Ar +	6	2		1538	7.87	bcc	0.1241	2+	0.074
										fcc	0.1269	3+	0.064
Cobalt	Co	27	58.93	Ar +	7	2		1495	8.83	hcp	0.125	2+	0.072
Nickel	Ni	28	58.71	Ar +	8	2		1453	8.90	fcc	0.1246	2+	0.069
Copper	Cu	29	63.54	Ar +	10	1		1084	8.93	fcc	0.1278	1+	0.096
Zinc	Zn	30	65.38	Ar +	10	2		420	7.13	hcp	0.139	2+	0.074
Germanium	Ge	32	72.59	Ar +	10	2	2	937	5.32	*	0.1224	4+	—
Arsenic	As	33	74.92	Ar +	10	2	3	816	5.78	—	0.125	3+	—
Krypton	Kr	36	83.80	Ar +	10	2	6	-157	—	fcc	0.201	Inert	—

Electron configuration subshells: 4d, 5s, 5p

Element		Z	At. wt.	Config	4d	5s	5p	MP (°C)	Density	Structure	Radius†	Valence	Ionic radius‡
Silver	Ag	47	107.87	Kr +	10	1		961.9	10.5	fcc	0.1444	1+	0.126
Tin	Sn	50	118.69	Kr +	10	2	2	232	7.17	bct	0.1509	4+	0.071
Antimony	Sb	51	121.75	Kr +	10	2	3	630.7	6.7	—	0.1452	5+	—
Iodine	I	53	126.9	Kr +	10	2	5	114	4.93	ortho	0.135	1-	0.220
Xenon	Xe	54	131.3	Kr +	10	2	6	-112	2.7	fcc	0.221	Inert	—

Electron configuration subshells: 4f, 5d, 6s, 6p, 6d, $7s^2$

Element		Z	At. wt.	Config	4f	5d	6s	6p	6d	$7s^2$	MP (°C)	Density	Structure	Radius†	Valence	Ionic radius‡
Cesium	Cs	55	132.9	Xe +			1				28.6	1.9	bcc	0.265	1+	0.167
Tungsten	W	74	183.9	Xe +	14	4	2				3410	19.25	bcc	0.1367	4+	0.070
Gold	Au	79	197.0	Xe +	14	10	1				1064.4	19.3	fcc	0.1441	1+	0.137
Mercury	Hg	80	200.6	Xe +	14	10	2				-38.86	—	—	0.155	2+	0.110
Lead	Pb	82	207.2	Hg + $6p^2$				2			327.4	11.38	fcc	0.1750	2+	0.120
Uranium	U	92	238.0	Rn + $5f^3$					6d	$7s^2$	1133	19.05	—	0.138	4+	0.097

* Diamond cubic

† One half of the closest approach of two atoms in the elemental solid. For noncubic structures, the average interatomic distance is given; e.g., in hcp, the atom is slightly ellipsoidal.

‡ Radii for CN = 6; otherwise, $0.97\,R_{CN=8} \approx R_{CN=6} \approx 1.1\,R_{CN=4}$. Patterned after Ahrens.

APPENDIX C

Properties of Selected Engineering Materials (20°C)*

Material	Density Mg/m³ (=g/cm³)	Thermal conductivity, $\frac{\text{watts}}{\text{mm}^2} / \left(\frac{°C}{\text{mm}}\right)$ †	Linear expansion, °C⁻¹‡	Electrical resistivity, ρ ohm·m§	Average modulus of elasticity, $\bar{E}$	
					MPa	psi
Metals						
Aluminum (99.9+)	2.7	0.22	22.5×10^{-6}	29×10^{-9}	70,000	10×10^6
Aluminum alloys	2.7(+)	0.16	22×10^{-6}	$\sim\!45 \times 10^{-9}$	70,000	10×10^6
Brass (70 Cu–30 Zn)	8.5	0.12	20×10^{-6}	62×10^{-9}	110,000	16×10^6
Bronze (95 Cu–5 Sn)	8.8	0.08	18×10^{-6}	$\sim\!100 \times 10^{-9}$	110,000	16×10^6
Cast iron (gray)	7.15	—	10×10^{-6}	—	140,000($\pm$)	$20 \times 10^{6}\pm$
Cast iron (white)	7.7	—	9×10^{-6}	660×10^{-9}	205,000	30×10^6
Copper (99.9+)	8.9	0.40	17×10^{-6}	17×10^{-9}	110,000	16×10^6
Iron (99.9+)	7.88	0.072	11.7×10^{-6}	98×10^{-9}	205,000	30×10^6
Lead (99+)	11.34	0.033	29×10^{-6}	206×10^{-9}	14,000	2×10^6
Magnesium (99+)	1.74	0.16	25×10^{-6}	45×10^{-9}	45,000	6.5×10^6
Monel (70 Ni–30 Cu)	8.8	0.025	15×10^{-6}	482×10^{-9}	180,000	26×10^6
Silver (sterling)	10.4	0.41	18×10^{-6}	$\sim\!18 \times 10^{-9}$	75,000	11×10^6
Steel (1020)	7.86	0.050	11.7×10^{-6}	169×10^{-9}	205,000	30×10^6
Steel (1040)	7.85	0.048	11.3×10^{-6}	171×10^{-9}	205,000	30×10^6
Steel (1080)	7.84	0.046	10.8×10^{-6}	180×10^{-9}	205,000	30×10^6
Steel (18 Cr–8 Ni stainless)	7.93	0.015	9×10^{-6}	700×10^{-9}	205,000	30×10^6
Ceramics						
Al₂O₃	3.8	0.029	9×10^{-6}	$>10^{12}$	350,000	50×10^6
Brick						
Building	2.3($\pm$)	0.0006	9×10^{-6}	—	—	—
Fireclay	2.1	0.0008	4.5×10^{-6}	1.4×10^6	—	—
Graphite	1.5	—	5×10^{-6}	—	—	—
Paving	2.5	—	4×10^{-6}	—	—	—
Silica	1.75	0.0008	—	1.2×10^6	—	—

Concrete	2.4(±)	0.0010	13×10^{-6}	—	14,000	2×10^6
Glass						
Plate	2.5	0.00075	9×10^{-6}	10^{12}	70,000	10×10^6
Borosilicate	2.4	0.0010	2.7×10^{-6}	$>10^{15}$	70,000	10×10^6
Silica	2.2	0.0012	0.5×10^{-6}	10^{18}	70,000	10×10^6
Vycor	2.2	0.0012	0.6×10^{-6}	—	—	—
Wool	0.05	0.00025	—	—	—	—
Graphite (bulk)	1.9	—	5×10^{-6}	10^{-5}	7,000	1×10^6
MgO	3.6	—	9×10^{-6}	10^3 (1100°C)	205,000	30×10^6
Quartz (SiO$_2$)	2.65	0.012	—	10^{12}	310,000	45×10^6
SiC	3.17	0.012	4.5×10^{-6}	0.025 (1100°C)	—	—
TiC	4.5	0.030	7×10^{-6}	50×10^{-8}	350,000	50×10^6
Polymers						
Melamine-formaldehyde	1.5	0.00030	27×10^{-6}	10^{11}	9,000	1.3×10^6
Phenol-formaldehyde	1.3	0.00016	72×10^{-6}	10^{10}	3,500	0.5×10^6
Urea-formaldehyde	1.5	0.00030	27×10^{-6}	10^{10}	10,300	1.5×10^6
Rubbers (synthetic)	1.5	0.00012	—	—	4–75	600–11,000
Rubber (vulcanized)	1.2	0.00012	81×10^{-6}	10^{12}	3,500	0.5×10^6
Polyethylene (L.D.)	0.92	0.00034	180×10^{-6}	10^{13}–10^{16}	100–350	14,000–50,000
Polyethylene (H.D.)	0.96	0.00052	120×10^{-6}	10^{12}–10^{16}	350–1,250	50,000–180,000
Polystyrene	1.05	0.00008	63×10^{-6}	10^{16}	2,800	0.4×10^6
Polyvinylidene chloride	1.7	0.00012	190×10^{-6}	10^{11}	350	0.05×10^6
Polytetrafluoroethylene	2.2	0.00020	100×10^{-6}	10^{14}	350–700	50,000–100,000
Polymethyl methacrylate	1.2	0.00020	90×10^{-6}	10^{14}	3,500	0.5×10^6
Nylon	1.15	0.00025	100×10^{-6}	10^{12}	2,800	0.4×10^6

* Data in this table were taken from numerous sources.

† Alternatively, W/mm·K. Multiply by 1.92 to get Btu/(ft^2·s)/(°F/in.).

‡ Or, K^{-1}; divide by 1.8 to get $°F^{-1}$.

§ Multiply ohm·m by 39 to get ohm·in.

APPENDIX D

Unified Numbering System (UNS) of Commercial Alloys

Series	Alloy types
Axxxxx	Aluminum and aluminum alloys
Cxxxxx	Copper and copper alloys
Exxxxx	Rare Earth and similar metals and alloys
Fxxxxx	Cast irons
Gxxxxx	AISI and SAE carbon and alloy steels. (See the footnote to Table 10–9.1.)
Hxxxxx	AISI and SAE H-steels
Jxxxxx	Cast steels (except tool steels)
Kxxxxx	Low melting metals and alloys
Mxxxxx	Miscellaneous non-ferrous metals and alloys
Nxxxxx	Nickel and nickel alloys
Pxxxxx	Precious metals and alloys
Rxxxxx	Reactive and refractory metals and alloys
Sxxxxx	Heat and corrosion resistant (stainless) steels, valve steels, and iron-base superalloys
Txxxxx	Tool steels, wrought and cast
Wxxxxx	Welding filler metals
Zxxxxx	Zinc and zinc alloys

Standards	Sources
AA	Aluminum Association
AISI-SAE	American Iron and Steel Institute/Society of Automotive Engineers
AMS	SAE/Aerospace Materials Specification
ASME	American Society of Mechanical Engineers
ASTM	American Society for Testing and Materials
AWS	American Welding Society
CDA	Copper Development Association
QQ & WW	Federal Specifications
MIL	Military Specifications
—	Metal and Alloy Index
—	Lead Industries Association
—	Zinc Institute

Index

Boldface page numbers refer to ''Terms and Concepts'' at the end of each chapter.